W9-CRV-126

Optics of Excitons in Confined Systems

Acknowledgements

The International Meeting on Optics of Excitons in Confined Systems has been held thanks to the support of

Università degli Studi di Messina
Opera Universitaria di Messina
Seconda Università degli Studi di Roma 'Tor Vergata'
Istituto di Metodologie Avanzate Inorganiche del Consiglio Nazionale delle Ricerche, Roma
Ecole Polytechnique Fédérale de Lausanne
Comitato Fisica del Consiglio Nazionale delle Ricerche
Gruppo Nazionale di Struttura della Materia del Consiglio Nazionale delle Ricerche
Assessorato al Turismo della Regione Siciliana
Cassa di Risparmio 'Vittorio Emanuele' di Messina

We also acknowledge the contribution to the organization of the meeting of

Istituto di Struttura della Materia, Università degli Studi di Messina
Dipartimento di Fisica, Università degli Studi di Messina
Istituto di Fisica Teorica, Università degli Studi di Messina
Dipartimento di Fisica, Seconda Università degli Studi di Roma 'Tor Vergata'
Institut de Physique Théorique, Ecole Polytechnique Fédérale de Lausanne
UNIPOINT, Centro Apple Education, Messina

Optics of Excitons in Confined Systems

Proceedings of the International Meeting, Giardini Naxos, Italy, 24–27 September 1991

Edited by A D'Andrea, R Del Sole, R Girlanda and A Quattropani

Institute of Physics Conference Series Number 123
Institute of Physics, Bristol, Philadelphia and New York

CODEN IPHSAC 123 1-354 (1992)

British Library Cataloguing-in-Publication Data are available

ISBN 0-85498-413-5

Library of Congress Cataloguing-in-Publication Data are available

Published under the Institute of Physics imprint by IOP Publishing Ltd
Techno House, Redcliffe Way, Bristol BS1 6NX, England
IOP Publishing Inc., The Public Ledger Building, Suite 1035, Independence Square, Philadelphia, PA 19106, USA

Printed in Great Britain by Galliard (Printers) Ltd, Great Yarmouth, Norfolk

Preface

This volume contains the Proceedings of the International Meeting on Optics of Excitons in Confined Systems, held in Giardini Naxos, Italy, from the 24th to the 27th of September 1991.

The first International Meeting on Excitons in Confined Systems was held in Rome, in April 1987, and was organized by two members of our committee. The aim of the meeting was to discuss the theoretical and experimental aspects of excitons in different confinement conditions, from semi-infinite solids to quantum wells. Some space was also devoted to the topic of excitons in quantum wires, which was, at that time, at its very beginning. The enormous growth and the increasing sophistication of the research work since then has led us to the decision of organizing a second meeting, whose proceedings are here reported. This second meeting is more concerned with confined structures, especially with the more recent ones, such as quantum wires and dots. Its aim is to provide an overview of the state of the art in semiconductors which exhibit resonance enhanced optical nonlinearities in the frequency range close to the valence-conduction band gap. In this range of energies a key role is usually taken by exciton dynamics, strong nonlinearities and fast optical switching which is markedly connected to electron–hole spatial confinement. The growth and control of the exciton spatial confinement are achieved by new crystal-growth techniques. In establishing the program, the organizers have profited by the broad experience of the other members of the Program Committee, A Baldereschi, L Banyai, B Gil and A Mysyrowicz.

It is always difficult to classify the material of a conference. We tried to establish an order going from simple and fundamental topics to those more complex and applied. The first chapter deals with the very fundamental aspects of exciton–photon interaction in confined systems, whose theory is not yet complete. The main point of debate, treated here by different authors, is how to calculate the radiative decay times of the excitations. The next three chapters deal with excitons in different confined systems, namely wells, quantum wires and dots, and superlattices. Nonlinear optical properties are treated in the fifth chapter. The more complex and application-oriented topic of impurities and external fields is treated in the sixth and last chapter.

The enthusiastic response of the invited speakers, contributors and participants has prompted the continuation of this series of conferences. The Third International Meeting on Optics of Excitons in Confined Systems will be organized in Montpellier, France, in September 1993, by M Voos, B Gil, G Bastard and J L Robert.

Giardini Naxos
September 1991

A D'Andrea
R Del Sole
R Girlanda
A Quattropani

Contents

Section 4: Superlattices

Section 5: Nonlinear Optical Properties of Confined Systems

Inst. Phys. Conf. Ser. No 123
Paper presented at the International Meeting on Optics of Excitons in Confined Systems, Giardini Naxos, Italy, 1991

Electrodynamics of excitons in two-dimensional systems

V M Agranovich

The Institute of Spectroscopy, USSR Academy of Sciences, Troitsk, Moscow obl., 142092 USSR

ABSTRACT: Peculiarities of optical linear and nonlinear properties of $2D$ excitons are discussed. New methods for calculation of optical response in the region of excitonic resonances, which, in particular, allow consistent account of the spontaneous emission of polariton states in $2D$ systems, are considered. Calculations of the linear polarizability of a monolayer are given as an example and the results are compared with those for a $3D$ crystal. Superradiance of an excitonic molecule in $2D$ systems and spectra of optically anisotropic molecular superlattices are discussed.

1. RETARDATION EFFECTS IN THE DIELECTRIC OPTICAL RESPONSE

In the study of optical properties of condensed media the equations of macroscopic electrodynamics are usually used complemented with so called material equations. These equations relate the displacement vector $\mathbf{D}(\mathbf{r}, t)$ with the electric field strength $\mathbf{E}(\mathbf{r}, t)$ (for simplicity we consider here nonmagnetic media and assume the fields $\mathbf{E}$ to be weak compared to atomic ones) and, generally speaking, they are nonlinear. Following Bloembergen(1965), nonlinearity of the material relation between $\mathbf{D}$ and $\mathbf{E}$ can be taken into account by the introduction of so called linear and nonlinear polarizabilities $\chi^{(n)}$. Most calculations of linear and nonlinear optical polarizabilities in condensed phases neglect the effects of retarded radiative intermolecular interactions (Flytzanis 1975, Madden and Kivelson 1984, Keldysh 1989). Consequently, the optical susceptibilities are expressed in terms of equilibrium material correlation functions which are calculated in the absence of the transverse electromagnetic field. These theories do not incorporate, therefore, spontaneous emission properly. In addition, at low temperature in crystals or monolayers the radiation field and matter (polarization) modes are strongly correlated and form new quasiparticles, polaritons. Conventional theories of polarizabilities neglect polariton effects and in some cases fail to give an adequate description of optical processes. Consider first the linear susceptibility in the vicinity of an exciton resonance (neglecting spatial dispersion):

$$4\pi\chi^{(1)}(\omega) \equiv \frac{\omega_p^2 f/\omega_0}{\omega_0 - \omega - i\gamma(\omega)}.$$

Here ω_p and ω_0 denote the plasma and the $k = 0$ exciton frequencies, respectively, f is the oscillator strength, and $\gamma(\omega)$ is the dissipative width resulting from exciton-phonon scattering. Since this expression is usually calculated without taking into account the retarded interaction, the function $\gamma(\omega)$ depends only on exciton dispersion $\omega_0(k)$. That is why the resulting function $\gamma(\omega)$ fails to give a correct description of the long-wavelength edge of excitonic absorption lines. In the region $\omega \leq \omega_0$ for excitonic transitions with sufficiently large oscillator strength f, retarded interactions strongly modify the exciton dispersion, and for $\omega \leq \omega_0$ this effect may be important. Correct calculations of the exciton-phonon scattering rate require the incorporation of polariton effects (Agranovich and Konobeev 1961). The same effect and for the same frequencies takes place also for two or three photon absorption (Gale et al 1986, Denisov et al 1987, Stevenson et al 1988, Fröhlich et al 1989). Another problem arises in calculating linear and nonlinear susceptibilities of monolayers. As was shown by Hopfield (1958) and Agranovich (1959), in perfect 3 D crystals the spontaneous emission rate of polaritons vanishes, so in calculating the imaginary part of the susceptibilities only damping mechanisms due to exciton-phonon scattering contribute to the dissipative rate $\gamma(\omega)$. This is not the case, however, for systems with restricted geometries such as monolayers. Even for perfect monolayers, as was shown by Agranovich and Dubovsky (1966), the account of retarded interaction strongly changes 2 D exciton spectrum. Instead of a single branch, two polariton branches appear and one of them has a giant radiative width (Fig.1). (Various aspects of this effect were considered also by Philpott and Sherman (1975) and in many other papers, see the references in (Agranovich and Mukamel 1990)).

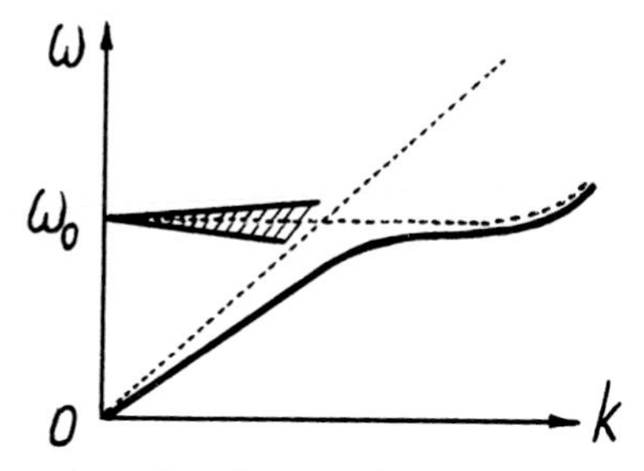

Fig.1. Dispersion curve for 2 D polaritons (the molecular transition dipole moment is perpendicular to the plane).

The radiative width Γ depends on the orientation of the transition dipoles and the wave vector and its typical value is

$$\Gamma \simeq \gamma(\frac{\lambda}{2\pi a})^2; (\frac{\lambda}{2\pi a})^2 \gg 1,$$

where γ is the radiative width of an isolated molecule, a is the lattice constant, $\lambda = 2\pi c/\omega_0$, ω_0 is the transition frequency and c is the velocity of light. For example, for the lowest transition in an Anthracene monolayer (see below) the radiation width Γ is about $10^4 s^{-1}$. It should be mentioned also that this result, and more generally, the influence of retardation on the spectrum of elementary excitations, is quite different from the case of a 3 D crystal, where retarded interactions contribute only to strong resonance renormalization of the dispersion curve for excitons (polaritons), but a radiative decay is totally absent. Ultrafast decay in Anthracene monolayers was first observed in picosecond meausurements carried out by Aaviksoo et al.(1987). The lowest electronic excited state in the surface monolayer in Anthracene is blue shifted with respect to the lowest exciton in the bulk by $\sim 204 cm^{-1}$ and is clearly seen in emission at $1.8K$. In the monolayer next to the surface it is blue shifted by $\sim 10cm^{-1}$ and in the following one by $\sim 2cm^{-1}$. These shifts are now well understood (Turlet et al 1983). Consequently, for temperatures low compared with the blue shift, the surface layer acts as an isolated monolayer and is an ideal system for investigations of 2 D excitons. Relative quantum yield measurements of the bulk and the surface emission indicate that the decay in the monolayer is purely radiative with a very small contribution of relaxation to the bulk. The picosecond timescales observed in these experiments are a beautiful example of superradiance of 2 D excitons. Thus, theoretical estimates and experiments with Anthracene crystal surfaces indicate that the radiative width may be of the order of a few tenths of a wavenumber, what, of course, should be taken into account in the calculation of susceptibilities at sufficiently low temperatures. An interesting demonstration of the necessity of properly incorporating retardation is provided also by the decay kinetics of excitonic gratings at the large exciton transition oscillator strength. At low temperatures, and for grating excitation with $\omega_1 > \omega_0$, following a fast relaxation ($\tau \sim 1ps$) a grating of the polariton density is created. The decay of this grating is related to the diffusion of polaritons rather than to the diffusion of excitons (Agranovich et al 1987, Agranovich and Leskova 1988). In this case the quantitative description of diffraction of the probe pulse requires the calculation of $\chi^{(3)}$, and polaritons as zero order states, and not the excitons should be incorporated in the calculation of $\chi^{(3)}$ (Knoester and Mukamel 1991). These comments suggest that for calculating linear and nonlinear optical susceptibilities, particularly in the region of a $2D$ excitonic transition $\omega \sim \omega_0$, it would be desirable to have exact formal expressions for the optical susceptibilities in terms of equilibrium correlation functions, which properly take into account retarded electromagnetic interactions. An example of such expression for bulk media is provided by the formula for the dielectric tensor $\varepsilon_{ij}(\omega)$, obtained by Dzyaloshinsky and Pitaevsky (1958), and the expressions for the second order susceptibilities $\chi^{(2)}_{ijl}$, obtained by Agranovich et al(1966) and by Obukhovsky and Strizhevsky (1966). Since in the derivation of these results the Maxwell equations were used, and not only the material relation between the displacement vector $\mathbf{D}$ and the electric field $\mathbf{E}$, the resulting formulae were expressed through correlation functions of the type $< EE >$, $< EEE >$, etc., E being the Maxwell field operator. These results were used later by Agranovich and Konobeev (1963) to calculate the tensor $\varepsilon_{ij}(\omega)$ in the excitonic region. The resulting expressions for $\chi^{(1)}$ and $\chi^{(2)}$ are given in terms of the macroscopic Maxwell field $\mathbf{E}$. In order to use these expressions the $\mathbf{E}$ field operator needs to be expressed in terms of the dynamical variables of the system and the Green function of the electromagnetic (Maxwell) fields should be also found. This is in general a complicated problem, particularly for an inhomogeneous medium.

2. FORMULA FOR LINEAR DIELECTRIC SUSCEPTIBILITY

In the note (Agranovich and Mukamel 1991) a new exact formula for the linear dielectric response was derived. This formula, in particular, allows us to take into account the effects of retardation and can be used naturally for 2 D systems. It relates the response to the cross correlation function of the polarization and electric field operators, and it reduces to formulae of the conventional (nonretarded) theories, when the radiation-matter interaction is treated perturbatively. The dielectric function $\varepsilon(\mathbf{k},\omega) \equiv 1 + 4\pi\chi^{(1)}(\mathbf{k},\omega)$

is given by

$$\varepsilon_{ij}(\mathbf{k},\omega) = \delta_{ij} + 4\pi M_{il}(\mathbf{k},\omega)[N(\mathbf{k},\omega)]^{-1}_{lj},$$

where

$$M_{il}(\mathbf{k},\omega) = \int d\mathbf{r} \int_{-\infty}^{+\infty} dt < \hat{P}_i(\mathbf{r},t)\hat{E}_l(\mathbf{r}_0,t_0) > \exp[-i\mathbf{k}(\mathbf{r}-\mathbf{r}_0) + i\omega t,] \tag{1}$$

$$N_{jl}(\mathbf{k},\omega) = \int d\mathbf{r} \int_{-\infty}^{+\infty} dt < \hat{E}_j(\mathbf{r},t)\hat{E}_l(\mathbf{r}_0,t_0) > \exp[-i\mathbf{k}(\mathbf{r}-\mathbf{r}_0) + i\omega t]. \tag{2}$$

It is more convenient sometimes, in particular when the Coulomb gauge is used, to use (1) and (2) with the operator of the vector-potential $\mathbf{A}(\mathbf{r}_0, t_0)$ instead of the operator $\mathbf{E}(\mathbf{r}_0, t_0)$. That alternative expression is also correct and shall be used in Section 4.

3. LINEAR RESPONSE FOR 3 D EXCITONS

In this Section we discuss the peculiarities of the use of the relation for $\chi^{(1)}_{ij}$ given in the previous Section. We shall consider here the simplest situations, e.g. cubic symmetry for 3 D excitons, the presence of the inversion centre, etc. Moreover, we shall first make no account for dissipative processes caused, e.g. by the interaction of excitons with phonons, scattering by impurities and so on. In 3 D crystals in this approximation the Hamiltonian of the crystal $\hat{H}$ is

$$\hat{H} = \sum_{\rho k} \mathcal{E}_\rho(\mathbf{k})\xi^\dagger_\rho(\mathbf{k})\xi_\rho(\mathbf{k}),$$

where $\mathcal{E}_\rho(\mathbf{k})$ is the energy of the $(\rho\mathbf{k})$ polariton, (ρ is the branch number) and $\xi^\dagger$ and ξ are its creation and annihilation operators. The quantities $\mathcal{E}_\rho(\mathbf{k})$ can be found within the microtheory framework, e.g.(Hopfield 1958, Agranovich 1959) by diagonalization of the Hamiltonian for excitons with due regard for retarding interaction. In the linear theory the operators $\xi^\dagger$, ξ are linear combinations of creation and annihilation operators of excitons and transverse photons (here the Coulomb gauge is implied). The dipole moment operator of a unit volume $\hat{P}(\mathbf{r},t)$, as well as the operator of the electric field strength $\hat{E}(\mathbf{r},t)$, can also be expressed in terms of the creation and annihilation operators of polaritons. For these operators the relations

$$\hat{P}(\mathbf{r},t) = \sum_{\rho k} \mathbf{P}(\mathbf{k},\rho)\tilde{\xi}_\rho(\mathbf{k},t)\exp(i\mathbf{k}\mathbf{r}) + h.c.,$$

$$\hat{E}(\mathbf{r},t) = \sum_{\rho k} \mathbf{E}(\mathbf{k},\rho)\tilde{\xi}_\rho(\mathbf{k},t)\exp(i\mathbf{k}\mathbf{r}) + h.c.,$$

are valid, where $\mathbf{P}(\mathbf{k},\rho)$ and $\mathbf{E}(\mathbf{k},\rho)$ are the amplitudes of the corresponding fields in a polariton $(\rho,\mathbf{k})$, and where the operators $\tilde{\xi}_\rho(\mathbf{k},t)$ are already written in the Heisenberg representation. Neglecting the effects of spatial dispersion and using the relations (1) and (2) we have $\chi_{ij} = 0$ at $i \neq j$ and the component $\chi_{11}(\mathbf{k},\omega) = \chi_{22}(\mathbf{k},\omega) = \chi_{33} = \chi(\mathbf{k},\omega)$ is determined by the relation

$$\chi(\mathbf{k},\omega) = \frac{\sum_{\rho=\rho_\perp} P(\mathbf{k},\rho)E^*(\mathbf{k},\rho)\delta(\omega-\omega_\rho(\mathbf{k}))}{\sum_{\rho=\rho_\perp} |E(\mathbf{k},\rho)|^2\delta(\omega-\omega_\rho(\mathbf{k}))}.$$

If at a given **k** the frequency ω coincides with some frequency $\omega_\rho(k)$, then for χ we obtain

$$\chi(\mathbf{k},\omega) = \frac{P(\mathbf{k},\rho)}{E(\mathbf{k},\rho)}, \omega = \omega_\rho(\mathbf{k}).$$

But if $\omega \neq \omega_{\rho\perp}(\mathbf{k})$, then in the above relation for $\chi(\mathbf{k},\omega)$ we obtain indeterminacy of the type of a zero divided by zero. In order to reveal this indeterminacy we consider in more details the equation for $\chi(\mathbf{k},\omega)$. It follows from this equation that the quantity $\chi(\omega)$ is given at a set of points $\omega = \omega_\rho(\mathbf{k})$. As the number of these points form an infinite denumerable set, we can use the known uniqueness theorem about analytical continuation (see, e.g. Shabat 1976). According to this theorem the function of the complex variable given on an infinite denumerable (contaning the limit point) set of points is uniquely determined by analytical continuation. In our case the use of this theorem means that as an expression for $\chi(\mathbf{k},\omega)$ at any ω we can use the expression

$$\chi(\mathbf{k},\omega) = \frac{P(\mathbf{k},\omega)}{E(\mathbf{k},\omega)}, \tag{3}$$

where k and ω are already not related by the dispersion equation $\omega = \omega_\rho(\mathbf{k})$ and where $P(\mathbf{k},\omega)$ and $E(\mathbf{k},\omega)$ functions should be obtained from the expressions for $P(\mathbf{k},\rho)$ and $E(\mathbf{k},\rho)$ in which the frequency $\omega_\rho(\mathbf{k})$ has been replaced by the frequency ω.

To proceed with calculations of the quantity $\chi(\mathbf{k},\omega)$ it is necessary to use explicit expressions for $P(\mathbf{k},\rho)$ and $E(\mathbf{k},\rho)$. This can only be done basing on the use of some microscopic model. We turn to the work of Agranovich and Konobeev (1963) in which the amplitudes $P(\mathbf{k},\rho)$ and $E(\mathbf{k},\rho)$ were calculated. Using the results of these calculations, we find that

$$\frac{P(\mathbf{k},\rho)}{E(\mathbf{k},\rho)} = \sum_\mu \frac{2E_\mu(\mathbf{k})|P(\mathbf{k},\mu)|^2}{E_\mu^2(\mathbf{k}) - \mathcal{E}_\rho^2(\mathbf{k})},$$

i.e. that (see (3))

$$\chi(\mathbf{k},\omega) = \sum_\mu \frac{2E_\mu(\mathbf{k})|P(\mathbf{k},\mu)|^2}{E_\mu^2(\mathbf{k}) - \hbar^2\omega^2}.$$

This relation can also be obtained by means of ordinary relations for χ_{ij}. Thus, the linear response of the crystal is, apart from a possible change of the damping constant (see Section 1), not affected by taking into account retardation effects. As shown in the next Section, the situation is different for $2D$ excitons.

4. LINEAR RESPONSE FOR 2 D EXCITONS

As already noted in Section 1, description of optical properties of a film within the framework of the phenomenological Maxwell electrodynamics requires the knowledge of its polarizability per unit area of the film. Therefore, in order to use the relations for polarizability given in Section 2 it is necessary to express the dipole moment operator of the unit area of the film $\hat{\mathbf{P}}(\mathbf{r})$ in terms of the variables of the system. Moreover, the same as for $3D$ excitons, it is also necessary to know the operator of the electric macrofield strength $\hat{\mathbf{E}}(\mathbf{r})$. In order to stress in the simplest way the problems arising in the determination of a polarizability we consider the contribution to a polarizability of Frenkel's excitons in an organic monolayer with a single molecule in the unit cell. The dipole moment operator $\hat{\mathbf{P}}$ in terms of molecular Bose creation and annihilation operators in this case is:

$$\hat{\mathbf{P}}(n) = \frac{\mathbf{P}^{0f}}{\Delta}(B_n + B_n^\dagger), \tag{4}$$

where Δ is the area of the unit cell, $\mathbf{P}^{0f}$ is the matrix element of the $0 \to f$ transition of the dipole moment of the molecule, the excited state f is assumed to be nondegenerated. Upon a transition to the Coulomb excitons

$$B_n = \frac{1}{\sqrt{N}} \sum_k B_k \exp(i\mathbf{kn}), \tag{5}$$

and therefore,

$$\hat{\mathbf{P}}(n,t) = \frac{\mathbf{P}^{0f}}{\Delta\sqrt{N}} \sum_k [B_k(t) + B_{-k}^\dagger(t)] \exp(i\mathbf{kn}). \tag{6}$$

In the case of a monolayer being discussed $n \equiv \mathbf{n}(n_1, n_2)$, $\mathbf{k} \equiv (k_1, k_2)$. In the determination of the operator of the macroscopic electric field $\hat{\mathbf{E}}(\mathbf{r}, t)$ it should be taken into account that in the Coulomb gauge the electric field strength $\hat{\mathbf{E}}(\mathbf{r}, t) = \hat{\mathbf{E}}^{\|}(\mathbf{r}, t) + \hat{\mathbf{E}}^{\perp}(\mathbf{r}, t)$, where

$$\hat{\mathbf{E}}^{\perp} = -\frac{1}{c}\frac{\partial \hat{\mathbf{A}}}{\partial t}.$$

and where the operator

$$\hat{\mathbf{A}}(\mathbf{r}, t) = \sum_{kj} \left(\frac{2\pi\hbar c^2}{Vkc}\right)^{1/2} \mathbf{e}_{kj} [a_{kj}(t) + a_{-k_j}^\dagger(t)] \exp(i\mathbf{kr}).$$

In this relation $\hat{a}_j^\dagger(\mathbf{q})$, $\hat{a}_j(\mathbf{q})$ are the Bose creation and annihilation operators of photons, $\mathbf{q}$ is the wave vector of the photon, and $j(=1,2)$ labels its polarizations. To obtain the operator of longitudinal part of electric field we recall that if $\mathbf{E}(\mathbf{r}, n)$ is the electric field at the point $\mathbf{r}$, created by the dipole in the site n, then

$$E_i(\mathbf{r}, n) = -T_{ij}(\mathbf{r} - \mathbf{n}) P_j(n),$$

where

$$T_{ij}(\mathbf{r} - \mathbf{n}) = \frac{\delta_{ij}}{|\mathbf{r} - \mathbf{n}|^3} - 3\frac{(\mathbf{r} - \mathbf{n})_i(\mathbf{r} - \mathbf{n})_j}{|\mathbf{r} - \mathbf{n}|^5} = -\nabla_i \nabla_j \frac{1}{|\mathbf{r} - \mathbf{n}|}.$$

Consequently,

$$\hat{E}_i^{\|}(\mathbf{r}) = -\sum_{nj} T_{ij}(\mathbf{r} - \mathbf{n}) \hat{P}_j(\mathbf{n}),$$

or, if one takes into account (4)-(6),

$$\hat{\mathbf{E}}^{\|}(\mathbf{r}, t) = \sum_k \mathbf{a}(\mathbf{r}, \mathbf{k})[B_k(t) + B_{-k}^\dagger(t)],$$

where

$$a_i(\mathbf{r}, \mathbf{k}) = \frac{1}{\Delta\sqrt{N}} \nabla_i \nabla_j \sum_n \frac{\exp(i\mathbf{kn})}{|\mathbf{r} - \mathbf{n}|} P_j^{0f}.$$

As we are interested only in the long-wave part of the field, we should replace the sum over $\mathbf{n}(n_1, n_2)$ by the integral over $\rho(x, y)$. Therefore, assuming $\mathbf{r} \equiv (\mathbf{r}_{\|}, z)$ we find that

$$a_i(\mathbf{r}, \mathbf{k}) = \frac{1}{\Delta^2\sqrt{N}} \nabla_i \nabla_j \{\exp(i\mathbf{k}\mathbf{r}_{\|}) \int d\rho \frac{\exp(i\mathbf{k}\rho)}{\sqrt{z^2+\rho^2}}\} P_j^{0f}.$$

As

$$\int d\rho \frac{\exp(i\mathbf{k}\rho)}{\sqrt{z^2+\rho^2}} = \frac{2\pi}{k} e^{-kz},$$

we obtain finally:

$$a_i(\mathbf{r}, \mathbf{k}) = \frac{2\pi}{\Delta^2\sqrt{N}} \nabla_i \nabla_j \{e^{-kz} \exp(i\mathbf{k}\mathbf{r}_{\|})\} P_j^{0f}.$$

For definiteness we consider the vector $\mathbf{P}^{0f}$ to be directed along the axis x and, moreover, we are interested below in the polarizability χ_{11}. To find this component of the tensor $\hat{\chi}$ we need to determine only the x- component of the operators $\hat{\mathbf{E}}$ and $\hat{\mathbf{P}}$. In particular, as in the case being discussed

$$a_1(\mathbf{r}, \mathbf{k}) = -\frac{2\pi P^{0f} k_1^2}{\Delta^2\sqrt{N} k} e^{-kz} \exp(i\mathbf{k}\mathbf{r}_{\|}),$$

the operator

$$\hat{E}_1^{\|}(\mathbf{r}, t) = -\frac{2\pi P^{0f}}{\Delta^2\sqrt{N}} \sum_k k_1^2 e^{-kz} \frac{\exp(i\mathbf{k}\mathbf{r}_{\|})}{k} [B_k(t) + B_{-k}^{\dagger}(t)]. \tag{7}$$

We can now proceed with calculation of the matrices $M_{ij}(k, \omega)$ and $N_{ij}(k, \omega)$. Substituting the sum $\mathbf{E} = \mathbf{E}^{\|} + \mathbf{E}^{\perp}$ into the relations (1) and (2) we find that

$$M_{ij}(k, \omega) = M_{ij}^{\|}(k, \omega) + M_{ij}^{\perp}(k, \omega),$$

$$N_{ij}(k, \omega) = N_{ij}^{\|,\|}(k, \omega) + N_{ij}^{\perp,\perp}(k, \omega) + N_{ij}^{\|,\perp}(k, \omega) + N_{ij}^{\perp,\|}.$$

The matrices introduced are determined, the same as before, by the relations (1) and (2). In these relations instead of the total operator $\hat{E}$ its either longitudinal or transverse part is used. Thus, e.g. $M_{ij}^{\|}(k, \omega)$ is determined by the correlator $< P_i E_l^{\perp} >$, $N_{ij}^{\perp,\|}$ by the correlator $< E_i^{\perp}, E_j^{\|} >$, etc. In our case, however, we do not need to calculate all these matrices as the longitudinal part of the macrofield created by a two-dimensional lattice of dipoles is small compared to the transverse part of the field. Actually, from the relation (7) it follows that the longitudinal part of the field at $z = 0$ is small (of the order of $ka \ll 1$). Thus, we have to calculate below only two quantities: $M_{ij}^{\perp}$ and $N_{ij}^{\perp,\perp}$. The correlation functions they contain can be expressed in terms of causal Green's functions for the same operators. Therefore, in what follows we concentrate on the calculation of the appropriate Green functions. In the nonrelativistic approximation the operator $\hat{H}_{int}$ can be written as

$$\hat{H}_{int} = \hat{H}_{int}^{(1)} + \hat{H}_{int}^{(2)},$$

where

$$\hat{H}_{int}^{(1)} = \sum_{q,\mu k} T_{\mu k}^{q} (\hat{a}_q + \hat{a}_{-q}^{\dagger})(\hat{B}_{\mu(-k)} - \hat{B}_{\mu k}^{\dagger}),$$

$$T_{\mu k}^{q} = -i\left(\frac{2\pi N}{Vq\hbar c}\right)^{1/2} E_{\mu k}(\mathbf{P}_{\mu k}\mathbf{e}_q). \tag{8}$$

$$\hat{H}_{int}^{(2)} = \sum_{q,q'} \Pi_{q'}^{q} (\hat{a}_q + \hat{a}_{-q}^{\dagger})(\hat{a}_{-q'} + \hat{a}_{q'}^{\dagger}),$$

$$\Pi_{q'}^{q} = \frac{\pi e^2 S \hbar N}{mVc} \frac{(\mathbf{e}_q \mathbf{e}_{q'})}{\sqrt{qq'}}. \tag{9}$$

In (8), (9) N is the total number of unit cells in a crystalline system; $\mathbf{P}_{\mu k}$ is the matrix element of the operator of the dipole moment betwen the ground and $\mu\mathbf{k}$ exciton states; $\mathbf{e}_q$ are the orts of polarization of photons; V is the volume of cyclicity. Neglecting local field corrections to the transverse part of the field $\mathbf{E}^{\perp}$ we shall assume that $T_{\mu k}^{q}$ is nonzero only if the tangential component of the wavevector $\mathbf{q}$ is equal to $\mathbf{k}$, i.e. $\mathbf{q} = (\mathbf{k}, q_n)$, where q_n is the normal to the surface component of the photon wave vector. The prime on the summation indicates that $\mathbf{q}_{\|} = \mathbf{q}'_{\|}$, S is the total number of electrons in the unit cell of the crystal. We introduce now the photon and the exciton causal Green functions, respectively

$$D_{q,q'}(t-t') = -i < T\{(\hat{a}_q + \hat{a}_{-q}^{\dagger})_t (\hat{a}_{-q'} + \hat{a}_q^{\dagger})_{t'}\} >,$$

$$G_{\mu k}(t-t') = -i < T\{(\hat{B}_{\mu k} - \hat{B}_{\mu(-k)}^{\dagger})_t (\hat{B}_{\mu(-k)} - B_{\mu k}^{\dagger})_{t'}\} > .$$

where <> denotes the averaging over the ground state. For vacuum photon and exciton fields the corresponding Fourier transforms of these functions are represented as follows:

$$D_{q,q'}(\omega) = D_q^{(0)}(\omega)\delta_{qq'} = \frac{2qc}{\omega^2 - q^2c^2 + i\eta}\delta_{qq'},$$

$$G_{\mu k}(\omega) = -\frac{2\omega_{\mu k}}{\omega^2 - \omega_{\mu k}^2 + i\eta},$$

$$\omega_{\mu k} \equiv \frac{E_{\mu k}}{\hbar}, \eta \to +0.$$

The Dyson equation for the function $D_{qq'}(\omega)$ is:

$$D_{q,q'} = D_q^{(0)}\delta_{qq'} + D_q^{(0)} \sum_{q_1} M_{q_1}^{q} D_{q_1,q'}$$

The mass operator M is:

$$M_{q_1}^{q}(\omega) = \frac{2}{\hbar}\Pi_{q_1}^{q} + \frac{1}{\hbar^2}\sum_{\mu} T_{\mu k}^{q} T_{\mu k}^{q_1} G_{\mu k}^{(0)}(\omega).$$

This expression can be simplified considerably using the sum rule which follows directly from the commutation relations for the operators of the coordinate and of the momentum of charges

$$\sum_{\mu} (\mathbf{P}_{\mu k})_i (\mathbf{P}_{\mu k})_j E_{\mu k} = \frac{\hbar^2 e^2 S}{2m}\delta_{ij}$$

Using this relation we reduce the mass operator to the following form

$$M_{q_1}^{q}(\omega) = \sum_{\mu} T_{\mu k}^{q} T_{\mu k}^{q_1} (\frac{\omega}{\omega_{\mu k}})^2 G_{\mu k}^{(0)}(\omega). \tag{10}$$

Below we restrict ourselves by consideration of the spectral region near an isolated exciton band and omit the summation in (10). In this approximation $M(\omega)$ is represented in a factorized form

$$M_{q_1}^{q}(\omega) = M_q(\omega) M_{q_1}(\omega),$$

$$M_q(\omega) = \frac{\omega}{\omega_{\mu k}} T^q_{\mu k} \{G^{(0)}_{\mu k}\}^{1/2}$$

and the Dyson equation can be solved exactly. Multiplying both sides of this equation by M_{q_1} and making summation over q_1 we obtain

$$\Phi_q \equiv \sum_{q1} M_{q_1} D_{q_1,q} = \frac{D^{(0)}_q M_q}{1 - \sum'_{q_1} D^{(0)}_{q_1} M^2_{q_1}(\omega)}$$

The poles of the function $\Phi_q(\omega)$, which determines both the exciton and the photon Green functions, give the spectrum of elementary excitations of the system. Actually, the photon Green function $D_{q,q'}$ in the assumed approximation can be written as follows:

$$D_{q,q'}(\omega) = D^{(0)}_q(\omega)\delta_{qq'} + D^{(0)}_q(\omega) M_q(\omega) \Phi_{q'}(\omega).$$

The exciton Green function $G_{\mu k}(\omega)$ can be written in the form

$$G_{\mu k}(\omega) = G^{(0)}_{\mu k}(\omega) + [G^{(0)}_{\mu k}(\omega)]^2 \sum_{q,q'} T^q_{\mu k} T^{q'}_{\mu k} [D^{(0)}_q(\omega)\delta_{qq'} + D^{(0)}_q(\omega) M_q(\omega) \Phi_{q'}(\omega)].$$

Substituting the expression for $\Phi_q(\omega)$ into the relations obtained we draw a conclusion that the poles of $\Phi_q(\omega)$ are also the poles of $D_{q,q'}$ and $G_{\mu k}(\omega)$. Using the explicit expressions for $D^{(0)}_q(\omega)$ and $M_q(\omega)$ we obtain the dispersion equation in the form

$$\omega^2 - \omega^2_{\mu k} = \sum_q \frac{8\pi N \omega_{\mu k} (\mathbf{P}_{\mu k} \mathbf{e}_q)^2}{\hbar V} \frac{\omega^2}{\omega^2 - q^2 c^2 + i\eta}, \quad q^2 = k^2 + q^2_n. \tag{11}$$

This equation was investigated by different authors for various cases of polarization of an excitonic transition. We shall not repeat this analysis and turn directly to calculation of the polarizability χ_{11} that is of interest. From the results presented in Section 2 it follows that this polarizability equals the ratio of spectral intensities of the functions $\int d\mathbf{r}_\| < 0 \mid \hat{P}_1(\mathbf{r}_\|, 0; t), \hat{A}_1(0,0;0) \mid 0 > \exp(i\mathbf{k}_\| \mathbf{r}_\|)$ and $\int d\mathbf{r}_\| < 0 \mid \hat{E}^\perp_1(\mathbf{r}_\|, 0; t), \hat{A}_1(0,0;0) \mid 0 > \exp(i\mathbf{k}_\| \mathbf{r}_\|)$. With the use of Eq.(6) and the expression for $\hat{\mathbf{E}}^\perp$ the expression for the sought polarizability can be written in the form

$$\chi_{11}(\omega, \mathbf{k}) = \frac{\tilde{M}(\omega)}{\tilde{N}(\omega)},$$

where

$$\tilde{M}(\omega) = -\frac{i}{\omega \Delta \sqrt{N}} \sum_{q_n j} \frac{P^{0f}_1 (\mathbf{e}_{qj})_1}{\sqrt{k^2 + q^2_n}} \psi_{k,q}(\omega),$$

$$\Psi_{k,q}(\omega) = \int_{-\infty}^{+\infty} dt e^{i\omega t} \{< 0 \mid \frac{d}{dt} (\hat{B}_k(t) + \hat{B}^\dagger_{-k}(t))(\hat{a}_{-q}(0) + \hat{a}^\dagger_q(0) \mid 0 >\},$$

while

$$\tilde{N}(\omega) = -i\omega \left(\frac{2\pi\hbar}{Vc}\right)^{1/2} \sum_{q_n, q'_n, j, j'} \frac{(\mathbf{e}_{qj})_1 (\mathbf{e}_{q',j'})_1}{\sqrt{qq'}} \phi_{q,q'}(\omega),$$

$$\phi_{q,q'}(\omega) = \int_{-\infty}^{+\infty} dt < 0 \mid (\hat{a}_q(t) + \hat{a}^\dagger_{-q}(t))(\hat{a}_{-q'}(0) + \hat{a}^\dagger_{q'}(0)) \mid 0 > \exp(i\omega t).$$

Let us remind that up to now all the operators here are in the Heisenberg represanta-tion. Using the expansion for the S-matrix (Abrikosov et al 1963) and the interaction representation it can be shown that the Fourier transform

$$< 0 \mid \frac{d}{dt}(\hat{B}_k(t) + \hat{B}^{\dagger}_{-k}(t))(\hat{a}_{-q}(0) + \hat{a}^{\dagger}_q(0) \mid 0 >_{\omega} = \frac{i\omega^2}{E_{\mu k}} G^{(0)}_{\mu k}(\omega) \sum_{q',j'} T^{q'}_{\mu k} \phi_{q,q'}.$$

Using the relationship between the spectral intensities and the imaginary part of the Green functions (Zubarev 1960) the function $\phi_{q,q'}$ is given by

$$\phi_{q,q'}(\omega) = -2Im\{D_{q,q'}(\omega)\}.$$

Thus

$$\chi_{11}(\omega, \mathbf{k}) = \frac{(P_1^{0f})^2}{\Delta\hbar} \frac{\alpha(\mathbf{k},\omega)}{\beta(\mathbf{k},\omega)},$$

where

$$\alpha(\mathbf{k},\omega) = G^{(0)}_{\mu k}(\omega) \sum_{q_n,q'_n,j,j'} \frac{(\mathbf{e}_{qj})_1(\mathbf{e}_{q'j'})_1}{\sqrt{qq'}} Im\{D^k_{qq'}(\omega)\};$$

while

$$\beta(\mathbf{k},\omega) = \sum_{q_n,q'_n,j,j'} \frac{(\mathbf{e}_{qj})_1(\mathbf{e}_{q'j'})_1}{\sqrt{qq'}} Im\{D^k_{qq'}(\omega)\}.$$

The nominator as well as the denominator are both nonzero only for the frequencies satisfying the dispersion relation (11). For these frequencies

$$\frac{\alpha(\mathbf{k},\omega)}{\beta(\mathbf{k},\omega)} = G^{(0)}_{\mu k}(\omega).$$

If at the frequency ω the nominator and denominator both turn to zero, the ratio $\frac{\alpha}{\beta}$ be-comes undetermined. Thus, in this case the same as in the preceding Section, analytical continuation of the expression obtained into the region of arbitrary real ω is needed. In contrast with the situation of the preceding section the solutions of the dispersion rela-tion (11) are complex and are determined by two functions $\omega_\rho(\mathbf{k}) + i\Gamma[\omega_\rho(\mathbf{k})]$. Taking this into account and making the analytical continuation we obtain

$$\chi_{11}(\mathbf{k},\omega) = \frac{2(P_1^{0f})^2\omega_{\mu k}/\hbar\Delta}{\omega^2_{\mu k} - \omega^2 - 2i\omega\Gamma(\omega,\mathbf{k})},$$

where

$$\Gamma(\omega,\mathbf{k}) = Im\{\sum_{q_n} \frac{4\pi N\omega_{\mu k}(P^{0f})^2 e_q^2}{\hbar V} \frac{\omega}{\omega^2 - (k^2 + q_n^2)c^2 + i\eta}\}$$

Of course, this result could already be expected on the basis of totally qualitative con-siderations and the analogous result should probably take place also for small clusters (with the size $d < \lambda$, λ is the wavelength). The quantity $\Gamma(\omega,\mathbf{k})$ turns out to be nonzero only if the point $(\omega,\mathbf{k})$ lies above the light straight line in the $(\omega,\mathbf{k})$ diagram. This function has been calculated in the works mentioned before and namely this quantity for the Anthracene has been measured by Aaviksoo et al (1987). As Γ is proportional to the oscillator strength of the excitonic transition in the expression obtained above the exciton-photon interaction is taken into account in all orders of the perturbation theory.

The details of the derivation presented above will be published elsewhere (Agranovich et al 1992). Moreover, in this work a new representation (in terms of the retarded Green functions) for linear and nonlinear polarizabilities is developed. In the next Section we shall outline the main idea of the derivation and the results.

5. LINEAR AND NONLINEAR POLARIZABILITIES IN TERMS OF THE RETARDED GREEN FUNCTIONS

In what follows we shall use the gauge with the scalar potential equal to zero. In this case the operators of the electric field strength $\hat{\mathbf{E}}$ and magnetic field $\hat{\mathbf{H}}$ are related to the operator of the vector-potential $\hat{\mathbf{A}}$ by

$$\hat{\mathbf{E}} = -\frac{\partial \hat{\mathbf{A}}}{\partial t}, \hat{\mathbf{H}} = curl \hat{\mathbf{A}}.$$

In the presence of external currents the Hamiltonian of the system has the form $\hat{H} + \hat{H}^{ext}$, where $\hat{H}$ is the Hamiltonian of the medium (with regard for retardation) and

$$\hat{H}^{ext}(t) = -\frac{1}{c} \int \mathbf{A}(\mathbf{r}, t) \hat{\mathbf{j}}^{ext}(\mathbf{r}, t) d\mathbf{r}.$$

Then, since

$$e^{-\frac{i}{\hbar} t(\hat{H} + \hat{H}^{ext})} = e^{-\frac{i}{\hbar} t \hat{H}} S_{ext}(t),$$

where the operator

$$S_{ext}(t) = T_t \exp\{-\frac{i}{\hbar} \int_{-\infty}^{t} \hat{H}^{ext}(t')dt'\}$$

the averages of the operators of the polarization $\hat{\mathbf{P}}$ and the vector-potential $\hat{\mathbf{A}}$ induced by the external currents are

$$< \hat{\mathbf{P}}(\mathbf{r}, t) > = < S_{ext}^{-1}(t) \hat{\mathbf{P}}(\mathbf{r}, t), S_{ext}(t) > .$$

$$< \hat{\mathbf{A}}(\mathbf{r}, t) > = < S_{ext}^{-1}(t) \hat{\mathbf{A}}(\mathbf{r}, t), S_{ext}(t) > .$$

where $\hat{\mathbf{P}}(\mathbf{r}, t) = e^{it\hat{H}} \hat{\mathbf{P}}(\mathbf{r}) e^{-it\hat{H}}$ and $\hat{\mathbf{A}}(\mathbf{r}, t)$ is defined analogously. Expansion of S_{ext} in powers of $\hat{H}^{ext}$ gives the expansions of the averages of the operators $\hat{P}$ and $\hat{A}$ in powers of $\mathbf{j}^{ext}$. The coefficients of these expansions turn out to be expressed in terms of different types of the retarded Green functions. For example, the Fourier component of the polarization induced by external currents is given by

$$P_i(k) = -\psi_{il}(k) j_l^{ext}(k) - \int dk_1 dk_2 \psi_{ilm}(k, k_1, k_2) j_l^{ext}(k_1) j_m^{ext}(k_2) \delta(k - k_1 - k_2) + ... \quad (12)$$

and analogously the vector-potential

$$A_i(k) = -\phi_{il}(k) j_l^{ext}(k) - \int dk_1 dk_2 \phi_{ilm}(k, k_1, k_2) j_l^{ext}(k_1) j_m^{ext}(k_2) \delta(k - k_1 - k_2) + ... \quad (13)$$

where $k \equiv (\mathbf{k}, \omega)$, $dk \equiv d\omega d\mathbf{k}$ and $\psi_{il}(k)$, $\phi_{il}(k)$ are the Fourier transforms of the Green functions

$$\psi_{il}(\mathbf{r}, t; \mathbf{r}_1, t_1) = -\frac{i}{\hbar c} \theta(t - t_1) < [P_i(\mathbf{r}, t) A_l(\mathbf{r}_1, t_1)] >,$$

$$\phi_{il}(\mathbf{r},t;\mathbf{r}_1,t_1) = -\frac{i}{\hbar c}\theta(t-t_1) < [A_i(\mathbf{r},t)A_l(\mathbf{r}_1,t_1)] >,$$

and so on. Substituting the expansions (12),(13) into the nonlinear optics expansion of the polarization $\mathbf{P}$ in powers of $\mathbf{E}$,

$$P_i(k) = \chi_{il}^{(1)}(k)E_l(k) + \int dk_1 dk_2 \chi_{ilm}^{(2)}(k,k_1,k_2)E_l(k_1)E_m(k_2)\delta(k-k_1-k_2) + ...$$

taking into account that $E_i(\mathbf{k},\omega) = \frac{i\omega}{c}A_i(\mathbf{k},\omega)$ and equating the terms of equal powers of $\mathbf{j}^{ext}$ we can determine the polarizabilities $\chi^{(n)}$ in terms of the retarded Green functions ψ_{il}, ϕ_{il}, ψ_{ilm}, ϕ_{ilm} and so on. For instance, for the linear polarizability $\chi_{ij}^{(1)}(\mathbf{k},\omega)$ we obtain

$$\chi_{ij}^{(1)}(\mathbf{k},\omega) = -i\frac{c}{\omega}\psi_{il}(\mathbf{k},\omega)\{\phi^{-1}(\mathbf{k},\omega)\}_{lj}. \tag{14}$$

Thus, new expressions for the optical susceptibility are obtained in terms of the equilibrium correlation functions which properly take into account the interaction between the matter and the radiation. These expressions are convenient for calculations of optical response of homogeneous media and, also, for example, thin films. It seems also, that the procedure of calculations of polarizabilities proposed here is formally more simple and direct than the commonly used method which requires transformation of the optical response to the external field into the optical response to the internal macrofield. However, in view of the wellknown "meanness conservation" law in the present formulation a similar problem arises for an arbitrary inhomogenious medium when inverting the matrix $\phi(\mathbf{k},\omega)$ (see Eq.(14)). The expressions for the polarizabilities of higher orders and also the transition to the arbitrary gauge are presented in (Agranovich et al 1992). Let us point out also that the use of Eq.(14) for $\chi^{(1)}$ for the excitonic region of the spectrum for 3 D and 2 D crystals gives the same results which have been established in Sections 3 and 4.

6. ON SUPERRADIANCE OF BIEXCITONS: CAN FAST RADIATIVE DECAY COMPENSATE WEAKNESS OF NONLINEARITY?

In this Section we point out a new possibility for observing biexcitons in two-dimensional microstructures. The main idea (see Agranovich and Mukamel 1990) is that in two-dimensional systems (interfaces, thin films, quantum wells or molecular monolayers) at low temperature a biexciton will exhibit a superradiant radiative decay, very similar to what was first observed by Aaviksoo et al (1987) for ordinary 2 D excitons. This may allow the observation of fluorescence in which the entire biexciton energy is transferred to a photon $\hbar\omega = E_2(k)$. This channel can therefore compete with the decay to an exciton and a photon with $\hbar\omega \approx E_2(k) - E_1(k')$ which is normally observed in semiconductors. In molecular crystals fast superradiant radiative decay may compete with exciton-exciton annihilation. For simplicity we discuss below a case of molecular crystals with Frenkel excitons. The molecular crystal Hamiltonian which includes exciton-exciton interactions was derived in (Agranovich 1968) (see also Agranovich and Galanin 1982).

Adopting a two level model for each molecule (with the ground state 0 and the excited state f, with gas phase electronic transition energy $\Delta\epsilon_f$), the Hamiltonian is given by

$$\hat{H} = \epsilon_0 + \hat{H}_1 + \hat{H}_2 + \hat{H}_3 + \hat{H}_4, \tag{15}$$

where ϵ_0 is the ground state energy of the crystal, and

$$\hat{H}_1 = \sum_{n,m}{}' < f0 \mid \hat{V}_{nm} \mid 00 > (P^\dagger_{nf} + P_{nf}),$$

$$\hat{H}_2 = \sum_n (\Delta\epsilon_f + \mathcal{D}_f) P^\dagger_{nf} P_{nf} + \sum_{n,m}{}' M^f_{nm} [P^\dagger_{mf} P_{nf} + \frac{1}{2}(P^\dagger_{nf} P^\dagger_{mf} + P_{nf} P_{mf})],$$

$$\hat{H}_3 = \sum_{n,m}{}' (< 0f \mid \hat{V}_{nm} \mid ff > - < f0 \mid \hat{V}_{nm} \mid 00 >)(P^\dagger_{nf} + P_{nf}) P^\dagger_{mf} P_{mf},$$

$$\hat{H}_4 = \frac{1}{2} \sum_{n,m} \Phi_{nm} P^\dagger_{nf} P_{nf} P^\dagger_{mf} P_{mf},$$

where

$$\Phi_{nm} =< ff \mid \hat{V}_{nm} \mid ff > + < 00 \mid \hat{V}_{nm} \mid 00 > -2 < f0 \mid \hat{V}_{nm} \mid f0 >, \qquad (16)$$

$$\mathcal{D}_f = \sum_n (< 0f \mid \hat{V}_{nm} \mid 0f > - < 00 \mid \hat{V}_{nm} \mid 00 >),$$

$$M^f_{nm} =< 0f \mid \hat{V}_{nm} \mid f0 > .$$

Here $P^\dagger_{nf}$ (P_{nf}) is the creation (annihilation) operator for an excitation on the nth site. These operators satisfy the Pauli commutation relations. $\hat{V}_{nm}$ is the intermolecular interaction and its matrix element is determined by

$$< \alpha\beta \mid \hat{V}_{nm} \mid \gamma\delta >= \int \varphi^\alpha_n \varphi^\beta_m \hat{V} \varphi^\gamma_n \varphi^\delta_m d\tau_n d\tau_m,$$

where φ^α_n is the state where the site n is in the state α. The operator of the number of the elementary excitations is $\hat{N} \equiv \sum_n P^\dagger_{nf} P_{nf}$. If in Eq.(15) we neglect $\hat{H}_1$ and $\hat{H}_3$, we obtain a zero-order Hamiltonian which conserves the number of excitons (i.e. commutes with $\hat{N}$):

$$\hat{H}_0 \equiv \epsilon_0 + \hat{H}_2 + \hat{H}_4.$$

The last term in this Hamiltonian is responsible for the appearance of a bound two-exciton state (biexciton). We suppose below that the exciton-exciton interaction energy (Eq.(16)) Φ_{nm} is sufficiently large, so that a bound two-exciton state does exist. We will take into account the mixing effect of states with different numbers of excitons. The single-particle and two-bound-particles eigenstates and eigenvalues of $\hat{H}_0$ with wave vector $\mathbf{k}$ are

$$\psi^0_1(\mathbf{k}) = \sum_n \psi_n(\mathbf{k}) P^\dagger_{nf} \mid 0 >, E \equiv E_1(\mathbf{k}),$$

$$\psi^0_2(\mathbf{k}) = \sum_{n,m} \psi_{nm}(\mathbf{k}) P^\dagger_{nf} P^\dagger_{mf} \mid 0 >, E \equiv E_2(\mathbf{k}),$$

where ψ_n and $\psi_{nm} = \psi_{mn}$ are the wave functions of the exciton and the biexciton, respectively, in the coordinate representation. Typically the matrix elements of $\hat{H}_1$ and $\hat{H}_3$ are much smaller than the spacing between the eigenstates of $\hat{H}_0$. (In Anthracene, for example, the former are $\sim 1000 cm^{-1}$, whereas the latter is $\sim 25000 cm^{-1}$ so that their ratio is $\sim 10^{-2}$). This justifies treating $\hat{H}_1$ and $\hat{H}_3$ as a small perturbation. We thus define the perturbed two-particle bound state:

$$\psi_2(\mathbf{k}) = \psi_2^0(\mathbf{k}) + \sum_s \frac{< \psi_2^0(\mathbf{k}) \mid \hat{H}_1 + \hat{H}_3 \mid \psi_s^0(\mathbf{k}) >}{E_2(\mathbf{k}) - E_s(\mathbf{k})}.$$

Therefore (the details of derivation see in (Agranovich and Mukamel 1990)),

$$\psi_2(\mathbf{k}) \approx \psi_2^0(\mathbf{k}) + \frac{D}{E_1(\mathbf{k})}\psi_1^0,$$

where D is of the order of the gas phase-condensed matter energy shift of the electronic excitation. It follows from this expression that the transition dipole of the biexciton to the ground state will be related to that of the exciton by

$$< \psi_2(\mathbf{k}) \mid P \mid 0 > \approx < \psi_1(\mathbf{k}) \mid P \mid 0 > \frac{D}{E_1(\mathbf{k})}$$

Since $|D/E_1(k)| \ll 1$, the biexciton state will have a weak borrowed oscillator strength to the ground state due to the mixing with ψ_1^0 (assuming that the single-exciton transition is allowed). This simple and well known argument explains why in three-dimensional semiconductor crystals biexciton decay proceeds through the channel $E_2 = E_1 + \hbar\omega$, where in the final state we have one exciton and one photon. However, for two-dimensional crystals and for small microcrystallites the situation may be quite different. Let us consider two-dimensional crystals (a molecular monolayer). We should bear in mind that the state of biexcitons, like the state of single excitons, is determined by a single wave vector $\mathbf{k}$ and not by two wave vectors $\mathbf{k}_1$, $\mathbf{k}_2$ as is the case for two free excitons. This is why all the results obtained for superradiance of two-dimensional excitons, may be directly used also for two-dimensional biexcitons. This means that biexcitons in a two-dimensional system will show a superradiant decay with the radiative width Γ_{BE}:

$$\Gamma_{BE} \approx \gamma(\frac{D}{E_1(\mathbf{k})})^2(\frac{\lambda}{2\pi a})^2. \tag{17}$$

For anthracene $D/E_1(\mathbf{k}) \sim 10^{-2}$, so that without superradiance $\Gamma_{BE} \approx 10^{-4}\gamma$. The quantum yield for this channel of direct biexciton emission is, therefore, very small. The third factor in Eq.(17) however gives superradiance. For Anthracene $(\lambda/2\pi a)^2 \sim 10^4$. In this case we see compensation of the smallness of $(D/E_1)^2$ by the large values of the superradiance factor $(\lambda/2\pi a)^2$. The radiative decay rate, Γ_{BE} is therefore in this case comparable in magnitude to the single exciton decay γ. This, of course, need not always be the case. The anharmonicity factor $(D/E_1)^2$ and the superradiant enhancement $(\lambda/2\pi a)^2$ are two independent parameters and their relative magnitude may be very different. Our main point is that in two-dimensional systems there exists a new possibility for observing biexcitons which should be taken into account (the influence of superradiance and two- dimentionality on other main channels of biexciton decay see in (Agranovich and Mukamel 1990)). We would like to point out also that the arguments presented here apply to any geometrically restricted microstructure such as J aggregates, and small clusters (smaller than the optical wavelength). Note, in addition, that the study of nonlinear properties of $2D$ excitons in Anthracene has already started (Kuwata 1987).

7. EXCITONS IN ANYSOTROPIC MOLECULAR SUPERLATTICES

In conclusion we would like to draw attention to new possibilities of investigations of excitons in confined molecular systems. Very recently molecular beam epitaxial methods have been successfully applied for growing multiple quantum well structures consisting of alternating layers of two crystalline organic semiconductors PTCDA and NTCDAC (So et al 1990). The individual layer thickness in the multilayer samples varied from $10\AA$ up to $200\AA$. It was discovered that there is a strong structural ordering in all layers, as well as across large spatial distances along the sample surface though the PTCDA and NTCDA crystal structures are incommensurate. The first spectroscopic investigations which gave information on a considerable dependence of excitonic transition frequencies on the layer thicknesses have also been performed. It was found from the optical absorption spectra that the lowest energy PTCDA singlet exciton line shifts to the higher energy with decreasing layer thicknesses. The maximal shifts, i.e. the difference $E(d_1 = 10\AA) - E(d_1 = 200\AA)$, was found to be $\sim 160cm^{-1}$. It seems tempting to explain this shift by the dependence of the superlattice dielectric tensor on the ratio $\frac{d_1}{d_1+d_2}$, where d_1 and d_2 are the layer thiknesesses, $L = d_1 + d_2$ is the superlattice constant (Agranovich 1991). However, in the case under discussion for all samples studied the layer thicknesses were the same $d_1 = d_2$, so the ratio $\frac{d_1}{d_1+d_2}$ was also the same for each sample. It is highly possible that the observed singlet exciton line shift is due to the effect which is well studied for the Frenkel excitons in Anthracene crystals and has already been mentioned above in the Section 1. As has already been pointed out, the excitonic transition in the first (or the last) anthracene monomolecular plane adjacent to vacuum turns out to be blue shifted to $\sim 200cm^{-1}$; for the second plane this shift is $\sim 10cm^{-1}$ and so on. It is clear that for PTCDA these shifts shoud be, generally speaking, different. In addition, for the multilayer systems the first (or the last) plane in the PTCDTA layer is adjacent to the NTCDA crystal rather than to vacuum. Nevertheless, the effect under discussion is possible since static multipoles of PTCDA and NTCDA molecules are different. In this case the electron structure of the interface can be of great importance. A high degree of layer ordering indicates, apparently, a strong molecular interaction between PTDCA and NTDCA molecules. When this interaction results in a charge transfer, then a grating of dipoles should be formed along the interfaces. The electric field caused by these dipoles will induce the Stark effect, so the analysis of the dependence of excitonic transition frequencies on the layer thicknesses becomes more complicated. Problems arising in this case are of a great interest, and the use of a new type of structures, in particular, with doping by different impurities, seems to be rather promising.

8. REFERENCES

Aaviksoo Ya, Lippmaa Ya, Reinot T 1987 Opt.Spectrosc.(USSR) 62 419

Abrikosov A A, Gorkov L P, Dzyaloshinsky I E 1963 Methods of Quantum Field Theory in Statistical Physics (New York:Prentice- Hall,Englewood).

Agranovich V M 1959 Zh.Eksp.Teor.Fiz. 37 340

Agranovich V M 1968 Theory of Excitons (Moscow:Nauka)(in Russian)

Agranovich V M 1991 Solid.State Commun. 78 747

Agranovich V M and Konobeev Yu V 1961 Sov.Phys.Solid State 3 260
Agranovich V M and Konobeev Yu V 1963 Sov.Phys.Solid State 5 1858
Agranovich V M and Dubovsky O A 1966 JETP Lett. 3 223
Agranovich V M and Galanin M D 1982 Electronic excitation energy transfer in condensed matter (Amsterdam: North-Holland).
Agranovich V M and Leskova T A 1988 Solid State Comm. 68 1029
Agranovich V M and Mukamel S 1990 Phys.Lett. A, 147 155
Agranovich V M and Mukamel S 1991 Solid State Comm.80 85
Agranovich V M and Leskova T A 1991 Laser Optics of Condensed Matter vol.2 The Physics of Optical Phenomena and Their Use as Probes of Matter Garmire E, Maradudin A A, Rebane K K (New York and London: Plenum Press) pp145-156
Agranovich V M, Ovander L N, Toshich B C 1966 Sov.Phys.JETP 23 885
Agranovich V M, Rupasov V I, Chernyak V Ya 1982 Fiz.Tverd.Tela, 24 2992
Agranovich V M, Ratner A M, Salieva M 1987 Solid State Commun. 63 329
Agranovich V M, Chernyak V Ya, Leskova T A 1992 (will be published)
Blombergen N 1965 Nonlinear Optics (New York:Benjamin)
Denisov V N, Mavrin B N, Podobed'ov V B 1987 Phys.Rep. 151 1
Dzyaloshinsky I E, Pitaevsky L P 1959 Sov.Phys.JETP 9 1282
Flytzanis C 1975 Quantum Electronics VI Rabin H, Tang C L(New York: Academic Press)pp 1-80
Fröhlich D, Kirchhoff S, Kohler P, Nieswand W 1989 Phys.Rev. B 40 1976
Gale G M, Vallee F, Flytzanis C 1986 Phys.Rev.Lett. 56 1867
Hopfield J J 1958 Phys.Rev. 112 1555
Keldysh L V 1989 The Dielectric Function of Condensed System, eds L V Keldysh, D A Kirzhniz and A A Maradudin (Amsterdam: North-Holland) pp 1-39
Knoester J, Mukamel S 1991 Physics Reports (in press).
Kuwata M 1987 J.Lummin.38 247
Madden P and Kivelson D 1984 Adv.Chem.Phys. 56 467
Obukhovsky V V, Strizhevsky V L 1966 Sov.Phys.JETP 23 91
Philpott M R and Sherman P G 1975 Phys.Rev.B 12 5381 Shabat B V 1976 Introduction to Complex Analysis (Moscow: Nauka)(in Russian)
Shen Y R 1984 The Principles of Nonlinear Optics (New York: Wiley)
Stevenson S H, Connolly M A and Small G J 1988 Chem.Phys. 128 157
So F F, Forrest S R, Shi Y Q, Steier W H 1990 Appl.Phys.Lett. 56 674
Turlet J M, Kottis Ph, Philpott M R 1983 Adv.Chem.Phys. 54 303
Zubarev D N 1960 Sov.Phys.-Usp. 3 320

Inst. Phys. Conf. Ser. No 123
Paper presented at the International Meeting on Optics of Excitons in Confined Systems, Giardini Naxos, Italy, 1991

Nonlocal theory of linear and nonlinear optical responses of confined excitons

Kikuo Cho, Hajime Ishihara, and Takeshi Okada

Department of Material Physics, Faculty of Engineering Science,
Osaka University, Toyonaka, 560 Japan

ABSTRACT: On the basis of a general formulation of nonlocal optical response, three topics have been discussed about linear responses of a slab and a two-level atom, and a nonlinear response of a slab.

1. INTRODUCTION

The subject of radiation-matter interaction has provided many attractive problems in physics from both fundamental and applicational viewpoints. The weight of interest is sometimes more on the nature of radiation or on that of matter. A good example of the latter is the role of spectroscopic data of atoms used in establishing quantum mechanics. The advent of laser belongs to the former example. Today, a considerable interest has been developed in the linear and nonlinear optical responses of confined electronic systems such as thin films, fine particles, quantum wells (wires, dots) and so on. One of the motivations in studying such systems is certainly an applicational one: Since we can now more or less control the size and shape of the materials, we may try to optimize the conditions to get the most effective devices from such materials. One of the major problems in this category is the optical responses in resonance with the exciton and/or biexciton levels confined in such mesoscopic systems. The characteristic point of this type of optical response is the nonlocality exhibited by such an optical medium. The nonlocal response is characterized by susceptibility function dependent on various different coordinates. In the simplest case of linear response, the induced polarization $\boldsymbol{P}(\boldsymbol{r},t)$ and the source field $\boldsymbol{E}(\boldsymbol{r},t)$ are related with one another as

$$\boldsymbol{P}(\boldsymbol{r},t) = \int \mathrm{d}\boldsymbol{r}' \int \mathrm{d}t' \chi(\boldsymbol{r},\boldsymbol{r}';t-t')\boldsymbol{E}(\boldsymbol{r}',t') \tag{1.1}$$

The medium is called local, when $\chi(\boldsymbol{r}-\boldsymbol{r}') \propto \delta(\boldsymbol{r}-\boldsymbol{r}')$. All the other dependences on $(\boldsymbol{r},\boldsymbol{r}')$ correspond to the case of nonlocal medium. Namely the induced polarization is determined by the source field not only at the same position but also at spatially different positions. From the general form of χ obtained from the linear response theory, the $(\boldsymbol{r},\boldsymbol{r}')$ dependence of χ is determined by the wave functions of the excited states of the (confined) system (Cho 1991). If the frequency of light is resonant with a particular excited state, its wave function dominates in determining the $(\boldsymbol{r},\boldsymbol{r}')$ dependence of χ . Therefore, all the optical media are generally nonlocal, since any wave function of the system is more or less extended. Only under special conditions, one may treat optical response as a local one. Obvious examples are (A) the case of well localized impurity levels, and (B) the case of completely off-resonant response, where all the excited states contribute almost uniformly. One may add another example, (C) the case of very small sample size in comparison with the relevant wavelength of the light. (Strictly speaking, this long wavelength approximation, LWA, should be distinguished from the local response approximation.) Since there are many other cases of interests than those mentioned above, a general theory of optical response must be nonlocal. The local response, which has been a paradigm in describing optical properties of matters, is just a special case of the nonlocal response.

The system of interacting radiation and matter is described by Maxwell and Schrödinger equations. Namely, the motion of charged particles (electrons and nuclei) obeys Schrödinger equation which contains vector potential $\boldsymbol{A}(\boldsymbol{r}, t)$ representing radiation field. The motion of the field is determined by Maxwell equations which include polarization (or current) as the source term producing the Maxwell field. Therefore, a given strength of the radiation field determines the induced polarization of the charged particle system, and in turn the induced polarization fixes the magnitude of the radiation field, which must be equal to its initial value used in the Schrödinger equation. In this way the motion of the whole system is determined selfconsistently as the solution of coupled Schrödinger and Maxwell equations (Cho 1991). Although this principle sounds reasonable, solving the coupled equations in practice does not seem feasible, until one notices a very general simplifying situation: The (linear and nonlinear) susceptibility functions, which appear as integral kernels in Maxwell integro-differential equations, have separable forms in site representation. This characteristic feature allows us to rewrite the Maxwell (integro-differential) equations into a set of simultaneous polynomial equations. In the case of linear (n-th order nonlinear) response, these polynomials are linear (n-th order). The number of unknown variables is determined by that of relevant states of material and radiation. Although the latter number is infinite in principle, one can handle it in the case of simple models (Cho 1986, Ishihara and Cho 1990), or restrict it to a finite number from a consideration of those states contributing to the resonant process in consideration. In this note, we mention three topics, [1] size-dependent radiative decay of an exciton in a slab, [2] spontaneous emission in our nonlocal (semiclassical) scheme, and [3] calculation of size-dependent nonlocal response of a third-order nonlinear process.

2. RADIATIVE DECAY OF AN EXCITON IN A SLAB

In a system much smaller than the wavelength of relevant light, it is well known that the radiative life time of an excited state is inversely proportional to its oscillator strength f. This factor f is defined through the matrix element of the ($\boldsymbol{k} = 0$ Fourier component of) transition dipole moment operator $\boldsymbol{P}(\boldsymbol{k} = 0)$. For a mesoscopic system, the matrix elements for a size-quantized excited state show explicit sample-size dependence. Therefore it is natural to expect a corresponding size dependence of the radiative lifetime (or decay width) of the excited state. This picture, however, is valid only for a very small system. As the system size becomes larger, the argument only in terms of $\boldsymbol{P}(\boldsymbol{k} = 0)$ tends to be less reliable; we need to take other Fourier components into account. In other words, nonlocal description becomes more and more necessary. Since the optical response should not depend on the size in the limit of a bulk sample, a reasonable theory must be able to describe the crossover from mesoscopic to bulk like behavior.

In the case of an exciton in a slab, its nonlocal description has been made several times (Kiselev et al. 1977, Cho and Kawata 1985, Cho and Ishihara 1990). A typical quantity representing the radiative lifetime of the size quantized exciton is the width of the absorption spectrum, i.e., $1-T-R$, where T and R represent transmittance and reflectance spectrum respectively. In Fig.1, we show the thickness dependence of the absorption width of the lowest exciton peak in normal incidence. As the material parameters, we used those for CuCl and CdS. In both cases, the absorption width grows linearly with slab thickness, and then saturates. The value of the thickness where a deviation from the linear growth becomes discernible is about one tenth of the light wavelength in the material, i.e., the wavelength divided by the background refractive index (~160Å in both cases.) The initial linear growth can be understood as the effect of giant oscillator strength (Rashba and Gurgenishvili 1962, Henry and Nassau 1970) : The oscillator strength (per unit surface area) of a thin slab is proportional to its thickness, and therefore, as long as the concept of oscillator strength (or the argument in terms of the $\boldsymbol{k}=0$ Fourier component alone) is valid, the width of the absorption peak is proportional to the thickness. The result of the nonlocal theory in Fig.1 shows that the concept of oscillator strength is valid for sample size of about 1/10 of the light wavelength in the medium. The argument which connects the absorption peak width with the radiative lifetime is as follows. The boundary conditions (including additional ones, so-called ABC's, if necessary) gives a simultaneous linear equations for all the amplitudes of relevant (incident, reflected, refracted, internally reflected, and transmitted) light waves. One may write these equations in a form where the amplitude of the incident wave E_i is on the right, and all the others on the left hand side, namely

$$\sum_m S_{nm} E_m = c_n E_i \tag{2.1}$$

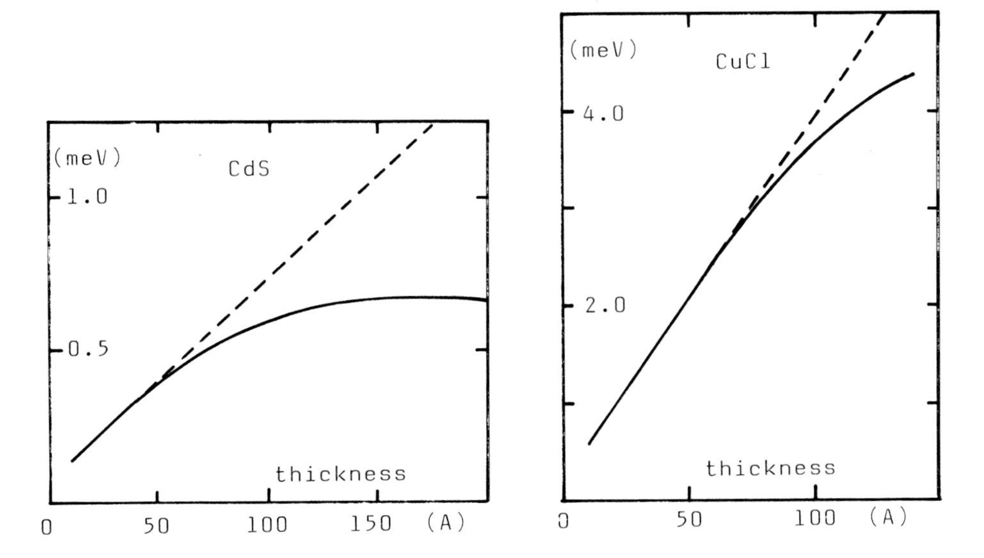

Fig. 1: The half-width of the lowest absorption peak in CdS and CuCl as a function of slab thickness. Parameters in the bulk dielectric function for CuCl (CdS) are $E_{1s}=3.2023$ (2.5530) eV, $\Delta_{LT}=5.7$ (1.46) meV, exciton mass=3.2 (0.78) m_e, background dielectric constant $\varepsilon_b=$ 5.6 (9.6), damping constant γ=0.05 (0.05) meV.

where the coefficients S_{nm} and c_n are the functions of material parameters, slab thickness, and incident angle, polarization and frequency. Inverting the matrix S, namely multiplying S^{-1} to the column vector (c), one gets the amplitudes for calculating reflectance R, transmittance T and so on. On the other hand, the condition for the existence of nontrivial solutions in the absence of the incident wave, i.e., for $E_i = 0$, is $\det(S) = 0$. The solutions give the dispersion (frequency versus parallel wave vector component $\boldsymbol{k}_{\|}$) of self-excitation modes. Among such modes, there are those which have complex frequency for a given $\boldsymbol{k}_{\|}$. They are radiatively decaying modes, as studied in detail by Fuchs et al.(1966) in the simple case of a slab of local medium. From the fact that S^{-1} includes $\det(S)$ in the denominator, the poles corresponding to the self-excitation modes contribute to the peak structure (position and width) of T and R (and consequently A, absorption spectrum). Therefore, a study of the width of absorption spectrum should provide the radiative width. As we show in section 3, the radiative width obtained from such a semiclassical theory can be shown to be same as that derived from a quantized field treatment.

3. RADIATIVE DECAY IN THE PRESENT NONLOCAL FRAMEWORK (TWO-LEVEL ATOM IN VACUUM)

As we mentioned in the introduction, the framework of the present nonlocal theory of radiation-matter interaction consists of solving Schrödinger and Maxwell equations simultaneously. The selfconsistency is required within the given order of describing induced polarization in terms of source field; namely, if one considers a problem of the n-th order (linear if $n = 1$, nonlinear if $n > 1$) response, the induced polarization is written as an n-fold integral of n field variables at different positions and times. (The kernel of the integral is nothing but the n-th order susceptibility.) Because of the selfconsistency, the states of the matter conceive radiation fields, and vice versa. Therefore, one may expect that a response field calculated by this theoretical framework should contain the effect of the radiative decay of the material system, although the scheme is a so-called semiclassical one. An application of the general scheme to the linear response of a two-level atom in vacuum has shown that the frequency dependence of the response field is given by the factor (Cho 1991)

$$1/(E_0 - \hbar\omega - i\gamma - q^2 A_1 - B_1) \tag{3.1}$$

where E_0 is the interval of the two levels, γ the decay constant (if any) in the susceptibility function, q the wave number of the light in vacuum, and

$$A_1 = \sum_{\xi} \iint \mathrm{d}\boldsymbol{r}\mathrm{d}\boldsymbol{r}' \rho_{\xi}(\boldsymbol{r}) \frac{\mathrm{e}^{iq|\boldsymbol{r}-\boldsymbol{r}'|}}{|\boldsymbol{r}-\boldsymbol{r}'|} \rho_{\xi}(\boldsymbol{r}')^* \tag{3.2}$$

$$B_1 = -\iint \mathrm{d}\boldsymbol{r}\mathrm{d}\boldsymbol{r}' \operatorname{div}\boldsymbol{\rho}(\boldsymbol{r}) \frac{\mathrm{e}^{iq|\boldsymbol{r}-\boldsymbol{r}'|} - 1}{|\boldsymbol{r}-\boldsymbol{r}'|} \operatorname{div}\boldsymbol{\rho}(\boldsymbol{r}')^* \tag{3.3}$$

The vector $\boldsymbol{\rho}$ is the transition dipole moment density of the two levels. The real and imaginary parts of $(q^2 A_1 + B_1)$ give the shift and broadening, respectively, of the transition energy between the two levels. In the LWA where we can expand the exponential factors in the above expression, the broadening is given by

$$\Gamma = (2/3)(\omega/c)^2 \left| \int \boldsymbol{\rho}(\boldsymbol{r})\mathrm{d}\boldsymbol{r} \right|^2 \tag{3.4}$$

Since this factor appears in the denominator of the field amplitude, the rate of the power decay is 2Γ, which is the well known expression of the spontaneous emission rate of a two level atom obtained from QED (quantum electrodynamics). In this way, it can be shown that the effect of radiative decay is correctly included in our semiclassical, nonlocal theory. It is quite interesting to make a further comparison between the semiclassical theory and (non-relativistic) QED. It seems to be possible to show the equivalence of $(q^2A_1 + B_1)$ and the second order (complex) frequency shift in QED without employing the LWA.

Another important point in the above result is that the total decay is described by the sum of γ and Γ, while the susceptibility χ contains only the former (γ). This means that χ should contain only the non-radiative decay. If one includes the radiative one in χ, it leads to the double counting of Γ. This is quite reasonable, if one considers that, with the very process of solving Maxwell equations, not of calculating χ, one starts taking account of the interaction of matter with the outside field, which determines the magnitude of radiative decay.

4. NONLOCAL CALCULATION OF A NONLINEAR PROCESS

The role of nonlocality in the linear response of micro- and mesoscopic systems is somewhat implicit. Although the radiative width of the lowest exciton peak shows a size dependence (size-linear growth and saturation) as in Fig.1, $\chi^{(1)}$ does not show an explicit size dependence in LWA. In contrast, $\chi^{(3)}$ in LWA does depends on size explicitly. This size dependence, or size enhancement, has recently been a major problem in the nonlinear optics of mesoscopic systems (Hanamura 1988, Takagahara 1989, Banyai et al. 1988, Spano and Mukamel 1989, 1991, Ishihara and Cho 1990, 1991). Since the essential point of the problem is the nonlocality (the coherent spatial extension of the eigenstate, or the existence of transfer effect among the localized basis states), one must study its manifestation both in $\chi^{(3)}$ and in response field.

The size enhancement of $\chi^{(3)}$ (per unit volume) is based on that of transition dipole moment which was discussed in sec.2. This idea is a simple extension of giant oscillator strength first introduced in the discussion of shallowly bound excitons (Rashba and Gurgenishvili 1962). The only difference is the mechanism of confinement; impurity potential in one case, and a surface potential wall in the other. A new problem, which was practically non-existent in the bound exciton problems, is the consistent description of the size dependence from meso- to macroscopic regime. Since $\chi^{(3)}$ should not depend on sample size in the bulk region, a reasonable theory should describe the size enhancement, its saturation and its approach to the bulk value with the change in the sample size in a single model. In our recent publications (Ishihara and Cho 1990, 1991), we have shown that a consistent picture of size enhancement and saturation is obtained by the detailed consideration of cancellation among various terms in $\chi^{(3)}$. Using a simple model of non-interacting Frenkel excitons in a periodic chain, we have shown the detailed dependence of $\chi^{(3)}$ on the size of chain (N), transfer energy (b), and damping constants. In the presence of the damping mechanism, the saturation of $\chi^{(3)}$ is shown to be derived in a natural way. This means that the damping constant in $\chi^{(3)}$ does provide a "coherence length" which sets the maximum size for LWA to be valid. Therefore, it is redundant and could possibly be inconsistent to introduce an extra coherence length in theory for the purpose of indicating the validity region of LWA.

One of the important remaining problems is the nonlocal calculation of nonlinear response field. Since we are interested in the response of a finite system (not a uniform infinite one), the calculation of $\chi^{(3)}$ in addition to $\chi^{(1)}$ does not at all mean the goal of the work. In view of the size range of our interests, it is essential to make a nonlocal calculation of the response field. The theoretical scheme to do it is an extension of that used for the calculation of the nonlocal linear response: Making use of the separable form of $\chi^{(3)}$ and $\chi^{(1)}$, we rewrite each frequency component of Maxwell equations into a set of simultaneous cubic equations. The numerical solution of these equations determines the response. One of the points we wanted to check is the validity region of LWA in nonlinear response. For this purpose, we took a model of a thin film consisting of linear chains perpendicular to its surface. On each chain we can excite non-interacting Frenkel excitons with a given transfer energy (b). No transfer is allowed between the chains. We consider a normal-incidence pump-probe spectroscopy for this system. In the presence of a pump field of a given frequency ω_p and intensity, we calculate the reflectance R and transmittance T of the probe light with frequency ω, and define the normalized absorption as $A_n = (1 - R - T)/(1 - R)$. In terms of this quantity, we compare the fully nonlocal calculation and the one with a partial use of LWA. For the latter, we take $\chi^{(3)}$ in LWA, and use it as a frequency dependent background susceptibility in the nonlocal calculation of the linear part of the response. In the fully nonlocal calculation, we diagonalize all the one- and two-particle excited states, and in terms of such eigenstates, $| a >$ and $| b >$, we define

$$F_{ab}(\Omega) = \int \mathrm{d}\boldsymbol{r} < a \mid \boldsymbol{\mu}(\boldsymbol{r}) \mid b > \boldsymbol{E}(\boldsymbol{r};\Omega) \tag{4.1}$$

where $\boldsymbol{\mu}(\boldsymbol{r})$ is the dipole density operator and $\boldsymbol{E}(\boldsymbol{r};\Omega)$ the Ω-Fourier component of the field at the position $\boldsymbol{r}$. We can rewrite Maxwell equations into simultaneous cubic equations of $F_{ab}(\Omega)$. The number of unknown quantities is $N_f(N_1 + N_1 N_2)$, where N_f is the number of relevant frequencies, and N_1, (N_2) that of one- (two-) particle states. For a linear chain of N sites, $N_1 = N$, $N_2 = N(N-1)/2$. Therefore the size of simultaneous cubic equations grows as N^3, which rapidly brings about a difficulty in numerical work as we increase the value of N. Fig.2 shows an example of the spectrum for the difference A_n(with pump) - A_n (without pump). The solid and dashed curves correspond, respectively, to the fully nonlocal calculation and the one with the partial use of LWA. The difference of the two curves is visible, though it is still rather small for this size ($N = 20$) of the sample. A direct calculation for larger values of N is already rather critical. The next best policy to go farther is to scale the resonant frequency and transfer energy for fixed value of N. By this scaling the wavelength of the resonant light is changed for a fixed value of sample size. Since our aim of this calculation is to see the validity limit of LWA, this scaling will be useful. Using this type of argument, we have calculated similar curves as Fig.2 for different relative values of D =[sample size / light wavelength]. This shows that, the difference between solid and dashed curves of Fig.2 becomes appreciable for $D > 0.1$. Similar scaling argument for background dielectric constant demonstrates that the light wavelength used for the definition of D must be that, not in vacuum, but in the medium. In this way, we have reached the conclusion that the validity condition for LWA both in linear and nonlinear responses is that $D < 0.1$, where the light wavelength is defined by $2\pi c/\omega\sqrt{\varepsilon_b}$.

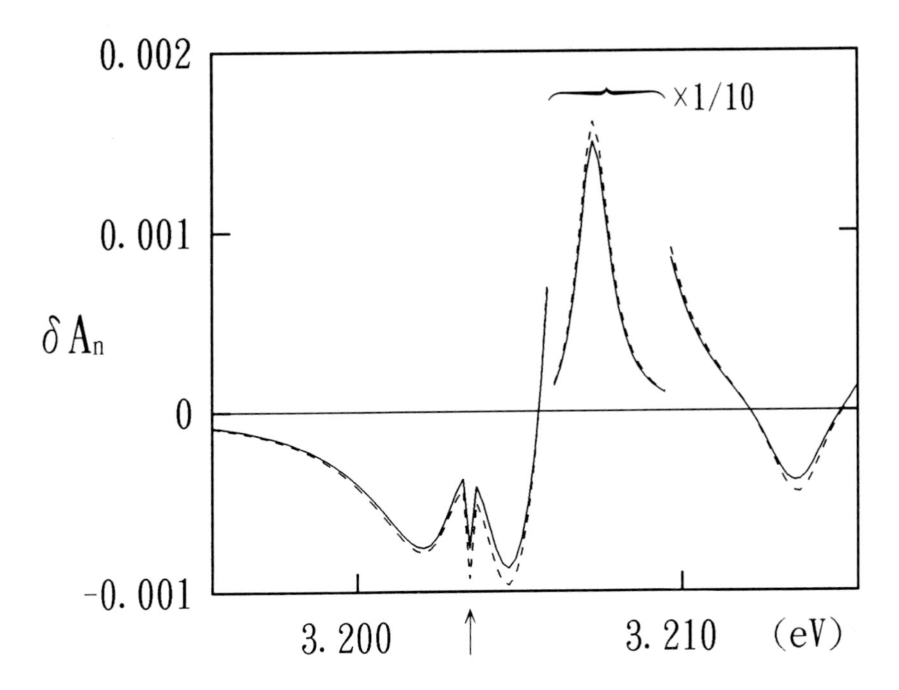

Fig. 2: The spectrum of $\delta A_n = A_n$(with pump) $- A_n$(without pump) as a function of probe light frequency. Parameters are; $N = 20$, $b = 57$ meV, lattice constant = 5.4 Å, ε_B =5.6, Δ_{LT} =5.7 meV, E_{1s} =3.2023 eV, transverse (longitudinal) damping constant = 0.60 (0.02) meV, Field strength of incident probe (pump) light = 1.0×10^{-6} (2.0×10^{-4}) V/cm. The arrow indicates the frequency of pump light (3.20374 eV), which is resonant with the first quantized exciton level.

ACKNOWLEDGMENTS

This work is supported in part by the Yamada Science Foundation and the Grant-in-Aid for Scientific Research on Priority Area "Electron Wave Interference Effects in Mesoscopic Structures" from the Ministry of Education, Science and Culture. The authors are grateful to Dr. Y. Ohfuti for useful discussions.

REFERENCES

Banyai L, Hu Y Z, Lindberg M, and Koch S W 1988 Phys. Rev. B37 8142
Cho K and Kawata M 1985 J. Phys. Soc. Jpn. 54 4431
Cho K 1986 J. Phys. Soc. Jpn. 55 4113
Cho K and Ishihara H 1990 J. Phys. Soc. Jpn. 59 754
Cho K 1991 Prog. Theor. Phys. Suppl. in print
Fuchs R, Kliewer K L, and Pardee W J 1966 Phys. Rev. 150 589
Hanamura E 1988 Phys. Rev. B37 1273

Henry C H and Nassau K 1970 Phys. Rev. B1 1628
Ishihara H and Cho K 1990 Phys. Rev. B41 1424
Ishihara H and Cho K 1990 Phys. Rev. B42 1724
Ishihara H and Cho K 1991 J. Nonlinear Opt. Phys. 1 in print
Kiselev V A, Makarenko I V, Razbirin B S, and Ural'tsev I N 1977 Sov. Phys. - Solid State 19 1374
Rashba E I and Gurgenishvili G E 1962 Sov. Phys. - Solid State 4 759
Spano F C, and Mukamel S 1989 Phys. Rev. A40 5783
Spano F C, and Mukamel S 1991 Phys. Rev. Lett. 66 1197
Takagahara T 1989 Phys. Rev. B39 10206

Inst. Phys. Conf. Ser. No 123
Paper presented at the International Meeting on Optics of Excitons in Confined Systems, Giardini Naxos, Italy, 1991

Polaritons in confined systems

L C Andreani,[a] F Bassani[b], F Tassone[b]

[a]IRRMA, PHB-Ecublens, CH-1015 Lausanne, Switzerland
[a]Scuola Normale Superiore, Piazza dei Cavalieri, I-56126 Pisa, Italy

ABSTRACT: The exchange interaction and the polaritonic effect of quantum well excitons are reviewed. The interaction between excitons and photons gives rise to stationary surface modes, and to radiative modes with a finite lifetime. Due to thermalisation processes, the effective lifetime of free quantum well excitons is found to rise linearly with temperature.

1. INTRODUCTION

Excitonic polaritons are the elementary excitations resulting from the interaction of excitons with the electromagnetic field (Hopfield 1958). In bulk crystals, polaritons are stationary states as long as interactions with phonons or impurities are neglected. The exciton-photon interaction alone does not produce optical absorption, nor a radiative decay of the exciton. The dispersion relations of polaritons have been verified in a variety of optical experiments. A fundamental parameter describing the dispersion of polaritons is the longitudinal-transverse (LT) splitting, which is a measure of the interaction strength between excitons and photons. This quantity, which is proportional to the oscillator strentgh per unit volume, can be calculated from the long-range part of the electron-hole exchange interaction.

The purpose of this paper is to describe how the above concepts are modified in quasi two-dimensional systems like quantum wells (QWs). It was already shown by Agranovich and Dubovskii (1966) that the interaction between excitons and photons in two-dimensional crystals gives rise to two kinds of modes: (i) surface modes, which are nonradiative and undamped, and (ii) oscillatory modes, which have a finite lifetime. The properties of two-dimensional polaritons were further investigated in the context of monomolecular layers (Philpott and Sherman (1975); Agranovich and Mukamel (1990) and references therein).

In the context of quantum wells, Nakayama (1985) and Nakayama and Matsuura (1986) calculated the dispersion of surface polaritons for $k_{\|}L \ll 1$, where $\mathbf{k}_{\|}$ is the exciton wavevector in the planes and L is the well width. Andreani and Bassani (1990) extended this calculation to all values of $k_{\|}L$, and showed that the polariton dispersion in the unretarded limit can also be obtained from a microscopic calculation of the exchange interaction. The radiative lifetime of free QW excitons was calculated by Hanamura (1988), Tassone et al (1990) and Andreani et al (1991).

In Sec. 2 we describe some of the features of the exchange interaction in QW excitons. In Sec. 3 we discuss the dispersion relations of surface modes, the radiative lifetime of oscillatory modes, and the normal-incidence reflectivity. Section 4 contains concluding remarks.

2. EXCHANGE INTERACTION

In direct-gap semiconductors with the zincblende structure and point group T_d, the lowest $\Gamma_6 \otimes \Gamma_8$ exciton has multiplicity $\Gamma_3 \oplus \Gamma_4 \oplus \Gamma_5$, and the spherical part of the exchange interaction separates the threefold-degenerate dipole-active state Γ_5 from the other states (Cho 1976, Rössler and Trebin 1981). In addition, the exchange interaction contains a nonanalytical part, which splits the optically active states Γ_5 into a longitudinal and two transverse excitons.

In a QW (point group D_{2d}) the ground state heavy- and light hole excitons have multiplicities $\Gamma_1 \oplus \Gamma_2 \oplus \Gamma_5$ and $\Gamma_3 \oplus \Gamma_4 \oplus \Gamma_5$ respectively. Optically active states are the twofold degenerate state Γ_5 (x, y polarisations) and the nondegenerate state Γ_4 (z polarisation, where z is the growth direction). Optically forbidden states are not affected by the exchange interaction, if small nonspherical terms are neglected (Rössler et al 1990). Here we are interested in the effect of the exchange interaction on dipole-active states, as a function of the in-plane exciton wavevector $\mathbf{k}_{\|}$. Let us take $\mathbf{k}_{\|} = k_{\|}\hat{x}$ along the x axis. Then optically active states can be classified as L-mode (polarisation vector $\boldsymbol{\epsilon} \parallel \hat{x}$), T-mode ($\boldsymbol{\epsilon} \parallel \hat{y}$), and Z-mode ($\boldsymbol{\epsilon} \parallel \hat{z}$). The Z-mode exists only for the light hole (LH) exciton.

The exchange interaction of QW excitons has been calculated by Andreani and Bassani (1990) in the effective mass approximation. The long-range part of the exchange energy is shown in Fig. 1 for the LH exciton in a 60 Å wide GaAs-$Ga_{0.6}Al_{0.4}As$ quantum well. Similar results are found for the L and T modes of the HH exciton.

The most surprising result in Fig. 1 is that the splitting between longitudinal and transverse excitons is linear in $k_{\|}$ for $k_{\|}L \ll 1$, and vanishes at $\mathbf{k}_{\|} = 0$. This can be interpreted in terms of the fact that the oscillator strength of QW excitons is *per unit area*, whereas the LT splitting is proportional to the oscillator strength *per unit volume*. There is no contradiction with reflectivity experiments (Ivchenko et al 1989), which use a different definition of the LT splitting. This issue is discussed in Sec. 3.3.

The vanishing exchange splitting between longitudinal and transverse excitons is characteristic of a two-dimensional system. In multiple quantum wells (MQWs) and superlattices, even with thick barriers, the LT splitting is again finite, due to the long-range nature of the Coulomb interaction (Wang and Birman 1991).

At $\mathbf{k}_{\|} = 0$, the Z mode of the LH exciton is split from the L, T modes by ~ 1 meV. This splitting can be interpreted as the energy difference between two planar sheets of dipoles oriented perpendicular and parallel to the plane, respectively. The ZT (or Γ_5-Γ_4) splitting has been recently observed experimentally (Berz et al 1991, Fröhlich et al 1991).

The exchange interaction depends strongly on $\mathbf{k}_{\|}$. However, this dependence is very difficult to observe, at least in GaAs-$Ga_{1-x}Al_xAs$ QWs, because of the much larger effect of spatial dispersion due to the center-of-mass motion of the exciton. The effects described in Fig. 1 might be more easily observable in CuCl quantum wells.

3. INTERACTION WITH RADIATION

In bulk crystals, an exciton with wavevector $\mathbf{k}$ interacts with one photon with the same wavevector and polarisation. The mutual interaction does not give rise to a radiative recombination, since there is no density of states for the decay: rather, it gives rise to stationary

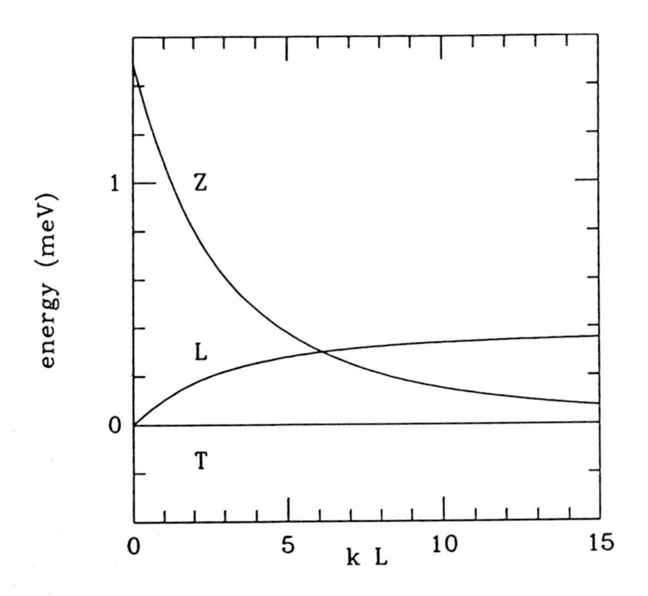

Figure 1: Long-range exchange energies of the T, L, and Z modes for the ground-state LH exciton in a 60 Å wide GaAs-$Ga_{0.6}Al_{0.4}As$ quantum well.

polariton states. The concept of polariton is thus strictly linked with the conservation of **k**. In fact, radiative decay is possible for bound excitons, for which **k**-conservation does not hold.

In a QW, translational symmetry along the growth direction is lost. An exciton with wavevector $\mathbf{k}_{||}$ interacts with photons with the same in-plane wavevector but with all possible values of k_z. Thus there is a one-dimensional density of states for radiative decay:

$$\rho(\mathbf{k}_{||},\omega) = \frac{V}{\pi S}(\frac{n}{\hbar c})^2 \frac{\hbar\omega}{\sqrt{k_0^2 - k_{||}^2}}\theta(k_0 - k_{||}), \tag{1}$$

where V (S) is the sample's volume (area), n is the index of refraction, and $k_0 = n\omega/c$. Only excitons with $k_{||} < k_0$ can decay radiatively. The density of states vanishes for excitons with $k_{||} > k_0$, which, therefore, remain stationary states. Thus the interaction of the exciton with the electromagnetic field gives rise to two kinds of states:

States with $k_{||} > k_0$. They do not decay radiatively, and in this sense they represent the analog of bulk polaritons. The electric field far from the quantum well is exponentially damped with decay constant $\alpha = ({k_{||}}^2 - k_0^2)^{1/2}$. These states are surface modes, lie on the right of the photon line in the $k_{||}$-ω plane, and do not couple to incident light propagating along the growth direction.

States with $k_{||} < k_0$. They have a finite radiative lifetime. They lie on the left of the photon line, and the electric field far from the quantum well oscillates with a wavevector $k_z = (k_0^2 - k_{||}^2)^{1/2}$ in the z-direction. These are the states which are seen in usual absorption or reflectivity experiments.

3.1. Nonradiative modes

There are three kinds of surface polaritons, according to polarisation, namely T-mode (or TE mode, s-polarisation) and L, Z-modes (or TM modes, p-polarisation). The dispersion relations

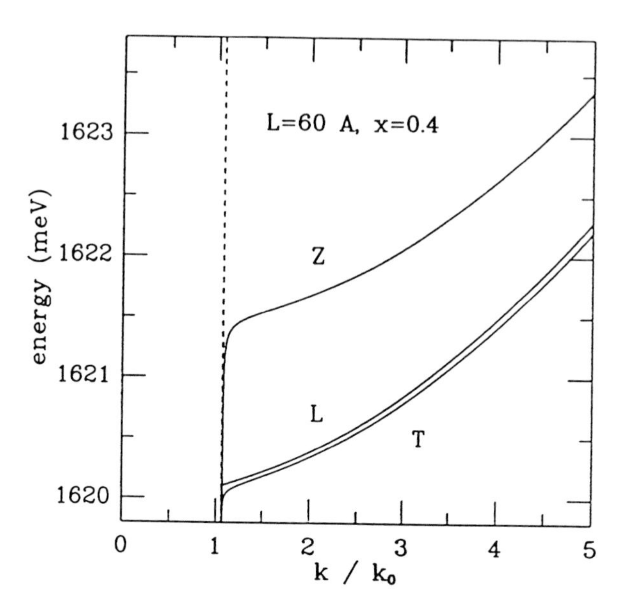

Figure 2: Dispersion of the LH surface polaritons in a 60 Å-wide GaAs-$Ga_{0.6}Al_{0.4}As$ quantum well. The unit of k vectors is $k_0 = 2.5 \times 10^5$ cm^{-1}. The dashed curve denotes the photon line.

can be calculated by solving Maxwell equations with the appropriate boundary conditions (Andreani and Bassani 1990, Tassone et al 1990). In the instantaneous limit, the dispersion relations reduce to the dispersion of the exciton, provided the long-range part of the exchange interaction is taken into account. This is an example of the equivalence between exchange interaction and interaction with the instantaneous electromagnetic modes, which is a general and powerful concept (Ehara and Cho 1982). In particular, the energy difference between unretarded L and T modes vanishes linearly in $k_{\|}$ as $k_{\|} \to 0$, in agreement with the exchange interaction described in Sec. 2. An example of the surface polariton dispersion is shown in Fig. 2 for the LH exciton in a GaAs-$Ga_{0.6}Al_{0.4}As$ QW. Interaction with the electromagnetic field is seen to modify the dispersion of the unretarded modes only in a narrow region around k_0. For $k_{\|} \gg k_0$, effects of retardation rapidly become negligible. Since in general $k_0 L \ll 1$, this means that effects of the $\mathbf{k}_{\|}$-dependent exchange interaction (see Sec. 2) occur at a much larger scale of $\mathbf{k}_{\|}$-vectors than polariton effects. In fact, the results of Fig. 2 show a very small LT splitting and a sizeable ZT splitting, which is characteristic of the exchange interaction at $\mathbf{k}_{\|} = 0$. The dispersion of the L-mode is analogous to that of a surface plasmon. Luminescence from the L-mode has been observed by growing a grating on a MQW sample (Kohl et al 1988, 1990), and the T and Z modes have been detected in a time-of-flight measurement (Ogawa et al 1990). Another possibility for observing surface modes would be the use of attenuated total reflection (Tassone at al 1991).

3.2. Radiative modes

The radiative lifetime of free QW excitons can be calculated from the dipole interaction $\mathbf{A} \cdot \mathbf{p}$ using Fermi's Golden Rule and the density of states (1) (Andreani et al 1991a). The result depends on the wavevector $k_{\|}$ and the polarisation of the exciton. We obtain, for $k_{\|} < k_0$,

$$\Gamma_{\mathrm{T}}(k_{\|}) = \frac{\pi}{n} \frac{e^2}{m_0 c} f_{xy} \frac{k_0}{k_z},$$

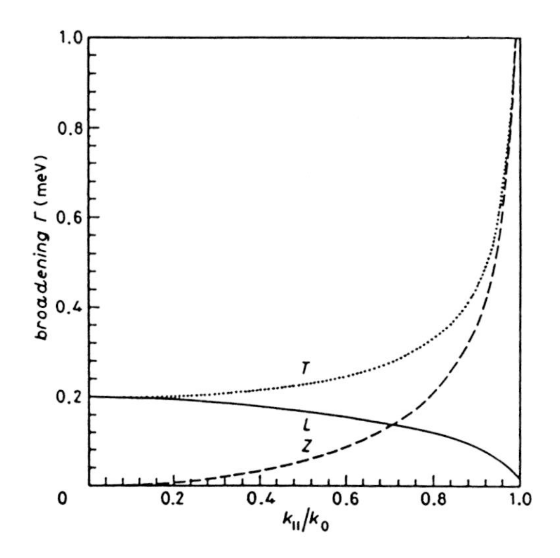

Figure 3: Radiative decay rate of excitons in a 20 Å wide CuCl quantum well.

$$\Gamma_{\rm L}(k_{\parallel}) = \frac{\pi}{n}\frac{e^2}{m_0 c} f_{xy}\frac{k_z}{k_0}, \tag{2}$$

$$\Gamma_{\rm Z}(k_{\parallel}) = \frac{\pi}{n}\frac{e^2}{m_0 c} f_z\frac{k_{\parallel}^2}{k_0 k_z},$$

where $k_z = (k_0^2 - k_{\parallel}^2)^{1/2}$, m_0 is the free electron mass, and f_{xy},f_z are the oscillator strengths per unit area. For the LH exciton, $f_z = 4f_{xy}$ (Andreani and Pasquarello 1990). The radiative width of the three modes is shown in Fig. 3 for a 20 Å wide CuCl quantum well. Note that the decay rate diverges as $k_{\parallel} \to k_0$ for the T and Z modes. This divergence, which can be traced back to the density of states (1), is integrable and disappears when the thermal average is taken. A slight broadening of the wavevector due to anelastic scattering would wash out this divergence, which we believe to have no physical consequences.

At $\mathbf{k}_{\parallel} = 0$, L and T modes have the same radiative width $\Gamma_0 = (\pi/n)(e^2/(m_0 c))f_{xy}$. Taking an oscillator strength $f_{xy} = 50 \cdot 10^{-5}$ Å^{-2}, which is appropriate for the HH exciton in GaAs-$Ga_{1-x}Al_xAs$ QWs of about 100 Å (Andreani and Pasquarello 1990), we obtain $\hbar\Gamma_0 = 0.026$ meV, or $\tau_0 = 1/\Gamma_0 = 25$ ps. This lifetime is an order of magnitude longer than found by Hanamura (1988). The discrepancy can be attributed to two reasons. First, Hanamura does not consider the index of refraction n. Second, he assumes the two-dimensional limit for the exciton, which considerably overestimates the oscillator strength. The quantitative difference is very important in comparing the radiative lifetime with the thermalisation time.

The radiative recombination mechanism calculated here is not expected to be relevant for too low a temperature, when excitons are bound to impurities and interface fluctuations. At finite temperature, thermalisation must be taken into account. Thermalisation is due to anelastic scattering with acoustic phonons (Schultheis et al 1986), which changes the exciton wavevector (scattering with optical phonons is not expected to be relevant for temperatures smaller than 36 meV, which is the LO phonon energy in GaAs). The scattering rate by acoustic phonons is measured (Schultheis et al 1986) to be linear in T with a coefficient $\gamma = 5$ μeV$/K$, for a QW of 135 Å width. A behavior linear in T is also calculated (Hanamura 1988), and the coefficient γ is determined to be 8 μeV$/K$ for $L = 100$ Å. Both values show that thermalisation processes are faster than radiative decay for $T > 5\ K$. There is also other evidence for this conclusion, coming from a high-energy thermal broadening of exciton lines. In the following we shall assume as a working assumption that thermalisation processes are much faster than radiative recombination, i.e., that excitons always have a thermal distribution while decaying radiatively.

In this assumption, the decay rate of the luminescence is given by the thermal average of the decay rate (2). The two characteristic energies are the thermal energy KT, and the kinetic energy of excitons which decay radiatively: the latter is at most $\hbar^2 k_0^2/(2M)$, where M is the exciton mass. Using $M = 0.25\ m_0$, we find $\hbar^2 k_0^2/(2M) \approx 1.1\,K$. This means that, for $T \gg 1\,K$, only a small fraction of excitons occupy the states with $k_{\|} < k_0$, which can decay radiatively. Performing the average over the Boltzmann distribution for $T \gg 1\,K$ gives, for the HH exciton,

$$\tau(T) = \frac{3}{2}\frac{2MKT}{\hbar^2 k_0^2}\tau_0, \tag{3}$$

where $\tau_0 = 1/\Gamma_0$ is the radiative lifetime at $k_{\|} = 0$ (we have also averaged over the forbidden triplet states). In practice, the radiative decay rate (2) is multiplied by the small fraction of excitons with $k_{\|} < k_0$, and is reduced accordingly. Equation (3) predicts an effective radiative lifetime which rises linearly with temperature. Numerically, for a 100 Å GaAs-$Ga_{0.6}Al_{0.4}As$ QW, we find $\tau(T) = 34\,T(K)$ ps. For $T \gg 1\,K$, this is clearly much longer than the "bare" radiative lifetime at $\mathbf{k}_{\|} = 0$.

The difference between our results and those of Hanamura (1988) should be emphasised. Hanamura calculates a radiative lifetime of 2.8 ps, and from a comparison with thermalisation times concludes that radiative recombination should dominate. This does not agree with experiment, as observed lifetimes are much longer. We calculate a lifetime of 25 ps, which is longer than typical thermalisation times, and the average over thermal distribution makes the effective lifetime even longer.

An effective lifetime rising linearly with temperature is also predicted by Feldmann et al (1987). The difference with our theory is that they do not specify the microscopic mechanism for the decay, and thus do not calculate the slope of the linear behaviour. The average over the thermal distribution is performed in a similar way in the two theories.

A radiative lifetime rising linearly with temperature is observed experimentally (Feldmann et al 1987). Our calculated lifetime are lower than those measured by Feldmann et al by about a factor of three. More recent measurements (Colocci et al 1990, 1991) show that the measured lifetime can be different in samples with the same nominal parameters. Moreover, thermal equilibrium between excitons and free electron-hole pairs is observed to change the temperature dependence of the lifetime. It is fair to say that there is still no complete understanding of the observed exciton lifetime in QWs, and even more elaborate models (Ridley 1990) are not in quantitative agreement with experiment.

Our theory is based on three assumptions: free excitons, conservation of $\mathbf{k}_{\|}$, and rapid thermalisation. At low temperatures, the decay mechanism is different, since the exciton is localised at interface fluctuations and the density of states of photons becomes three-dimensional. At high temperatures, when interaction with optical phonons becomes dominant, the observed lifetime reaches a maximum and then decreses as a function of temperature.

For well widths $L \gg a_B$, where a_B is the exciton radius, the closely spaced quantised levels of the center-of-mass motion of the exciton must be taken into account in the thermal average.

This can be shown to change the temperature dependence of the lifetime for a T-linear behaviour to a $T^{3/2}$-law (Andreani et al 1991b). This clearly holds only for $L \ll \lambda$, where λ is the wavelength of light, otherwise the bulk polariton effect comes into play and the radiative recombination mechanism disappears.

3.3. Reflectivity

Recent reflectivity experiments on MQW and superlattice samples (Ivchenko et al 1989) have been interpreted as yielding the LT splitting, which is found to be greatly increased as compared to the bulk. In order to clarify the relation between the experimental results and our microscopic theory, here we calculate the normal-incidence reflectivity from a QW surrounded by a barrier of thickness D on one side and by a semi-infinite barrier on the other side. The index of refraction is assumed to be the same throughout the structure. The dielectric response of the exciton is described by a nonlocal susceptibility $\chi(z, z')$, which is computed in linear response theory: we consider an isolated exciton resonance, the effect of all other resonances being lumped into the background dielectric constant. The reflectivity is then calculated by solving Maxwell equations with the usual boundary conditions. This microscopic method (which is similar to the one described by Cho (1986)) takes full account of polariton effects. No additional boundary conditions are needed, since the microscopic details of the exciton wavefunction are contained in the nonlocal susceptibility. In order to describe broadening, we introduce a finite width γ, which takes into account all *nonradiative* relaxation processes. Neglecting multiple reflections, we obtain

$$R(\omega) = (\frac{n-1}{n+1})^2 - \frac{8n(n-1)}{(n+1)^2}\frac{(\gamma+\Gamma_0)\cos(k_0 l) + (\omega_0-\omega)\sin(k_0 l)}{(\omega_0-\omega)^2+(\gamma+\Gamma_0)^2}\Gamma_0, \tag{4}$$

where $l = L + 2D$, ω_0 is the resonance frequency, $k_0 = n\omega/c$, and $\Gamma_0 = (\pi/n)(e^2/(m_0 c))f_{xy}$ is the radiative linewidth at $\mathbf{k}_{\|} = 0$. The exciton reflectivity appears as a weak structure on the background reflectivity due to the barrier-vacuum interface. It is important to note that the radiative linewidth Γ_0 automatically appears from the solution of Maxwell equations in the radiative region. In the resonance denominator, the radiative linewidth adds to the nonradiative width γ and can usually be neglected. However, the radiative linewidth also appears in the numerator and governs the size of the reflectivity. For $k_0 l \ll 1$, the reflectivity has a Lorentzian lineshape with a peak value of $\sim \Gamma_0/\gamma$.

Reflectivity data are usually analysed in terms of a local dielectric model, in which the QW layer is described by the dielectric function

$$\varepsilon(\omega) = \varepsilon_\infty(1 + \frac{\tilde{\omega}_{LT}}{\omega_0 - \omega - i\tilde{\gamma}}). \tag{5}$$

The parameters $\omega_0, \tilde{\omega}_{LT}, \tilde{\gamma}$ are obtained from a fit of the experimental data. Calculating the reflectivity within the local dielectric model, we find that the result (4) is reproduced only if the width $\tilde{\gamma}$ is much larger than $\tilde{\omega}_{LT}$. Under this condition (which is usually verified experimentally), the relation between $\tilde{\omega}_{LT}$ and Γ_0 is found to be $\tilde{\omega}_{LT} = 2\Gamma_0/(k_0 L)$ (or $\tilde{\omega}_{LT} = 2\Gamma_0/(k_0(L_w + L_b))$, in the case of a MQW). Thus we see that the effective parameter $\tilde{\omega}_{LT}$ is not the exchange splitting between longitudinal and transverse excitons, as it is in the bulk.

4. CONCLUSIONS

The exchange splitting between longitudinal and transverse excitons is found to vanish linearly in $k_{\|}$ for $\mathbf{k}_{\|} \to 0$, in agreement with the dispersion of polaritons in the unretarded limit. The LT splitting defined from the exchange interaction does not coincide with the effective parameter $\tilde{\omega}_{LT}$ of the local dielectric model used to analyse reflectivity experiments. In general, effects of the exchange interaction are small in GaAs/AlGaAs structures, except for the upward shift of the z-polarised LH exciton. Interaction with radiation gives rise to stationary surface modes, and to a finite lifetime for radiative modes with $k_{\|} < k_0$. The radiative lifetime is found to be of the order of 25 ps at $k_{\|} = 0$. Since this is longer than the thermalisation time, the observed decay rate is given by a thermal average, yielding an effective lifetime which rises linearly with temperature.

Acknowledgments - This work was supported in part by the European Community through contract No. ST2J-0254. One of us (LCA) acknowledges partial support from the Swiss National Science Foundation under Grant No. 20-5446.87.

Agranovich V M and Dubovskii O A 1966, Pis'ma Zh. Eksp. Theor. Fiz. 3, 345 [JETP Lett. 3, 223]
Agranovich V M and Mukamel S 1990, Phys. Lett. A 147 155
Andreani L C and Bassani F 1990, Phys. Rev. B 41, 7536
Andreani L C and Pasquarello A 1990, Phys. Rev. B 42, 8928
Andreani L C, Tassone F and Bassani F 1991a, Solid State Commun. 77, 641
Andreani L C, D'Andrea A and Del Sole R 1991b, to be published
Berz M W, Andreani L C, Steigmeier E F, and Reinhart F K 1991, Solid State Commun., submitted.
Cho K 1976, Phys. Rev. B 14, 4463
Cho K 1986, J. Phys. Soc. Jpn. 55, 4113
Colocci M et al 1990, Europhys. Lett. 12, 417
Gurioli M et al 1991, Phys. Rev. B 44, 3115
Ehara K and Cho K 1982, Solid State Commun. 44, 453
Feldmann J et al 1987, Phys. Rev. Lett. 59, 2337
Fröhlich D et al 1991, to be published
Hanamura E 1988, Phys. Rev. B 38, 1228
Hopfield J J 1958, Phys. Rev. 112 1555
Ivchenko et al 1989, Solid State Commun. 70, 529
Kohl M, Heitmann D, Grambow P and Ploog K 1988, Phys. Rev. B 37, 10927
Kohl M, Heitmann D, Grambow P and Ploog K 1990, Phys. Rev. B 42, 2941
Nakayama M 1985, Solid State Commun. 55, 1053
Nakayama M and Matsuura M 1986, Surf. Sci. 170, 641
Ogawa K, Katsuyama T and Nakamura H 1990, Phys. Rev. Lett. 64, 796
Philpott M R and Sherman P G 1975, Phys. Rev. B 12, 5381
Ridley B K 1990, Phys. Rev. B 41, 12190
Rössler U, Jorda S and Broido D 1990, Solid State Commun. 73, 209
Rössler U and Trebin H R 1981, Phys. Rev. B 23, 1961
Tassone F, Bassani F and Andreani L C 1990, Il Nuovo Cimento D 12, 1673
Tassone F, Bassani F and Andreani L C 1991, Phys. Rev. B, submitted
Wang B S and Birman J L 1991, Phys. Rev. B 43, 12458

Inst. Phys. Conf. Ser. No 123
Paper presented at the International Meeting on Optics of Excitons in Confined Systems, Giardini Naxos, Italy, 1991

Investigation of resonant polaritons in quantum wells

D. Fröhlich, P. Köhler, E. Meneses-Pacheco, G. Khitrova[1], and G. Weimann[2]

Institut für Physik, Universität Dortmund, 4600 Dortmund 50, FRG
[1] Optical Science Center, University of Arizona, Tucson, USA
[2] Walter Schottky Institut, TU München, 8046 Garching, FRG

ABSTRACT: Resonant polaritons are optically excited in GaAs-$Al_xGa_{1-x}As$ multiple quantum wells. The radiation is coupled via a ZnSe prism into the sample which allows measurements with a rather large in-plane wavevector $\mathbf{K}_{\|}$. The shift of the **Z** mode to higher energies as compared to the **L**, **T** modes is clearly resolved for the light-hole resonance. The experimental results are compared with theoretical predictions.

1. INTRODUCTION

The investigation of excitons in quantum well structures by linear and nonlinear optical methods has attracted great interest by experimentalists and theorists as well. For recent reviews we refer to Schmitt-Rink et al. (1989) and Cingolani and Ploog (1991). Due to the confinement of the electron-hole wave function in the growth direction, the symmetry is lowered from T_d to D_{2d}. This leads to a splitting of the four-fold valence band (Γ_8-symmetry in T_d) into two two-fold bands (Γ_6 and Γ_7 in D_{2d}). One thus gets two exciton series termed heavy hole (HH) and light hole (LH). The detailed calculation of such important exciton parameters like binding energy, oscillator strength and electron-hole exchange interaction has to include valence band mixing, the nonparabolicity of the conduction band, Coulomb coupling between excitons of different subbands and even the difference in dielectric constant between well and barrier materials. We refer to Andreani and Pasquarello (1990a) for an accurate theory of excitons in GaAs-$Al_xGa_{1-x}As$ quantum wells and previous literature.

Another important parameter, which is expected to be influenced by the confinement of the carriers, is the exchange interaction. Chen et al. (1988) have calculated exchange effects on excitons in quantum wells. They show that both short- and long-range parts of the exchange interaction exhibit an important enhancement. One thus expects an increase of the splitting between ortho- and paraexcitons (short range exchange interaction) and an increase of the longitudinal-transverse splitting ΔE_{LT}^{QW} compared to excitons in bulk GaAs. Andreani and Bassani (1990b), however, clearly show, that ΔE_{LT}^{QW} vanishes linearly in K as K→0, although the optical oscillator strength is enhanced. Besides this rather surprising fact there are other qualitative and quantitative new aspects of polaritons in QW structures. Nakayama (1985) and Nakayama and Matsuura (1986) have calculated the dispersion of surface polariton modes. For the excitation of surface polariton modes special techniques like attenuated total reflection or the presence of a grating on the sample are needed. The coupling of photons to nonradiative exciton-polaritons in structured materials has been demonstrated bei Kohl et al. (1988 and 1990). Tassone et al. (1990) give a detailed calculation of the polaritons which result from coupling with travelling waves in the barrier. They explicitely distinguish between "resonant polaritons" and "surface polaritons". The distinction between these two types of modes is given by the value of the in-plane wavevector $K_{\parallel}$. For $K_{\parallel}<K_0$ one deals with resonant polariton modes, whereas for $K_{\parallel}>K_0$ surface polariton modes are excited. $K_0=n\omega/c$ is the wavevector of the radiation in the material. For the resonant polariton they calculate the dispersion of the three modes, termed **T**,**L** and **Z** mode, which are distinguished by their polarization direction relative to $\mathbf{K}_{\parallel}$ and the growth direction (z-direction). For the **T** and **L** modes the polarization is in the QW-plane perpendicular and parallel to $\mathbf{K}_{\parallel}$, respectively. For the **Z** mode the polarization is parallel to the growth direction. As proposed by Andreani and Bassani (1990b) the splitting between the **Z** and **T** modes should be directly observable in the LH-exciton resonance, if a sufficiently large $K_{\parallel}$ is achieved in the experimental configuration. In the HH-exciton resonance this effect can not be observed since the oscillator strength vanishes for polarization parallel to the growth direction. In this contribution we present experimental results of the splitting between the T and **Z** modes for samples of different well-width between 50 and 150 Å. After a short description of the experimental details we briefly cite the relevant formulas from the theory of Andreani and Bassani (1990b). We then present our experimental results, which are compared with the theory.

2. EXPERIMENTAL DETAILS

For the excitation of the **Z**-mode one needs $K_{\parallel}$ as large as possible. Because of the large refractive index ($n\sim3.3$) one gets a maximum $K_{\parallel}\sim0.3K_0$ if the light is directly coupled from air into the QW. A larger value of $K_{\parallel}$ ($\sim0.7K_0$) can be obtained with the use of a ZnSe prism ($n=2.4$) which is in optical contact with the sample (Fröhlich et al. 1987). In Fig.1 the schematics of the optical coupling is shown for the polarization in the plane of incidence (electric field $\mathbf{E_m}$, m stands for mixed polarization) and in the x-y plane, which is termed $\mathbf{E}_{\parallel}$, since the electric field is parallel to the QW-plane.

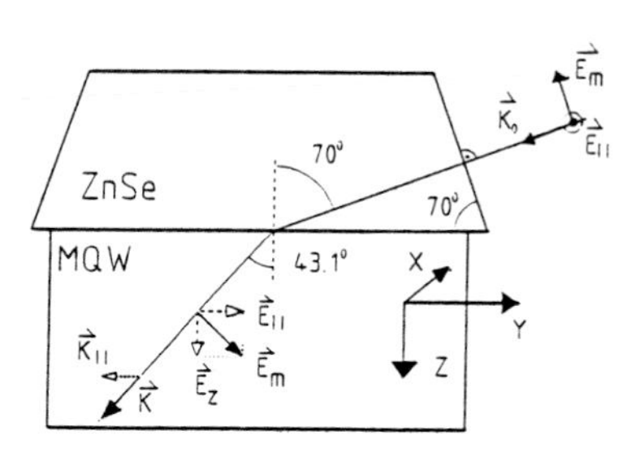

Fig.1.Schematics of the optical coupling of the laser pulse into the MQW

For the decomposition of the spectrum of the mixed polarization ($\mathbf{E_m}$) into the $\mathbf{E}_{\parallel}$ and $\mathbf{E_z}$ components we have used the Fresnel formulas and the relations of the oscillator strength $f_z/f_{xy}=4$ and $f_z=0$ for the LH and HH resonances, respectively. The spectra were measured in absorption and by photoluminescence excitation (PLE) with the use of a tunable titanium-sapphire laser which was pumped by an Ar-ion laser. The spectrometer was set to the low energy side of the free exciton luminescence of the HH exciton. With the use of a half-wave plate the spectra were taken for both polarizations in the same scan. The measurements were done at a temperature of 1.5K on different samples as listed in Table 1.

3. THEORY AND EXPERIMENTAL RESULTS

In this section we follow the treatment of Andreani and Bassani (1990b). The "resonant exciton polariton" ($K_{\parallel}<K_0$) can be found (in the unretarded limit) by a microscopic

calculation of the exchange interaction of the exciton in the QW. It was shown that this exchange interaction follows two different regimes depending on the value of $K_{\parallel}L$. Before showing our experimental findings we present some theoretical results of Andreani and Bassani (1990b). The long range exchange energy of the **L**,**T** and **Z** modes for the light-hole exciton is given by

$$\langle J_{LR}^{\mu\parallel(x,y)} \rangle = g(\Gamma_5) \frac{2\pi}{\varepsilon_\infty} \mu^2 |F_{QW}(0)|^2 |\hat{\mu}\cdot\hat{K}_{\parallel}|^2 K_{\parallel} P(K_{\parallel}) \qquad (1)$$

$$\langle J_{LR}^{\mu\parallel z} \rangle = g(\Gamma_4) \frac{2\pi}{\varepsilon_\infty} \mu^2 |F_{QW}(0)|^2 [2\mathcal{D} - K_{\parallel} P(K_{\parallel})] \qquad (2)$$

where $\hat{\mu}$ and $\hat{K}$ are unit vectors and $\mu = \langle u_c|er|u_v\rangle$ is the dipole matrix element between bulk Bloch functions. $g(\Gamma_5)$, $g(\Gamma_4)$ are spin-orbit factors of the HH(LH) for the x,y and z polarization, respectively. $\mathcal{D} = \int dz |c(z)|^2 |v(z)|^2$, where c(z), v(z) are envelope functions of conduction and valence subbands, respectively. $F_{QW}(\rho)$ is the envelope function for the relative electron-hole motion in the x-y plane and

$$P(K_{\parallel}) = \int dz \int dz' c^*(z) v(z) v^*(z') c(z') e^{-k_{\parallel}|z-z'|} \; .$$

In the two-dimensional limit $\mathcal{D} = 3/(2L)$ (Andreani and Bassani 1990b), the function $P(K_{\parallel})$ can be calculated for infinite barriers (Andreani 1989) and the dependence of the oscillator strength on the well width was taken from Andreani and Pasquarello (1988). Using eqs. (1),(2) one can calculate the dependence of the **L**,**T**,**Z** modes on the product $K_{\parallel}L$ for a given well width. Our results cover only a small region $K_{\parallel}L < 1$ (see Table 1). The PLE of the sample (c) is shown in Fig.2. The spectrometer was set to the heavy-hole exciton energy (E=1.555 eV). The LH exciton spectrum for the mixed polarization ($\mathbf{E_m}$) is shifted to the high energy side in comparison to the LH exciton for the polarization in the x-y plane ($\mathbf{E_{\parallel}}$). The shift of the Z-mode can be obtained from the curves for $I_{\parallel}$ and I_m making use of the Fresnel formulas and the oscillator strength relation $f_z/f_{xy}=4$ and $f_z=0$ for the LH and HH excitons, respectively.

Table 1. Samples data

sample	well width L(Å)	Al fraction x	barrier width L_B(Å)	$K_{\parallel}L$
(a)	50	0.35	150	0.093
(b)	66	0.31	104	0.12
(c)	90	0.43	1000	0.16
(d)	100	0.45	180	0.18
(e)	130	0.35	150	0.23
(f)	150	0.26	1000	0.26

For sample (e) we have also measured the HH exciton in PLE. The HH exciton for the z-polarization after unfolding vanishes. The disappearance of the HH exciton has also been seen by Ogawa and Katsuyama (1990).

In Fig.3 we show our experimental results for all samples (a)-(f) together with the theory. The agreement between theory and experiment is quite good. The experimental values, however, seem to be systematically larger by a small amount as compared to the calculated values.

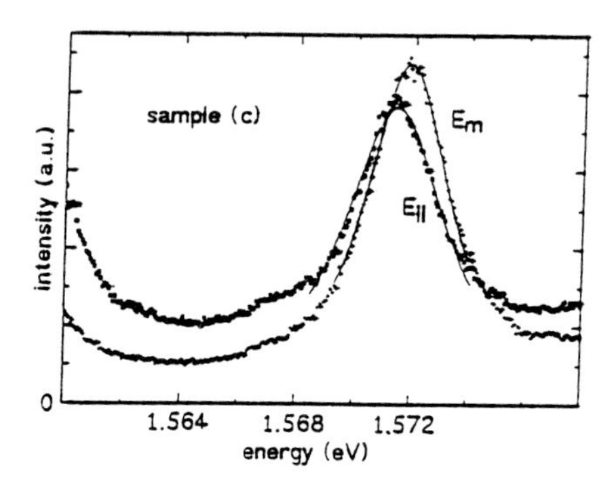

Fig.2. PLE of the LH exciton for sample (c). E_m, $E_{||}$ denote electric field in the plane of incidence (mixed polarization) and in the xy-plane, respectively. Solid line: fit to a Lorentzian.

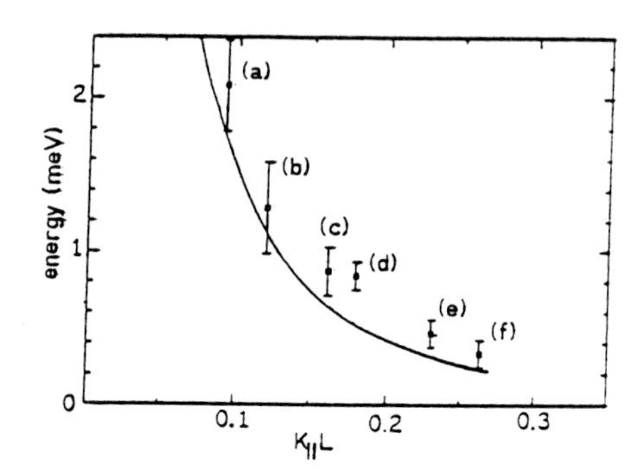

Fig.3 . Splitting between **Z** and **T** mode as function of $K_{||}L$ for samples (a)-(f). Solid line, theory; full squares, experiment.

4. SUMMARY

We have observed a shift of the light-hole exciton to the high energy side for the electric field polarization in the direction perpendicular to the layer plane (z-direction) in comparison to the case when the polarization lies in the layer plane (x-y plane) . This is a characteristic property of polaritons in quantum wells. The good agreement with a recent theory is a strong indication in favor of this interpretation.

5. ACKNOWLEDGMENTS

We would like to thank Dr. L.C. Andreani for many helpful discussions. One of us (E.M.P.) is financially supported by CAPES (Brazil).

6. REFERENCES

Andreani L.C. and Pasquarello A. 1990a Phys. Rev. B 42 8928
Andreani L.C. and Bassani F. 1990b Phys. Rev. B 41 7536
Andreani L.C. and Pasquarello A. 1988 Europhys. Lett. 6 259
Andreani L.C. 1989 Ph.D. thesis, Scuola Normale Superiore, Pisa (unpublished)
Chen Y., Gil B., Lefebvre P. and Mathieu H. 1988 Phys. Rev. B 37 6429
Cingolani R. and Ploog K. 1991 Advances in Physics (to be published)
Fröhlich D., Wille R., Schlapp W. and Weimann G. 1987 Phys. Rev. Lett. 59 1748
Kohl M., Heitmann D., Grambow P. and Ploog K. 1988 Phys. Rev. B 37 10927
Kohl M., Heitmann D., Grambow P. and Ploog K. 1990 Phys. Rev. B 42 2941
Nakayama M. 1985 Solid State Commun. 55 1053
Nakayama M. and Matsuura M. 1986 Surf. Sci. 170 641
Ogawa K. and Katsuyama T. 1990 J. of Luminesc. 45 205
Schmitt-Rink S., Chemla D.S. and Miller D.A.B. 1989 Advances in Physics 38 89
Tassone F., Bassani F. and Andreani L.C. 1990 IL Nuovo Cimento 12 D 1673

Inst. Phys. Conf. Ser. No 123
Paper presented at the International Meeting on Optics of Excitons in Confined Systems, Giardini Naxos, Italy, 1991

Optical spectroscopy of excitons in CdTe/CdZnTe II–VI heterostructures

Hélène Tuffigo

DRFMC / SP2M / PSC, CEN-Grenoble, 85X 38041 Grenoble cedex France.

ABSTRACT: This paper presents the spectroscopic results recently observed for CdTe/CdZnTe strained heterostructures. The peculiar band structure of this system allows us to observe experimentally unusual optical properties: (i) the quantization of the exciton centre-of-mass motion for quantum wells of thickness larger than 200Å, (ii) the mixed nature of the electronic band structure (type I for heavy-hole excitons and type II for light-hole excitons).

1. INTRODUCTION

Though emitting almost in the same energy range as the much studied GaAs/AlGaAs system (the bandgap of CdTe is 1.606eV), the CdTe/CdZnTe system is in fact very different as concerns the Coulomb interaction (compared to the carrier confinement energies), the lattice mismatch induced strain and the value of the valence band offset. Our aim is to emphasize the influence of these factors on the excitonic properties of this system.

In section 2, we will present the specificity of the CdTe/CdZnTe system, that is the order of magnitude of three physical parameters, the valence band offset (and hence the confinement potentials), the mismatch induced strain (and hence the influence on the band structure) and the electron-hole Coulomb interaction. This system appears as a good candidate for the experimental observation of an unusual "exciton centre-of-mass quantization" and this subject will be detailed in section 3, where we will present the experimental evidence for two types of quantization, exciton centre-of-mass quantization versus separate carrier quantization, the adequate description depending on the well thickness.

The rest of the paper will deal with the influence of the strain on the nature of the exciton in this system. Because of the small lattice parameter mismatch and the small natural valence band offset, the configuration of the valence band is very sensitive to the internal strain; consequently, the nature of the exciton is easily "tailorable" depending on the structural characteristics of the heterostructure. In the case of {100}-oriented CdTe/$Cd_{1-x}Zn_xTe$ strained superlattices (section 4), an optical study as a function of the superlattice period points out the mixed nature of the band structure (i.e. direct or indirect for heavy or light-hole exciton); the experimental values of exciton binding energy compares favorably with a variational calculation taking into account the specific nature of the valence band (i.e. very small valence band offset).

An additional consequence of the internal strain in our system is the very large piezoelectric effect observed for polar axis-oriented CdTe/$Cd_{1-x}Zn_xTe$ heterostructures, which strongly modifies their optical properties [André et. al. 1990; Cibert et. al. 1991]; this effect will not be described in this paper.

2. INTRODUCTION TO THE $CdTe/Cd_{1-x}Zn_xTe$ SYSTEM

The CdTe/ZnTe system has a lattice mismatch of 6%; therefore, despite the small values of x (in the $CdTe/Cd_{1-x}Zn_xTe$ heterostructures) that we are dealing with (about 10%), the structures experience a large biaxial strain. The bandgap difference ΔE_g between CdTe and $Cd_{1-x}Zn_xTe$ is small (dE_g/dx =0.525eV for low values of the zinc content x) and occurs mainly in the conduction band (as will be shown later on); in other words, the zero-strain valence band offset is negligeable. In such a situation, strain does much more than just perturb the band structure: instead, it totally governs the confinement potentials for holes.

Most of the heterostructures studied consist of CdTe and $Cd_{0.92}Zn_{0.08}Te$ layers deposited by molecular beam epitaxy on a $Cd_{0.96}Zn_{0.04}Te$ {001}-oriented substrate; in this case, if the whole structure remains coherent with the substrate, the CdTe and $Cd_{0.92}Zn_{0.08}Te$ layers experience a biaxial stress of opposite sign. Therefore, the strain-induced splitting of the Γ_8 band between the heavy and light holes produces an unusual band configuration [Mariette et. al. 1988] (see Figure 1) since the heavy and the light holes are confined in the CdTe and the CdZnTe layers respectively.

Compared to the III-V compounds, II-VI semiconductors are generally characterized by a larger exciton binding energy and a larger electron effective mass (e.g. respectively 2.5 and 1.5 larger in bulk CdTe than in bulk GaAs). In II-VI quantum wells, the electron-hole correlation is strong enough to maintain the electron-hole Coulomb interaction along the growth axis down to relatively small values of the well thickness; moreover, due to the larger value of the electron mass and the shallow confinement potentials, the electron is less confined than in III-V systems. Therefore, a correct description of the exciton state in II-VI quantum well heterostructures must take into account the concept of "exciton centre-of-mass quantization", often ignored in the work on III-V quantum wells (Centre of mass quantization in GaAs/GaAlAs quantum wells has been considered by [Schultheis et. al. 1984], [D'Andrea et. al. 1988] and [Kusano 1989]).

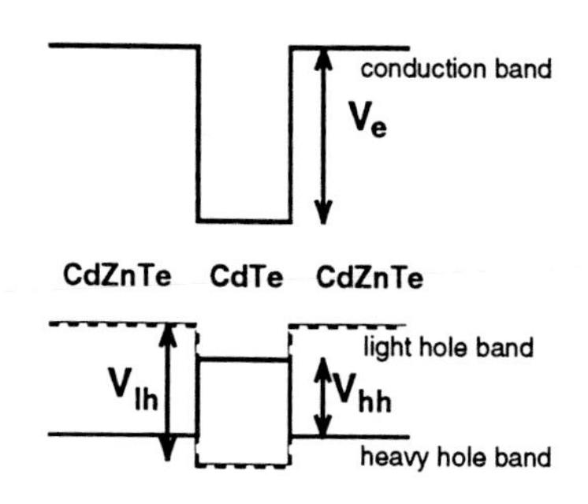

Figure 1: band structure of a $CdTe/Cd_{0.92}Zn_{0.08}Te$ heterostructure deposited on a $Cd_{0.96}Zn_{0.04}Te$ {100}-oriented substrate.

3. EXPERIMENTAL EVIDENCE OF TWO TYPES OF QUANTIZATION

Depending on the relative values of the exciton binding energy and the carrier kinetic energies, we have been able to distinguish two types of quantization which occur in two different quantum well thickness (L_z) ranges: $L_z \leq 200$Å and $L_z \geq 200$Å.

3.1 Separate carrier quantization

This corresponds to the well-known description of excitons states usually labelled e_ih_j, each constructed from a single pair of separately quantized 2D electron and hole states e_i and h_j [Dingle et. al. 1974]. This description is valid when the energy separation between the carrier subbands is large enough to neglect the mixing of the different 2D carrier subbands by the electron-hole Coulomb interaction, that is to say, when the quantum well is narrow and deep

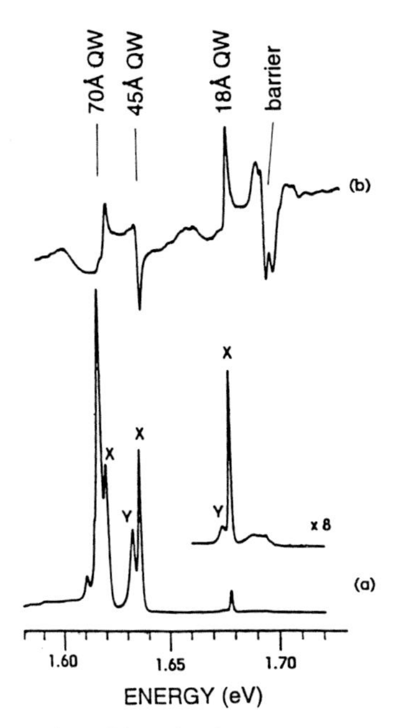

Figure 2: Photoluminescence and reflectivity spectra of a CdTe/CdZnTe heterostructure with three embedded quantum wells (of thickness 18Å, 45Å and 70Å).

enough to give rise to large confinement energies.

Figure 2 shows the photoluminescence (PL) and reflectivity spectra of a heterostructure constitued by three CdTe quantum wells of thicknesses 70Å (22 monolayers), 45Å (14 ML) and 18Å (5.5 ML) embedded between $Cd_{0.82}Zn_{0.18}Te$ barriers. These thicknesses are given acccurately by RHEED intensity oscillations during growth [Lentz et. al. 1989]. The signature of each quantum well is given by a single line observed in the reflectivity spectrum associated with a line called X in the photoluminescence spectrum; this line corresponds to the recombination of the intrinsic e_1h_1 exciton. The low lying lines Y in PL involve bound exciton recombinations [Kheng et. al. 1991]. This sample allows to visualize the increase of the carrier confinement energy when the QW thickness decreases from 70Å down to 18Å. This increase is well explained by a simple one band envelope function approximation [Bastard 1988]. The structure observed at 1.695eV in the reflectivity spectrum corresponds to the excitonic transition in the barriers and the buffer layer (of the same composition). This feature corresponds to a broad and weak line in the luminescence spectrum: the low intensity shows the efficiency of the transfert between barriers and wells.

It should be noted that (i) no appreciable Stokes shift is observed between emission and absorption associated with line X and that (ii) the line width (1-1.5 meV) does not vary significantly with the well width contrary to with what is observed for GaAs/GaAlAs systems [Tanaka et. al. 1987]. This can be explained by a large delocalization of the exciton wavefunction in the alloy due to the small barrier heights in these heterostructures [Tuffigo et. al. 1990]; consequently, the contribution of interface defects is reduced and the exciton is mainly perturbed by alloy fluctuations.

3.2 Quantization of the exciton centre-of-mass motion

Figure 3 presents (on a relative scale) PLE spectra obtained for "broad" quantum wells (with thicknesses varying between 220Å and 1000Å). The most intense line of each spectrum (taken as the zero of energy) is attributed to the ground state of the free exciton in the well [Tuffigo et. al. 1988]. On the high energy side of this line, one observes a series of well-resolved excited state lines, N=2, 3, ... (on an energy scale of 10-20meV); the number, the spacings and relative intensity of these lines depend strongly on L_z. These lines have been attributed to quantized states of the exciton's centre-of-mass motion along z (the growth axis) [Merle d'Aubigné et. al. 1987; Tuffigo et. al. 1988]. In the present case, the confinement energies for an electron and a hole (e.g. 1.3meV and 0.2meV respectively for a 500Å-thick CdTe quantum well) are small compared to the exciton binding energy so that the electron-hole Coulomb interaction completely mixes the subbands and maintains the electron-hole correlation along z and this down to surprisingly small values of the well thickness. In the simplest description of such states (which assume, in particular, infinitely high barriers), the line positions follow

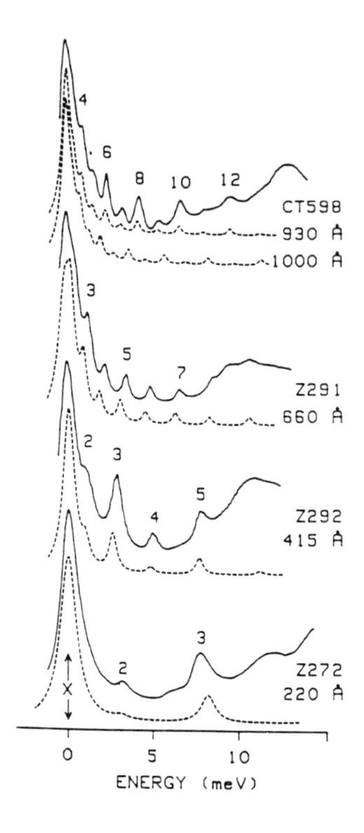

Figure 3: PLE spectra (continuous line) obtained at 1.8K for CdTe/CdZnTe quantum wells of thickness 220Å, 415Å, 660Å and 1000Å; the energy scale of this figure is relative (to the gound state of the free exciton). Luminescence is recorded on the low energy side of the free exciton line. The dotted-line spectra below each PLE spectrum are the calculated absorption spectra for these L_z values. For sample CT598, the real thickness is less precisely known, and we also show a calculation using the value L_z=930Å that best fits the line spacings.

$$E(N,L_z) = E_T + \frac{\hbar^2\pi^2}{2M^*L_z^2} N^2 \quad \text{(eq. 1)},$$

where M^* is the exciton translational mass $=m_e^*+m_h^*$ and E_T is the transverse energy. To predict the optical absorption spectrum more accurately, we have done a complete calculation [Tuffigo et. al. 1990] of the propagation of light waves and spatially dispersive exciton waves through the multi-layer sample, including evanescent waves in the barriers (whose height is just the optical gap difference, taken to be 45 meV, the approximate value for most of the samples). The calculated line positions relative to the lowest energy peak (see Figure 4) fit the experimental peaks quite accurately down to $L_z \cong 200$Å (taking $m_e^*=0.099m_0$ and $m_h^*=0.513m_0$ from [Neumann et. al. 1988] and ω_{LT}=0.49meV [Merle et. al. 1984]). The finite barrier height gives a slight downward curvature in plots against $1/L_z^2$ as compared to the straight lines given by eq. 1, and such a curvature appears in the data of Figure 4.

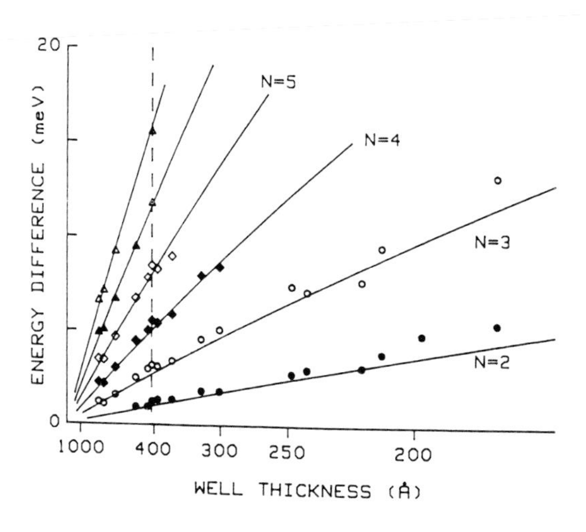

Figure 4: Experimental energy positions (relative to the ground state N=1 line) of the lines observed in PLE for 15 wells (180Å<L_Z<660Å) (the different peak for a given sample are represented by different symbols on one vertical line); the energies are plotted against $1/L_Z$. The curves represent the calculated variation of the absorption line energies assuming an average barrier height of 45meV.

Performing the complete calculation of the wave propagation through the structure presents the advantage of yielding the shape of the spectra; in Figure 3, for three of the samples, the only fitting parameter is the damping coefficient adjusted to fit the linewidths. By comparing the spectra obtained for the different thicknesses, one notices some striking features of the relative intensities of the excited states lines (which have alternating intensities for samples Z292 and CT598 but decrease regularly with N for sample Z291). This results from interference effects in the well, being determined mainly by the ratio of the wavelength of the photon-like wave (λ_{ph}=2500Å in CdTe), to the wavelength of the quantized exciton-like wave.

3.3 Transition between the two types of quantization

In the centre-of-mass quantization description, the exciton wavefunction is a linear combination of 3D wavefunctions of electrons and holes. When the well thickness decreases, giving rise to discrete energy levels whose energy separation exceeds the exciton binding energy, the Coulomb interaction can not mix anymore the electron (or hole) wavefunctions and we recover the usual description of a 2D exciton as being constructed from a single quantized electron state e_i and a single quantized hole state h_i. Because $M^*=m_e^*+m_h^*\sim m_h$, the theoretical N=1, 2, 3... energy differences are close to the separation between the h_1, h_2, h_3...hole subbands. In other words, the N=2, 3... exciton states univocally correspond to the "forbidden" e_1h_2, e_1h_3... excitons of the better known separate carrier quantization. Prediction of the thickness range in which the transitions occurs presents a difficult problem, especially when taking into account the finite barriers effect [D'Andrea et. al. to be published]. (Theoretical predictions of the transition range assuming infinitely high barriers have been obtained by [D'Andrea et. al. 1988] and [Platero et. al. 1990].)

We have experimentally observed for samples having L_z<200Å the e_2h_2 exciton simultaneously with the N=1, 2, 3... (or respectively e_1h_1, e_1h_2, e_1h_3...) excitons [Mariette et. al. 1991]; this is of great interest since the e_2h_2 exciton does not exist in the centre-of-mass quantization description and is therefore an unambiguous signature of the separate carriers approach. The experimental estimate of the thickness range in which the transition occurs for our system is consequently around L_z=200Å.

4. STRAIN-INDUCED MIXED TYPE I-TYPE II SUPERLATTICE SYSTEM

4.1 Influence of the period

We now would like to discuss the importance of the strain on the band structure. We have studied a set of CdTe/$Cd_{0.92}Zn_{0.08}Te$ superlattices with equal thicknesses of CdTe and $Cd_{0.92}Zn_{0.08}Te$ deposited on a $Cd_{0.96}Zn_{0.04}Te$ substrate (the average deformation of such superlattices is therefore equal to zero). The use of various complementary spectroscopic techniques (photoluminescence, PLE, reflectivity, polarized PLE) has allowed us to identify unambiguously the heavy and light hole exciton transitions (e_1h_1 and e_1l_1 respectively), see Figure 6. Piezoreflectivity measurements performed on the same samples have confirmed this identification [J. Calatayud, J. Allegre, private communication]. Then, assuming a zero valence band offset in the absence of strain, we have calculated the optical heavy- and light-hole gaps (see [Tuffigo et. al. 1991]) E_1H_1 and E_1L_1, and compared them with the experimental excitonic energies (see Figure 5.a). The difference between the calculated gaps and the experimental excitonic energies (schematized by the arrows in Figure 5.a) can be considered as "semi-experimental values" of the exciton binding energies. As can be seen in Figure 5.a, the heavy hole exciton binding energy determined in this way remains roughly constant over the whole range of periods whereas on the contrary, the light hole exciton binding energy decreases drastically as the period decreases. This results is fully consistent with the hypothesis of a direct heavy hole exciton and an indirect light hole exciton (the words "direct" and "indirect" apply to real space always in this paper), as represented in Figure 1; the larger the period, the smaller the overlap between the electron and the light hole wavefunctions, and therefore, the smaller the indirect exciton binding energy.

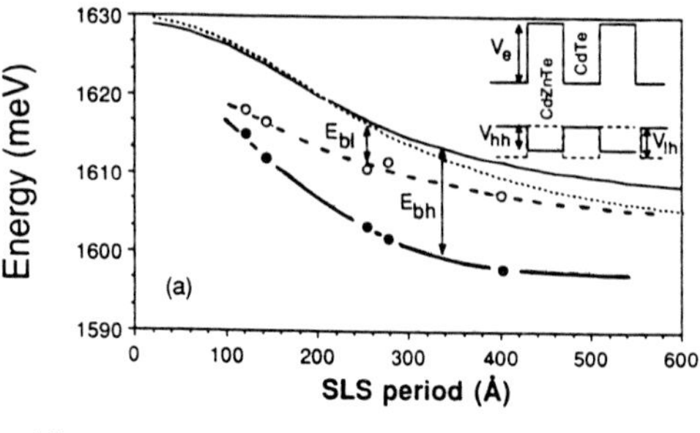

Figure 5.a: Influence of the SL period on the calculated gaps E_1H_1 (thin solid line) and E_1L_1 (thin dashed line) for $CdTe/Cd_{1-x}Zn_xTe$ SL (x=0.08, L(CdTe)=L($Cd_{1-x}Zn_xTe$), grown on a $Cd_{0.96}Zn_{0.04}Te$ substrate) and on the experimental excitonic energies of e_1h_1 (•) and e_1l_1 (o) (the thick solid and dashed lines drawn through the experimental points are only a guide for the eye).

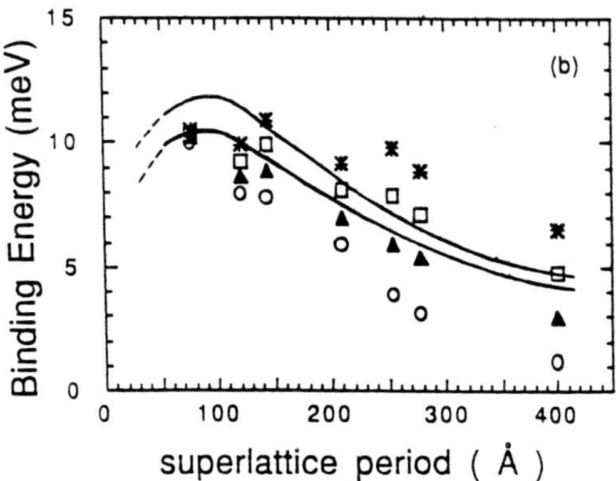

Figure 5.b: Plot of the light hole (indirect) exciton binding energy as a function of the superlattice period.
experimental points: circles, triangles, squares and stars correspond to α=0, α=0.05, α=0.1 and α=0.15 respectively.
theoretical curves: the lower one is obtained with the masses taken from [Neumann et. al. 1988], whereas the upper one is obtained for an infinite in-plane hole mass.

We now compare these results with a more complete calculation of the exciton binding energy taking into account the effect of the Coulomb interaction on the band structure ([Berroir et. al. to be published; Peyla et. al. to be published]). This effect is important in our system since the hole confinements are mainly due to this Coulomb interaction; in this respect, it has been shown [Deleporte et. al. 1991] that the Coulomb interaction can even localize the light hole in the center of the CdTe well, in the case of certain separate confinement heterostructures. In this calculation the exciton binding energy is defined as the difference between the hole energy with and without Coulomb interaction. In Figure 5.b, the calculated light hole exciton binding energy (quasi independent on the valence band offset in the range of interest $0.1<\alpha<0.1$, see [Tuffigo et. al. 1991]), is compared -for two limit values of the light hole mass- to the semi-experimental values of the exciton binding energies obtained for different values of the valence band offset α. This comparison inclines us to think that α=0.1 (which means $\Delta Ec=0.9\Delta Eg$ and $\Delta Ev=0.1\Delta Eg$) is the better compromise.

4.2 Influence of the deformation

These results confirm that the potentials for heavy and light holes are almost entirely determined by the superlattice's strain state; therefore, by tailoring the internal strain experienced by each layer of the superlattice (CdTe and CdZnTe), it is possible to change the superlattice from type I (the direct heavy hole exciton being at lower energy) to type II (the indirect light hole exciton being at lower energy). This was achieved by growing a superlattice of a given period ($CdTe/Cd_{0.92}Zn_{0.08}Te$ 65Å/65Å) on $Cd_{1-z}Zn_zTe$ buffer layers of various zinc concentrations (z varying between 0 and 0.08): changing the buffer's lattice parameter modifies the average superlattice deformation.

Figure 6 shows typical spectra obtained for two SLs in different strain states. One is deposited directly on a $Cd_{0.96}Zn_{0.04}Te$ substrate (same configuration as before); the comparison of the PL, reflectivity and polarized PLE spectra (Figure 6 a-c) confirms that in

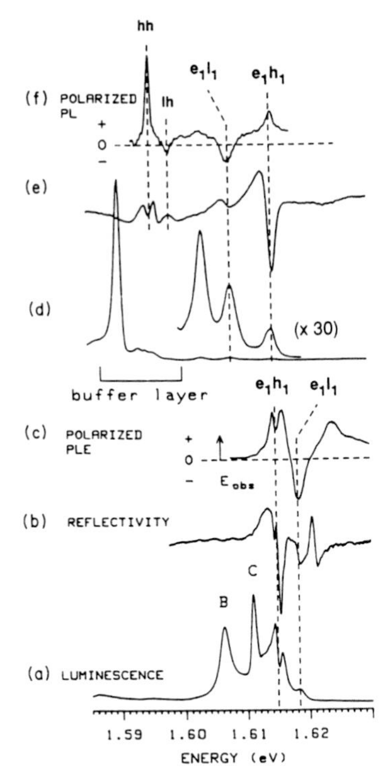

Figure 6: Photoluminescence (a and d), reflectivity (b and e) and polarization spectra (c and f) obtained for a type I (a,b,c) an a type II (d,e,f) superlattice; the strong CdTe emission at 1.59eV shows efficient energy transfert to the buffer layer.

Figure 7: Influence of $e^{||}$, the superlattice's average in-plane strain, on the calculated gaps E_1H_1 and E_1L_1 (continuous lines) and on the experimental excitonic energies e_1h_1 (●) and e_1l_1 (o).

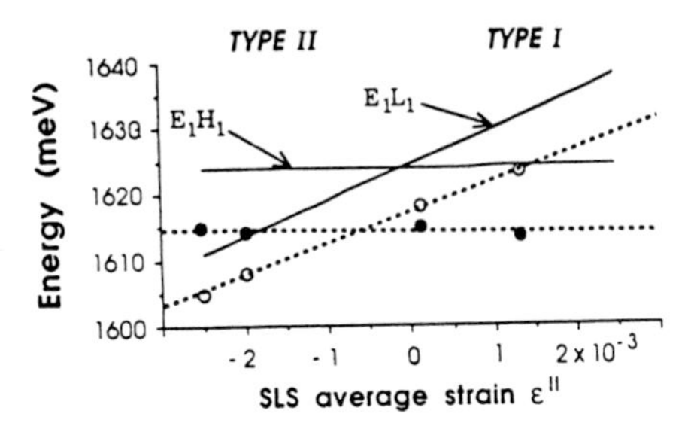

this case, the direct heavy hole exciton lies at lower energy than the indirect light hole exciton: we are dealing with a type I SL. (The negative polarization signal is the signature of the light hole character). The other one is deposited on a CdTe buffer layer; an important change is observed in the optical spectra, as shown in Figure 6 d-f: the relative energy position of intrinsic lines e_1h_1 and e_1l_1 is reversed, the indirect light hole exciton lying now at lower energy.

In Figure 7, we plot the experimental positions of e_1h_1 and e_1l_1 as a function of the deformation of the superlattice for four samples. Assuming that the elastic constants and deformation potentials are insensitive to the zinc concentration (in the range of interest), we can consider that the hole confinement potentials are unchanged when one varies the buffer composition: the only effect of such a variation is to shift the heavy and light hole potentials (of Figure 1) as a whole relative to eachother. Then, the inversion of the optical type in the superlattices observed in Figure 7 is entirely due to the strain changes (in contrast to other systems, see for example [Gershoni et. al. 1989]).

6. CONCLUSION

In this paper, we have emphasized the main optical characteristics of CdTe/CdZnTe heterostructures. First, this system has allow to give the experimental evidence of two types of quantization, depending on the well thickness: exciton quantization versus separate carrier quantization. Second, we have stressed the importance of the lattice-mismatch induced strain and showed how it is possible to play with. For {100}-oriented heterostructure, we have shown that it is possible to tailor the valence band configuration (due to its mixed nature) and consequently observe the transition from a type I superlattice to a type II superlattice. Moreover, for polar axis oriented heterostructures, large piezoelectric effects that modify considerably the optical characteristics are also observed. These last two results make this system especially attractive to study the influence of an electric field on the basic electro-optic properties of II-VI heterostructures.

ACKNOWLEDGEMENTS

I am particularly indebted to R.T. Cox, N. Magnea and H. Mariette for their helpful contribution throughout this work. I would like to thank Ph. Peyla for giving me his theoretical results prior to publication. The results reported in this review article were obtained in collaboration with J. Cibert, P. Gentile, G. Lentz, Le Si Dang, Y. Merle d'Aubigné, J.L. Pautrat, Ph. Peyla and A. Wasiela. This work was performed within the CEA-CNRS joint group for II-VI semiconductor microstructures.

REFERENCES

R. André, C. Deshayes, J. Cibert, Le Si Dang, S. Tatarenko and K. Saminadayar, Phys. Rev. B42 (1990) 11392.
G. Bastard, "Wave mechanics applied to semiconductor heterostructures", les éditions de physique, 1988.
J.M. Berroir et. al. to be published.
J. Calatayud, J. Allegre et. al., private communication.
J. Cibert, R. André, C. Deshayes, G. Feuillet, P.H. Jouneau, R. Mallard, K. Saminadayar, Proceedings of the Fifth International Conference on II-VI Compounds, Tamano (Japan) 1991.
A. D'Andrea and R. Del Sole in Excitons in Confined Systems (edited by R. Del Sole, A. D'Andrea and A. Lapiccirella), Proceedings in Phys. 25, Springer (1988) 102.
A. D'Andrea, R. Del Sole, K. Cho and H. Ishihara, Proc. 19th Int. Conf. on Physics of Semiconductors, Warsaw, 1988, Ed. W. Zawadzki.
A. D'Andrea, N. Tomassini, R. Del Sole, R.T. Cox, H. Tuffigo, to be published.
E. Deleporte, G. Peter, J. M. Berroir, G. Bastard and C. Delalande, to appear in the Proceedings of EP2DS9-MSS5, Nara 1991.
R. Dingle, W. Wiegmann and C.H. Henry, Phys. Rev. Lett. 33, 827 (1974).
D. Gershoni and H. Temkin, J. Lumin. 44 (1989) 381.
K. Khengh, R.T. Cox, S. Tatarenko, F. Bassani, K. Saminadayar, this conference.
J. Kusano, Y. Segawa, M. Mihara, Y. Aoyagi and S. Namba, Solid State Commun. 72, 215 (1989).
G. Lentz, A. Ponchet, N. Magnea and H. Mariette, Appl. Phys. Lett. 55, 2733 (1989).
H. Mariette, F. Dal'Bo, N. Magnea, G. Lentz and H. Tuffigo, Phys. Rev. B38, 12443 (1988).
H. Mariette, N. Magnea, H. Tuffigo, to appear in Physica Scripta 1991.
Y. Merle d'Aubigné, Le Si Dang, A. Wasiela, N. Magnea, F. Dal'Bo and A. Million, J. Phys. 48 (1987) C5-363.
C. Neumann, A. Nöthe and N.O. Lipari, Phys. Rev. B37 (1988) 922.
Ph. Peyla et. al. to be published.
G. Platero and M. Altarelli, Proceedings of 20th Int. Conf. on the Physics of Semiconductors (edited by E.M. Anastassakis and J.D. Joannopoulosa) World Scientific (1990) 1489.
L. Schultheis and K. Ploog, Phys. Rev. B29 (1984) 7058.
D.L. Smith and C. Mailhiot, Rev. Mod. Phys. 62 (1990) 173.
M. Tanaka and H. Sakaki, J. Cryst. Growth 81(1987) 153.
H. Tuffigo, R.T. Cox, N. Magnea, Y. Merle d'Aubigné and A. Million, Phys. Rev. B37 (1988) 4310.
H. Tuffigo, R.T. Cox, G. Lentz, N. Magnea, H. Mariette, J. of Cryst. Growth 101 (1990) 778.
H. Tuffigo, N. Magnea, H. Mariette, A. Wasiela, Y. Merle d'Aubigné, Phys. Rev B43 (1991) 14629.
H. Tuffigo, N. Magnea and H. Mariette, Materials Science and Engineering, B9 (1991) 185.

Inst. Phys. Conf. Ser. No 123
Paper presented at the International Meeting on Optics of Excitons in Confined Systems, Giardini Naxos, Italy, 1991

Conduction-band offset measurements on GaAs/GaAlAs heterostructures

C B Sørensen and E Veje

Physics Laboratory, H.C. Ørsted Institute,
Universitetsparken 5, DK-2100 Copenhagen, Denmark

ABSTRACT: We are studying the optical spectra emitted from GaAs/GaAlAs heterodiodes which are forward biased and also photoexcited with a laser. Due to the forward bias, electrons are injected ballistically from the active GaAlAs layer into the GaAs buffer, where they can recombine radiatively with carbon acceptors. We observe spectral features which we interprete as resulting from such recombinations, and from their positions, the conduction-band offset can be deduced. Our data for a mole fraction x = 0.29 confirm a recently published result.

1 INTRODUCTION

The conduction-band offset is defined as that fraction of the total band offset which occurs for the conduction band at a heterojunction. Although the conduction-band offset for GaAs/ GaAlAs junctions has been widely studied, the results are still inconclusive, as discussed by Petersen et al (1989) and by Goossen et al (1987). The majority of previous experimental results seem to show that the conduction-band offset is approximately 65% of the total band offset, independent of the mole fraction x, of the $Ga_{(1-x)}Al_xAs$ layer. However, the few recent data by Petersen et al (1989) and by Goossen et al (1987) may indicate that the conduction-band offset is a larger fraction of the total band offset for small values of x. Because of this, and since the size of the conduction-band offset is of importance in basic research as well as for practical purposes like design of devices, we are initiating a systematic study to elucidate the problem for GaAs/GaAlAs heterostructures as a function of x. Our method is similar to that used by Petersen et al (1989) in the sense, that hot electrons are injected ballistically from an active GaAlAs layer into a GaAs buffer, where they may recombine radiatively with acceptors. However, to obtain a better signal-to-noise ratio, we assist this process with soft laser-excitation of the samples.

2 EXPERIMENTAL PROCEDURE, DATA TREATMENT AND RESULTS

Our samples were grown in a Varian Modular Gen II machine using molecular beam epitaxy (MBE). From top, they consist of an n-type GaAs cap, doped with $2x10^{17}$ silicon atoms per cm^3, an n-type $Ga_{(1-x)}Al_xAs$ active layer, also doped with $2x10^{17}$ silicon atoms per cm^3, a thin, undoped $Ga_{(1-x)}Al_xAs$ spacer and a thick unintentionally doped GaAs buffer, which actually is slightly p-type due to incorporation of carbon atoms during the growth. Consequently, such a structure can be regarded (from bottom) as a p-i-n diode, and by applying a forward bias to the structure, electrons are ejected ballistically from the GaAlAs layer to the GaAs buffer, where radiative recombination may take place. This is sketched in Fig. 1, which schematically gives a presentation of the energy properties of the GaAlAs/GaAs heterojunction. The conduction-band offset is denoted ΔE_C.

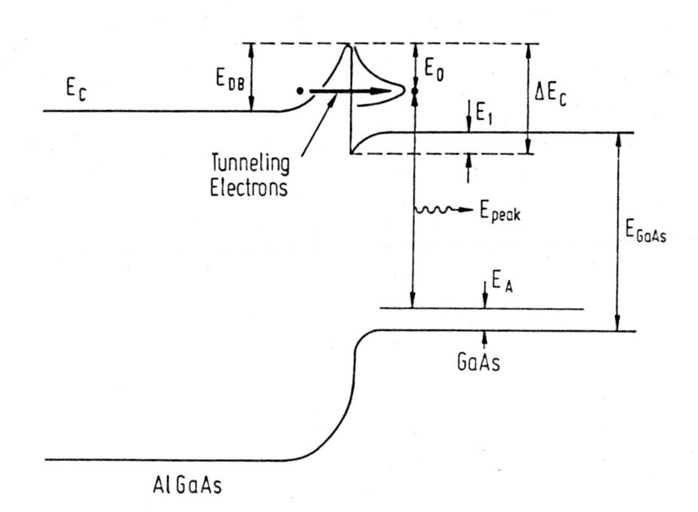

Fig. 1. The low-temperature forward-bias diode band bending. The depletion barrier on the GaAlAs side (not to scale) and associated tunnel current are shown.

The optical measurements were carried out at our photoluminescence facility, which has been described in detail by Veje (1990). Here it suffices to mention that photons emitted from radiative recombination processes were analyzed with a one-meter scanning spectrometer with high resolution and were subsequently detected by a photomultiplier using single-photon counting.

Spectral scans were carried out with, as well as without a positive bias across the sample, and with or without illuminating the sample with a HeNe laser or an argon-ion laser.

The mole fraction, x, of the $Ga_{(1-x)}Al_xAs$ layer was determined from the RHEED oscillation periods related to the MBE growth, as well as with the use of photoluminescence, namely from the position of the bulk GaAlAs exciton features at 10 K and also from the position of the band-to-band transitions at room temperature. The three determinations yielded the same results.

During data taking, the samples were cooled to 10 K by a closed-cycle helium refrigeration system.

In the optical spectrograms, we naturally observed the well-established exciton transitions in the near-bandgap regions, and in addition also a weaker spectral feature, similar to that seen by Petersen et al (1989) occured. We interprete it as resulting from the recombination process depicted in Fig. 1 and from the peak position, the transition energy, E_{peak}, was deduced. In passing, we mention that besides of the main peak, also phonon replicas with energy spacings of 36 meV, corresponding to the longitudinal-optical phonon energy, were observed.

From Fig. 1 is seen, that

$$\Delta E_C = E_{peak} + E_A + E_0 + E_1 - E_{GaAs}$$

from which ΔE_C can be deduced. In the data evaluation, we used for E_0 and E_1 the values given by Petersen et al (1989) and with a mole fraction x = 0.29, we obtained a ratio of the conduction-band offset to the total band-gap offset of 0.67 which equals the value reported by Petersen et al (1989), and this is encouraging. At the time of writing, work is in progress for other values of x.

The intensities of the various spectral features were measured as functions of the potential drop across the sample, see Fig. 2. As can be seen from this figure, the exciton-line intensities decrease rapidly with increasing potential drop. This is clearly related to the low binding energies of the excitons, so that they are easy to ionize. Contrary to this, the intensity of the light ascribed to the above-mentioned radiative recombination process depends much less on the potential drop. This is to be expected for recombinations where the initial state is unbound. Thus, the data in Fig. 2 confirm our interpretation.

3 ACKNOWLEDGMENTS

The equipment has been granted from the Danish Natural Science Research Council, the Danish National Agency of Technology, The Carlsberg Foundation, Direktør Ib Henriksens Foundation, and the NOVO Foundation. Skillful assistance from Ms. Mette B. Jensen is also highly appreciated.

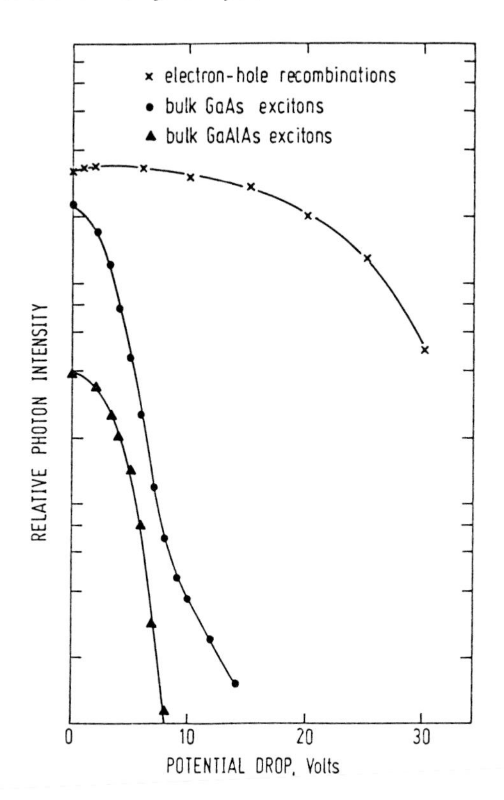

Fig. 2. Relative photon intensities for electron-hole recombinations, bulk GaAs excitons, and bulk GaAlAs excitons are plotted semilogarithmically versus the potential drop across the sample.

4 REFERENCES

Goossen K W, Lyon S A and Alavi K 1987 Phys. Rev. B36 9370
Petersen C L, Frei M R and Lyon S A 1989 Phys. Rev. Lett. 63 2849
Veje E 1990 Proc. 14th Nordic Semiconductor Meeting, ed O Hansen (Århus:Århus University Press)pp. 29-32

Inst. Phys. Conf. Ser. No 123
Paper presented at the International Meeting on Optics of Excitons in Confined Systems, Giardini Naxos, Italy, 1991

Fine structure and dispersion of quantum well excitons

S Jorda*, U Rössler* and D Broido†

* Institut für Theoret. Physik, Universität Regensburg, D-8400 Regensburg
† Department of Physics, Boston College, Chestnut Hill, Mass. 02167, USA

ABSTRACT: We present a systematic formulation of the exchange interaction for excitons in Quantum Wells and perform numerical calculations of the exchange splitting and dispersion for $Al_{0.3}Ga_{0.7}As$ Quantum Wells with L = 50 and 150 Å.

1 INTRODUCTION

The dispersion of excitons for small wave vector is determined by the fine-structure splitting, caused by the electron-hole exchange interaction, and by the dispersion of conduction and valence bands (Rössler 1979). For quantum wells the situation differs from that of bulk material due to the lowering of the symmetry and the heavy-hole light-hole mixing, which on one side gives a more complex splitting pattern and on the other side mirrors the intricate subband dispersion. In the formulation of the exchange interaction for excitons in quantum wells we take into account the heavy-light hole mixing for the subbands and relate the exchange splitting of quantum well excitons to the exchange parameters of the bulk excitons. We perform numerical calculations of the exchange splitting for $Al_{0.3}Ga_{0.7}As/GaAs$ quantum wells with L = 50 and 150 Å within the axial model, i.e. neglecting the warping of the hole subbands. For evaluating the exciton dispersion we exploit the ambiguity in choosing the relative and center-of-mass coordinates to formulate the hole subband states (and hence the Coulomb interaction) independent of the exciton total in-plane wave vector. Instead it appears in a linear coupling with the relative momentum. This coupling is taken into account in solving the exciton integral equation.

2 THE EXCHANGE HAMILTONIAN

The lowest electronic excitations in a quantum well system are the excitons formed by an electron in the first conduction subband (C1) and a hole in the first heavy (HH1) or light (LH1) hole subband. Taking into account the spin of the particles these exciton states span an eightfold space. In a quantum well structure with (001) as growth direction the symmetry group is D_{2d} and the electron-hole exchange Hamiltonian can be represented as an 8x8 matrix (Rössler et al 1990), which in angular momentum representation $|M_h, M_e\rangle$ falls into four blocks corresponding to different values of $|M_e + M_h|$. The eigenvalues of this exchange Hamiltonian agree with the splitting pattern obtained from the decomposition of $\Gamma_8 \otimes \Gamma_6$ (T_d) in the symmetry group D_{2d}: Heavy hole exciton states $\Gamma_6 \otimes \Gamma_6 = \Gamma_1 \oplus \Gamma_2 \oplus \Gamma_5$, light hole exciton states $\Gamma_7 \otimes \Gamma_6 = \Gamma_3 \oplus \Gamma_4 \oplus \Gamma_5$. The dipole allowed states are Γ_4 ($\mathbf{e} \parallel z$) and Γ_5 ($\mathbf{e} \parallel x, y$). The formulation of the Hamiltonian for quantum well excitons and of the corresponding eigenvalue problem as an integral equation was given by Broido et al (1986) and Zhu et al (1987) without taking into account the exchange interaction. Here we add the formulation of the exchange contribution whose general form is

$$V^X_{\lambda_e\lambda_h\lambda'_e\lambda'_h} = \int dx_1 \int dx_2\, \psi^\dagger_{\lambda_e}(x_1)\, \psi^\dagger_{\lambda'_h}(x_2)\, \frac{e^2}{|\boldsymbol{r}_1 - \boldsymbol{r}_2|}\, \psi_{\lambda_h}(x_1)\, \psi_{\lambda'_e}(x_2). \tag{1}$$

The single-particle wavefunctions for electrons and holes are

$$\begin{aligned} \psi_{\lambda_e}(x_e) &= \sqrt{\frac{\mathcal{V}}{\mathcal{A}}}\, e^{i\vec{k}_e\cdot\vec{\rho}_e} \sum_{k_z} C_{n_e}(k_z) e^{ik_z z_e} u_{c\boldsymbol{k}\sigma_e}(x_e) \\ \psi_{\lambda_h}(x_h) &= \sqrt{\frac{\mathcal{V}}{\mathcal{A}}}\, e^{i\vec{k}_h\cdot\vec{\rho}_h} \sum_{k'_z}\sum_{M_J} C^j_{n_h\vec{k}_h M_J}(k'_z) e^{ik'_z z_h} u_{v\boldsymbol{k}' M_J}(x_h) \end{aligned} \tag{2}$$

where $C(k_z)$ are the Fourier transforms of the corresponding subband functions, $\vec{k}_{e,h}, \vec{\rho}_{e,h}$ the in-plane wave vector and particle coordinates, respectively and $\boldsymbol{k} = (\vec{k}_e, k_z)$, $\boldsymbol{k}' = (\vec{k}_h, k'_z)$ the three-dimensional wave vectors. $\mathcal{A}$ and $\mathcal{V}$ are the normalization area and volume. $u_{c\boldsymbol{k}\sigma_e}(x_e)$ and $u_{v\boldsymbol{k}'M_J}(x_h)$ are the periodic parts of the bulk Bloch functions of the electron and hole, respectively, which depend on space and spin coordinates x and are assumed to be the same for well and barrier material. The sum over M_J in Eq. 2 is due to heavy-hole light-hole mixing for finite in-plane wave vector $\vec{k}_h$.

In performing the integrals in Eq. 1 we first consider the spin part of the Bloch functions to obtain $\boldsymbol{J}\cdot\boldsymbol{\sigma} - \frac{3}{2}$. In a second step the products of the periodic parts of the electron and hole Bloch functions are represented as Fourier series

$$u^*_{c\boldsymbol{k}}(\boldsymbol{r})\, u_{v\boldsymbol{k}'}(\boldsymbol{r}) = \sum_{\boldsymbol{G}} D_{cv\boldsymbol{k}\boldsymbol{k}'}(\boldsymbol{G})\, e^{i\boldsymbol{G}\cdot\boldsymbol{r}} \tag{3}$$

where $\boldsymbol{G} = (\vec{G}_\parallel, G_z)$ are three-dimensional reciprocal lattice vectors, in accordance with the assumption that the $u's$ are the same for well and barrier material. By integrating over all space coordinates we obtain

$$\begin{aligned} V^X_{\lambda_e\lambda_h\lambda'_e\lambda'_h} &= -\sum_{M_J M'_J} \frac{2}{3}(\boldsymbol{J}\cdot\boldsymbol{\sigma} - \frac{3}{2})^{\sigma_e M_J}_{\sigma'_e M'_J} \sum_{\boldsymbol{q}} \frac{4\pi e^2}{\mathcal{V}q^2}\mathcal{V}^2 \sum_{\substack{k_z k'_z \\ k''_z k'''_z}} C^*_{n_e}(k_z) C_{n'_e}(k'_z) \times \\ &\times\; C^{j'^*}_{n'_h \vec{k}'_\parallel M'_J}(k'''_z) C^j_{n_h \vec{k}_\parallel M_J}(k''_z) \sum_{\boldsymbol{G}\boldsymbol{G}'} D_{cv\boldsymbol{k}\boldsymbol{k}''}(\boldsymbol{G}) D^*_{c'v'\boldsymbol{k}'\boldsymbol{k}'''}(\boldsymbol{G}') \times \\ &\times\; \mathcal{L}\,\delta_{k_z,k''_z+q_z+G_z}\, \mathcal{L}\,\delta_{k'_z,k'''_z+q_z+G'_z}\, \delta_{\vec{q}_\parallel,\vec{Q}_\parallel-\vec{G}_\parallel}\, \delta_{\vec{q}_\parallel,\vec{Q}_\parallel-\vec{G}'_\parallel} \end{aligned} \tag{4}$$

where $\vec{k}_\parallel = \vec{k}_h$, $\vec{k}'_\parallel = \vec{k}'_h$, $Q_\parallel = \vec{k}_e - \vec{k}_h$ is the exciton in-plane center-of-mass wave vector and $\mathcal{L}$ is a periodicity length.

Let us first consider the case $\boldsymbol{G}, \boldsymbol{G}' \neq \boldsymbol{0}$, which corresponds to the analytic exchange contribution. Because $\vec{Q}_\parallel$ and all $\vec{k}'s$ are small compared to the length of reciprocal lattice vectors, they can be neglected in $D_{cv\boldsymbol{k}\boldsymbol{k}'}(\boldsymbol{G})$ and we have $\boldsymbol{G} = \boldsymbol{G}'$. For the two-band model ($c = c' = \Gamma_6$, $v = v' = \Gamma_8$) this part of the exchange matrix reads

$$\begin{aligned} V^{X,A}_{\lambda_e\lambda_h\lambda'_e\lambda'_h} &= -\sum_{M_J M'_J} \frac{2}{3}(\boldsymbol{J}\cdot\boldsymbol{\sigma} - \frac{3}{2})^{\sigma_e M_J}_{\sigma'_e M'_J} \sum_{\boldsymbol{G}\neq\boldsymbol{0}} \frac{4\pi e^2}{\mathcal{V}G^2} |\mathcal{V}\, D_{cv\boldsymbol{o}\boldsymbol{o}}(\boldsymbol{G})|^2 \times \delta_{G_z 0} \\ &\times\; \mathcal{L}^2 \sum_{q_z k_z} C^*_{n_e}(k_z) C^j_{n_h\vec{k}_\parallel M_J}(k_z - q_z) \sum_{k'_z} C_{n'_e}(k'_z) C^{j'^*}_{n'_h\vec{k}'_\parallel M'_J}(k'_z - q_z). \end{aligned} \tag{5}$$

The contribution for $\boldsymbol{G} = \boldsymbol{G}' = \boldsymbol{0}$, which is called nonanalytic in the bulk, has to be evaluated by taking into account the differences in the wave vectors $\boldsymbol{k} - \boldsymbol{k}'' = \boldsymbol{k}' - \boldsymbol{k}''' = -(\vec{Q}_{\|}, q_z)$ of the periodic parts of the Bloch functions in a $\mathbf{k}\cdot\mathbf{p}$ expansion, which gives

$$D_{cv\boldsymbol{k}\boldsymbol{k}''}(\boldsymbol{0}) = \frac{1}{\mathcal{V}}\frac{\hbar}{m}\frac{\langle c|(\boldsymbol{k}-\boldsymbol{k}'')\cdot\boldsymbol{p}|v\rangle}{E_c - E_v} \tag{6}$$

where c, v refer to the conduction and valence band edge of the bulk material. The exchange coupling between different pairs of conduction and valence bands results in a two-band model as a background screening of the exchange term (Rössler and Trebin 1981). We obtain for the nonanalytic exchange part

$$\begin{aligned} V^{X,N}_{\lambda_e\lambda_h\lambda'_e\lambda'_h} &= -\sum_{M_J M'_J}\frac{2}{3}(\boldsymbol{J}\cdot\boldsymbol{\sigma} - \frac{3}{2})^{\sigma_e M_J}_{\sigma'_e M'_J}\frac{4\pi e^2\hbar^2}{\varepsilon_b \mathcal{V} E_g^2 m^2} \times \sum_{q_z}\eta(\vec{Q}_{\|}, q_z) \\ &\times \mathcal{L}^2 \sum_{k_z} C^*_{n_e}(k_z) C^j_{n_h \vec{k}_{\|} M_J}(k_z - q_z) \sum_{k'_z} C_{n'_e}(k'_z) C^{j'^*}_{n'_h \vec{k}'_{\|} M'_J}(k'_z - q_z) \end{aligned} \tag{7}$$

where

$$\eta(\vec{Q}_{\|}, q_z) = \begin{cases} \dfrac{q_z^2}{Q_{\|}^2 + q_z^2} & \text{for } |v\rangle = |z\rangle \\ \dfrac{Q_{\|}^2}{Q_{\|}^2 + q_z^2} & \text{for } |v\rangle = |x\rangle \text{ (if } \vec{Q}_{\|} \parallel \vec{e}_x) \\ 0 & \text{for } |v\rangle = |y\rangle \end{cases} \tag{8}$$

The dependence of $\eta(\vec{Q}_{\|}, q_z)$ on the relative orientation between $(\vec{Q}_{\|}, q_z)$ and the momentum matrix element $\langle c|p|v\rangle$ gives rise to the longitudinal-transverse splitting. At this point we emphasize that the k-dependence of the periodic parts of the electron and hole Bloch functions is crucial for this part of the exchange interaction. Andreani and Bassani (1990) in their evaluation of the electron-hole exchange interaction have neglected this k-dependence. To our opinion, they obtain a LT-splitting only due to an approximate evaluation of lattice sums (Andreani et al 1988).

3 EVALUATION OF THE DISPERSION

The exchange interaction obtained in section 2 is now considered by first order perturbation theory using the exciton functions calculated within the axial model (Broido et al 1986 and Zhu et al 1987). The only off-diagonal term between degenerate states occurs between the Γ_3 and Γ_4 light hole excitons. In order to be able to identify longitudinal and transverse exciton states, we perform in the two Γ_5-blocks a transformation from the angular momentum eigenstates to states which transform as x and y. The exchange parameters $A = \sum_{\boldsymbol{G}} |\mathcal{V} D_{cv\boldsymbol{oo}}(\boldsymbol{G})|^2 \frac{4\pi e^2}{\mathcal{V}G^2}$ and $N = \frac{4\pi e^2 P^2}{\mathcal{V}\varepsilon_b E_g^2}$ are taken to be the bulk values $\mathcal{V}A = 0.1775 \cdot 10^5\ meV$ and $\mathcal{V}N = 0.71 \cdot 10^5\ meV$ which correspond to the splittings reported by Ekardt et al (1979) and Ulbrich et al (1977). Our choice of $\vec{k}_{\|} = \vec{k}_h$ and $\vec{Q}_{\|} = \vec{k}_e - \vec{k}_h = \vec{k}'_e - \vec{k}'_h$ in section 2 has the advantage that the hole-subband states with their entanglement do not depend on $\vec{Q}_{\|}$. However, we have to take into account the coupling term $\frac{\hbar^2}{m_e}\vec{k}_{\|}\cdot\vec{Q}_{\|}$ deriving from the parabolic electron subband dispersion. This is done by considering this coupling between s and p excitons in the solution of the exciton integral equation.

In Fig. 1 we show results for light hole exciton states for a 50 Å quantum well. The energies represent sums of hole subband-, binding energies and $\vec{Q}_{\|}$ dependent contributions of dispersion and exchange terms for small in-plane total exciton wave vectors. Fig. 2 shows similar results for a 150 Å quantum well, the exchange contribution being given on an extended wave vector scale (see inset). The main features are the vanishing LT-splitting for $Q_{\|} \rightarrow 0$ (as already found by Andreani and Bassani (1990)), the dominance of the anisotropy splitting between Γ_4 and $\Gamma_3 \oplus \Gamma_5$ exciton states caused by electron-hole exchange interaction and the strongly nonparabolic exciton dispersion in Fig. 2, which is caused by the corresponding light-hole subband dispersion. More detailed results, also for heavy-hole excitons, will be shown in a forthcoming publication (Jorda et al 1991).

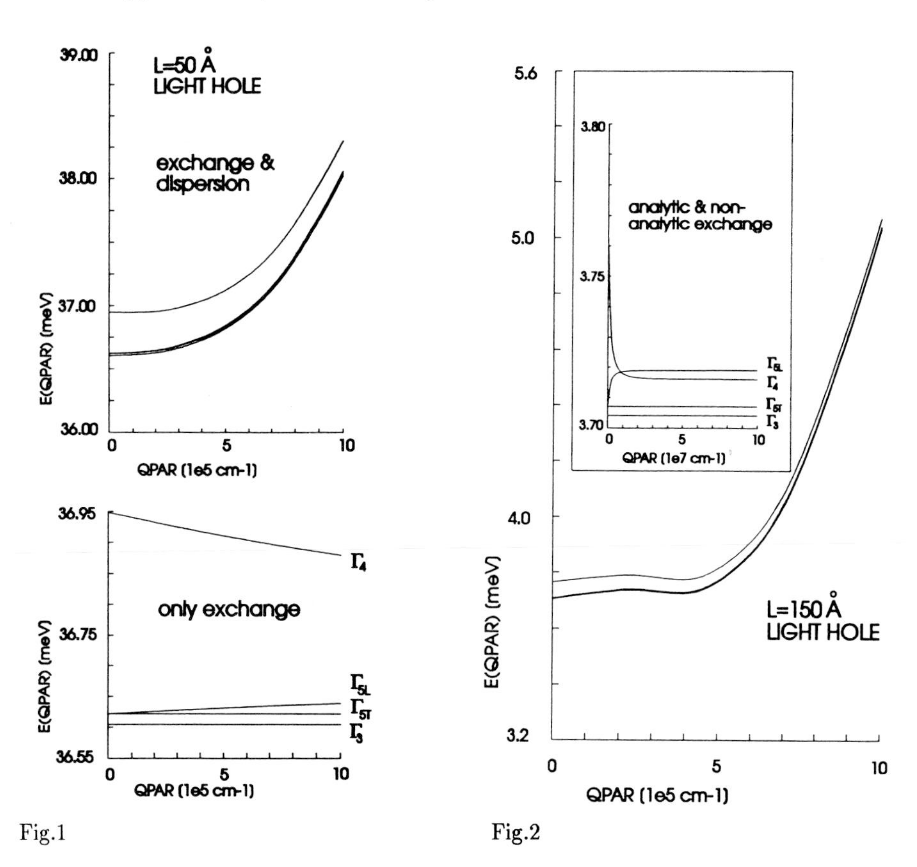

Fig.1 Fig.2

REFERENCES

Andreani L C, Bassani F and Quattropani A 1988 Nuovo Cimento D 10 1473
Andreani L C and Bassani F 1990 Phys. Rev. B 41 7536
Broido D A and Sham L J 1986 Phys. Rev. B 34 3917
Ekardt W, Lösch K and Bimberg D 1979 Phys. Rev. B 20 330
Jorda S, Rössler U and Broido D to be published.
Rössler U 1979 Festkörperprobleme XIX/Advances in Solid State Physics, ed J Treusch (Vieweg, Braunschweig 1979) p 77
Rössler U and Trebin H R 1981 Phys. Rev. B 23 1961
Rössler U, Jorda S and Broido D 1990 Solid State Commun. 73 209
Ulbrich R G and Weisbuch C 1977 Phys. Rev. Lett. 38, 865
Zhu B and Huang K 1987 Phys. Rev. B 36 8102

Inst. Phys. Conf. Ser. No 123
Paper presented at the International Meeting on Optics of Excitons in Confined Systems, Giardini Naxos, Italy, 1991

Exciton reflectivity and photoreflectivity in quantum well structures

V.P.Kochereshko, E.L.Ivchenko, I.N.Uraltsev.

A.F.Ioffe Physico-Technical Institute of the USSR Academy of Sciences, 194021, Sankt-Petersburg.

ABSTRACT: We have used the recently found effect of giant photoreflectivity in GaAs/AlGaAs quantum well structures [1] to create the light-induced grating. Interference of two coherent laser beams has been used to form a periodical structure of illuminated and unilluminated areas on the structure surface to form a periodical modulation of the refractive index. We have observed the resonance diffraction of the probe beam on this light-induced lateral grating. The giant diffraction efficiency has been observed in the vicinity of the quantum well exciton resonances.

1. INTRODUCTION

In the recent years there has been an impressive activity in investigation of the low-dimensional systems by reflectivity methods [2,3,4]. Reflectance experiments have been successively used to determine the excitonic parameters as: the resonance frequencies, oscillator strengths and dampings of excitons in quantum well structures and short-period superlattices [5]. Using this technique we have documented the transition from the 2D character of the exciton wave function in quantum wells to the 3D character in short-period superlattices [6].

We have previously observed the effect of photoinduced modification of the refractive index in the vicinity of the exciton resonances in quantum well structures. We have found the strong increase of reflectance in the resonance region at the above barrier illumination of extremely low intensity $\leqslant 10\ mW/cm^2$. We have demonstrated that this effect is connected with photoinduced decreasing of an exciton scattering, and is caused by a neutralization of barrier impurities or defects under an illumination [1].

We have used the photoreflectance effect to create a light-induced grating. Two coherent laser beams have been used to form aperiodical structure of illuminated and unilluminated areas onthe structure surface to produce periodical variations of theexciton damping and, consequently, of the refractive index.We have observed the resonance diffraction of the probe beamon this light-induced lateral grating. The giant diffraction efficiency $R_{\pm 1}/R_0 \approx 10^{-3}$ has been observed in the vicinity of the light- and heavy-hole exciton resonances.

2. LIGHT-INDUCED LATERAL GRATING

2.1 Theory

The impressive reduction in the exciton damping, Γ, is expected under pumping above the barrier bandgap to create periodic modulation of Γ in a SQW.

Let us assume that a SQW structure is illuminated by two coherent light beams impinging on the sample at the angles φ_1 and φ_2 with respect to the normal of the interface, as shown schematically in Fig.1a. Due to the interference of the beams the resulting pump intensity, J, is modulated along the x-axis laying in the incidence plane:

$$J(x)=J_o + J_1\cos(2\pi x/\Lambda) \quad . \qquad (1)$$

The modulation period:

$$\Lambda=\lambda|\sin\varphi_1 - \sin\varphi_2|^{-1} \qquad (2)$$

is supposed to be much longer than the light wavelength, λ, and the diffusion length of photocarriers which takes part in the neutralization of impurity centers [1]. It follows then that a grating of the neutralized centers is formed near the SQW and, as a result, the exciton damping Γ varies periodically along the x-axis. The variation $\Gamma(x)$, can be demonstrated in experiments on diffraction of the probe lightwave.

The diffraction efficiently can be evaluated from the relation between the electric fields of the reflected and incident waves

$$E(x) = [\, r_{01} + \frac{t_{01}\, t_{10}\, i\tilde{\omega}\, e^{i\Phi}}{\omega_o - \omega - i\Gamma(x,T)} \,]\, E_o(x) \quad , \qquad (3)$$

where T is the pumping duration and allowance is made for the x-dependence of Γ. For the simplest model of deep-level filling by photocarriers, $\Gamma(x,T)$ can be written as

$$\Gamma(x,T)=\Gamma_o + \Delta\Gamma \exp[-\xi J(x)T] \quad , \qquad (4)$$

where ξ is a constant, J(x) is given by Eq.(1) , $\Delta\Gamma$ is the difference between the equilibrium and steady-state values of the damping. The angle φ_d, at which the first order diffraction peak is observed, is expressed in terms of φ_1, φ_2 and the probe incidence angle, φ_o, as follows

$$\sin\varphi_d = \sin\varphi_o \pm \frac{\lambda_{probe}}{\lambda_{pump}}(\sin\varphi_1 + \sin\varphi_2) \quad . \qquad (5)$$

Assuming for the sake of simplicity that the oscillating part of Γ is small in comparison with Γ_o we obtain for the diffracted intensity:

$$R_{\pm 1}=\left|\frac{E_{\pm 1}}{E_o}\right|=[t_{01}t_{10}I_1(u_1)e^{-u_o}\frac{\Delta\Gamma_1\ \tilde{\omega}}{(\omega_o-\omega)^2+\Gamma_o^2}]^2 \quad . \quad (6)$$

Here $u_i=\xi J_i T$ $(i=0,1)$, the intensities J_0 and J_1 are introduced in Eq.(1), $I_\nu(z)$ is the modified Bessel function of order ν. It should be mentioned that the spectrum $R_{\pm 1}(\omega)$ is independent on the phase Φ and has the lineshape with a peak at $\omega = \omega_o$ and the effective halfwidth $\Gamma_o/\sqrt{2}$. In the GaAs/AlGaAs heterostructures the spectrum consists of two peaks corresponding to the 1s-exciton e1-hh1 and e1-lh1. Notice that $\Delta\Gamma$ can be different for the two exciton states.

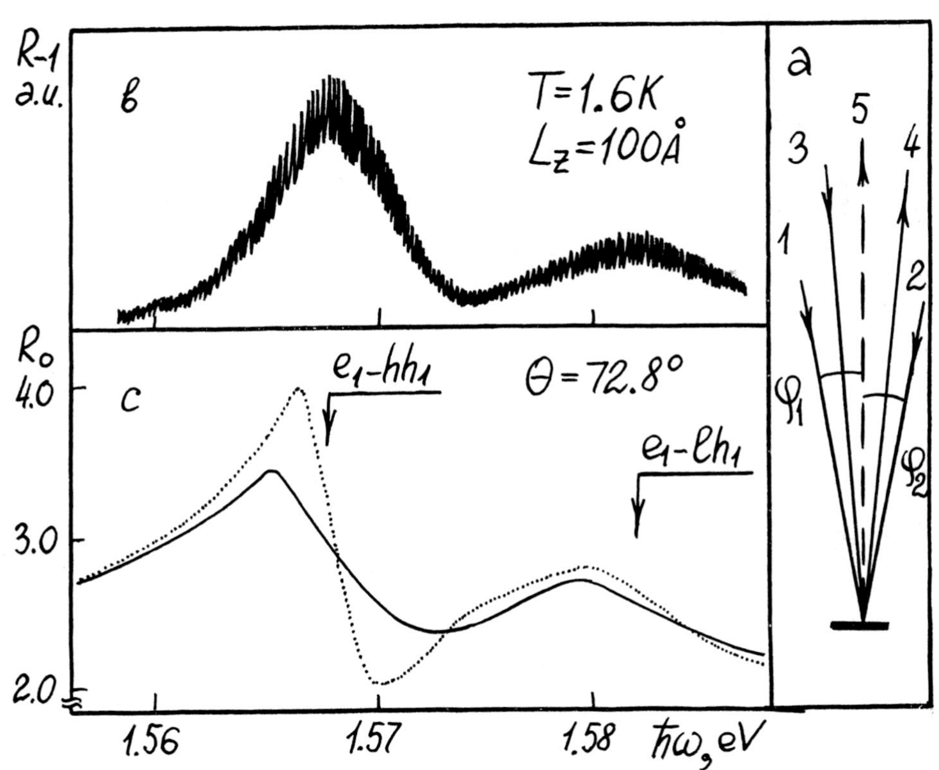

Fig.1 a) Experimental scheme. 1,2-laser pump beams, 3-probe beam, 4-reflected-R_o, 5-first order diffraction R_{-1} ;

b) First order diffraction R_{-1} on light-induced grating;

c) Reflection p-polarized spectra taken at oblique incidence from sample without (solid) and with additional illumination (dotted). We have extracted for solid and dotted curves the same set of parameters: ε= 12.5, $\hbar\omega_o^h$= 1.568 eV, $\hbar\omega_o^l$= 1.583 eV, $\hbar\omega_{LT}^h$=0.48 meV, $\hbar\omega_{LT}^l$=0.25 meV, except $\hbar\Gamma_h$=2.4 meV, $\hbar\Gamma_l$=3.5 meV for solid curve and $\hbar\Gamma_h$=1.5 meV, $\hbar\Gamma_l$ =2.9 meV for dotted curve.

2.2 Experiment

Fig.1a shows the scheme of diffraction of probe beam on light-induced grating. Experiments have been made at 1.6 K on $GaAs/Al_{0.3}Ga_{0.7}As$ SQW structure with L_z=100 A .The diffraction spectrum Fig.1b has been measured in the "backward" configuration when the two interfering pump beams (1) and (2) as well as the probe beam (3) are propagating forward and only the diffracted light propagates in a backward direction (4). In order to create the lateral grating, the He-Ne laser beam of the intensity 10 mW/cm^2 was split into two which made to coincide on the sample. The probe beam was the quasi-monochromatic light output from a 70 W quartz-halogen lamp through a 0.5 m double monochromator. Pump and probe beams were polarized perpendicular to the incidence of plane. To exclude the photoluminescence influence on the measured signal a delay of registration was used to be sufficiently longer than the photoluminescence lifetime ($\propto$1 ns) and shorter than the decay time of the photoreflectivity ($\propto$1 ms) [1].

Fig.1c shows the reflectivity spectrum of the structure registred at Brewster angle geometry.

As expected the diffraction signal is observed only in the vicinity of exciton resonance frequencies. It follows from the exciton parameters of our sample that

$$(\Delta\Gamma_l\tilde{\omega}_l)^2 \ll (\Delta\Gamma_h\tilde{\omega}_h)^2 \quad .$$

The first order diffraction signal related to the reflectance R amounts 10^{-3} which is a typical value for the diffraction by light-induced grating in bulk crystals. Taking into account the microscopic thickness of the investigated SQW (L≈100 A) we conclude that their diffraction efficiencies are extremely high.

3. REFERENCES

1. Ivchenko E.L.,Kochereshko V.P.,Uraltsev I.N., Yakovlev D.R., Phys. Stat. Sol. (b) 161,217 (1990)
2. Ivchenko E.L., Kop´ev P.S., Kochereshko V.P., Uraltsev I.N., Yakovlev D.R., Ivanov S.V., Meltser B.Ya.,Kalitievskii M.A., Sov. Phys. Semicond. 22,497 (1988)
3. Andreani L.C.,Tassone F., Bassani F., Solid State Commun. 77,529 (1989)
4. Dignam M.M., Sipe T.E., Phys. Rev. B41,N5,2865 (1990)
5. Ivchenko E.L.,Kochereshko V.P., Kop´ev P.S., Kosobukin V.A., Uraltsev I.N.,Yakovlev D.R., Solid State Comm.,70,529 (1989)
6. Uraltsev I.N., Kochereshko V.P., Kop´ev P.S., Vasiliev A.M., Yakovlev D.R., Surface Science 229,459 (1990)

Inst. Phys. Conf. Ser. No 123
Paper presented at the International Meeting on Optics of Excitons in Confined Systems, Giardini Naxos, Italy, 1991

The exciton formation time in a GaAs quantum well

R.Eccleston, R.Strobel, and J.Kuhl
Max-Planck-Institut für Festkörperforschung
Heisenberstraße 1, 7000 Stuttgart 80, Germany

K.Köhler
Fraunhofer Institut für Angewandte Festkörperphysik,
7800 Freiburg, Germany

ABSTRACT: The dynamics of exciton formation and the exciton formation time are determined in a GaAs asymmetric double quantum well tunneling structure using the non-linear photoluminescence (PL) cross-correlation technique.

1.INTRODUCTION

GaAs-$Al_{0.3}Ga_{0.7}As$ asymmetric double quantum well (ADQW) structures consist of a wide and narrow quantum well (10nm and 5nm thick respectively in these experiments) separated by a thin barrier through which tunneling from the narrow well may occur, and have been the subject of many investigations (Nido (1990)) of resonant and non-resonant tunneling times by means of time-resolved photoluminescence (TRPL) . We report here strong changes in the time-integrated PL efficiency of the narrow well of such structures as the photo-excited free electron-hole pair density, n_{ex}, is varied. Fig. 1 shows the time and spectrally integrated PL intensity at 8K emitted from the n=1 heavy-hole exciton of the narrow well of two ADQW's as a function of n_{ex}. For a 20nm barrier sample (crosses), through which negligible tunneling from narrow to wide well can occur, a completely linear dependence on density is observed. However, for a 4nm barrier sample (open circles), where TRPL measurements indicate a tunneling time of 10ps, the intensity increases with n_{ex}^2. A linear dependence is however observed in the wide well of the same 4nm barrier sample (dots in Fig.1) where tunneling is also absent. We attribute the dramatically different dependence in the narrow well of the tunneling structure to competition between n_{ex}-independent tunneling of free carriers out of the well and n_{ex}-dependent formation of excitons from free carriers in a bimolecular process (ie. at a rate proportional to the product of free electron and free hole densities) . At low n_{ex}, most electrons tunnel before excitons can be formed and so relatively little excitonic PL from the narrow well is observed. At higher density, the tunneling probability is unchanged, but the exciton formation probability is increased. Proportionately more excitonic PL is then observed, leading to the measured non-linear dependence. The role of the exciton formation has been rarely considered in the dynamics of photoexcited carriers in quantum well systems, and the influence on conventional experiments is often weak. For example, the risetime of excitonic TRPL transients is usually dominated by exciton cooling rather than formation (Damen 1990). In ADQW's however, the PL efficiency is strongly sensitive to exciton formation. In this work, this is exploited to measure the exciton formation rate using the time-resolved non-linear PL cross-correlation technique.

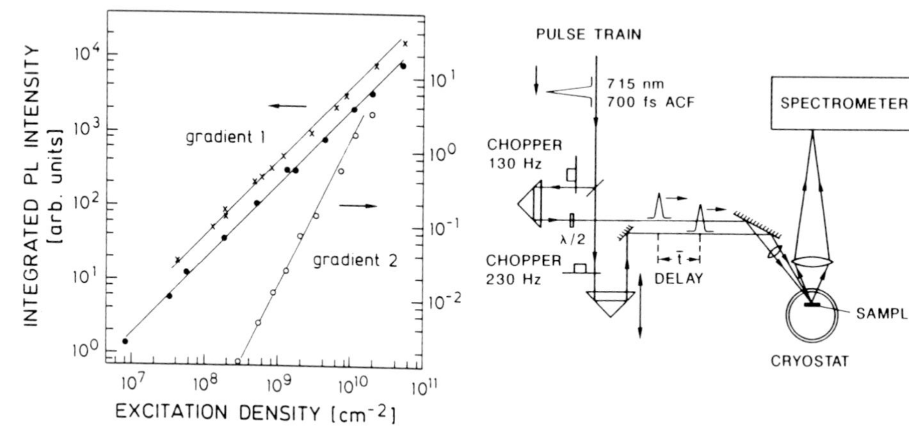

Fig.1 Integrated PL (arb. units) vs. n_{ex} at 8K for: 4nm barrier/narrow QW (circles), 4nm barrier/wide QW (dots) and 20nm barrier/narrow QW (crosses).

Fig. 2 The non-linear PL cross-correlation measurement.(ACF - auto-correlation function).

2.EXPERIMENTAL

In the non-linear PL cross-correlation measurement (Fig.2), two 700fs pulses from a hybrid mode-locked Pyridine 1 dye laser operating at 715nm are overlapped on the ADQW sample which is maintained at 8K in a helium cryostat. The time-integrated narrow well exciton PL originating from the photocreated free carriers is detected using a cooled photomultiplier. Beam 1 is chopped at 230Hz and beam 2 at 130Hz and the photoluminescence signal is detected at the sum frequency, 360Hz. A signal is generated at the sum frequency only because the PL emission is a non-linear function of n_{ex}. As discussed above, this arises from the enhanced probability of exciton formation due to the mutual interaction of the two carrier populations created by the two pulses. By varying the delay between the two pulses, the decay of the non-linear change in PL output may be measured. The decay occurs due to depletion of the carrier population created by the leading pulse via electron and hole tunneling or exciton formation. The times associated with these 3 processes can be extracted by modeling the cross-correlation signal using a system of rate equations to describe the carrier dynamics in the narrow well:

$$\frac{dn}{dt} = -Cnp - \frac{n}{\tau_e} + \frac{N}{\tau_h} + G(t) + G(t - \bar{t}) \tag{1}$$

$$\frac{dp}{dt} = -Cnp - \frac{p}{\tau_h} + \frac{N}{\tau_e} + G(t) + G(t - \bar{t}) \tag{2}$$

$$\frac{dN}{dt} = Cnp - \frac{N}{\tau_e} - \frac{N}{\tau_h} - \frac{N}{\tau_{LT}} \tag{3}$$

where n, p, and N are the density of free electrons, free holes, and excitons, respectively; τ_e and τ_h are the tunneling times for electrons and holes; τ_{LT} is the exciton radiative lifetime; $G(t)$ and $G(t-\bar{t})$ are carrier generation terms given by two Gaussian pulses of appropriate FWHM, delayed by $\bar{t}$. The bimolecular generation rate of excitons is given by the term Cnp, where C is the bimolecular exciton formation coefficient; n/τ_e, p/τ_h are the tunneling rates for free electrons and holes. Tunneling of an electron or hole bound in an exciton ('excitonic' electrons or holes) is given by N/τ_e and N/τ_h. The excitonic tunneling terms N/τ_e and N/τ_h not only decrease the density of excitons, but also give a positive contribution to the free hole and free electron densities, respectively. The radiative recombination term, N/τ_{LT}, determines the photoluminescence intensity emitted from the sample. Time integration of $N(t,\bar{t})/\tau_{LT}$ gives the total time-integrated PL as a function of $\bar{t}$. Because exciton formation is a bimolecular process, the growth of the exciton population is non-exponential and has no single characteristic time. But, from (1)(2)(3), we can define a density-dependent exciton formation time, τ_{ex}, equal to the 1/e time for a photoexcited population of electrons and holes to form excitons in the absence of tunneling, given by $\tau_{ex} = (e-1)/Cn_{ex}$. Note that calculation of the total PL intensity vs. n_{ex} in the limit of $\tau_{ex} << \tau_e$ using (1)(2)(3) gives the n_{ex}^2 dependence measured in Fig. 1.

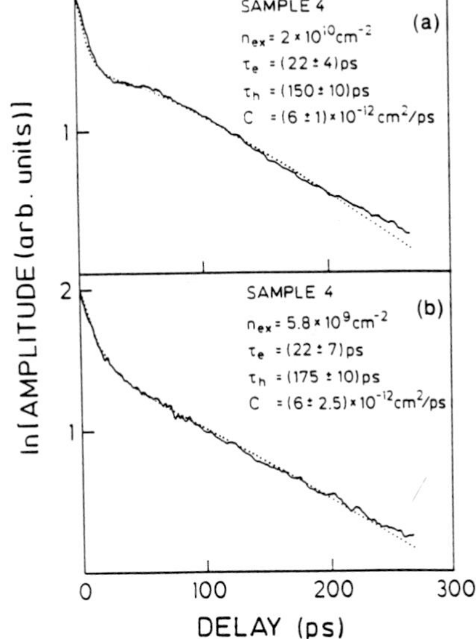

Figure 3. PL cross-correlation signal vs. delay for 4nm barrier/narrow QW at (a) n_{ex}=2 $\times 10^{10} cm^{-2}$ and (b) n_{ex}=5.8 $\times 10^{9} cm^{-2}$. Also shown is the modeled fit.

3.RESULTS AND DISCUSSION

Figure 3 shows non-linear PL curves from a 4nm barrier sample at two different excitation densities and the best fit using Equs. (1)(2)(3). In the low density case (curve (b)), the signal closely approximates to two exponential decays given by τ_e (=22ps) and τ_h (=160ps). The same result is obtained from both experiment and model if n_{ex} is reduced by a further factor of 10. However, at higher n_{ex} (curve (a)), the faster decay becomes even faster and a non-exponential 'shoulder' region is observed. The faster decay corresponds to the case where the carriers from the first pulse are depleted faster by exciton formation than by electron tunneling. The shoulder arises because as excitonic electrons tunnel, free holes are released which contribute further to the non-linearity. This provides direct evidence that the exciton is broken apart during the tunneling process. The value of C is determined most accurately in curve (a), because the fast decay is then given mainly by the exciton formation process, and corresponds to τ_{ex} =14ps at this value of n_{ex}, consistent with the upper limit of 20ps inferred from TRPL measurements (Damen (1990)). Shorter tunneling times ($\tau_e = 7ps$ and $\tau_h = 66ps$) are obtained, as expected from a thinner (3nm) barrier. The non-exponential shoulder is also absent at $n_{ex} = 2\times10^{10}cm^{-2}$ as expected because τ_e is now less than τ_{ex}. Note that a similar ratio, $\tau_e/\tau_h \approx 0.1$, is obtained for both the 3nm and 4nm samples. This ratio is in both cases much less than expected from semi-classical tunneling theory (for further details see: Strobel (1991)).

The good agreement between experiment and model as a function of barrier thickness and n_{ex} confirms that the non-linear PL originates from competition between exciton formation and tunneling, and demonstrates the bimolecular nature of exciton formation. It should be stressed that although the quantitative values are obtained from a three parameter fit, the accuracy of the values remains quite high since τ_h is relatively independent of the other parameters, and C and τ_e may be distinguished by varying n_{ex}. The results imply that the correlation of the photocreated 'geminate' electron-hole pair is rapidly broken and that excitons are subsequently formed purely from non-geminate pairs. This is consistent with the large initial excess energy of the carriers. Several LO-phonons are immediately emitted and separation of the geminate pair is likely to be very efficient. We find the same results for excess photon energies between 130meV (715nm) and 15meV above the exciton line. We therefore conclude that the exciton formation and tunneling dynamics are largely unaltered by different excess carrier energy within this regime.

4.CONCLUSIONS

The PL output of the narrow well of ADQW structures varies non-linearly with excitation density. This is due to the bimolecular nature of the exciton formation process, which makes the carrier dynamics in such structures strongly carrier density-dependent. Using PL cross-correlation measurements we determine a value of $6\times10^{-12}cm^2/ps$ for the bimolecular exciton formation coefficient, and also the electron and hole non-resonant tunneling times.

5. REFERENCES

M. Nido, M.G.W. Alexander, W.W. Rühle, T. Schweizer, and K. Köhler, Appl. Phys. Lett. **56**, 355 (1990), and the references therein.

D. von der Linde, J. Kuhl, and E. Rosengart, J. Luminescence **24/25**, 675 (1981).

T.C. Damen, J. Shah, D.Y. Oberli, D.S. Chemla, J.E. Cunningham, and J.M. Kuo, J.Luminescence **45**, 181 (1990).

R.Strobel, R.Eccleston, J.Kuhl, and K.Köhler, Phys. Rev. **B 43**, 12564, (1991).

Inst. Phys. Conf. Ser. No 123
Paper presented at the International Meeting on Optics of Excitons in Confined Systems, Giardini Naxos, Italy, 1991

The dynamics of the exciton–lattice interaction in a GaAs quantum well

R.Eccleston, R.Strobel, W.W.Rühle, J.Kuhl, B.F.Feuerbacher, and K.Ploog.
Max-Planck-Institut für Festkörperforschung, Heisenbergstrasse 1, 7000 Stuttgart 80, Germany

ABSTRACT: The initial value and time evolution of the heavy-hole (hh) exciton temperature is investigated in a high-quality 27nm GaAs quantum well for both light-hole (lh) exciton and free-carrier excitation by time-resolved photoluminescence (TRPL). The TRPL transient is modeled to obtain the exciton-LA phonon energy loss rate. The results suggest excitons are formed elastically.

1.INTRODUCTION

We report TRPL risetime measurements in a GaAs-$Al_{0.3}Ga_{0.7}As$ 27nm single QW with an extremely narrow heavy-hole exciton PL linewidth (0.2meV) (Ploog (1991)). Photo-excitation very close to the hh exciton PL line is then possible without direct resonant creation of hh excitons, allowing investigation of new excitation regimes. Specifically: for sufficiently low excess photon energy, the usual rapid LO-phonon emission by free carriers is forbidden. Cooling *or heating* of excitons then occurs via slower LA-phonon emission and absorption. The evolution of the exciton temperature then becomes slow enough to detect changes in the initial exciton temperature, induced by variation of the excess energy of the exciting photon, on a streak camera. Such changes influence the risetime of the TRPL signal due to the increase in the effective exciton oscillator strength as the exciton population cools (Feldmann (1987), Damen (1990)). Here the magnitude of the risetime is related to the exciton LA-phonon energy loss rate. Comparison with results for excitation of lh excitons (which in a 27nm QW also lie in the bandgap) can be used to estimate the contribution of exciton formation to exciton energy loss. The mechanism of lh-hh exciton transfer is also investigated. The sample was cooled in a helium flow cryostat and excited with pulses from a synchronously pumped mode-locked Styryl 8 laser. The PL was spectrally dispersed in a 32cm spectrometer and temporally resolved in a 2D synchroscan streak camera. The temporal resolution of the system was between 15ps and 25ps and the spectral resolution was 0.5meV.

2.LIGHT-HOLE EXCITON TO HEAVY-HOLE EXCITON SCATTERING

The inset of Fig.1 shows TRPL transients at two initial excited carrier densities, n_0, for excitation at the lh exciton line and detection at the hh exciton line at 15K. For large n_0 (curve a), an immediate fast rise of the hh exciton PL to 80% of the peak value is observed within the time resolution of the streak camera, indicating that scattering between lh and hh exciton bands occurs within 20ps at this density. The second, slower component of the risetime originates from the slow rise in excitonic PL due to hot exciton cooling described earlier. The scattering between the hh and lh exciton states therefore establishes an exciton population which is initially hotter than the lattice. At lower n_0, the fast rise in the PL disappears (curve b). The density-dependence of the risetime demonstrates that the lh-hh transfer mechanism is exciton-exciton scattering. Fig. 1 shows the variation of total risetime with n_0. At the lowest densities investigated, the PL risetime increase saturates, consistent with exciton-exciton scattering becoming weaker than (density independent) impurity or LA-phonon scattering. At the highest densities, the risetime is dominated by cooling. The filling of the hh exciton states is closely equivalent to the establishment of a thermal distribution within the two (strongly coupled) exciton bands. For free carriers this thermalisation process is generally considered to occur on the timescale of $\leq$1ps. The much longer times (at low n_0) measured here are consistent with the lower efficiency of exciton-exciton scattering (Kuhl (1989)).

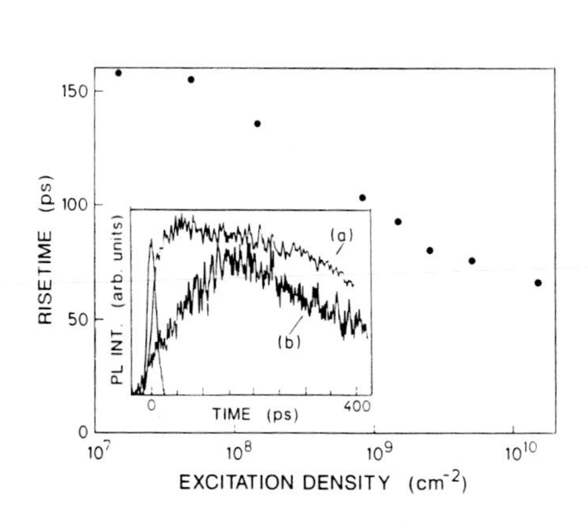

Fig. 1. PL risetime vs. n_0 for lh exciton excitation and hh exciton PL detection at 15K. The inset shows TRPL profiles at (a) $n_0 = 1.5 \times 10^{10} cm^{-2}$, (b) $n_0 = 1.5 \times 10^8 cm^{-2}$.

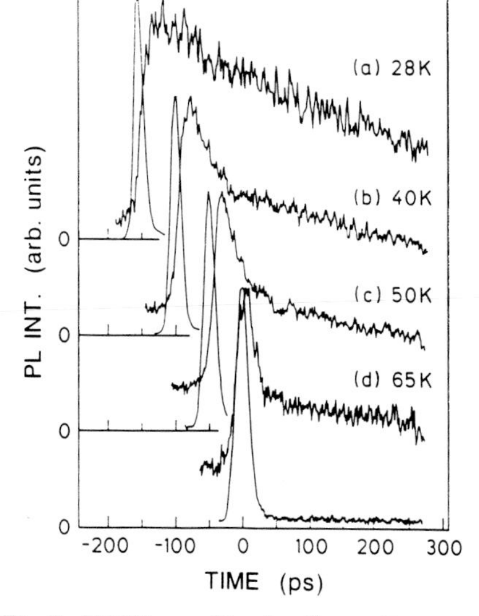

Fig.2. TRPL profiles for lh exciton excitation and hh PL exciton detection at different T_L and $n_0 = 1.5 \times 10^{10} cm^{-2}$. For clarity each curve is shifted by 50ps.

The lattice temperature, T_L, was varied to determine the temperature at which the heavy-hole exciton population was formed. As T_L was increased, the cooling component of the risetime decreased and near 28K (Fig.2a) was close to the time-resolution limit, indicating that hh excitons are formed at an initial temperature which is close to 28K. This value is in agreement with the temperature expected after thermalisation to a Maxwellian distribution in 2D from an initial delta-function energy distribution at an excess energy of 2.4meV (the lh-hh line splitting). This is consistent with exciton-exciton scattering since the thermalisation occurs with no energy loss from the exciton distribution. For higher T_L (Figs.2b-d), an initial fast decay is observed, which becomes faster as T_L increases. This fast decay is the exact inverse of the increase in effective exciton oscillator strength due to warm exciton cooling observed at low temperature. The exciton population internally thermalises to a temperature near 28K and is then heated to the lattice temperature with a corresponding fall in the radiative recombination rate. This also shows that lattice-related processes such as phonon scattering cannot contribute significantly to the filling of the hh exciton states, because the hh excitons can be formed initially colder than the surrounding lattice.

3. EXCITON LA-PHONON EMISSION

Now we investigate the change in initial exciton temperature (at T_L=15K) as the excess photon energy, ΔE, above the hh exciton line is varied. Fig.3a shows the previous resonant lh exciton pumping result (i.e. ΔE=2.4meV). Fig.3b is obtained at ΔE =6.9meV, i.e. only slightly above the electron-hh bandedge, located at ΔE=5.7meV (Ploog (1991)). In Figs.3a-d, a progressive increase in the PL risetime is observed, indicating a higher initial exciton temperature. However, beyond $\Delta E \approx$ 10meV the TRPL profile no longer varies strongly with ΔE (compare Figs.3d and 3e), implying no further increase in the initial exciton temperature. This is consistent with the onset of very rapid energy loss via LO-phonon emission within the streak camera time-resolution. No further increase in the TRPL risetime with ΔE is therefore apparent beyond this threshold.

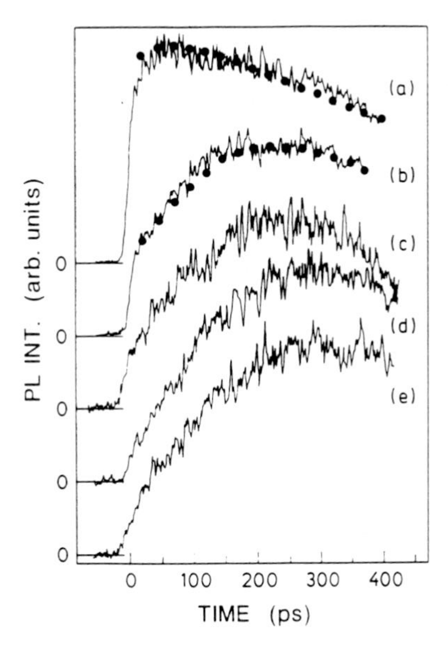

Fig. 3. TRPL profiles at 15K for detection at the hh exciton PL line for different excess photon energy, ΔE, at $n_0 = 1.5 \times 10^{10} cm^{-2}$. (a) ΔE =2.4meV, (b) ΔE =6.9meV, (c) ΔE =8meV, (d) ΔE =12.9meV, and (e) ΔE =20.8meV. Curves (a) and (b) also show the theoretical fit (dots).

The exciton LA-phonon energy loss rate and exciton temperature evolution may be extracted from the magnitude of the TRPL risetimes. The luminescence intensity, I(t), is proportional to the instantaneous exciton recombination rate, given in a QW by (Feldmann (1987)): $I(t) \propto -dN/dt = (1 - exp(-\Gamma/kT_{ex})N/K$, where T_{ex} is the exciton temperature, Γ is the homogeneous linewidth (0.2meV,(Feuerbacher (1990))),

N is the exciton density. K is a constant determined from the exponential tail of the luminescence beyond 500ps where $T_{ex}=T_L$. We consider here only the lh exciton and near-bandedge measurements (Figs.3a and 3b). LO phonon energy loss can then be completely neglected. We assume that excitons are formed elastically, i.e. the entire exciton binding energy is transfered into extra kinetic energy of the bound electron-hole pair. The 2D LA phonon deformation potential energy loss rate in eV/s for a Maxwell-Boltzmann distribution of excitons is:

$$\langle\frac{dE}{dt}\rangle_{LA} = -2.04 \times 10^4 (D_c - D_v)^2 \left(\frac{M_{ex}}{m_0}\right)^2 (T_{ex} - T_L) \qquad (1)$$

where M_{ex} is the total exciton mass $(0.7m_0)$, D_c and D_v are the deformation potentials of the conduction and valence band of GaAs given by 10.7eV and -3.4eV,respectively. Further details of the calculation are described elsewhere (Eccleston (1991)). Figs.3a and 3b show the calculated TRPL transients, which are in reasonable agreement with experiment. We therefore conclude that for ΔE near the bandedge, the exciton cooling rate determined from TRPL risetimes is quantitively consistent with LA-phonon emission. We find that T_{ex} falls to within 15% of T_L after 100ps for ΔE =2.6meV and after 300ps for ΔE =6.9meV. The good agreement with theory may be used to support the assumption that excitons are formed elastically. If significant energy was lost, e.g. by emission of a phonon in the formation process, a much smaller difference in risetime would be obtained between Figs. 3a and 3b. In fact an energy loss of no more than 1meV is permitted without generating significant differences from the observed transients.

4. CONCLUSIONS

Bandgap lh excitons scatter to hh exciton states via exciton-exciton scattering. The resultant hh exciton temperature corresponds to a value of kT close to the lh-hh exciton splitting energy. Thermal equilibrium with the lattice then occurs by either cooling or heating. Near the bandedge, the PL risetime increases with excess photon energy due to a larger initial exciton temperature. Above $\Delta E \approx$ 10meV, no further increase in risetime is observed due to the onset of rapid LO phonon emission. The magnitude of the TRPL risetime is consistent with the calculated LA phonon emission rate assuming quasi-elastic exciton formation.

5. REFERENCES

T.C.Damen, J.Shah, D.Y.Oberli, D.S.Chemla, J.E.Cunningham, and J.M. Kuo, J. Luminescence **45**, 181 (1990).

R.Eccleston, R.Strobel, W.W.Rühle, J.Kuhl, B.F.Feuerbacher, and K.Ploog, Phys. Rev. **B44**, 1395 (1991).

J.Feldmann, G.Peter, E.O.Göbel, P.Dawson, K.Moore, C.T.Foxon, and R.J.Elliot, Phys. Rev. Lett. **59**, 2337 (1987).

B.F.Feuerbacher, J.Kuhl, R.Eccleston and K.Ploog, Solid State Comm. **74**, 1279 (1990).

J.Kuhl, A.Honold, L.Schultheis, and C.W.Tu, Festkörperprobleme **29**, 157 (1989).

K.Ploog, A.Fischer, L.Tapfer, and B.F.Feuerbacher, Appl.Phys. **A52**, 135 (1991).

Inst. Phys. Conf. Ser. No 123
Paper presented at the International Meeting on Optics of Excitons in Confined Systems, Giardini Naxos, Italy, 1991

Resonant decay of electron–hole pairs into optical phonons in a 2D system

F Bechstedt and H Gerecke

Friedrich-Schiller-Universität, Max-Wien-Platz 1, O-6900 Jena, Germany

ABSTRACT: The Raman scattering excitation of electron–hole pairs in a two–dimensional hole plasma that occur in a p–modulation–doped MQW structure is studied. Particular interest concerns the resonant decay of the excited pairs into optical phonons and the resulting lineshapes.

1. INTRODUCTION

The inelastic light scattering has proved to be a powerful tool for the study of electronic and vibronic elementary excitations in semiconductor microstructures. Besides studies of zone–centre optical phonons the most papers in this respect are related to single–particle or collective inter– and intrasubband excitations of quasi–2D electron systems. Holes in such materials with a complicated valence band structure have been investigated only in few papers (Pinczuk et al 1985, 1986, Heiman et al 1987). If the energy of the electronic excitations comes into the region of the optical phonon energy the corresponding Raman spectra overlap. Recently Dahl et al (1990, 1991) and Kraus et al (1991) reported magneto–Raman studies of p–modulation–doped GaAs-GaAlAs multiple quantum wells that indicate the existence of coupled LO–phonon–hole intersubband excitations. These spectra show $h1 - h3$ electron–hole pair excitations in the heavy–hole subbands that strongly interact with GaAs–like confined LOn phonons. The question is if the lineshape of the two coupled excitations has to be interpreted in terms of a resonantly coupled two–level system or a Fano–type profile (Bechstedt and Peuker 1975).

2. THEORETICAL BASIS

We consider a p–modulation–doped multi–quantum–well structure consisting of two zincblende semiconductors, e.g. GaAs and GaAlAs. This system is studied by backscattering of incident light with frequency ω_i and polarization $\mathbf{e}_i$ into light of frequency ω_s and polarization $\mathbf{e}_s$ along the growth axis. For parallel polarizations $\mathbf{e}_i = \mathbf{e}_s$ the electronic Raman scattering is governed by single–particle intersubband (or interband) excitations whereas the simultaneous vibronic light scattering is restricted to processes induced by the polar intraband Fröhlich interaction between electrons and optical phonons. We assume that the photon energies $\hbar\omega_i$ and $\hbar\omega_s$ are in near resonance with the dipole–allowed E_0–transition between the fourfold–degenerated valence band maximum and the twofold–degenerated lowest conduction (c) band minimum at the Γ point of the bulk semiconductor forming the well. More strictly, as described in Fig. 1 the resonance enhancement is studied near the $hn_i \rightarrow cn_i$ transition (incoming resonance) with the occupied initial heavy–hole subband hn_i (n_i = 2 or 3 Heiman et al 1987, Dahl et al 1990, 1991, Kraus et al 1991) and the empty electron subband cn_i in the intermediate states. For the considered hole gas only the highest heavy–hole subband $h1$ is partially unoccupied.

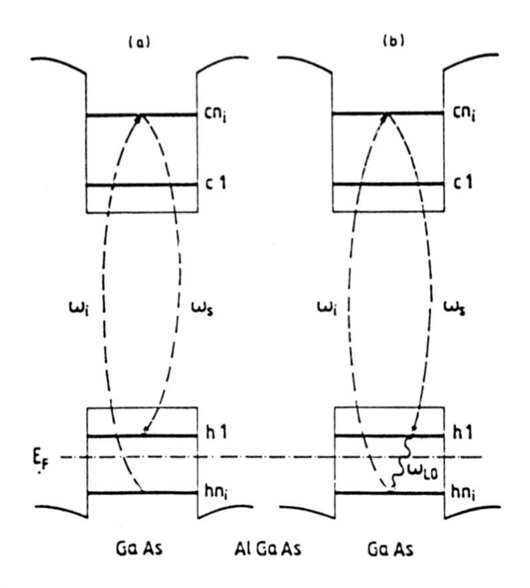

Figure 1: Electron and heavy–hole subbands and the most resonant Raman transitions in a p–modulation–doped GaAs–GaAlAs multiple–quantum–well structure. (a) Electronic $hn_i \rightarrow h1$ Raman transition, (b) Vibronic process including the coupling of $h1$ and hn_i subbands.

Consequently, the most pronounced electronic Raman transitions between occupied and empty valence states are $hn_i \rightarrow h1$ (Fig. 1a). The strongest vibronic Raman process is shown in Fig. 1b. It is nearly doubly resonant. As in the electronic case the incident light $\hbar\omega_i$ is in resonance with the $h1 \rightarrow cn_i$ optical transition in agreement with the assumption that the optical phonon energy $\hbar\omega_{LO}$ comes in the region of the energy of the $h1 - hn_i$ electron–hole pairs.

The total scattering is mainly determined by the imaginary parts of the electronic and vibronic Green's functions even if the strong coupling of the corresponding elementary excitations produces an additional dependence of the scattering amplitudes on the Raman shift $\omega = \omega_i - \omega_s$. The $h1 - hn_i$ electron–hole pair excited in the not completely filled heavy–hole subbands can be described by the bare two–particle Green's function. Apart from an occupation–dependent factor and neglecting the coupling of the hole motion parallel and perpendicular to the interfaces it can be written as

$$G_{1n_i}(\omega) = \frac{1}{\hbar\omega - i\Gamma_e - (\varepsilon_{hn_i} - \varepsilon_{h1})}, \tag{1}$$

where ε_{hn} is the quantization energy for the n–th heavy–hole subband and Γ_e represents a phenomenological inverse lifetime of the electron–hole pair. On the other hand, the Green's function for the dressed confined well–terminated LOn phonons obeys a Dyson equation of the form

$$G_{nn'}(\omega) = G^{(0)}_{nn'}(\omega) + \sum_{n'',n'''} G^{(0)}_{nn''}(\omega)\Sigma_{n''n'''}(\omega)G_{n'''n'}(\omega) \tag{2}$$

with the unperturbed Green's function

$$G^{(0)}_{nn'}(\omega) = \frac{\delta_{nn'}}{\hbar\omega - i\Gamma_p - \hbar\omega_{LO}} \tag{3}$$

and a generalized phonon self–energy operator

$$\Sigma_{nn'}(\omega) = pF_{1n_i}(n)F^*_{1n_i}(n')G_{1n_i}(\omega), \tag{4}$$

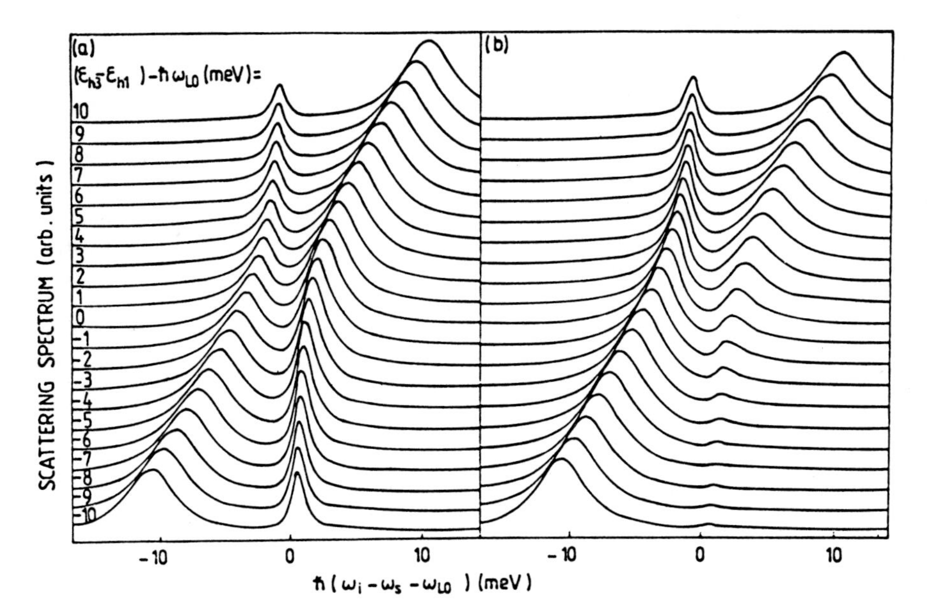

Figure 2: Scattering spectra for coupled electronic and vibronic elementary excitations.

where an inverse lifetime Γ_p of the phonons and the two–dimensional hole density p are introduced. The frequency of the LOn phonons is identified with that ω_{LO} of the corresponding bulk optical phonons. The intraband Fröhlich interaction is represented by the matrix elements $F_{1n_i}(n)$.

The Fröhlich interaction of the well–like LOn phonons with electrons and holes is governed by the macroscopic electrostatic potential $\Phi_n^{LO}(z)$ which can be represented by trigonometric and constant functions (Bechstedt et al 1991). Assuming strong hole confinement it therefore follows strict selection rules $F_{1n_i}(n) \sim \delta_{n,n_i\pm 1}$. Consequently, under the assumption of

$$\hbar\omega \approx (\varepsilon_{hn_i} - \varepsilon_{h1}) \approx \hbar\omega_{LO} \tag{5}$$

and infinite wells the $h1 - hn_i$ electron–hole pairs can resonantly decay into LO($n_i \pm 1$) phonons and vice versa. As a result of this strong coupling the Green's function shows a two–mode behaviour. The poles $\omega_\pm$ of its diagonal terms (considering the limit $\Gamma_e, \Gamma_p = 0$)

$$\hbar\omega_\pm = \frac{1}{2}\left\{\hbar\omega_{LO} + (\varepsilon_{hn_i} - \varepsilon_{h1}) \pm \left[[\hbar\omega_{LO} + (\varepsilon_{hn_i} - \varepsilon_{h1})]^2 + 4p\sum_{+-}|F_{1n_i}(n_i \pm 1)|^2\right]^{\frac{1}{2}}\right\} \tag{6}$$

exhibit electron–hole–pair–like or phonon–like character.

3. RESULTS

Typical scattering spectra are plotted in Fig. 2 for various detunings between the bare energies of electron–hole pairs $(\varepsilon_{hn_i} - \varepsilon_{h1})$ and phonons $\hbar\omega_{LO}$ and for two resonance conditions $E_{E_0} + \varepsilon_{cn_i} + \varepsilon_{hn_i} - \hbar\omega_i = 0$ (a) and $-10meV$ (b). We have applied accepted GaAs parameters (Bechstedt et al 1991), a well thickness of 10nm, $\Gamma_e = 2.5meV$, $\Gamma_p = 0.5meV$, and

$p = 2.4 \times 10^{-11} cm^{-2}$. In agreement with experiment (Dahl et al 1990, 1991, Kraus et al 1991) the initial heavy–hole subband is chosen to be $n_i = 3$. The two series of spectra show the typical behaviour of a resonantly coupled two–level system although one of these, the electronic one, is remarkably broadened. In the representation of Fig. 2 versus $\omega_i - \omega_s - \omega_{LO}$ the phonon–like peak practically exhibits no shift with the detuning between $(\varepsilon_{hn_i} - \varepsilon_{h1})$ and $\hbar\omega_{LO}$. Only the sign of the detuning determines the position of this line below or above the energy zero point. On the other hand, the electron–hole–pair related peak completely follows the detuning. Since the electronic Raman scattering only shows the incoming resonance it is not very sensitive to a small change of the incident photon energy $\hbar\omega_i$. This fact is in contrast to the behaviour of the doubly resonant phonon scattering. For $\hbar\omega_{LO} > (\varepsilon_{h3} - \varepsilon_{h1})$ and $\hbar\omega_i > (E_{E_0} + \varepsilon_{c3} + \varepsilon_{h3})$ the phonon related peak disappears more or less. In the presence of an external magnetic field the principal features of the l ineshape are conserved. The quantization of the electron and hole motions mainly influences the resonance behaviour. The corresponding logarithmic singularities change over into Lorentzian–like ones. Together with a reduction of the lifetime broadening of the virtual electron–hole pairs it gives rise to a strong resonance enhancement.

A Fano–like antiresonance behaviour can hardly be observed in the theoretical scattering spectra of Fig. 2. On the one hand this fact can be understood since the pure electronic scattering produces no continuous spectrum. On the other hand, because of the interface disorder (Kraus et al 1991) the electronic Raman peak is much broader than the vibronic one, $\Gamma_e \gg \Gamma_p$, and the scattering cross section can be formally related to a Fano–like profile. However, the complex Fano–lineshape parameter depends strongly on the scattering shift. For larger shifts its module is always greater than unity what destroys the antiresonance behaviour.

REFERENCES

Bechstedt F and Peuker K 1975 phys. stat. sol. (b) 72 743

Bechstedt F, Gerecke H and Kraus J 1991 Phys. Rev. B (submitted)

Dahl M, Kraus J, Müller B, Schaack G and Weimann G 1990 Proc. 3rd Int. Conf. Phonon Physics and 6th Int. Conf. Phonon Scattering in Cond. Matter, eds Hunklinger S, Ludwig W and Weiss G (Singapore: World Scientific) p 758

Dahl M, Ils P, Kraus J, Müller B, Schaack G, Schüller C and Weimann G 1991 Superlattices and Microstructures 9 77

Heiman D, Pinczuk A, Gossard A C, Fasolino A and Altarelli M 1987 Proc. 18th Int. Conf. Physics of Semiconductors, eds Engström O (Singapore: World Scientific) p 617

Kraus J, Ils P, Schüller C, Ebeling J K, Schlapp W and Weimann G 1991 J. Phys.: Condens. Matter (submitted)

Pinczuk A, Störmer H L, Gossard A C and Wiegmann W 1985 Proc. 17th Int. Conf. Physics of Semiconductors, eds Chadi J D and Harrison W A (New York: Springer) p 329

Pinczuk A, Heiman D, Sooryakumar R, Gossard A C and Wiegmann W 1986 Surface Sci. 170 573

Inst. Phys. Conf. Ser. No 123
Paper presented at the International Meeting on Optics of Excitons in Confined Systems, Giardini Naxos, Italy, 1991

Resonance width of the light-hole exciton in GaAs-$Ga_{1-x}Al_xAs$ quantum wells

A Pasquarello[a] and L C Andreani[b]

[a]Institut de Physique Théorique, EPFL, CH-1015 Lausanne, Switzerland
[b]IRRMA, CH-1015 Lausanne, Switzerland

ABSTRACT: The light-hole exciton in GaAs-$Ga_{1-x}Al_xAs$ quantum wells is degenerate with the continuum of the heavy-hole exciton for well widths $L < 130$ Å. The two exciton series are coupled, due to valence-band mixing. Thus the light-hole exciton acquires a resonance broadening. In this work this broadening is calculated. We present two methods which are both based on the diagonalisation of the Hamiltonian on a finite variational basis, chosen in such a way that the discrete as well as the continuum parts of the spectrum are well represented.

In GaAs-$Ga_{1-x}Al_xAs$ quantum wells (QW's), for well widths L smaller than about 130 Å, the difference in quantization energies of the light-hole (LH) and heavy-hole (HH) subbands makes the ground-state LH exciton to be degenerate with the continuum of the HH exciton. Due to valence band mixing, the ground-state LH exciton is coupled to the HH continuum. This coupling gives rise to a lifetime broadening which is known as Fano-resonance effect (Fano 1961). A further effect of the coupling is an energetic shift of the resonance position.

The calculation of the resonance width requires either a numerical integration of the Schrödinger equation (Broido and Yang 1990) or the application of the theory developed by Fano (1961). Both approaches are often not easy to apply, because of the difficulty of obtaining the eigenstates in the continuum. In this work we present two new methods to calculate the resonance width which are simple to apply. They are based on a diagonalisation of the Hamiltonian on a finite basis set. The key idea is that the basis functions are chosen in such a way to describe the discrete as well as the continuum parts of the spectrum. An advantage of our methods is that they can easily be applied as an extension of usual variational calculations.

We use a theory of quantum-well excitons which includes valence band mixing and a number of other effects (Andreani and Pasquarello 1988, 1990). In this theory, the valence-band mixing is described by the 4×4 Luttinger-Kohn (1955) matrix in the axial approximation. The exciton envelope function is labelled by a spin index s which runs from $-\frac{3}{2}$ to $\frac{3}{2}$ and is represented as

$$F^s(\rho, \theta, z_e, z_h) = \sum_j \int \frac{d\mathbf{k}_\parallel}{(2\pi)^{3/2}} g_{ij}(k_\parallel) e^{im\alpha} e^{i\mathbf{k}_\parallel \cdot \boldsymbol{\rho}} c_i(z_e) v^s_{j\mathbf{k}_\parallel}(z_h), \quad (1)$$

where $\mathbf{k}_\parallel = (k_\parallel, \alpha)$ is the Bloch vector of the subbands, $\boldsymbol{\rho} = (\rho, \theta)$ is the relative coordinate in the xy plane, $c_i(z_e), v_{j\mathbf{k}_\parallel}(z_h)$ are envelope functions of conduction and valence subbands, the index i (j) denotes the principal quantum number of the conduction (valence) subbands, $g_{ij}(k_\parallel)$ is the in-plane wavefunction in k-space, and m is a conserved quantum number, which can be interpreted as the projection of the total angular momentum along the growth direction z. Note that the basis set already contains valence band mixing. Trial states belonging to different subbands, but with the same value of m, are coupled by the Coulomb potential. It follows that s-states of the LH exciton are coupled to d-states of the HH-exciton. In this work we consider only exciton states associated to the first conduction subband. Coulomb coupling between excitons belonging to different conduction subbands is neglected.

The exciton levels are found by the variational method expanding the radial function $g_{ij}(k_{\|})$ into two types of wavefunctions $h_\nu(k_{\|})$: the first are decreasing exponentials in ρ-space, which in k-space have the form

$$h_\nu(k_{\|}) = \frac{2\alpha_\nu^2}{(k_{\|}^2 + \alpha_\nu^2)^{3/2}}, \tag{2}$$

and give a good description of discrete exciton states. The second kind of basis functions are gaussians in $\mathbf{k}_{\|}$-space:

$$h_\nu(k_{\|}) = e^{-(k_{\|}-k_\nu)^2/(2\sigma^2)}. \tag{3}$$

For a small width σ, states (3) describe delocalized states in ρ-space, and are well suited in order to represent states of the exciton continuum. By carefully choosing the fixed parameters k_ν, we can increase the density of variational eigenstates of the HH continuum in the energy region of the LH resonance. The resulting eigenvalues are equally spaced in energy, provided the k_ν^2 are taken to be equally spaced.

The first method to calculate the resonance broadening is an application of the theory developed by Fano (1961). The HH and LH exciton series are diagonalised separately neglecting Coulomb coupling, and the full Hamiltonian is written in the basis of the states so obtained. In this basis the off-diagonal elements describe the coupling between the two excitonic series. The resonance width of the LH exciton is given by

$$\Gamma(\overline{E}) = 2\pi|V(\overline{E})|^2\rho(\overline{E}) \tag{4}$$

where $\overline{E}$ is the energy of the resonance, $\rho(E)$ the density of states, and $V(E)$ is the coupling of the LH exciton to the HH excitonic series. Unlike Fano's original theory the states which represent the continuum are discrete in our model. As long as the density of states in the resonance region is high enough the density of states ρ and the coupling V at the energy $\overline{E}$ can easily be interpolated.

In the second method, which we will refer to as the "absorption" method, the Hamiltonian is diagonalised on the full variational basis. Then, the oscillator strengths of the single variational eigenstates are calculated. As an example, we plot in Fig. 1 the value of the oscillator strength versus energy for the variational eigenstates which fall near to the LH resonance in the case of a 80 Å wide GaAs-$Ga_{0.7}Al_{0.3}As$ QW. Finally, the resonance width is determined by a fitting procedure.

In order to test the accuracy of the previous methods we have applied them to a one-dimensional model, whose exact solution is known. The model Hamiltonian is given by

$$H = -\frac{\hbar^2}{2m}\frac{d^2}{dx^2} + V\delta(x - L), \tag{5}$$

where $x > 0$ and V gives the strength of the barrier potential. We consider as discrete state the ground state for $0 < x < L$ obtained in the case of an unpenetrable barrier ($V \to \infty$). When V is finite, this state is no longer an eigenstate and is coupled to a continuum of states,

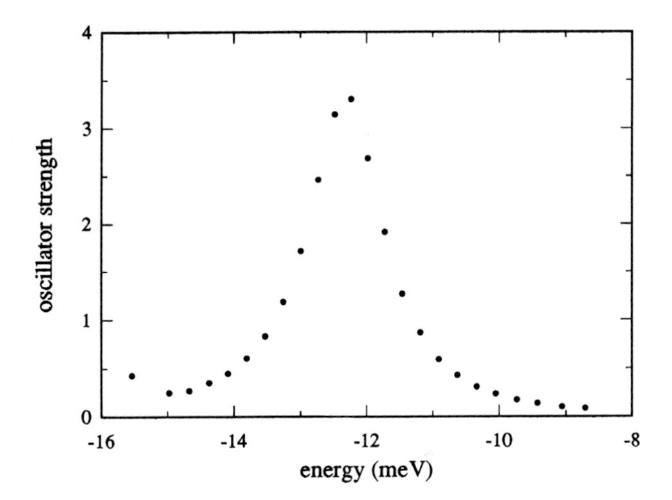

Fig. 1 Oscillator strength vs energy for variational eigenstates in the region of the LH resonance for a 80 Å wide GaAs-$Ga_{0.7}Al_{0.3}As$ QW. The energy is given with respect to the subband transition edge.

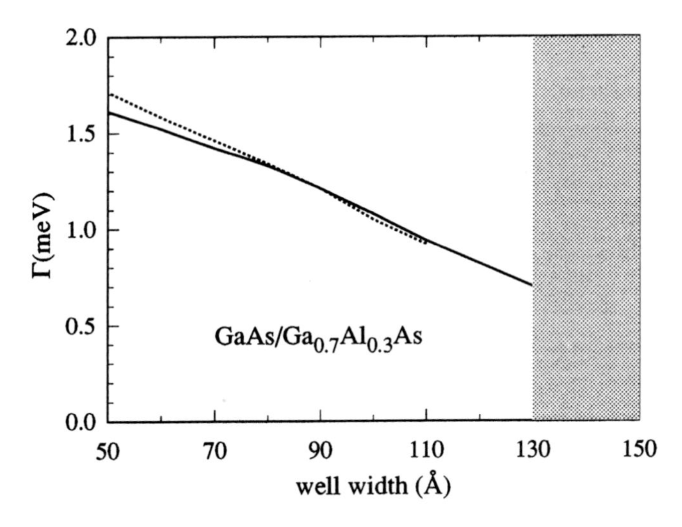

Fig. 2 Resonance width (FWHM) of the ground state LH exciton in GaAs-$Ga_{0.7}Al_{0.3}As$ quantum wells. Solid line, result of Fano theory. Dotted line, result from "absorption" calculation.

in analogy to the LH exciton problem. The conclusions of this study are that both methods give accurate results for the linewidth. The "absorption" method converges mores slowly as a function of the density of variational eigenvalues.

In Fig. 2 we show the calculated Fano width (full width at half-maximum) for the LH exciton in GaAs-$Ga_{0.7}Al_{0.3}As$ quantum wells. Our final result is the continuous curve, which results from the Fano calculation. The dotted curve represents the result of the absorption calculation. The difference between solid and dotted curve arises from lack of convergence in the absorption calculation. The Fano width starts from a finite value at the continuum edge and increases as the well width is reduced, reflecting the increase in $V(E)$ for increasing energies.

On the experimental side, few data are available so far. The Fano width must be disentangled from other contributions to the total linewidth. The most important of them are scattering from acoustic phonons and defects, which produce a homogenous broadening, and well-width fluctuations, which produce an inhomogeneous broadening. As the Fano width is found to be of the order of an meV and so comparable to the other sources of broadening, it should be possible to extract it from absorption spectra. Lineshape fits could resolve homogeneous from inhomogenous broadening. Assuming that that the other contributions to the broadening are the same for the HH and LH excitons, the Fano width can then be obtained as the difference of the homogeneous broadenings of the two excitons.

Andreani L C and Pasquarello A 1988, Europhys. Lett. 6 259
Andreani L C and Pasquarello A 1990, Phys. Rev. B 42 8928
Broido D A and Yang S R E 1990, Phys. Rev. B 42 11051
Fano U 1961, Phys. Rev. 124 1866
Luttinger J M and Kohn W 1955, Phys. Rev. 97 869

Inst. Phys. Conf. Ser. No 123
Paper presented at the International Meeting on Optics of Excitons in Confined Systems, Giardini Naxos, Italy, 1991

Temperature dependent studies of the radiative recombination in GaAs/AlGaAs multiple quantum wells

J P Bergman[1], P O Holtz[1], B Monemar[1], M Sundaram[2], J L Merz[2] and A C Gossard[2].

1) Department of Physics and Measurement Technology, Linköping University, S-581 83 Linköping, SWEDEN.
2) Center for Quantized Electronic Structures (QUEST), University of California, Santa Barbara, California 93106, USA.

ABSTRACT: Photoluminescence and photoluminescence decay measurements have been performed of the recombination from different GaAs/AlGaAs quantum well samples, in the temperature region from 2 K to 500 K. We have found that the measured decay is mainly due to radiative recombinations for temperatures below 200 K, while at higher temperatures it is dominated by non-radiative recombinations.

1. INTRODUCTION

The photoluminescence (PL) decay from the free exciton (FE) recombination in a quantum well (QW) has been the topic of several studies. These studies have mainly been concentrated on either low temperature or room temperature recombination. At low temperatures the decay is influenced by the localization of the FE at interface defects, as reported by Hegarty et al (1982). The temperature dependence of the FE recombination in the low temperature region (<40 K) has been investigated and explained by Feldman et al (1987). At room temperature the measured decay has been assigned to mainly depend on non-radiative recombinations, as reported by Sermage et al (1989) .

In this paper we present experimental results for the measured decay time in the temperature region between 2 K and 500 K for different GaAs/AlGaAs multi QW samples. Our results are compared with a model of the radiative recombination kinetics involving free carriers and excitons in QW systems, as presented by Ridley (1990). A similar study of the recombination in samples containing different single QW´s has recently been presented by Colocci (1991) . Their results differ however in some cases from our results.

2. SAMPLES AND EXPERIMENTAL PROCEDURES

The samples used in this study were grown by Molecular Beam Epitaxy (MBE) on a semi-insulating substrate. The epitaxial layer was grown with non-interrupted growth and consisted of a buffer layer with 4000 Å GaAs and 5 periods of 20 Å GaAs and 150 Å AlGaAs. This was followed by 50 QW's with width L_z and separeted by 150 Å AlGaAs, and finally a 100 Å GaAs cap layer. All layers were nominally undoped and the x-value was 0.3 in all $Al_xGa_{1-x}As$ layers.

The photoluminescence measurements were performed with a laser excitation photon energy of ~ 1.7 eV (7200 Å), which is below the AlGaAs bandgap at all temperatures of interest. The PL signal was detected by a GaAs photomultiplier tube or a cooled Ge detector. The PL decay

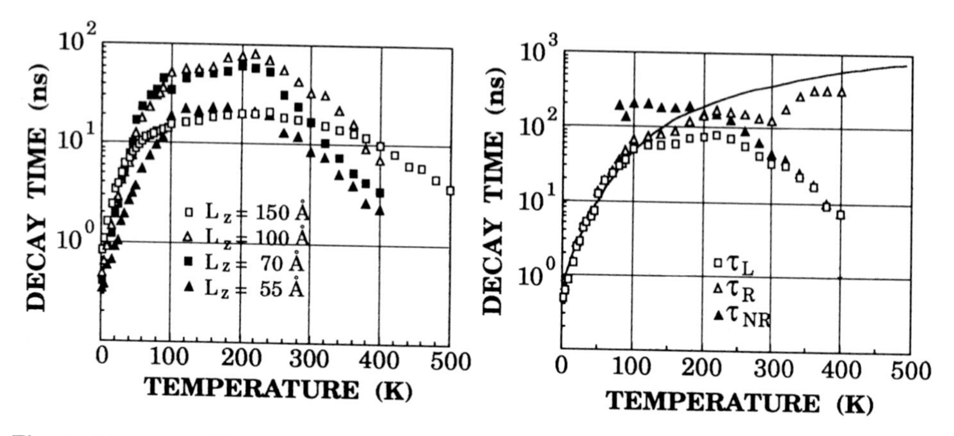

Fig. 1. Measured PL decay times as a function of temperature for four different MQW samples.

Fig. 2. The temperature dependence of the measured decay time (τ_L) and the estimated values for the radiative recombination (τ_R) and the nonradiative recombination (τ_{NR}) for a 100 Å wide QW. The solid line corresponds to the radiative decay as calculated from Ridley (1990).

measurements were made with a conventional time correlated photon counting system. For the excitation we used 5 ps pulses from a synchronously pumped dye laser, equipped with a cavity dumper. The excitation photon energy was also in this case around 1.7 eV.

3. RESULTS AND DISCUSSION

Fig. 1 shows the measured PL decay times as a function of temperature for multi QW samples with different QW width, L_Z. These values are obtained with low excitation intensity giving close to exponential decays. At higher excitation intensities we observe an increasing non exponentiality of the decay, caused by an additional fast component. This is most clearly seen in the temperature region between 100 and 250 K, where the decay time has its maximum value. At higher temperatures (>300 K), the measured decay is exponential irrespective of the excitation intensity. The nonlinearity is consistent with a model of the radiative decay time presented by Ridley (1990). It predicts an additional fast component when the photoexcited electron concentration n(t=0) exceeds the equilibrium density of holes p_o, which would be the case during high laser excitation. This effect will disappear at higher temperatures where the measured decay is governed by non-radiative recombination, which seems to be the case in all our samples as can be seen by the decrease of the measured decay time at high temperatures.

The temperature dependence of the measured decay time $\tau_L(T)$ is however similar to that observed by Colloci et al (1991) and Gurioli et al (1991), even if our the values have a maximum at a somewhat higher temperature. The highest measured values for the decay time, occuring at about 200 K, are longer than the previously reported high temperature values by Colloci et al (1991) and Matsueda et al (1989).

From the combined measurement of the measured decay time $\tau_L(T)$ and the integrated intensity $I_L(T)$ we can, in a similar way as Miller et al (1980) and Gurioli et al (1991), extract the non-radiative $\tau_{NR}(T)$ and the radiative $\tau_R(T)$ decay times as

$$\tau_R(T) = \tau_L(T)\,\frac{I_L(0)}{I_L(T)}$$

$$\tau_{NR}(T) = \tau_L(T)\,\frac{I_L(0)}{I_L(0)-I_L(T)}$$

The values of $\tau_R(T)$ and $\tau_{NR}(T)$ as well as the measured values of $\tau_L(T)$ are shown in Fig. 2, for a QW with L_z=100 Å.

Also shown in Fig. 2 is the calculated radiative decay time according to Ridley (1990). This is based on the assumption of thermal equilibrium between FE´s and free carriers in the well. The curve in Fig. 2 is calculated with the assumption of an equilibrium density of holes, p_o, of 1×10^{10} cm^{-2}. As can be seen in the figure we obtain good agreement between this model and our experimental results. The assignment of our measured decay time at temperatures up to about 100 K as the true radiative decay is further supported by a measurement of the integrated PL intensity as a function of laser excitation intensity. At 77 K we observe a linear increase when the excitation is changed over 4 decades, from 10^{-3} W/cm^2 to 10 W/cm^2. It should also be pointed out that Ridley´s model predicts a strong dependence of the radiative decay time with p_o, which could be one possible explanation for the differences between different samples in Fig.1 as well as in the measurements of Colocci et al (1991).

We have finally performed a lineshape fitting of the luminescence spectrum at different temperatures from 10 K up to and above room temperature. The measured PL spectra for the sample with Lz= 100 Å are shown in Fig. 3 as a function of photon energy and temperature. We have used the model presented by Christen and Bimberg (1990) and considered both the

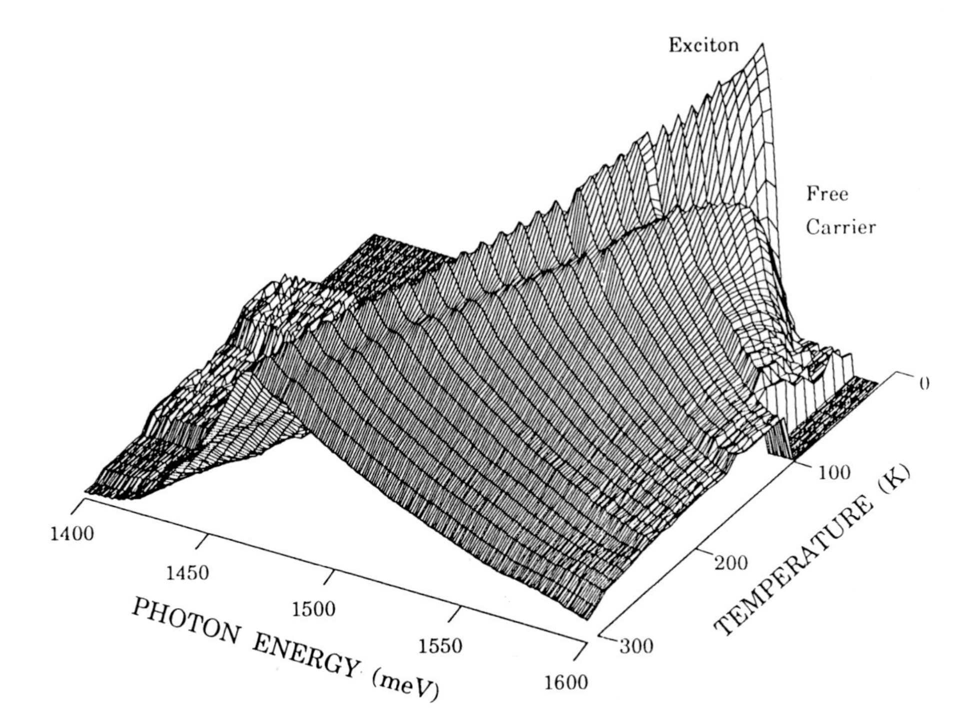

Fig.3. A compilation of measured PL spectra at different temperatures for the sample with L_z=100 Å. The intensity is shown with a logarithmic scale. The recombination due to free carriers can be seen as a high energy shoulder to the dominating exciton recombination.

ground state and the first excited state of the heavy- and ligth hole exciton recombination, together with the free carrier recombination. Fig. 4. shows the ratio between the integrated intensity of the free carrier recombination and the total exciton recombination, obtained from the fitting. From these measurements and the fittings we can clearly follow the relation between exciton and free carrier recombination and conclude that the FE recombination is present and dominating at room temperature, in agreement with the conclusions of Christen and Bimberg (1990). From Fig. 4 we observe that, at 300 K, the total intensity of the free carrier recombination is 77% of the exciton recombination. This corresponds to an amplitude ratio in the PL spectrum of 40%, between the free carrier and exciton recombination.

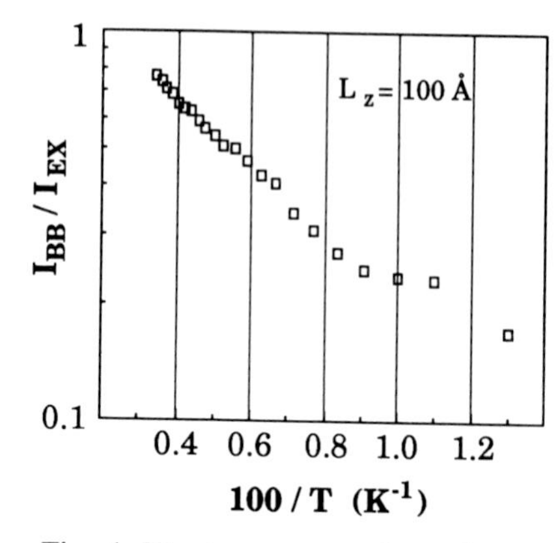

Fig. 4. The temperature dependence of the ratio between total exciton and free carrier (BB) recombination, obtained from the fitting of the spectra in Fig. 3.

4. CONCLUSIONS

The PL decay of the recombination in a GaAs/AlGaAs QW has been studied for samples with different width, L_Z. We conclude that the measured decay is dominated by the radiative decay for temperatures up to about 200 K. At higher temperatures the measured decay is dominated by non-radiative recombination processes. The maximum value of the measured decay time of 80 ns and 62 ns for the sample with width 100 Å and 70 Å, respectively, is to our knowledge higher than any previously obtained values in comparable studies. Our measurements are in agreement with the model for the radiative recombination in QW systems, assuming thermal equilibrium between excitons and free carriers, as presented by Ridley (1990).

In this study we have neglected any possible influence on the decay time due to reabsorbtion or recycling effects. In bulk GaAs these effects have been calculated by Asbeck (1977) to be of large importance for the measured decay time. To our knowledge no similar calculations have been performed for QW´s, and especially the temperature dependence of such effects would be of great interest.

5. REFERENCES

Asbeck P 1977 J. Appl. Phys. 48 820.
Colocci M, Gurioli M, Vinattieri A, Fermi F, Deparis C, Massies J and Neu G 1990 Europhys. Lett. 12 417.
Colocci M, Gurioli M and Vinattieri A 1991 Presented at 11th General Conference of the Condensed Matter Division (EPS) Exeter, England.
Christen J and D Bimberg 1990 Phys. Rev. B42 7213.
Feldmann J, Peter G, Göbel E O, Dawson P, Moore K, Foxon C and Elliott R J 1987 Phys. Rev. Lett. 59 2337.
Gurioli M, Vinattieri A, Colocci M, Deparis C, Massies J, Neu G, Bosacchi A and Franchi S 1991 Phys. Rev. B44 3115.
Hegarty J, Sturge M D, Weisbuch C, Gossard A C and Wiegmann W 1982 Phys. Rev. Lett. 930.
Matsueda H and Hara K 1989 Appl. Phys. Lett. 55 362.
Miller R C, Kleinman D A, Nordland J:r W A and Gossard A C 1980 Phys. Rev. B22 863.
Ridley B K 1990 Phys. Rev. B41 12190.
Sermage B, Alexandre F, Beerens J and Tronc P 1989 Superlattices and Microstructures 6 373.

Inst. Phys. Conf. Ser. No 123
Paper presented at the International Meeting on Optics of Excitons in Confined Systems, Giardini Naxos, Italy, 1991

VUV studies of quantum well excitons in RbCl-KBr-RbCl structures

A Ejiri and A Hatano

Department of Pure and Applied Sciences, University of Tokyo 3-8-1, Komaba, Meguro-ku Tokyo 153

ABSTRACT: RbCl-KBr-RbCl single well structures which are likely interface-mixture-free are studied via VUV absorption. An evidence that blue shifts on the exciton band due to the well layer are purely caused to the quantum confinement is observed. The blue shifts can be explained in terms of kinetic energies of the confined exciton in the square well. Effects of the layer-layer interface-mixture are also discussed.

1. INTRODUCTION

The one dimensional quantum well exciton (QWE) in alkali halides (AH) have been first investigated by the present authors (Ejiri et al 1989 1990a 1990b 1990c) via vacuum ultraviolet (VUV) absorption measurements of alternating multilayer structures (MLS); KCl(barrier)-KBr(well) and RbCl(barrier)-KBr(well) with the well thicknesses of 3-10 nm. The exciton absorption band due to the well layer (KBr) exhibits blue shift and steeper rise in comparison with the KBr bulk exciton band. The amounts of the observed blue shift were often much larger than theoretical values expected from the kinetic energy of QWE confined in square well potential with the infinite depth, which is given by

$$E = \frac{\hbar^2}{2m_z}\left(\frac{\pi}{L}\right)^2 (n+1)^2 + E_{x,y}, \tag{1}$$

where m_z is an effective exciton mass in the longitudinal ($z-$) direction, L a thickness of the well, n a positive integer including zero, and $E_{x,y}$ the energy due to its transverse motion (to be called the quasi-particle (QP) model).

On the other hand, Bastard et al (BMCE)(1982) used a realistic model-wave-function of exciton taking into account of the electron-hole distance in their theoretical work of GaAlAs-GaAs-GaAlAs structure, and indicated the confinement of the wave-function-extent significantly (about 30%) existing even in the wide well of $L \approx 10a_B$, where a_B is the Bohr radius of bulk exciton. In this model, exciton-binding-energy-shift due to the confinement is expected to be one order larger than the kinetic energy shift of the QP model. Since the a_B of the KBr bulk exciton is 4.2Å(Tomiki et al 1973), the BMCE theory seems to be applicable to the present results, and it is very interesting whether AH-QWE is attributed to the QP model or to the BMCE model.

The large blue shifts observed in MLS have been, however, successfully explained in terms of the interface-mixture (IM) formed between the evaporated layers in the MLS by the present authors (Ejiri et al 1990a 1990b 1990c) rather than the BMCE model, because IM plays a role to make the blue shift very large. Whereas the steep rise could be qualitatively explained (Ejiri et al 1990a 1990b) in terms of BMCE's exciton-wave-function confinement effect because narrowing of the wave-function-extent in the well makes the exciton band sharper.

Furthermore, it has been proposed by Ejiri et al (1990b) that possible occurrences of IM-free and also interface-mismatch-free RbCl-KBr-RbCl single well structure (SWS) could be realized through some improvements on sample preparation techniques of thermal annealing and of deposition rate control. This is probably because annealed layers become hard to mix with next-coming another material, and also because there is a very small difference of the lattice constants less than 1% between RbCl and KBr. Forbidden band gap (E_g) and lattice constants

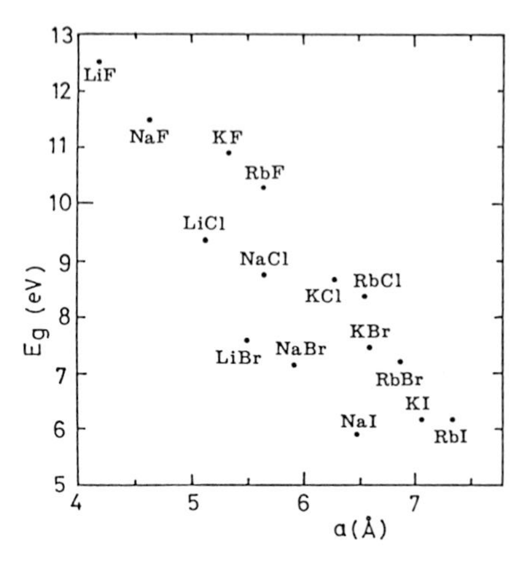

Fig.1. The band gap energy E_g of several alkali halides versus the lattice constant a.

(a) of several alkali halides are illustrated in Fig.1.

In the present work, it is intended to reveal the true blue shift due to the quantum confinement using such the IM-free SWS. Experimental procedure will be described briefly in Section 2, and successfully observed exciton bands of the IM-free SWS will be demonstrated and their blue shifts will be discussed in Section 3.

2. EXPERIMENTAL

RbCl-KBr-RbCl SWSs were fabricated on cleaved surface of NaCl single crystals by means of evaporation in vacuum of 1×10^{-6} Torr. NaCl has a large band gap enough to find the KBr exciton doublet and its lattice constant is rather close to RbCl's one. Deposition rate was always kept in less than 0.1 nm/s and its average was kept about 1-2 nm/min. Thermal annealing was performed at 180-200°C for about 15 minutes or more in the vaccum just after deposition of each layer consisting SWS. However,the third layer (RbCl) of SWS was often not annealed for avoiding IM formation due to annealing. The substrate crystals were always thermally annealed in similar condition before depositions to make its surface clean. KBr single layers were also fabricated on KCl single crystal substrates in similar manners in order to take reference spectra for comparing with SWS spectra.

It was already pointed out by Ejiri et al (1989) that structural features of KBr thin layers deposited on collodion film substrates were polycrystalline and bulky in the two-dimensional direction even in KBr 5nm layer through the transmission electron microscopy techniques. Moreover, Saiki et al (1991) observed that a KCl film covered a KBr single-crystal-substrate uniformly even at its one unite layer and occured two dimensional layer growth in the similar deposition conditions with ours. Therefore, epitaxial growth of each layer can be expected in the present SWS. VUV absorption measurements of these samples were performed with a 1m Seya-Namioka monochromator under a resolution of about 1000 at 80 K as well as at room temperature in the synchrotron radiation facility, ISSP-SRL (TOKYO). Wavelength reproducibility of the monochromator is kept in the range of less than 0.1 nm.

3. RESULTS AND DISCUSSION

The absorption spectra of IM free RbCl-KBr-RbCl SWSs on NaCl substrates where the well thickness are 3-10 nm are successfully observed (Ejiri et al 1990c). Typical results are shown in Fig.2. Since the IM free RbCl-KBr-RbCl SWS with KBr10nm has been already revealed to exhibit no blue shift comparing with the bulky KBr20nm single layer structure (Ejiri et al 1990b), the 1st exciton bands of these SWSs are compared with that of the SWS with KBr10nm. The KBr5nm and KBr6nm bands indicate no blue shift from the KBr10nm band as a usual manner, but seem to slightly exhibit a red shift although in the range of experimental error. The bands of SWSs with KBr3nm are, however, exhibit blue shift as large as 0.01-0.015 eV.

When layer-layer IMs are formed in the multilayer structure, absorption band due to the IMs appears at an energy region upper than the well exciton band (Ejiri et al 1989-1990c), and the well exciton band decreases its intensity and often exhibits a large blue shift. Two origins

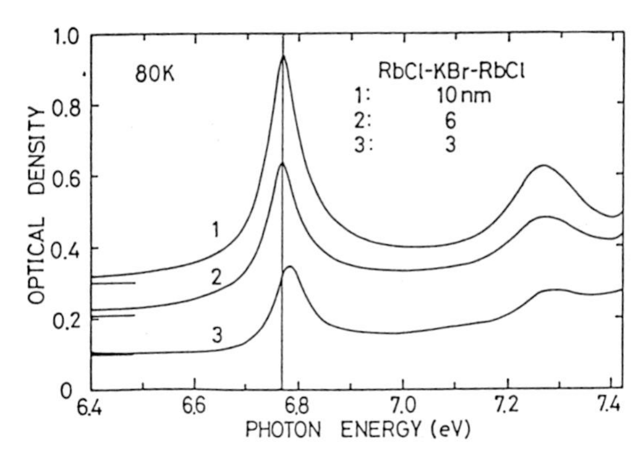

Fig.2. Typical absorption spectra around the KBr exciton region in the RbCl-KBr-RbCl single well structures with KBr 10nm, 6nm, and 3nm.

for the large blue shift can be imagined. One is that the well thickness becomes very thin due to the IM formation, and the other is that IM works to enlarge the effective band gap appreciably (Ejiri et al 1990a). Both effects likely occur in the real structures containing IM. In the SWS fabrication process, it has been revealed that IM is easily formed between the noannealed KBr layer and the next coming RbCl layer but hard to be formed between the annealed layer and the next coming layer.

On the absorption spectra obtained in the present study, such IM effects are not indicated, and the absorption band due to IM is disappeared. The 1st exciton band in the RbCl-KBr-RbCl SWS usually becomes more sharper than the single-KBr-layer-band. This has been ascribed to weak interface-mismatch in the structure (Ejiri et al 1990b). The band width, however, depends also on a complex of several conditions in the fabrication. The most narrow width of the first band of 66 meV (FWHM) has been achieved yet in a SWS with KBr5nm, whereas that of the annealed KBr single layer on KCl crystal is usually about 90 meV. Annealing temperatures for the first, second, and third layer of this RbCl20nm-KBr5nm-RbCl10nm SWS were 200°C, 190°C, and 180°C respectively, and deposition speed for each layer was kept in 1 nm/min.

Observed characters of the first exciton band in these SWSs are summarized in Table I, and blue shifts observed are illustrated in Fig.3. The solid line is a theoretical curve of the QP model where the effective electron mass (Hodoy 1971) and effective hole mass (Overhof 1971) for the free exciton are used. The observed points are removed lower by 0.84 nm on the abscissa as surface dead layers $2a_B$ of exciton in the well. Binding energy shift derived from BMCE model (Bastard et al 1982) is also traced in Fig.3 (broken line) for comparison. Consequently, it can be stated that the blue shifts of KBr QWE in the IM-free SWS is closely fitted with the QP model, and is attributed to pure quantum confinement.

Table I: Observed characters of the 1st exciton band in RbCl-KBr-RbCl single well structures.

Sample No.	Layer thickness (nm) RbCl-KBr-RbCl	Blue shift(meV)	FWHM(meV)
1	20 - 3 - 10	10	–
2	20 - 3 - 10	15	74
3	20 - 5 - 10	-3	–
4	20 - 5 - 10	-4	66
5	20 - 6 - 10	-3	74
6	20 - 10 - 20	0	70
7	20 - 10 - 20	0	78
8	20 - 10 - 20	0	80
9	20 on KCl	0	91
10	20 on KCl	0	91

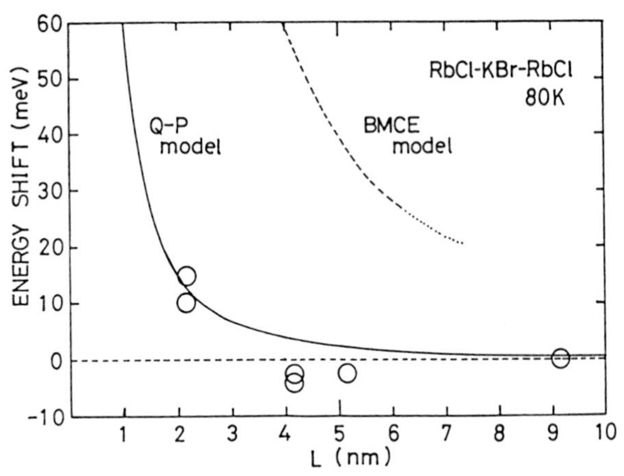

Fig.3. Observed blue shifts of the single well structures RbCl-KBr-RbCl (open circle). Solid line indicates a theoretical values of the quasi-particle (QP) model and broken line is of Bastard et al's (BMCE) model.

The 3nm thickness in KBr corresponds to about its 5 molecular layers or less. After subtraction of the dead layers, remains are about 3 molecular layers and become about 2 nm. The kinetic energy for the 2nm well becomes 0.015 eV which is very close to the observed values as can be seen in Fig.3. The intensity of the band is significantly larger than the expected one as for the 3nm well thickness. However, in such a super thin layer, a mesoscopic enhancement is also expected in the optical transition of exciton (Hanamura 1989), then, the absorption intensity isn't likely linear but rather nonlinear for the well thickness in this case.

4. CONCLUSIONS

In the successfully fabricated IM-free RbCl-KBr-RbCl SWS on NaCl substrate, it is well demonstrated that the blue shift of the QWE band can be explained in terms of the quasi-particle model. The AH-QWE presently observed is well understood as a physical picture of the small exciton confined in the rather wide well. RbCl-KBr-RbCl likely gives a good quantum well structure, however, further studies will be required for the small red shift observed at the KBr5-6nm SWS.

5. ACKNOWLEDGEMENTS: The authors are indebted to M. Fujisawa and other staffs of SRL-ISSP (The University of Tokyo) for their generous help to use synchrotron radiation facility. They are also grateful to Prof. K. Nakagawa of Kobe University for his help to the measurements and Prof. E. Hanamura of The University of Tokyo for suggestive discussions. This work was supported by a Grant-in-Aid for scientific research B (No.63460028) and C (No.02640252) from the Ministry of Education.

REFERENCES

Bastard G, Mendez E E, Chang L L, and Esaki L 1982 Phys. Rev. **B26** 282.
Ejiri A and Nakagawa K 1989 J. Phys. Soc. Jpn. **58** 2669.
Ejiri A, Nakagawa K, and Hatano A 1990a Physica Scripta **41** 95.
Ejiri A and Nakagawa K 1990b Solid State Commun. **73** 849.
Ejiri A, Hatano A and Nakagawa K 1990c *Proc. 4th Asia-Pacific Phys. Conf.* eds S H Ahn, S H Choh, Il-T Cheon and C Lee (World Scientific) **1** pp405-408.
Hanamura E 1988 Phys. Rev. **B38** 1228.
Hodoy J W J 1971 J. Phys. **C4** L8.
Overhof H 1971 Phys. Stat. Sol. **43** 575.
Saiki K, Nakamura Y and Koma A 1991 Surf. Sci. **250** 27.
Tomiki T, Miyata T, and Tsukamoto H 1973 J. Phys. Soc. Jpn. **35** 495.

Inst. Phys. Conf. Ser. No 123
Paper presented at the International Meeting on Optics of Excitons in Confined Systems, Giardini Naxos, Italy, 1991

Indirect transitions in GaAs/(Al,Ga)As single quantum wells bordered by an ultra-thin AlAs layer

M. Leroux, B. Chastaingt, N. Grandjean, G. Neu, C. Deparis and J. Massies

Laboratoire de Physique du Solide et Energie Solaire, CNRS, 06560 Valbonne (France)

ABSTRACT: Under hydrostatic pressure, GaAs/$Al_{1-x}Ga_xAs$ quantum wells bordered by one or two AlAs monolayers exhibit intense type II excitonic luminescence which allows an analysis of the X_z electronic band structure in ultra thin AlAs layers. The results are well interpreted in the framework of the envelope function formalism.

1. INTRODUCTION

Recently, a novel approach for the control of excitonic transition energies in the GaAs/(Al,Ga)As quantum well (QW) system has been proposed (Neu et al. 1991). It consists in the insertion of ultra-thin AlAs layers between the GaAs well and the $Al_xGa_{1-x}As$ barrier, resulting in a Double Barrier Quantum Well (DBQW) structure. As expected, such structures display increased confinement energies, together with an increase of the optical exciton oscillator strength for a given well width. Though the structures studied involved monomolecular or bimolecular AlAs layers, the GaAs QW direct Γ transition energies were well accounted for using the envelope function formalism. This paper deals with doubly indirect transitions between the X-like electrons in AlAs mono- or bilayers asymmetric wells and the GaAs Γ hole states. Hydrostatic pressure, driving X levels down and Γ levels up, allows the AlAs wells to supply the electron ground state of such single quantum well structures, in a staggered configuration similar to those observed in type II superlattices or multiquantum wells. High pressure has already proven to be a useful tool in studying quantum confined systems, for instance in the measurement of band offsets (Venkateswaran et al. 1986, Wolford et al. 1986, Wilkinson et al. 1991) or in the study of Γ-X mixing effects (Skolnick et al. 1989).

This work reports a high pressure photoluminescence (PL) study at low temperature of GaAs/$Al_{1-x}Ga_xAs$ quantum wells of various thicknesses bordered by 0, 1 or 2 AlAs monolayers. In the case of DBQWs, very efficient type II excitonic luminescence is observed. The results are well accounted for using the envelope function approximation, under the various band structure configurations high pressure allows to design. The confinement energy of electrons in AlAs mono- or bilayers has been obtained.

2. EXPERIMENTAL DETAILS

The three samples studied were grown by molecular beam epitaxy at 610 °C on (001) GaAs substrates. They consist of a 1 μm thick GaAs buffer followed by 3 QWs of width 7, 14 and 28 GaAs monolayers (20, 40 and 80 Å respectively) with 500 Å wide $Ga_{0.7}Al_{0.3}As$ barriers. One sample is a reference one, in the two others each well is bordered on both sides by one monolayer of AlAs (sample DBQW(1ML)) or 2 monolayers (sample DBQW(2ML)).

The samples were thinned to about 30 μm and placed in a diamond anvil cell cooled at liquid helium temperature. The pressure transmitting medium was Ar and pressure was monitored through the red shift of the R1 line of ruby. The luminescence was excited using the 4880 Å line of an Ar laser, and detected with a cooled GaAs photomultiplier.

3. RESULTS AND DISCUSSION

Figures 1 and 2 show the pressure dependence of PL transition energies for the reference sample and sample DBQW(1ML) respectively. The high pressure luminescence of GaAs/AlGaAs single QWs has already been studied by other groups (Venkateswaran et al. 1986, Wolford et al. 1986, Masumoto et al. 1989). The results may be roughly summarized as follows: The type I QW luminescence energies increase with pressure, essentially following the direct gap increase, up to the crossovers between the confined Γ electron levels and the AlGaAs X band gap, a point where the QW luminescence is strongly quenched. Sometimes some low intensity indirect transition are observed (Venkateswaran et al. 1986). Our results are very similar. The three QWs luminescences quench at a given pressure (this is indicated by dotted lines in Fig. 1), until only GaAs buffer PL persists (transitions labelled D_Γ correspond to direct gap excitons in GaAs). Some low intensity staggered indirect excitonic transitions are also observed around the crossover pressures (open symbols in Fig. 1).

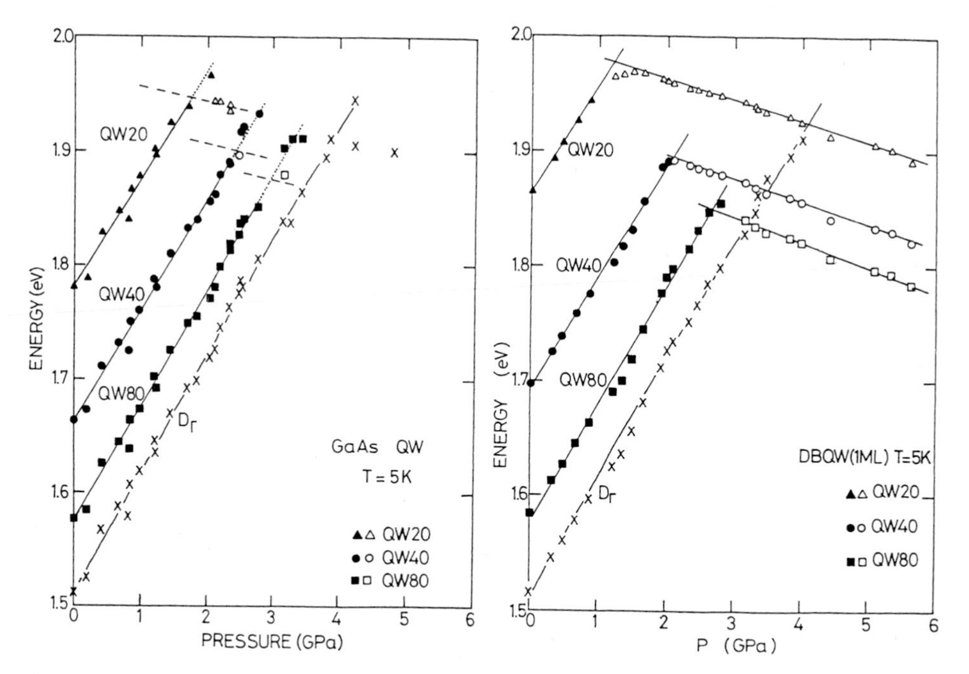

Figure 1 : Pressure dependence of quantum wells PL energies in the reference sample.

Figure 2: Pressure dependence of the PL energies of QWs bordered by one AlAs monolayer.

In the case of the double barrier quantum well (Fig. 2, sample DBQW(1ML)), the situation appears quite different. Due to the AlAs monolayer induced increase of the DBQW confinement energies , direct to indirect crossovers occur at lower pressures than in the previous case, as expected. But very intense indirect luminescence is now observed, persisting in the whole pressure range studied (up to 6 GPa). These indirect no phonon excitonic recombinations are accompanied by weak phonon repliqua with phonon energies of 34 and 47 meV. Such PL line shapes are characteristic of X_z electron confinement in the barrier as observed in type II GaAs/AlAs superlattices or multi quantum wells (Lefebvre et al. 1989, Scalbert et al. 1989, Dawson 1990). In fact, in sample DBQW(1ML), the envelope function calculation to be discussed below shows that the X_{xy} levels are not confined in the AlAs monolayer.

The high pressure PL results of sample DBQW(2ML) are not shown. Though the transition energies are different from those of sample DBQW(1ML) (at a given pressure, type I direct transitions are higher in energy, whereas type II indirect ones are lower in energy) the spectra are qualitatively the same, i.e. intense indirect excitonic transition can be observed in the whole pressure range studied.The atmospheric pressure value of the direct ($E_\Gamma(0)$) and extrapolated indirect ($E_X(0)$) excitonic transition energies, together with their linear pressure coefficients α, for the three samples studied are listed in Table 1.

Sample/well width (Å)	$E_\Gamma(0)$ (eV)	α_Γ (meV/GPa)	$E_X(0)$	α_X(meV/GPa)
QW/20	1.782	95		
QW/40	1.663	98		
QW/80	1.575	101		
DBQW(1ML)/20	1.865	93	2.002	-19
DBQW(1ML)/40	1.691	98	1.939	-21
DBQW(1ML)/80	1.574	103	1.909	-22
DBQW(2ML)/20	1.918	95	1.969	-20
DBQW(2ML)/40	1.726	95	1.888	-21
DBQW(2ML)/80	1.588	107		

Table 1: Pressure dependence of the confined exciton luminescence energies in the samples studied.

Electron and hole confinement energies have been computed using the transfer matrix approach of the envelope function formalism. A fractional band offset of 0.32 has been used. The calculation assumes Γ and X electron states to be confined in the GaAs wells and AlAs layers respectively. Figure 3 shows the Rydberg energy (E_{Ry}) of confined Γ excitons in our samples, as calculated following the method of Leavitt and Little (1990). It ranges from 8 to 12 meV for the QWs studied. Similarly, the Rydberg energy of confined X excitons is taken to be about 10 meV by analogy with its value in type II GaAs/AlAs superlattices (Duggan and Ralph 1987). Allowance for exciton localisation is also made, since PL excitation spectroscopy at atmospheric pressure has evidenced a Stokes shift of 5 to 20 meV, depending on the well studied.

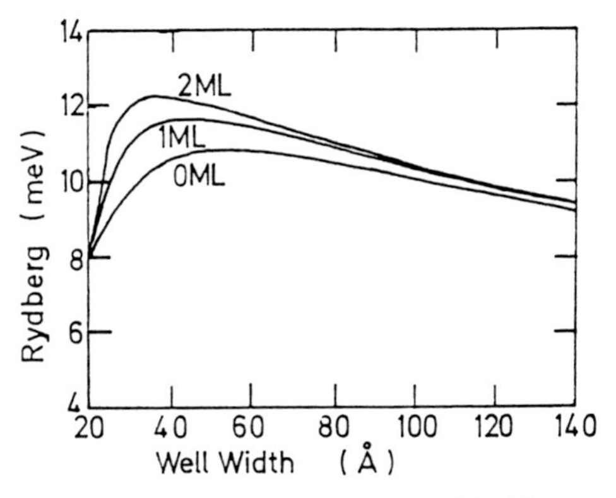

Figure 3 : Well width dependence of the Γ exciton Rydberg for wells bordered by 0,1 or 2 AlAs monolayers.

Using this procedure, the type I-type II crossover pressures and energies are calculated to be 1.25 GPa and 1.970 eV, 2.17 GPa and 1.892 eV, 2.75 GPa and 1.853 eV for the 20, 40 and 80 Å wide QWs respectively in sample DBQW(1ML). These values are in good aggreement with the experimental ones: 1.2 GPa and 1.978eV, 2.1 GPa and 1.895 eV, 2.7 GPa and 1.849 eV respectively. Such a good agreement supports the assumption that the observation of PL from indirect excitons is due to electron localisation in the AlAs layer. Moreover, since the same electron state is involved in each well in the indirect transitions, the difference of PL energy between two wells directly reflects the difference of heavy hole confinement energy (assuming similar exciton Rydberg and localisation energy). The experimental differences, at a given pressure, between the type II PL of wells 20 and 40 Å wide, and between wells 40 and 80 Å wide, are 63 and 30 meV respectively in sample DBQW(1ML), once again comparing well with the calculated values of 58 and 26 meV.

Due to the good agreement between experiment and theory, a direct estimation of the electron confinement energy in the pressure induced AlAs mono- and bilayer wells can be obtained. Indeed, this confinement energy is given by:
$E_e^X = E_{lum}(0Gpa) - E_g^X + \Delta V - E_{hh}^\Gamma + \Delta E$ where $E_{lum}(0GPa)$ is the extrapolated 0 pressure value of the luminescence energy, E_g^X is the AlAs energy gap, E_{hh}^Γ the heavy hole confinement energy, ΔV the valence band offset and ΔE the sum of the exciton coulombic and localisation energies. In the case of sample DBQW(1ML), the data of figure 2 lead to E_e^X-ΔE values of 181, 175 and 172 meV for the 20, 40 and 80 Å wide wells respectively. Assuming ΔE to be in the 10-15 meV range (Duggan and Ralph 1987), we get $E_e^X = 189\pm7$ meV, to be compared with 190 meV, as given by the envelope function calculation. Similarly, in sample DBQW(2ML), the confinement energy of electrons in 2 AlAs monolayers is found to be $E_e^X = 144\pm9$ meV, in agreement with the theoretical estimate of 150 meV. These experimental determination of X-type electron confinement energy in ultra thin assymetric AlAs wells could be helpful for instance in the design of resonant tunneling structures.

4. CONCLUSIONS

Efficient excitonic radiative recombination of type II is observed under pressure in GaAs/$Ga_{1-x}Al_xAs$ single quantum wells bordered by ultrathin AlAs layers. In accordance with our calculation, this is attributed to excitons formed by holes in GaAs QWs and electrons in the ultra-thin AlAs QWs. Separate hole and electron confinement energies have been obtained. We suggest that such structures, allowing on a single sample to study under pressure (i.e. for various Γ and X energy levels configurations) and especially on various well width (i.e. for various values of Γ-X overlap), should prove useful in the study of the origin of Γ-X mixing in microstructures, for instance by selectively excited and (or) time resolved high pressure photoluminescence.

5. ACKNOWLEDGMENTS

We are grateful to Y. Chen (Laboratoire de Microstructures et Microélectronique, Bagneux) for is contribution in the starting stage of this work.

6. REFERENCES

Dawson P 1990 Optical and Quantum Electronics 22 S231
Duggan G and Ralph H I 1987 Phys. Rev. B 35 4152
Leavitt R and Little J W 1990 Phys. Rev. B 42 11774
Lefebvre P Gil B Mathieu H Planel R 1989 Phys. Rev. B 39 5550
Masumoto Y Kinoshita Y Shimomura O and Takemura K 1989 Phys. Rev. B 40 11772
Neu G Chen Y Deparis C and Massies J 1991 Appl. Phys. Lett 58 2111
Scalbert D Cernogora J Benoit a la Guillaume C Maaref M Charfi F F and Planel R 1989 Solid State Comm. 70 945
Skolnick M S Smith G W Spain I L Whitehouse C R Herbert D C Whittaker D M and Reed L J 1989 Phys. Rev. B 39 11191
Venkateswaran U Chandrasekhar M Chandrasekhar H R Vojak B A Chambers F A and J M Meese 1986 Phys. Rev. B 33 8416
Wilkinson V A Prins A D Dunstan D J Howard L K and Emeny M T 1991 J. Electronic Materials 20 509
Wolford D J Kuech T F Bradley J A Gell M A Ninno D and Jaros M 1986 J. Vac. Sci. Technol. B 4 1043

Inst. Phys. Conf. Ser. No 123
Paper presented at the International Meeting on Optics of Excitons in Confined Systems, Giardini Naxos, Italy, 1991

Excitons in asymmetrical double quantum wells

P Lefebvre, P Bonnel, Y Chen* and B Gil

Groupe d'Etudes des Semiconducteurs, Université des Sciences et Techniques du Languedoc - CP 074. 34095 Montpellier Cedex 5, France.

*Laboratoire de Microstructures et Microélectronique, Centre National de la Recherche Scientifique, 196 Avenue Henri Ravera, 92260 Bagneux Cedex, France.

ABSTRACT : The lack of symmetry plane, in Asymmetrical Double Quantum Wells (ADQW's) leads to novel optical properties, such as the constitution of excitons out of spatially separate electrons and holes, favoured by peculiar resonances of valence subbands, for particular cases of GaAs-(Ga,Al)As ADQW's. The electronic structure of such ADQW's is analyzed, within the Envelope Function Formalism. Interwell probability transfers of holes are predicted, either when the in-plane wave vector differs from zero, or when external uniaxial stress is applied. In this last case, the resulting excitonic mixings are experimentally observed.

1. INTRODUCTION

The physics of excitons confined in Asymmetrical Double Quantum Wells (ADQW's) recently received an increasing interest, due to the possibility to produce novel effects. The asymmetry of the structures makes it possible to obtain the so-called *barbell excitons* (Peeters and Golub 1991), which are constituted by electrons and holes located in each of the two wells, and coupled by the coulombic interaction across the intermediate barrier. Such excitons are expected to exhibit so much larger lifetimes than excitons simply confined in single wells so that peculiar thermodynamical behaviors should be observed. For instance, recent experimental findings on the temperature behavior of excitons in double quantum wells (Fukuzawa et al 1990, Kash et al 1991) were explained by invoking, first, the Bose-Einstein Condensation of excitons and, second, a statistic broadening of the emission line, due to the short-range electric dipole repulsion between excitons trapped on interfacial inhomogeneities. Observations of "indirect" excitons by reflectance measurements are allowed (Bonnel et al 1990) when confined states originating from either well come into incidental (Leopold and Leopold 1990, Roussignol et al 1991) or externally induced resonance. The corresponding envelope functions then become delocalized through the whole ADQW, so that crossed (or *indirect*) transitions become possible. Necessary conditions can be achieved by application of external perturbations such as electric fields (Charbonneau et al 1988) or mechanical stresses (Lefebvre et al 1991).
In a more general way, the characteristics of confined excitons (binding energies, optical oscillator strengths...) are strongly correlated to

numerous details of the real quantum well environment, such as the complexity of the in-plane dispersion relations of the valence subbands. These details have been included in a recent theoretical work, by Andreani et al (1990), together with the Coulomb coupling between different quantized states, in the case of a single quantum well, and non-negligible effects were found, improving the agreement with experimental data. In the case of ADQW's, the situation is made moreintricate by the coupling between the quantum wells, which complicates the in-plane band structure of the system. Moreover, Coulomb coupling effects between different subbands, turning out to be very close in energy, should be strongly enhanced.

This paper presents experimental low-temperature reflectance studies on (Ga,Al)As-GaAs ADQW's, submitted to externally applied uniaxial stress along the (110) direction. By its selective influence on light-hole and heavy-hole states, the uniaxial stress generates a variety of *inter-well* and *intra-well* anticrossings between valence states. The observed energy shifts of the excitonic reflectance structures are compared to the calculated energy shifts of the corresponding subband-to-subband transitions. Evidence is made of stress-induced couplings between *excitonic* states.

2. THEORETICAL BACKGROUND

The model used in order to compute the quantized states -more particularly, valence states- in a (001)-grown ADQW, and their variation under (110) in-plane uniaxial stress, is described and discussed extensively elsewhere (Lefebvre et al 1991). It is inspired with the model proposed by Andreani et al (1987) which was dedicated to single quantum wells, possibly submitted to internal biaxial strain fields, having the same symmetry as the quantizing square potential. The method consists in an exact solution of a system of equations, which account for the continuity conditions of the envelope functions and of the probability current density, at each interface. By doing so, given a value of the in-plane wave vector $\mathbf{k}_{\perp}$, a set of eigenenergies can be found, as well as the corresponding eigenstates, expanded over the four basis states $|3/2,\pm3/2\rangle$ and $|3/2,\pm1/2\rangle$ (heavy-hole and light-hole states). For evident reasons of symmetry, mixings between different kinds of states can only be obtained for nonvanishing values of the in-plane wave

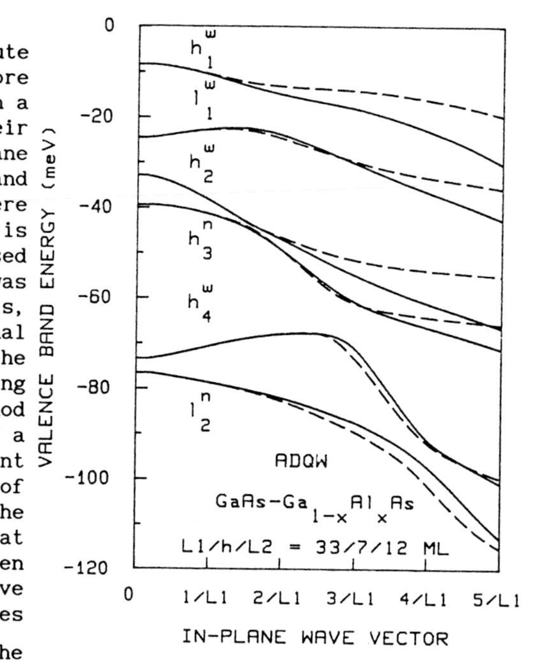

Fig. 1. *In-plane dispersion curves for a GaAs-(Ga,Al)As ADQW, with an aluminum mole fraction of x = 0.30 in the barriers. Solid lines show the dispersion relation for a wave vector along the (100) axis, and dashed lines represent the dispersion along the (110) axis.*

vector. This model was here adapted to the calculation of in-plane dispersion relations for ADQW's and to the application of in-plane uniaxial stress, which can produce mixings between light- and heavy-hole states, even at the centre of zone, by addition of the adapted strain Hamiltonian. Figure 1 displays an example of in-plane dispersion relation, obtained for the valence levels of a $GaAs-Ga_{0.7}Al_{0.3}As$ ADQW with respective layer thicknesses $L_1/h/L_2$ = 33/7/12 monolayers (ML). These curves resemble the mere superimposition of the dispersion curves of both single quantum wells, but some novel anticrossings occur, at certain points of the Brillouin zone, between bands originally related to different wells. A first consequence is the modification of the curvatures of the dispersion relations, which is strongly dependent on the degree of coupling between the two wells, *i.e.* the geometry of the ADQW. The properties of the excitons in the structure are thus modified, since the in-plane "effective masses" of the holes are changed.

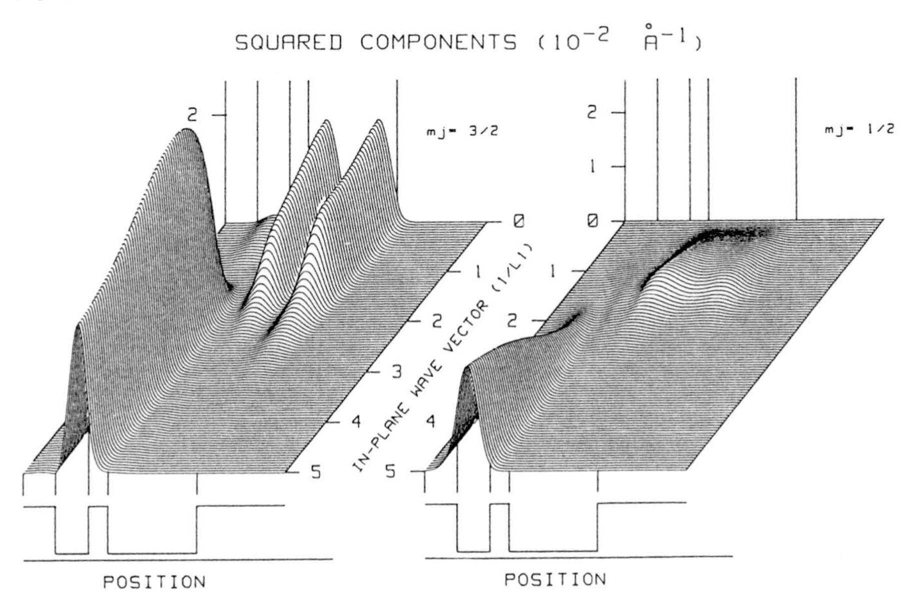

Fig. 2. *Heavy-hole character (m_J = ±3/2, left-hand) and light-hole character (m_J = ±1/2, right-hand), shown by each squared component of the h_2^w state, versus the in-plane vector* **k**. *At* $|\mathbf{k}| \sim 2/L_1$, *the spatial distibution of the state transfers from the wide well into the narrow one, keeping a heavy-hole character, except for large values of* $|\mathbf{k}|$, *where an intrawell mixing between light and heavy holes occurs.*

The second consequence is an alteration of the spatial repartition of the wave functions. This is illustrated by the probability distributions shown in Figure 2 : as the vector $\mathbf{k}_\perp$ increases, the second heavy-hole state of the wider well anticrosses the first heavy-hole level of the narrower well, and an *inter-well* probability transfer occurs. It is clear that, depending on which proportion of the Brillouin zone effectively contributes to excitonic wave functions, this effect should have a crucial incidence on the characteristics of the excitons in the ADQW. Moreover,

the possible modifications of these characteristics are again heavily correlated to the design of the structure, which rules, in particular, the energetic distancy between the various levels.
Comparable spatial transfers can also be obtained, for states at the centre of zone, by application of an in-plane uniaxial stress. When a compressive stress is applied along the (110) direction, light-hole subbands present a larger stress-induced energy shift than the heavy-hole ones, as shown in Figure 3. As a consequence, the valence states present a series of *anticrossings*, instead of crossings, because of the lack of symmetry plane in the ADQW. The stress then modifies not only the energies of the various subbands, at the zone center, but also the curvatures of their in-plane dispersion relations and, consequently, the distribution of the related wave functions upon the basis valence states |3/2,±3/2> and |3/2,±1/2>, as illustrated in Figure 4.

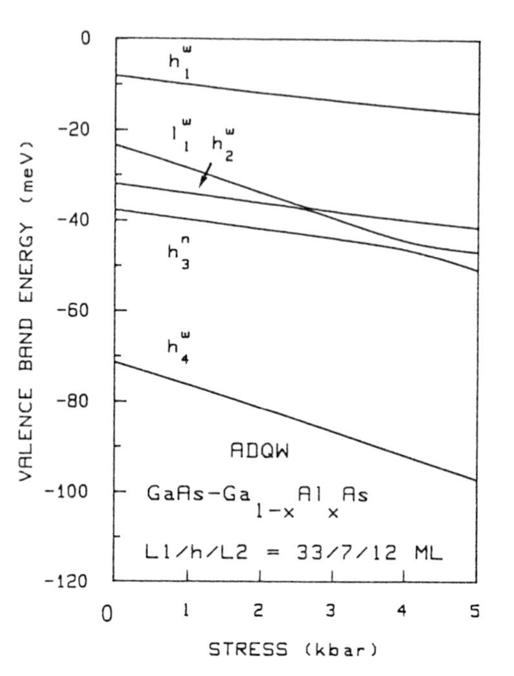

Fig. 3. Energy shifts of the five upper valence subbands, when the above ADQW is submitted to a compressive stress along the (110) axis.

Concerning excitonic binding energies and oscillator strengths in double quantum wells, several tentative approaches have been proposed (Peeters and Golub 1990, Galbraith and Duggan 1989, Lefebvre et al 1990), but none of them clearly includes the above-mentioned effects, for evident reasons of simplicity. As an illustration, Figure 5 shows an example of calculation of the binding energies in a GaAs(Ga,Al)As ADQW with constant well thicknesses and for varying dimensions of the intermediate barrier. This result was obtained using a variational approach and assuming parabolic in-plane dispersion curves for valence states and, thus, no coupling between them. Previous studies by Bonnel et al (1990) showed that, when the coupling between the wells is increased, some states remain strongly localized in their original well, while some others, coming into incidental resonance, tend to become delocalized over the whole ADQW. The calculation shows that *intra-well* excitons exhibit a poor dependence on the coupling, while *inter-well* excitons, involving near-resonant valence states, present rapidly varying Rydberg energies. The oscillator strengths vary correspondingly as shown in previous works by Lefebvre et al (1990). We will show, in the next section, that the experimental findings cannot be totally interpreted by using a simple model, which does not account for the variations of the Rydbergs under stress. The characteristics of the observed excitons, derived from the comparison between experiments and band-to-band calculations, strongly suggest the appearance of peculiar exciton mixings, which were not observed in previous works (Gil et al 1988) on single quantum wells of comparable dimensions.

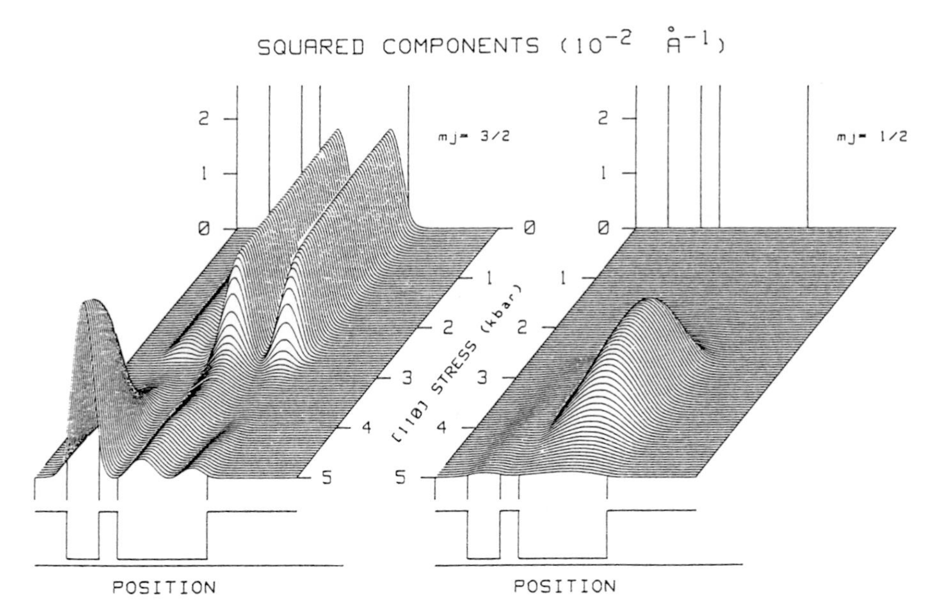

Fig. 4. Squared components of the h_2^w state of the same ADQW, taken at **k = 0**, *under (110) uniaxial stress. The heavy-hole character of the state is destroyed, at ~ 3 kbar, by an intrawell anticrossing with the l_1^w state as one can see on the right hand part of the figure. Under a uniaxial stress of ~ 4 kbar, the heavy-hole character is recovered, but the state is now mainly localized in the narrow well, due to the interwell anticrossing with the h_3^n level.*

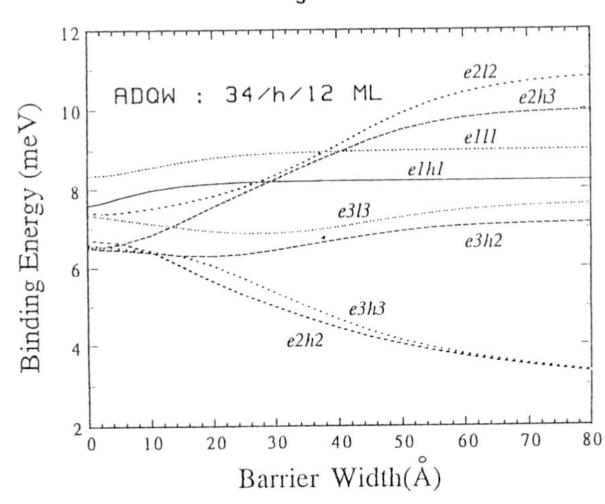

Fig. 5. Calculated binding energies of various excitons in a GaAs-(Ga,Al)As ADQW, with an aluminum mole fraction x = 0.30, in the barriers. The GaAs layers have respective widths of 34 and 12 ML, while the central barrier width is continuously varied.

3. EXPERIMENTAL RESULTS

Reflectance spectra were taken from samples grown by MBE on (001)-oriented GaAs substrates, immersed in a pumped liquid helium bath and submitted to a (110)-oriented uniaxial stress of varying magnitude. The latter was measured within ± 50 bar from the stress-induced energy shift and splitting of the reflectance signature of the light- and heavy-hole excitons, freely propagating in the GaAs substrate (Bonnel et al 1990). Detailed results taken on several samples will appear elsewhere (Gil et al 1992). Figures 6-A and 6-B just show the evolution of the low- (A) and high-energy (B) parts of the reflectance spectrum versus the applied (110)-stress, for a sample having the physical characteristics described above. The notations are the following : subscripts correspond to the quantum numbers of the levels, in the context of the global ADQW, and superscripts indicate whether a given level "originates" from the wide well (w) or from the narrow one (n).

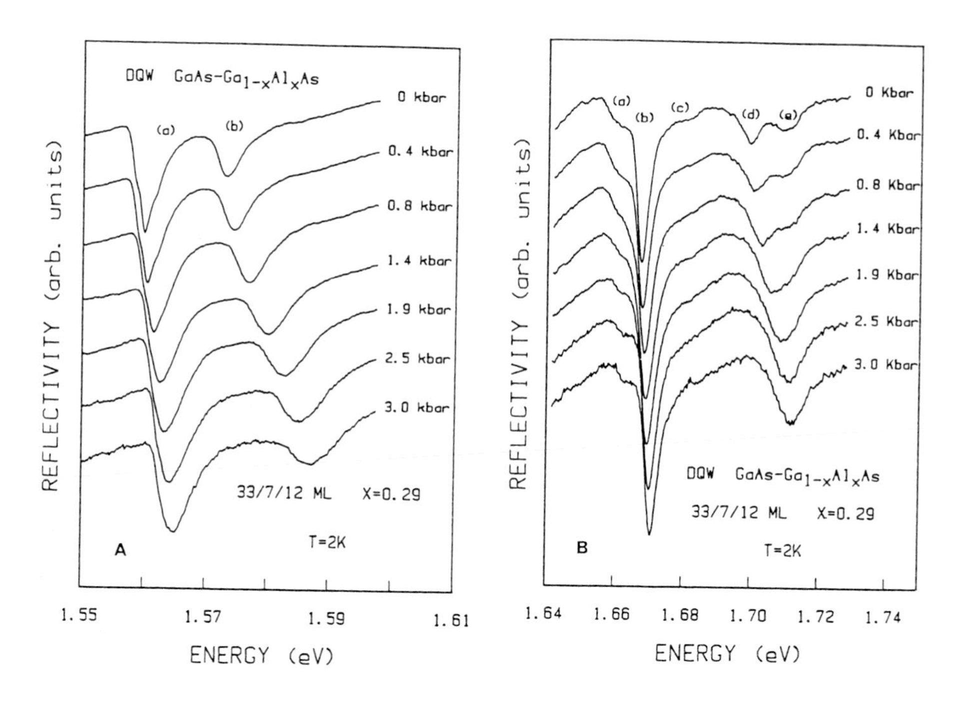

Fig. 6. Low-temperature reflectance spectra for a $GaAs\text{-}Ga_{0.71}Al_{0.29}As$ ADQW with the following layer thicknesses : 33/7/12 ML, and under varying magnitudes of the applied (110)-stress. (A) : low-energy spectrum, involving excitons mostly located in the wider well : (a) = $e_1^w\text{-}h_1^w$ and (b) = $e_1^w\text{-}l_1^w$. (B) : high-energy zone, where the excitons mainly involve states originating from the thinner well : (a) = $e_2^n\text{-}h_2^w$, (b) = $e_2^n\text{-}h_3^n$, (c) is difficult to assign, (d) = $e_2^n\text{-}l_2^n$, (e) = $e_3^w\text{-}h_3^n$ and/or $e_3^w\text{-}h_2^w$.

The high-energy spectra reveal rather fast changes in the shapes of the reflectance structures having different sensitivities to the stress, depending on the valence states involved. This was not observed in the case of single quantum wells (Gil et al 1988), for which the reflectance structures kept a constant shape, when the stress was changed. Figure 7 displays a plot of the measured transition energies versus the applied (110) stress. These measurements are compared to the calculated subband-to-subband transition energies, shifted, at zero stress, down to the experimental points, by quantities assumed equal to the binding energies of the corresponding excitons. This point is particularly important for the interpretation of the high-energy part of the spectra : the $e^n_2 l^n_2$ transition is identified as such from its slope. Now, the measured relative energies of $e^n_2 l^n_2$ and $e^w_3 h^n_3$ cannot be reproduced by any calculation, even by tentative variations of the layer widths -within a reasonable range- or of the barrier heights. It can be inferred that the $e^n_2 l^n_2$ exciton is pushed to lower energy than the $e^w_3 h^n_3$ one, only because of a stronger Rydberg energy (~ 10 meV). It is to note that the values of the different Rydbergs, obtained by comparison between the band-to-band calculations and the experimental points, are not in perfect agreement with the computations sketched in Figure 5, especially in the high-energy part of the spectra.

Concerning the behavior of the transitions under stress, the first constatation is an overall very good agreement between the experimental and the numerically computed slopes. Nervertheless, comparing with previous reflectance studies on GaAs-(Ga,Al)As single quantum wells (Gil et al 1988), new features are observed :

i) In the low-energy range, the $e^w_1 l^w_1$ transition exchanges its stress shift with the forbidden $e^w_1 h^w_2$, at their anticrossing point, near ~ 2.5 kbar, instead of keeping its original slope. Such an *intra-well* anticrossing, involving states mostly localized in the 33 ML-wide well, is only made possible by the presence of the adjacent 12 ML-wide well, which causes the asymmetry of the ADQW. However, the observed complete tranfer of oscillator strengths between both transitions cannot be accounted for by simple band-to-band calculations of the electron-hole envelope function overlap integrals, as it was done in a previous work by Bonnel et al (1990) : coulombic effects probably have to be correctly included, following the above arguments.

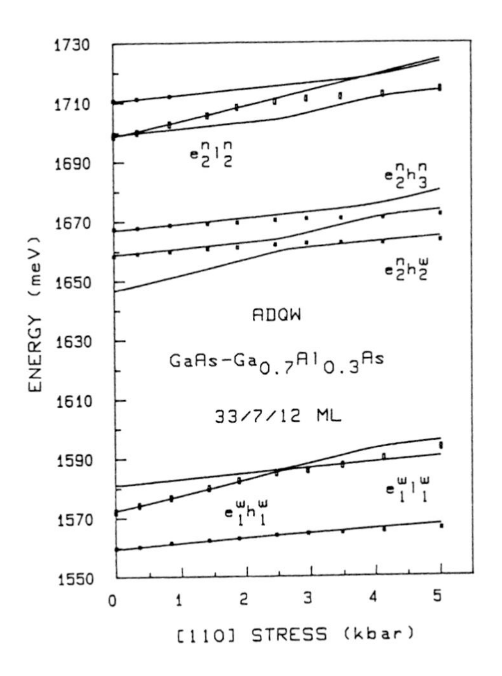

Fig. 7. *Stress-induced shift of the measured excitonic transitions versus the (110)-stress, marked by rectangles, compared to the shifts of the corresponding band-to-band computed transition energies.*

ii) In the high-energy range, where states related to the thinner well mainly participate, a complex system of anticrossings is calculated, whereas rather simpler experimental energy shifts are measured. Here again, the mechanisms of exchange between excitonic oscillator strengths, accounting for intricate mixings of states, should be invoked to explain the observations. A dramatic demonstration of this last point is provided by a careful examination of the stress behavior of the $e_2^n l_2^n$ exciton, whose energy shift is strongly bent down, just as if it was repelled by the higher-lying $e_3^w h_3^n$ exciton. The reflectivity structure related to the latter ends by "drawning" into the one of the former. This observation is absolutely inconsistent with the predictions of the calculation, which is not surprising, keeping in mind that the relative positions of both transitions are due to a strong difference between their Rydbergs. One can thus conclude that complex excitonic mixings are responsible for the measured energy shifts.

4. CONCLUSION

After a systematic comparison between theoretical predictions (Lefebvre et al 1991) and optical measurements (Gil et al 1992) on GaAs-(Ga,Al)As ADQW's, under (110) uniaxial stress, of which this paper presents a few examples, we conclude that the symmetry properties and measurable characteristics of excitons in these structures cannot be modeled in a straightforward way. It is very likely that a satisfactory description of the observations should include complex excitonic interactions together with the peculiar **k**-space dispersion of valence states, as it was already done (Andreani et al 1990), by heavy computations, in case of single quantum wells.

ACKNOWLEDGEMENTS

The authors are grateful to the staff at the LPSES-CNRS (Valbonne, France) - C. Deparis, J. Massies and G. Neu - for providing nice MBE samples, and to Prof. H. Mathieu, for his encouragements and fruitful discussions.

REFERENCES

Andreani L C, Pasquarello A and Bassani F 1987 Phys. Rev. B 36, 5887
Andreani L C, Pasquarello A 1990 Phys. Rev. B 42, 8928
Bonnel P, Lefebvre P, Gil B, Mathieu H, Deparis C, Massies J, Neu G and Chen Y 1990 Phys. Rev. B 42, 3435
Charbonneau S, Thewalt M L W, Koteles E S, Elman B 1988 Phys. Rev. B 38, 6287
Clérot F, Deveaud B, Chomette A, Regreny A and Sermage B 1990 Phys. Rev. B 41, 5756 (1990)
Fukuzawa T, Mendez E E and Hong J M 1990 Phys. Rev. Lett. 64, 3066
Galbraith I and Duggan G 1989 Phys. Rev. B 40, 5515
Gil B, Lefebvre P, Mathieu H, Platero G, Altarelli M, Fukunaga T and Nakashima H 1988 Phys. Rev. B 38, 1215
Gil B, Lefebvre P, Bonnel P, Mathieu H, Deparis C and Massies J 1992 (unpublished)
Kash J A, Zachau M, Mendez E E, Hong J M and Fukuzawa 1991 Phys. Rev. Letters 466, 2247 (1991)
Lefebvre P, Deparis C, Bonnel P, Gil B, Yong Chen, Massies J, Neu G and Mathieu H 1990 Superlattices and Microstructures 8, 187
Lefebvre P, Bonnel P, Gil B and Mathieu H 1991 Phys. Rev. B *(in press)*
Leopold D J and Leopold M M 1990 Phys. Rev. B 42, 11147
Peeters F M and Golub J E 1991 Phys. Rev. B 43, 5159
Roussignol P, Gurioli M, Carraresi L, Colocci M, Vinattieri A, Deparis C, Massies J and Neu G 1991 Superlattices and Microstructures 9, 151.

Inst. Phys. Conf. Ser. No 123
Paper presented at the International Meeting on Optics of Excitons in Confined Systems, Giardini Naxos, Italy, 1991

Excitons in GaAs/$Al_{0.3}Ga_{0.7}As$ symmetric double quantum wells studied by optical spectroscopy

Q X Zhao[1], T Westgaard[1,2], O Hunderi[1,2], B O Fimland[3], K Johannessen[4]

1) Department of Physics, Norwegian Institute of Technology, N-7034 Trondheim, Norway
2) SINTEF Applied Physics, N-7034 Trondheim, Norway
3) Department of Physical Electronics, Norwegian Institute of Technology, N-7034 Trondheim, Norway
4) SINTEF DELAB, N-7034 Trondheim, Norway.

ABSTRACT: GaAs/$Al_{0.3}Ga_{0.7}As$ symmetric coupled double quantum well (CDQW) structures have been studied by photoluminescence (PL) and photoluminescence excitation (PLE) spectroscopy. Exciton energy levels and ratios of absorption oscillator strengths of "antisymmetric" to "symmetric" exciton resonances are obtained. The oscillator strength ratios show a pronounced reduction with decreasing coupling between the wells. The experimental results can be explained within effective mass theory if mixing between the different exciton states is included in the calculations.

1. INTRODUCTION

Coupled symmetric double quantum wells structures have interesting electronic properties which can be utilized in optoelectronic devices (Miller et al 1986). In double quantum well structures with narrow barriers the energy levels in the two wells can be coupled. The electronic states corresponding to single well one-particle states split into symmetric and antisymmetric states. The exciton states constructed from products of either symmetric or antisymmetric hole and electron one-particle wavefunctions are denoted "symmetric" and "antisymmetric" excitons. Both types of excitons have even parity. The splitting of energy levels due to coupling between the two wells is very sensitive to barrier width, but it is also dependent on well widths. Previous experimental studies on CDQWs have been concentrated on strongly coupled structures (Yariv et al 1985, Chen et al 1987). Exciton effects and mixing between "symmetric" and "antisymmetric" exciton states have very often been neglected in the analysis of experimental results. A recent theoretical study of CDQW structures by Dignam and Sipe (1991) shows that mixing between "symmetric" and "antisymmetric" excitons can have a large effect on the electronic structure of a CDQW system.

In this presentation we report results from experimental and theoretical studies of CDQW structures covering a wide range of coupling strengths. Exciton energy levels and relative excitonic transition strengths are obtained from PLE spectra. Theoretical energy levels and oscillator strengths for the excitonic transitions have been calculated in a model which allows for mixing of exciton states. A similar model has previously been used by Kamizato and Matsuura (1989) to describe the lowest exciton state, but it has not been used to explain the behaviour of the other excitonic states.

2. SAMPLES AND EXPERIMENTAL SETUP

The samples were grown in a Varian Gen II modular MBE machine on semiinsulating substrates oriented in the [001] direction. For all samples the growth was started with a 500 Å thick n-doped ($5x10^{17}$ cm^{-3} Si) GaAs buffer layer and continued with an undoped GaAs buffer layer. The buffer layer was followed by a 20 period superlattice consisting of 20 Å GaAs and 20 Å AlAs layers. After this structure an 800 Å $Al_{0.3}Ga_{0.7}As$ barrier was grown. On top of this barrier three double quantum well structures with equal $Al_{0.3}Ga_{0.7}As$ barrier widths L_b and decreasing quantum well widths (L_z) were grown. The double wells were separated by 800 Å $Al_{0.3}Ga_{0.7}As$ barrier layers. The growth was ended with an 800 Å thick $Al_{0.3}Ga_{0.7}As$ barrier layer and a 200 Å thick n-doped ($5x10^{17}$ cm^{-3} Si) GaAs capping layer. Sample A contains three double wells with widths L_z = 150.0 Å, 99.1 Å and 59.4 Å , while sample B contains three double wells with widths L_z = 124.5 Å, 79.2 Å and 39.6 Å. All the double wells in both samples have barrier widths L_b =14.2 Å .

The samples were mounted on the coldfinger in a closed-cycle cryostat and cooled down to a temperature of 12 K. The excitation source was a tunable Ti:Sapphire laser pumped by an argon-ion laser. The luminescence emitted by the sample was dispersed by a 0.85 m double monochromator and detected by a cooled GaAs photomultipler using photon counting techniques.

3. EXPERIMENTAL RESULTS AND DISCUSSION

In the PL spectra only the transitions from the lowest excitonic states to the ground state of the CDQW structures appear. In Fig.1 the PLE spectra for the CDQWs are shown. From this figure we can see that the four distinct exciton peaks (labelled 1, 2, 3, 4) clearly appear for the thinner double quantum wells (for example for L_z =79.2 Å). With increasing well widths L_z the relative intensities of the excitonic transitions in the PLE spectra change. This variation of excitonic transition strengths in the PLE spectra can be explained within the effective mass theory if mixing between different exciton states is included in the calculations. The Hamiltonian for a CDQW system with electron-hole interaction (exciton effects) can be written:

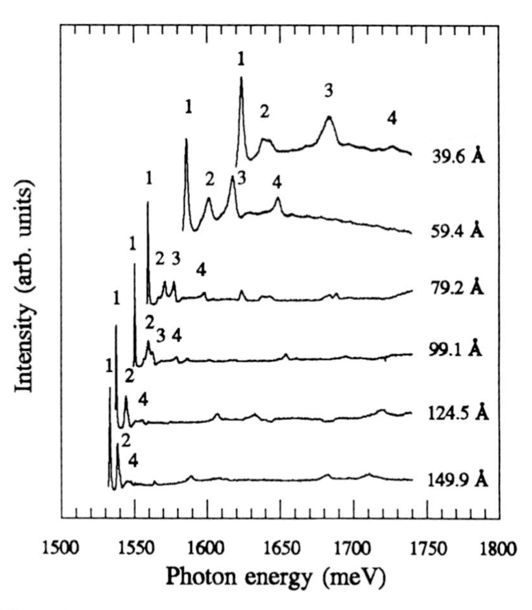

Fig.1. PLE spectra of CDQW structures from samples A and B. The well width corresponding to each curve is shown to the right.

$$H_{ex} = H_e + H_{hh/lh} + H_{eh} \quad (1)$$

where

$$H_{eh} = -\frac{\hbar^2}{2\mu_\pm}\frac{1}{\rho}\frac{\partial}{\partial\rho}(\rho\frac{\partial}{\partial\rho}) - \frac{e^2}{4\pi\varepsilon_0\varepsilon(\rho^2+(z_e-z_h)^2)^{1/2}}$$

H_e and $H_{hh/lh}$ are single-particle Hamiltonians for electrons and heavy/light holes in a CDQW system. $\mu_\pm$ are effective masses in the plane perpendicular to the growth direction for heavy holes (+) and light holes (-). The interaction between heavy and light hole bands is neglected in this calculation. The trial wavefunctions for the exciton Hamiltonian (1) are constructed from linear combinations of the eigenfunctions of the two lowest states of the effective mass Hamiltonians H_e and $H_{hh/lh}$:

$$\Psi_i = \Sigma\, a_{ij}\, \Phi_j \tag{2}$$

where $\Phi_j = N_j f_{e,j}(z_e)\, f_{h,j}(z_h)\, \varphi_j(\rho, |z_e - z_h|)$

with j =1,2. N_j are normalization constants. $f_{e,j}(z_e)$ and $f_{h,j}(z_h)$ are the eigenfunctions of the Hamiltonians H_e and $H_{hh/lh}$. If mixing between "symmetric" and "antisymmetric" exciton states is neglected, one has $a_{12}=a_{21}=0$ in equation (2). We approximate φ_j with a hydrogenic ground state wavefunction with the exciton radius R as a variational parameter. This approximation has been shown to work for SQW structures (Bastard et al 1982):

$$\varphi_j(\rho, |z_e - z_h|) = \exp(-(\rho^2 + (z_e - z_h)^2)^{1/2} / R) \tag{3}$$

Linnerud and Chao (1991) have used a trial wavefunction where φ_j has cylindrical symmetry to calculate CDQW exciton energy levels. This approach which is motivated by the anisotropy of the effective masses, introduces only small changes in the exciton energy levels, and it has no significant effect on the degree of mixing of exciton states. The eigenvalue equation with the exciton Hamiltonian (1) and the trial wavefunction (2) can be written as

$$\Sigma\, (H_{ij} - E\, S_{ij})a_{ij} = 0 \qquad j=1,2 \tag{4}$$

$H_{ij} = \langle\Phi_i | H_{ex} | \Phi_j\rangle$ and $S_{ij} = \langle\Phi_i | \Phi_j\rangle$ are matrix elements. The exciton eigenenergies can be obtained by solving the characteristic equation of (4) and varying the parameter R to minimize the eigenenergy of lowest state. When the eigenenergies and the corresponding parameters of the trial wavefunctions are obtained, the optical oscillator strengths related to excitonic absorption can be straightforwardly calculated (Zhu 1988).

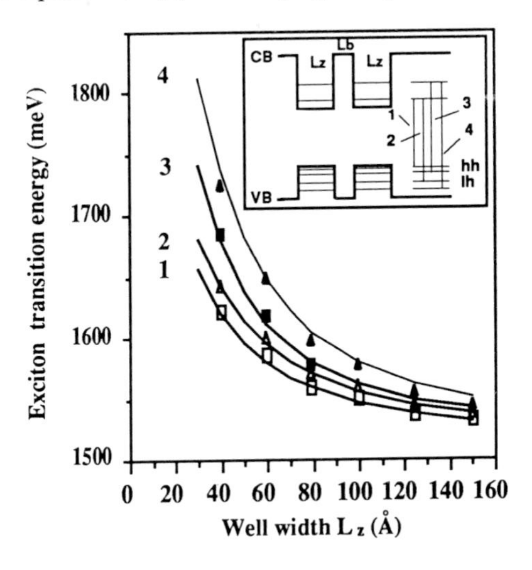

Fig.2. Exciton energy levels in CDQW structures with 14.2 Å barriers calculated in a model with mixing between "symmetric" and "antisymmetric" excitons. The experimental points are exciton resonances found in PLE measurements. The insert shows schematically the transitions in a CDQW structure.

The theoretical curves of energy levels shown in Fig. 2 have been calculated with the following parameters for GaAs ($Al_{0.3}Ga_{0.7}As$): electron effective mass 0.0665 (0.0916) m_0, heavy hole effective mass 0.34 (0.466) m_0 and light hole effective mass 0.094 (0.1069) m_0. The GaAs bandgap was set to 1.520 eV and the $Al_{0.3}Ga_{0.7}As$ bandgap was 1.957 eV. In the calculations a band offset parameter Q=0.65 was used, and the relative dielectric constant ε =12.5 of GaAs was used for the whole CDQW structures. The relative ratios of oscillator strengths between "antisymmetric" excitons and "symmetric" excitons are shown in Fig.3 for both heavy and light hole related excitons together with experimental values obtained from the PLE spectra. For comparison calculated oscillator strength ratios where mixing between different exciton states is neglected, are also shown. From Fig. 3 it is clear that only the theoretical values obtained for trial wavefunctions with mixing between different states can reproduce correct trends for the oscillator strength ratios obtained from the PLE data.

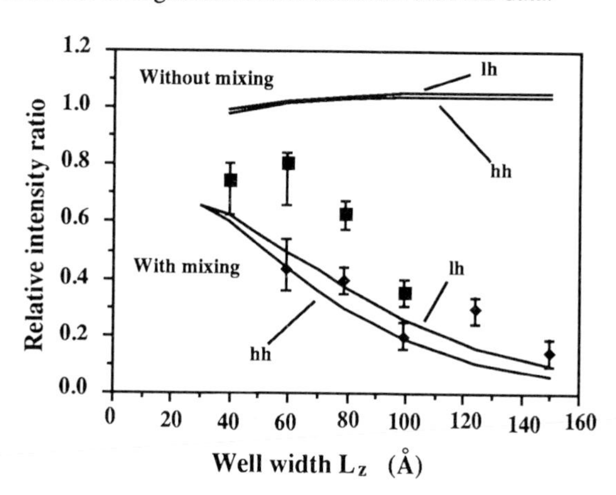

Fig. 3. Calculated ratios of the oscillator strengths between "antisymmetric" and "symmetric" excitons in CDQW structures with 14.2 Å barriers with mixing effects and without mixing effects. (The curves hh are related to heavy holes, lh to light holes). The experimental points are obtained as integrated peak intensities from the PLE spectra. Heavy hole exciton ratios: ■ . Light hole exciton ratios : ◆.

4. CONCLUSIONS

CDQW structures covering a wide range of coupling strengths have been studied both experimentally and theoretically. The results show that exciton effects and mixing between different exciton states are important for weakly coupled wells, i.e. where the energy splitting between symmetric and antisymmetric energy levels is comparable to single quantum well exciton binding energies.

REFERENCES

Bastard G, Mendez E E, Chang L L and Esaki L 1982 Phys. Rev. B 26 1974

Chen Y J, Koteles E S, Elman B S and Armiento C A 1987 Phys. Rev. B 36 4562

Dignam M M and Sipe J E 1991 Phys. Rev. B 43 4084

Kamizato T and Matsuura M 1989 Phys. Rev. B 40 8378

Linnerud I and Chao K A 1991 private communication. They have used an expression with three variable parameters:

$\varphi_j(\rho, |z_e\text{-}z_h|) = (1+ \alpha\, (z_e\text{-}z_h)^2)\exp(-(\rho^2 + \beta\, (z_e\text{-}z_h)^2)^{1/2}/R)$

Miller D A B, Weiner J S and Chemla 1986 IEEE J. Quantum Electron. QE-22 1816

Yariv A, Lindsey C and Siven U 1985 J. Appl. Phys. 58 3669

Zhu B 1988 Phys. Rev. B 37 1526

Inst. Phys. Conf. Ser. No 123
Paper presented at the International Meeting on Optics of Excitons in Confined Systems, Giardini Naxos, Italy, 1991

Optical studies on strained $Ga_xIn_{1-x}As/InP$ quantum wells

A Kux, C Wetzel, B K Meyer, R Meyer[*], D Grützmacher[+] and A Kohl[+]

Physik Department E 16, TU Munich, James Franck Str., 8046 Garching, F.R.G.
[*]Forschungszentrum Jülich, Institut für Schicht-und Ionentechnik, Leo Brandt Str., 5170 Jülich, F.R.G.
[+]Institute of Semiconductors Electronic, RWTH Aachen, Sommerfeldstr., 5100 Aachen, F.R.G.

ABSTRACT The effect of strain on $Ga_xIn_{1-x}As/InP$ heterostructures and Quantum Wells (QW) was investigated using photoluminescence, impact ionisation and transmission measurements. The respective Ga concentrations were in the range from $x=0.17$ to 1, whereas the QW width (L_z) varied from 5 nm to 1.5 nm. In contrast to recent investigations concluding that for $x>0.8$ a type I to type II superlattice transition occurs, we find strong excitonic luminescence at 1.1 eV for $x>0.8$ indicating that the $Ga_xIn_{1-x}As/InP$ strained layer superlattices remain type I also for $x>0.8$.

1. INTRODUCTION

Absorption, photoluminescence (PL) and photoconductivity measurements for quantum well structures in the $Ga_xIn_{1-x}As/InP$ system have been an active and rapidly developing area of investigation (Wang et al 1990, Gershoni et al 1988 Meyer et al 1991a). The motivation for this interest comes from the fields of application this ternary alloy system has utilizing ultrahigh speed devices with high electron mobilities (Reithmaier et al 1989, Grundmann et al 1990, Reihlen et al 1990. The fabrication of high electron mobility transistors takes advantage of modulation doping into the InP barrier, which creates a two-dimensional electron gas (Hardtdegen et al 1991). In this materials system quantum well heterostructures including the effect of non-lattice-matched structures allow to tailor the band gap energy over a very wide range of emission wavelength. For lattice mismatched QWs the layer thickness must be below a critical value to avoid the formation of misfit dislocations. In that case the strain is accommodated elastically and/or plastically. The detection of microwave induced changes in the photoluminescence, optically detected impact ionisation (ODII), gives the possibility to extract more information on the nature of the recombination than standard PL alone. This will be along with absorption and PC studies on strained $Ga_xIn_{1-x}As/InP$ quantum wells the subject of this paper.

2. EXPERIMENTAL

The $Ga_xIn_{1-x}As$ quantum well structures were grown in a low pressure MOVPE reactor with a growth temperature of 640^0 C, details of the apparatus and the growing sequence can be found in Meyer et al 1991a. PL was excited with a Kr^+ ion laser, the emitted light was detected using a single monochromator (SPEX) and a LN_2 cooled Ge detector (North Coast). The ODII experiments were performed in a 12 GHz microwave cavity, where changes induced by the chopped microwaves on the PL intensity were studied. Low microwave powers and high chopping frequencies were applied to avoid bolometric effects. The sample temperature was usually 1.9 K.

3. EXPERIMENTAL RESULTS AND DISCUSSION

In Fig.1 the PL spectrum of a 3 nm QW with Gallium content of x=0.48 is shown. At 944 and 934 meV excitonic recombination is observed, the two peaks are due to monolayer fluctuations, followed by a broad band at 888 meV. Whether it is due to excitonic transitions cannot be decided from PL alone. The advantage of ODII is that different radiative contributions give different signs in the ODII spectrum. It bases on the fact that free carriers are accelerated in the microwave electrical field present in the resonator. The gain in energy is used to impact ionise bound excitons and neutral donors. Excitons and free electrons are split off, respectively (Wang et al 1989, Meyer et al 1991b). In the PL this is reflected in an increase of free exciton recombination and free to bound transition and a decrease in the bound exciton and donor-acceptor recombination. Excitonic transitions due to monolayer fluctuations are usually observed with the same sign. This behaviour is apparent from the ODII spectrum in Fig.1.b. Whereas the second excitonic transition is only observable as a shoulder at lower energies (936 meV) the increase at lower energies shows the involvement of a free to bound transition. The acceptor binding energy with respect to low energy excitonic transition is of the order of 45 meV. It is attributed to Zn, which is the usual contaminant in the MOVPE reactor due to growth of laser structures.

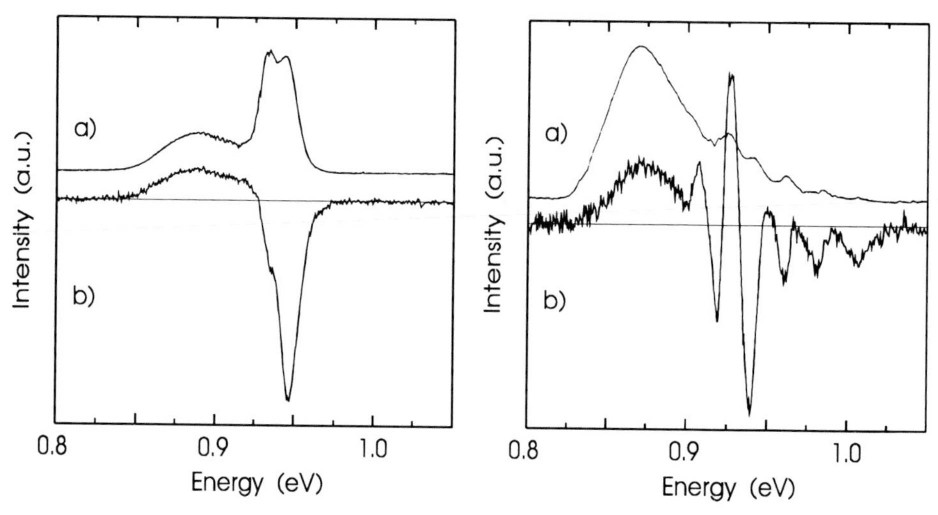

Fig.1: PL (a) and ODII (b) for a 3 nm QW in $Ga_{0.48}In_{0.52}As/InP$

Fig.2: PL (a) and ODII (b) for a 3 nm QW in $Ga_{0.11}In_{0.89}As/InP$

Changing the Gallium content causes quite dramatic changes in the PL spectrum (see Fig.2). For x=0.11 the dominating excitonic lines due to monolayer fluctuations of 7 and 8 monolayers (see below) can hardly be inferred from the PL spectrum. ODII now resolves 4 lines with negative sign. They might be of excitonic origin, however, hot electron luminescence cannot be ruled out (see discussion below). The first line starts at 1106 meV, they increase in intensity and the separation has a mean energy interval of 44 ± 2 meV. The most intense lines are centered at 918 and 956 meV and could be attributed to the H_1E_1 transitions originating from QWs having 6 and 7 monolayers. The strong and broad positive signal at 828 meV could be attributed to a free to bound transition with an approximate acceptor binding energy of 70 ± 10 meV. It is, however, known that the binding energy of the shallow acceptors decreases towards low Ga content (Goetz et al 1983) making this interpretation less likely. It can also be due to extended defects, if relaxation has already taken place.

For low Gallium content the critical layer thickness can be in the range of the QW width, i.e. 3 to 4 nm (People et al 1985). Excitonic recombination above the main QW recombination is an unusual fact. It appears, however, only in the x=0.11 (and partly in the x=0.18) samples, the separation is not a monotonic function of QW thickness. For d=1.5 nm we find 30 ± 1 meV, for d= 3 nm it is 44 ± 2 meV and for d=5 nm we observe 21 ± 1 meV (Meyer et al 1991b). Hot electron luminescence might cause these oscillations, where the spacings are given by the phonon energies of the LO-phonons. For the 3 nm wide well this is quite reasonable taking into account that the wave function of the first subband considerably leaks into the barrier and the barrier bulk LO-phonon energy (InP) is 43 meV. For wider wells (>5 nm) it should decrease to the InGaAs bulk value of 31 meV not far away from the experimental value of 21 meV. The 1.5 nm well shows spacings of 30 meV and at first sight outside the range of considerations expecting the largest value for the smallest well. However in very thin QWs the bulk phonon energies are not appropriate, apart from not considering polaron effects in 3 or 2 dimensions, the role of surface phonons are not taken into account. Experimentally for x=0.17 Raman measurements revealed phonon lines at 30 and 32 meV in agreement with the PL and ODII experiments. Whereas we feel that the results can partly be explained by the recombination of still hot carriers, the very low value of 30 meV for d=1.5 nm still needs more theoretical support.

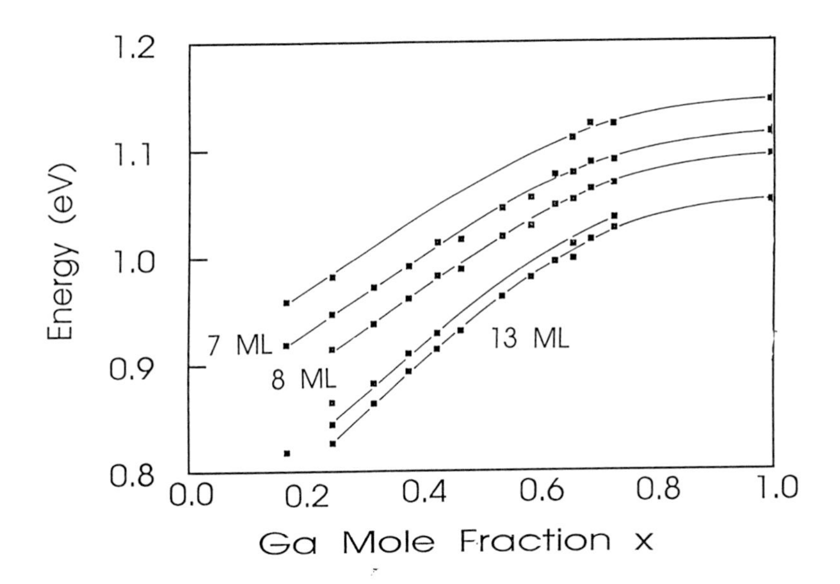

Fig. 3: Energy position of the E_1H_1 transition as obtained by PL as a function of the QW thickness and of the Gallium content

The PL measurements presented in Fig.4 showed how the composition changes affect the excitonic transition energies. The PL peak positions in the QWs with 1.5 and 3 nm width are assigned to the number of monolayers (ML) showing a width variation of one monolayer (i.e. 7 and 8 MLs for the 1.5 nm wells). It is worth noting that by increasing the Gallium content above x>0.8 still strong excitonic luminescence at 1.1. eV is observed. By changing the growth conditions (Meyer et al 1991a) we can rule out interface effects as origin and we therefore conclude in contrast to Gershoni et al 1988 that transitions from type I to type II superlattice do not occur, in line with Wang et al 1990.

The PL experiments are supported by transmission measurements. For x<0.47 and d=1.5 nm they are shown in Fig.4. They indicate,however, in contrast to PL, monolayer fluctuations of 6 and 7 monolayers only, when we calculated and attributed the H_1E_1 and L_1E_1 transitions correspondingly. A reason for that is, that due to exciton diffusion in PL normally recombination within a 8 ML QW is stronger compared to the 7 ML QW.

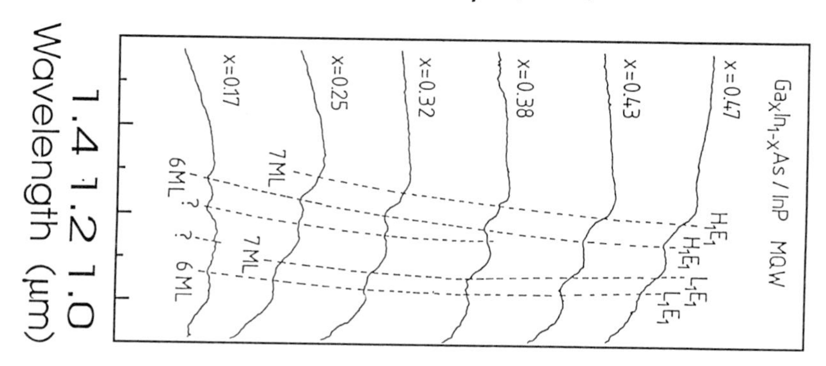

Fig.4 : Transmission spectra for 1.5 nm thick $Ga_xIn_{1-x}As$/InP Quantum Wells for different Gallium content (T=10 K)

4. CONCLUSIONS

We demonstrate the advantage of optically detected impact ionisation to resolve additonal features in the luminescence spectrum. Hot electron emission might be the origin of the multiple line spectra in the thin QWs with low Gallium content.

References

Gershoni D, Temkin H, Vandenberg J M, Chu S N G, Hamm R A and Panish M B 1988 Phys. Rev. Lett. **60** 448
Götz K H, Bimberg D, Jürgensen H, Selders J, Solomonov A V, Glinskii G F and Razeghi 1983 J. Appl. Phys. **54** 4543
Grundmann M, Lienert U, Christen J, Bimberg D, Fischer-Colbrie A and Miller J N 1990 J. Vac. Sci. Technol. **B8** 751
Hardtdegen H, Meyer R, Loken-Larsen H, Appenzeller J, Schäpers T and Lüth H 1991 to be published
Meyer R, Hardtdegen H, Carius R, Grützmacher D, Stollenwerk M, Balk P, Kux A and Meyer B K 1991a J. Electr. Mat. to be publ.
Meyer B K, Wetzel C, Grützmacher D and Ömling P 1991b Mater. Sci. Engin. **B9** 293
Reihlen E H, Birkedal D, Wang T Y and Sringfellow G B 1990 J. Appl. Phys. **68** 1750
Reithmaier J P, Cerva H and Lösch R 1989 J. Appl. Phys. **54** 46
Wang F P, Monemar B and Ahlström M 1989 Phys. Rev. **B39** 11195
Wang T Y and Stringfellow G B 1990 J. Appl. Phys. **67** 34

Inst. Phys. Conf. Ser. No 123
Paper presented at the International Meeting on Optics of Excitons in Confined Systems, Giardini Naxos, Italy, 1991

Exciton mixings in thin (Ga,In)As-AlAs strained-layer quantum wells

Bernard GIL[+], Karen J.MOORE[++], and Yong CHEN[+++]

+ Groupe d'Etudes des Semiconducteurs, Université de Montpellier II
Case Courrier 074, 34095 Montpellier Cedex 5, France.
++ Department of Mathematics and Physics, Manchester Polytechnic,
All saints, Manchester M15 6BH, England.
+++ Laboratoire de Microstructures et Microélectronique,
196 Avenue H.Ravera, 92260 Bagneux Cedex, France.

ABSTRACT: Model calculations of varying complexity are used to describe the band structure of semiconductor quantum wells and superlattices. However, the physics of the valence band usually neglects coupling with the Γ_7 split-off band. We report the first spectroscopic demonstration of the crucial influence of the light-hole and split-off hole valence band coupling on the energy of the light-hole exciton in (GaIn)As-AlAs strained-layer multiple quantum wells. In addition, we demonstrate the effects of including valence band coupling in calculations of the light-hole exciton binding energy.

1-INTRODUCTION

Model representations of increasing complexity have been used to describe excitons confined in quantum wells (Bastard et al. 1982, Grundmann and Bimberg 1988, Bauer and Ando 1988, Andreani and Pasquarello 1990). Most of them treat the $k_\perp = 0$ case. Their predictions essentially depend of the trial function used to minimize the exciton binding energy . Inclusion of valence band mixings as well as other important effects such as interband Coulomb couplings or difference in dielectric constants leads to complex calculations. The net result is a substantial modification of the Rydberg energies. These models restrict the valence band states to the upper valence band and do not include contribution of the spin-orbit split-off states. Our objectives are first to show that the deep valence band states may strongly change the light-hole confinement energies in quantum structures with valence potential of depth comparable to the spin-orbit coupling, and second to estimate how big the influence of these valence band states can be on the light-hole exciton energy . For the sake of the simplicity, we shall limit ourselves to $\mathbf{k}_\perp = 0$ valence band states. The electron hole exchange interaction which remains small although strongly enhanced with respect to the bulk case (Chen et al. 1988, Andreani and Bassani 1990, Rössler et al. 1990) will not be calculated.

2-EXCITON HAMILTONIAN AT $k_\perp = 0$.

The conduction and valence Bloch waves are noted $|\phi_e\rangle$, $|\phi_{hh}\rangle$, $|\phi_{lh}\rangle$, and $|\phi_{so}\rangle$ respectively. Then we write down the exciton hamiltonian as follows:

$$
H_{ij} = \begin{array}{c} \\ \\ \\ \end{array}
\begin{array}{ccc}
|\phi_e,\phi_{hh}\rangle & |\phi_e,\phi_{lh}\rangle & |\phi_e,\phi_{so}\rangle \\
\left|\begin{matrix} H^{hh}_{exc} + H^{hh}_{strain} \\ \\ 0 \\ \\ 0 \end{matrix}\right. &
\begin{matrix} 0 \\ \\ H^{lh}_{exc} + H^{lh}_{strain} \\ \\ H^{so-lh}_{exc} + H^{so-lh}_{strain} \end{matrix} &
\left.\begin{matrix} 0 \\ \\ H^{lh-so}_{exc} + H^{lh-so}_{strain} \\ \\ H^{so}_{exc} + H^{so}_{strain} \end{matrix}\right|
\end{array} \qquad (1)
$$

$$H^{hh}_{exc} = -\hbar^2/2\mu^\perp_+ \nabla^2_\rho - \hbar^2/2m_e \partial^2_{z_e} - \hbar^2/2m_{hh} \partial^2_{z_h} + V_{\Gamma_6}(z_e) + V_{\Gamma_8}(z_h) - e^2/\varepsilon r$$

$$H^{lh}_{exc} = -\hbar^2/2\mu^\perp_- \nabla^2_\rho - \hbar^2/2m_e \partial^2_{z_e} - \hbar^2/2m_{lh} \partial^2_{z_h} + V_{\Gamma_6}(z_e) + V_{\Gamma_8}(z_h) - e^2/\varepsilon r$$

$$H^{so}_{exc} = -\hbar^2/2\mu^\perp_{so} \nabla^2_\rho - \hbar^2/2m_e \partial^2_{z_e} - \hbar^2/2m_{so} \partial^2_{z_h} + V_{\Gamma_6}(z_e) + V_{\Gamma_7}(z_h) - e^2/\varepsilon r \ldots$$
$$\ldots + \Delta_o(z_h)$$

$$H^{lh-so}_{exc} = 2\sqrt{2}\, \hbar^2 \gamma_2 \partial^2_{z_h}$$

$$H^{hh}_{strain} = a\,(e_{xx} + e_{yy} + e_{zz}) + b\,(e_{zz} - e_{xx})$$

$$H^{lh}_{strain} = a\,(e_{xx} + e_{yy} + e_{zz}) - b\,(e_{zz} - e_{xx})$$

$H^{so}_{strain} = a'\,(e_{xx} + e_{yy} + e_{zz})$,and $H^{lh-so}_{strain} = \sqrt{2}b'\,(e_{zz} - e_{xx})$

Where a, b, a', b' are orbital and spin-dependent deformation potentials which where measured for GaAs by Chandrasekhar and Pollak (1977). The e_{ij} quantities represent the component of the built-in strain (if any).

$1/\mu^\perp_\pm = 1/m_e + 1/m^\perp_{hh}$; $\quad 1/\mu^\perp_{so} = 1/m_e + \gamma_1$; $\quad 1/m^\perp_{hh \atop lh} = \gamma_1 \pm \gamma_2$

$1/m_{hh} = \gamma_1 - 2\gamma_2$; $\quad 1/m_{lh} = \gamma_1 + 2\gamma_2$; $\quad 1/m_{so} = \gamma_1$

3-INFLUENCE OF THE Γ_8-Γ_7 MIXING ON THE LIGHT-HOLE ENERGIES

We have measured the experimental values of the splitting between the light-hole exciton and the heavy-hole exciton on a series of $Ga_{0.96}In_{0.04}As$-AlAs multiple quantum wells. Figure 1 represents the reflectance data taken at 2 K for these samples. The valence confinement energies have been calculated following the procedure detailed in the work of Gil et al.(1991).The dashed line on figure 2 represents the calculated splitting without taking into account the coupling between the light-hole and split-off hole wave functions whilst the full line represents the result of the two-band calculation which is obviously required to improve the agreement between the experiment (full ovals) and the band to band calculation.

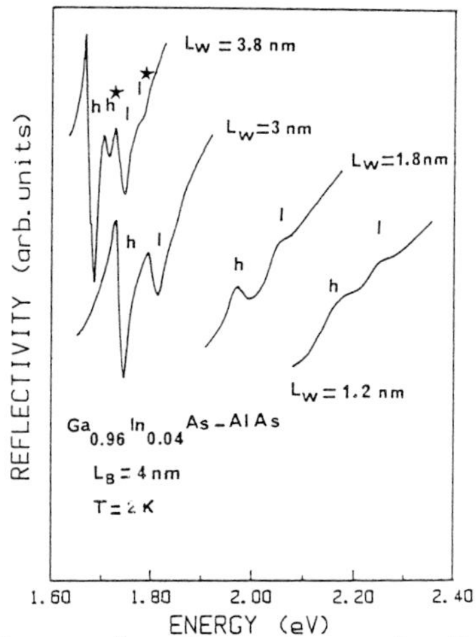

Figure 1: Reflectance spectra at 2K of $Ga_{0.96}In_{0.04}As$-AlAs superlattices with GaInAs well widths L_w, and AlAs barrier width L_B. h and l label the e_1hh_1 and e_1lh_1 excitonic features and h^* and l^* the corresponding hot excitons.

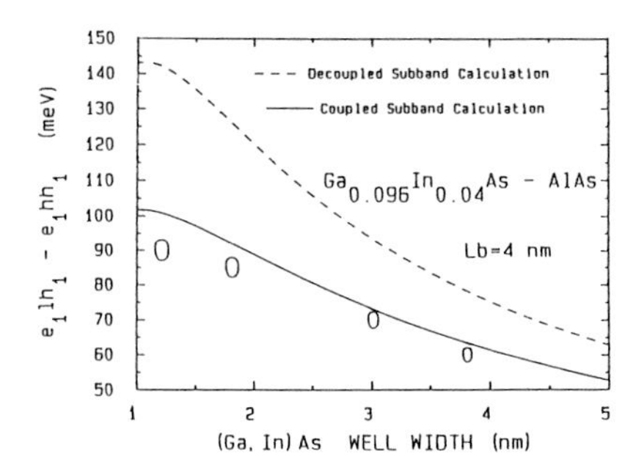

Figure 2: Comparison of the measured (solid ovals) energy difference between elhhl and ellhl and calculations including (solid line) and neglecting (dashed line) valence band coupling.

4-EXCITON BINDING ENERGIES

The coupling between the light-hole and the split-off valence band also modifies the exciton binding energy. To estimate the coupling-induced correction to the light-hole exciton binding energy, we consider a perturbative approximation and use the 2x2 block in the hamiltonian (1) . The two-component exciton envelope function can be written as:

$$\Psi_{exc} = \begin{vmatrix} \alpha\Phi_\alpha \\ \beta\Phi_\beta \end{vmatrix} = \begin{vmatrix} \alpha\phi(r)\kappa_e(z_e)\kappa_{lh}(z_h) \\ \beta\phi(r)\kappa_e(z_e)\kappa_{so}(z_h) \end{vmatrix}$$

where $\phi(r)$ is a quasi- 2D envelope function, $\kappa_e(z_e)$, $\kappa_{lh}(z_h)$, and $\kappa_{so}(z_h)$ are solutions of the square well problem for the electron and the holes respectively.
The exciton energies can be obtained by solving the following determinantal equation:

$$\det \begin{vmatrix} \langle\Phi_\alpha|H_{22} - E_{exc}|\Phi_\alpha\rangle & \langle\Phi_\alpha|H_{23}|\Phi_\beta\rangle \\ \langle\Phi_\beta|H_{32}|\Phi_\alpha\rangle & \langle\Phi_\beta|H_{33} - E_{exc}|\Phi_\beta\rangle \end{vmatrix} = 0$$

For the sake of simplicity, we shall use infinite potential barriers for the carrier confinement and drop all strain-dependent terms. We shall follow a variational approach and write the quasi-2D envelope function:

$$\phi(r) = \sqrt{2}/\sqrt{\pi}\ \lambda\ e^{-\lambda\rho}$$

where λ is a variational parameter which can be fixed by minimizing the energy of the system. In figure 3, we show the variation of the light-hole exciton binding energy as a function of the well width with (solid line) and without (dashed line) taking into account the valence band coupling. The heavy-hole exciton has been also represented. No significant difference between the two curves is observed for a large range of well-width. As the well width decreases, the valence band coupling-induced modification becomes more important, reflecting a significant mixing of the light-hole and split-off band exciton wave functions. Although our model is very simplified, the estimation gives a reasonable trend for the valence band coupling-induced corrections to the light-hole exciton binding energies.

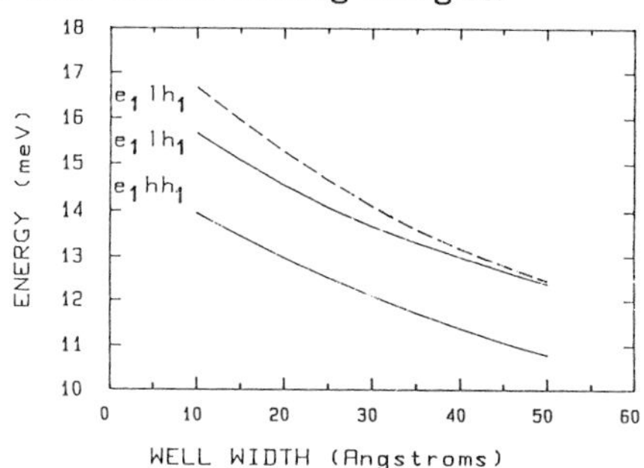

Figure 3: Variation of the heavy-hole exciton and light-hole exciton binding energies as a function of well width including (solid line) and neglecting (dashed line) valence band coupling.

5-REFERENCES

Andreani L.C. and Bassani F. 1990 Phys.Rev.B, **41**, 7536
Andreani L.C., and Pasquarello A.1990 Phys.Rev.B, **42**, 8928
Bastard G., Mendez E.E., Chang L.L., and Esaki L.1982 Phys.Rev.B, **26**,1974
Bauer G.E.W., and Ando T. 1988 Phys.Rev.B,**38**,6015,(1988)
Chandrasekhar M., and Pollak F.H. 1977 Phys.Rev.B, **15**, 2127
Chen Y., Gil B., Lefebvre P., and Mathieu H. 1988 Phys Rev.B, **37**, 6429
Gil B., Lefebvre P., Boring P., Moore K.J., Duggan G., and Woodbridge K. 1991 Phys.Rev.B **44**, 1942
Grundmann M., and Bimberg D. 1988 Phys.Rev.B, **38**, 13486
Rössler U., Jorda S., and Broido D. 1990 Solid State commun. **73**, 209

Inst. Phys. Conf. Ser. No 123
Paper presented at the International Meeting on Optics of Excitons in Confined Systems, Giardini Naxos, Italy, 1991

Spin-polarized excitons in pseudomorphic, strained $In_{0.16}Ga_{0.84}As/GaAs$ and $In_{0.16}Ga_{0.84}As/Al_{0.29}Ga_{0.71}As$ quantum wells on GaAs

M. Kunzer, G. Hendorfer, U. Kaufmann, K. Köhler, and W.W. Rühle*

Fraunhofer-Institut für Angewandte Festkörperphysik,
Tullastrasse 72, 7800 Freiburg, Germany

* Max-Planck-Institut für Festkörperforschung
Heisenbergstrasse 1, 7000 Stuttgart 80, Germany

ABSTRACT: Optical pumping techniques and time -resolved photoluminescence have been used to study spin and exciton lifetimes in two pseudomorphic quantum well systems as a function of well width. From excitation spectroscopy with circularly polarized light, band offsets for the $In_{0.16}Ga_{0.84}As/Al_{0.29}Ga_{0.71}As$ system are deduced.

1. INTRODUCTION

Pseudomorphic epitaxial layers grown on a bulk substrate have attracted great interest since the lattice -mismatch induced strain in these layers can be utilized as an additional parameter for tailoring the electrical and opto-electronic properties of quantum well (QW) devices [1,2]. In this work we have studied the two pseudomorphic QW systems $In_{0.16}Ga_{0.84}As/GaAs$ and $In_{0.16}Ga_{0.84}As/Al_{0.29}Ga_{0.71}As$ grown on SI-GaAs. Using optical pumping techniques [3,4] and time-resolved photoluminescence (PL) we have determined electron spin-flip times τ_s (spin lifetimes) and optical decay times τ (exciton lifetimes) of photoexcited carriers as a function of well width. We also briefly discuss results of PL excitation with circularly polarized (cp) light from which band offsets for $In_{0.16}Ga_{0.84}As/Al_{0.29}Ga_{0.71}As$ are derived.

2. EXPERIMENTAL DETAILS

The magneto-optical setup used for the present steady state PL, PL excitation and Hanle depolarization [3,4] experiments has been described previously [5]. Excitation is achieved with cp light from a tunable Ti-sapphire laser. The degree of circular polarization of the excitonic QW luminescence can be analyzed with a photoelastic modulator. The luminescence decay measurements were performed on a system [6] which uses ≈ 5 ps dye laser pulses for excitation ($h\nu$ = 1.49 eV) and a streak camera for temporal dispersion (time resolution ≈ 20 ps).

The two QW structures investigated have been grown by molecular-beam-epitaxy on (001) oriented SI-GaAs substrates. Each structure contains five undoped $In_{0.16}Ga_{0.84}As$ QWs with widths of 20, 50, 100,

130 and 160 Å as schematically shown in the insert of Fig. 1b. The wells are separated from each other by 300 Å wide undoped barriers which consist of GaAs for structure 1, and of $Al_{0.29}Ga_{0.71}As$ for structure 2. The QWs are separated from the substrate by a GaAs/AlGaAs superlattice buffer and are protected by a GaAs capping layer. Independent studies have shown [7] that the wells investigated here are pseudomorphic but begin to relax for widths above 160 Å.

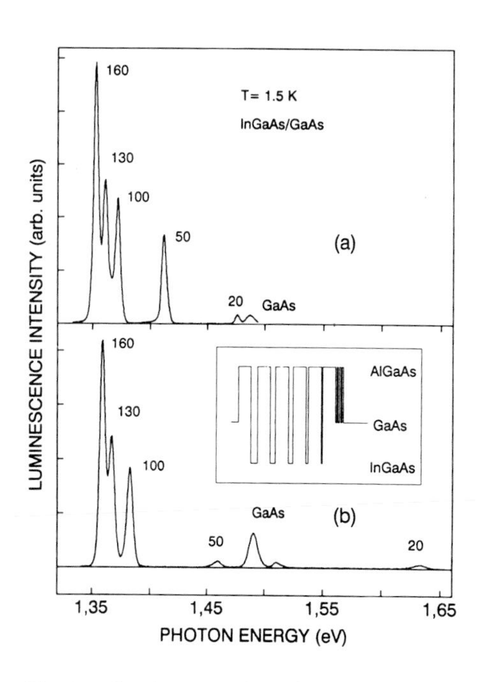

Fig. 1 Luminescence intensity versus photon energy for InGaAs/GaAs (a) and InGaAs/AlGaAs (b) QWs. The numbers associated with the peaks correspond to QW widths in Å. Excitation power density ≈ 1 W/cm^2.

3. BACKGROUND

Optical pumping, i.e. excitation of PL with cp light, offers the advantage that either the total luminescence $I^+ + I^-$ or its degree of circular polarization $P = (I^+ - I^-)/(I^+ + I^-)$ can be measured. Here I^+ and I^- are intensities of right and left cp light. The polarization technique works particularly well for QWs where the uppermost valence band at the Γ point is split by confinement and strain. In the present case, the heavy hole (hh) band lies above the light hole (lh) band and it is possible to excite electrons selectively from the hh band to the conduction band. Due to angular momentum selection rules [3,4] the spin of the free carries thus created are oriented (spin-polarized) along the excitation direction (parallel to the observation direction). If this carrier polarization does not decay totally via spin-flip processes during the exciton lifetime, the resulting excitonic recombination radiation is circularly polarized too and P contains information about spin- and exciton lifetimes. Another advantage of optical pumping concerns PL excitation spectroscopy. When monitoring P of the fundamental excitonic QW recombination as a function of excitation energy, hh and lh excitons are readily distinguished by the sign of P [2].

4. RESULTS AND DISCUSSION

Total luminescence spectra of the InGaAs/GaAs and the InGaAs/AlGaAs QW structure are shown in Fig. 1a and Fig. 1b, respectively. All peaks labelled with numbers are due to the fundamental exciton transitions where an electron in the lowest InGaAs conduction subband recombines with a hole in the highest InGaAs hh subband. Confinement effects are clearly visible and the stronger blue shift of the exciton peaks of InGaAs/AlGaAs is understood in terms of the increased barrier heights for the confined carriers.

PL excitation spectra under optical pumping conditions were obtained for all InGaAs/AlGaAs wells of structure 2 by monitoring P of the individual peaks in Fig. 1b while scanning the laser excitation energy.

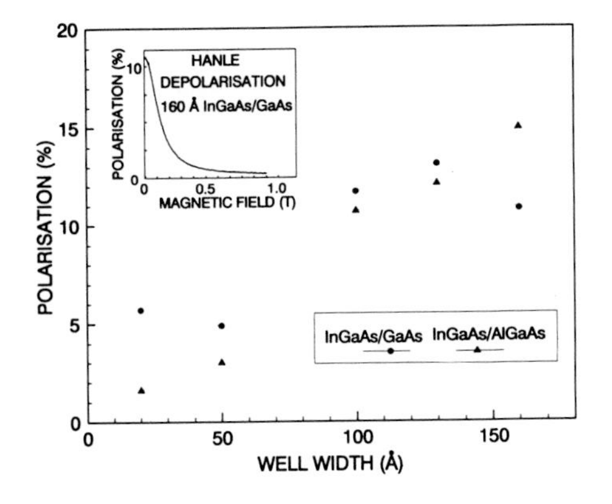

Fig. 2 Degree of circular polarization P of the excitonic QW luminescence peaks using cp light for excitation. The relative error in P is 15%. The insert shows a Hanle depolarization curve as a function of a transverse magnetic field.

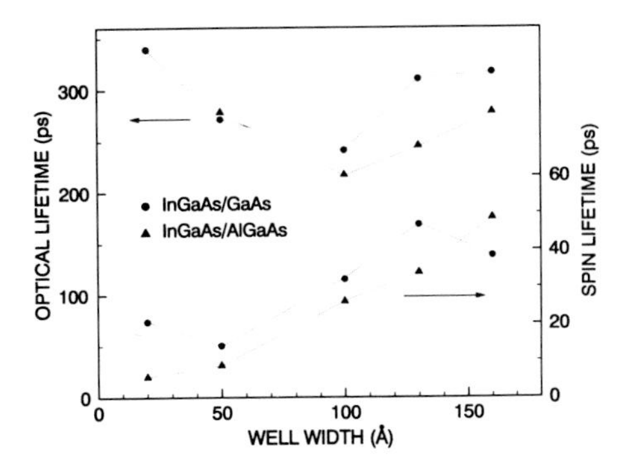

Fig. 3 Optical lifetime τ (upper traces) and spin lifetime τ_s (lower traces) versus well width at liquid He temperatures.

Heavy hole and light hole exciton features can be clearly distinguished and a reliable subband assignment of higher lying exciton peaks is possible. A square well model, described in detail elsewhere [5], was fitted to these data and a reliable value for the conduction band offset Q_c of $In_{0.16}Ga_{0.84}As/Al_{0.29}Ga_{0.71}As$ could be deduced, $Q_c = 0.70 \pm 0.05$. This is the first optically determined band offset for this heterostructure system. For comparison, the Q_c values for InGaAs/GaAs reported in the literature [9] scatter between 0.4 and 0.8. A recently reported value [10] of $Q_c \approx 0.60$ for In contents between 7% and 20% appears most reliable.

In measuring absolute values of P for the excitonic QW peaks in Fig. 1 we have used suitable $\lambda/4$ platelets and have chosen the excitation energies only slightly higher than the luminescence peaks in order to avoid lh exciton excitation. In addition it is difficult to suppress effectively diffuse scattered laser light (including polarized luminescence from the Ti sapphire laser) and this results in an undesired background polarization. We have separated the real excitonic polarization from this background by the Hanle depolarization technique [3,4]. In this method a transverse magnetic field B is applied to the sample. The field destroys the quantization axis defined by the excitation beam and the spins start to precess around the field directions thus leading to a reduction of the zero field P value. The theoretical shape of such a depolarization curve as a function of B is a Lorentzian. In principle g-values can be inferred from such curves [11] but this is beyond the scope of this paper. Here we have simply used the depolarization curves to obtain accurate P values which are plotted in Fig. 2 versus well width. Note that P decreases with decreasing width, the only deviation from this trend being one 160 Å well. We suspect that this is an indication for beginning, partial strain relaxation. The insert in Fig. 2 is an example of a Hanle curve. A line-shape analysis shows that it is a perfect Lorentzian.

In the upper part of Fig. 3 we have plotted the exciton lifetimes τ obtained from decay curves of the exciton peaks in Fig. 1 following picosecond pulse excitation. The relative error in τ is about 15%. There appears to be a minimum in τ for well widths around 100 Å

With the above τ values one can now evaluate the spin lifetimes τ_s since, *in the present case*, these two quantities are related to P by the equation

$$P = (1 + \tau/\tau_s)^{-1}$$

The τ_s values thus obtained are plotted in the lower part of Fig. 3. The relative error is estimated to be about 20%. Two significant properties of τ_s are evident from that plot. Firstly, a strong decrease of τ_s with increasing confinement for both QW structures and secondly, smaller values for structure 2. Again one 160 Å well behaves exceptional indicating once more that it is partially relaxed.

We suggest the following arguments to explain our results. i) The order of magnitude of τ_s (10 – 50 ps) indicates strong excitonic exchange coupling of the photogenerated carriers. ii) The decrease of τ_s with decreasing quantum well width can be understood in terms of the kinetic energy of the carriers. Stronger confinement implies a higher kinetic energy and as a consequence a higher rate of reflections of the carriers at the interface. Each reflection at the interface exhibits a certain probability for inducing a spin flip process due interactions of the carrier spin with paramagnetic defects in the interface region. Therefore the longer spin lifetimes in structure 1 indicate a better interface quality for InGaAs/GaAs as compared to InGaAs/AlGaAs. In the latter case a contribution of barrier alloy scattering to the spin flip rate may be an additional reason for the strongly reduced spin lifetime in the thinner wells.

In summary, we have performed optical and magneto-optical investigations on the pseudomorphic systems InGaAs/GaAs and InGaAs/AlGaAs. For the latter we have found a conduction band offset of 70% which exceeds that of the former by about 10%. The two systems are similar to each other as far as the optical decay times and their dependence on confinement is concerned. However, differences are found with respect to the polarization signals as well as the corresponding spin lifetimes and their confinement dependence. We believe that the interface quality is of major importance for these effects.

We thank P. Ganser for assistance in MBE growth, K. Sambeth for expert technical support, and J. Schneider for useful discussions.

5. REFERENCES

1. T.P. Pearsall in "Semiconductors and Semimetals" Vol. 32, Eds.: R.K. Willardson and A.C. Beer (Academic, New York 1990) p 1
2. G.C. Osbourn, P.L. Gourdley, I.J. Fritz, R.M. Biefeld, L.R. Dawson and T.E. Zipperian in "Semiconductors and Semimetals", Vol.24, Eds.: R.K. Willardson and A.C. Beer (Academic, New York 1987) p 459
3. see various chapters in "Optical Orientation" Eds.: F. Meier and B.P. Zakharchenya (North-Holland, Amsterdam 1984)
4. C. Hermann and G. Lampel and V.I. Safarov, Ann. Phys. Fr. **10**, 1117 (1985)
5. M. Kunzer, G. Hendorfer, U. Kaufmann and K. Köhler, submitted to Phys. Rev. B
6. M. Nido, M.G.W. Alexander, W.W. Rühle and K. Köhler, SPIE Proc. Vol. 1268, 177 (1990)
7. T. Schweizer, K. Köhler, P. Ganser, A. Hülsmann and P. Tasker, Appl. Phys. A **53**, 109 (1991)
8. C. Weisbuch, R.C. Miller, R. Dingle, A.C. Gossard and W. Wiegmann, Solid State Commun. **37**, 219 (1981)
9. M.J. Joyce, M.J. Johnson, M. Gal and B.F. Usher, Phys. Rev. B **38**, 10978 (1988)
10. J.-P. Reithmaier, R. Höger, H. Riechert, P. Hiergeist and G. Abstreiter, Appl. Phys. Lett. **57**, 957 (1990)
11. M.J. Snelling, G.P. Flinn, A.S. Plaut, R.T. Harley, A.C. Tropper, R. Eccleston and C.C. Phillips, submitted to Phys. Rev. B

Inst. Phys. Conf. Ser. No 123
Paper presented at the International Meeting on Optics of Excitons in Confined Systems, Giardini Naxos, Italy, 1991

One-dimensional excitons and centre-of-mass quantization of excitons in quantum well wires

D. Heitmann, H. Lage, M. Kohl, R. Cingolani*, P. Grambow, and K. Ploog**
Max-Planck-Institut für Festkörperforschung
Heisenbergstr. 1, 7000 Stuttgart 80, Germany

ABSTRACT: We review excitonic excitations in quantum well wire structures which were studied by photoluminescence spectroscopy of laterally microstructured $Al_xGa_{1-x}As$ – GaAs quantum well systems. It is found that for very narrow wires the concept of quasi one-dimensional excitons is an appropriate model to describe the behavior of the observed transitions. The 1D character of these excitons is manifested in the observation of several exciton resonances which are related to different energetically separated 1D quantum confined subbands, by a reduced diamagnetic shift and by an increased binding energy with respect to the 2D reference sample. For wider quantum well wire structures one also observes an up to threefold splitting of the lowest heavy hole exciton resonance. Here the splitting arises from a quantization of the excitonic center-of-mass motion perpendicular to the direction of the wires. Identifying the exciton wavevector K_{ex}^n with $n\frac{\pi}{L_x}$, where L_x is the active wire width, one can reconstruct the exciton dispersion.

1. INTRODUCTION

The recent progress in crystal growth techniques has made it possible to fabricate layered semiconductor heterostructures, quantum wells and superlattices with atomically flat interfaces. These novel systems have unique physical properties which arise from the quasi-two-dimensional (2D) behavior and, in superlattices, from coupling between these 2D layers. Optical spectroscopy, in particular photoluminescence (PL), is a powerful tool to study these systems. The PL spectra of quantum-well (QW) systems are governed by efficient intrinsic radiation of free excitons and exhibit a superior optical performance (Ref.[1], for a recent review see e.g. Ref.[2]). One of the challenging topics of current interest involves systems of further reduced dimensionality, namely 1D quantum wires and 0D quantum dots. Besides the expectation of novel physical phenomena it is hoped that the optical properties, i.e. oscillator strengths, nonlinear coefficients and others can be further improved with respect to the 2D systems to realize high performance optical devices. 1D and 0D systems have been prepared by starting from a 2D-layered system and employing modern micro and nanometer lithographic techniques[3–21] or by using novel growth techniques.[22–24] An excellent review has recently been given by Kash[25] where these different preparation techniques are described extensively. There is also an enormous amount of theoretical work on excitons in low-dimensional systems. We will refer here to some recent work which will be directly addressed in this review[26–35] and further references therein and to other contributions of this volume, in particular to the reviews by G.W. Bryant and S.W. Koch.

In this paper we will discuss experiments on AlGaAs-GaAs quantum well wire systems which have been prepared by deep-mesa-etching techniques.[10,11,13,14,19] We will demonstrate that with decreasing wire width the excitonic excitation spectrum exhibits a gradual change. For wide wires one observes a center-of-mass quantization of the excitons. For very narrow wire widths the resonances show the behavior of 1D excitons, i.e. excitons related to energetically separated 1D quantum confined subbands.

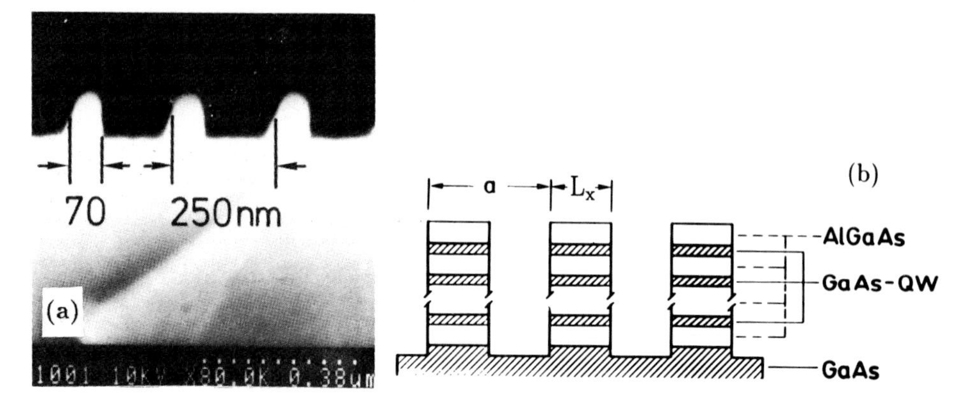

Fig. 1: (a) Scanning electron micrograph of the profile of a QW wire sample. The grating period is 250 nm, the wire width 70 nm. (b) Schematic cross-section of a deep-mesa-etched QW wire structure.

2. SAMPLE PREPARATION AND EXPERIMENTAL TECHNIQUES

The samples were grown by molecular beam epitaxy (MBE) at 610°C on semi-insulating (001) GaAs substrates. They consisted typically of multi-layered identical GaAs QW with well widths $L_z = 10 - 14$ nm separated by $L_B = 10 - 14$ nm thick $Al_xGa_{1-x}As$ barriers. The Al-concentration of the different samples was kept in the range $0.26 < x < 0.37$. A mask consisting of periodic photoresist lines was prepared by holographic lithography. The photoresist was exposed to the superposition of two expanded coherent laser beams (wavelength $\lambda = 458$ nm) which gives rise to a sinusoidal intensity modulation and, after the development, to a spatially modulated photoresist profile. The width of the photoresist lines was adjusted by the exposure and development time of the resist. With subsequent reactive ion etching in a $SiCl_4$ plasma nearly rectangular grooves were etched through all wells into the samples. The resulting geometrical wire widths, L_x, ranged down to 60 nm, which was determined by a scanning electron microscope. On each QW wire sample a small part was left unpatterned to take reference spectra. A scanning electron micrograph of an etched quantum well wire sample is shown in Fig. 1a, a sketch of the system in Fig. 1b. The important point of all the preparation techniques is to optimise all process parameters in such a way that even at very small wire widths there is still strong enough PL and that not all radiative transitions are quenched due to nonradiative processes. More details of our preparation process are given in Refs.[36,37].

The PL was excited with normally incident light. PL excitation (PLE) measurements were performed with a Styryl 9 - dye laser or a Titanium-Sapphire laser, which were pumped by an Ar^+-laser. The luminescence light was collected normally to the sample in a solid angle of 30° opening and analyzed with a 1 m monochromator and photon-counting techniques. The spectral resolution was set to 0.03-0.05 nm and the sample temperature was 4-8K if not otherwise indicated. For magneto-optic measurements the sample was placed in a superconducting split-coil magnet with windows or in an axial magnet which was coupled via fibers to the optical set up.

3. QUASI-ONE-DIMENSIONAL EXCITONS

Polarization-dependent PLE spectra obtained from a QW wire sample with $L_x = 70$ nm (etched from a structure with three QW with L_z=14 nm and L_B=10 nm) are shown in Fig. 2. The transitions in the reference spectrum are the e_1-hh_1 (heavy hole) transitions (808.525 nm, 808.755 nm) and the e_1-lh_1 (light hole) transitions (804.975 nm, 805.225 nm). From calculations we can ascribe the small splitting of these transitions to QW-thickness fluctuations of one monolayer.[38] The PLE spectrum of the QW wire sample exhibits a different behaviour, which

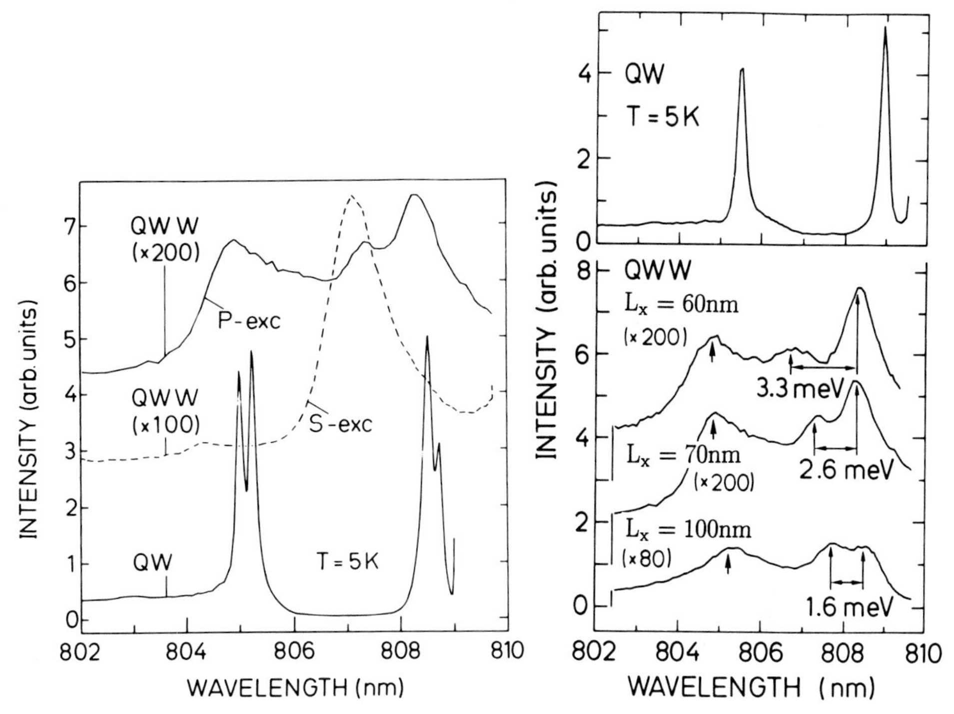

Fig. 2 (left): PLE spectra of a QW wire (QWW) with $L_x \approx 70$ nm and of the corresponding reference QW. The spectra are shifted vertically with respect to each other for clarity and show the dependence on the polarization of the exciting laser light. For s-polarized (p-polarized) light $\vec{E}$ is parallel (perpendicular) to the wires. The short (long) wavelength resonance of the reference QW represents a light (heavy) hole exciton excitation. The heavy hole excitons of the QW wire samples is split into two resonances which represent 1D excitons, i.e. excitons related to energetically separated quantum confined 1D subbands. (From Ref.[13]). Fig. 3 (right): PLE spectra of QW wire samples with three different geometrical wire widths L_x as indicated. The reference spectrum in the upper part is taken at a spot located directly near the $L_x = 100$ nm QWW structure on the wafer. The spectra are shifted vertically with respect to each other for clarity and show an increasing splitting of the 1D heavy hole excitons with decreasing wire widths L_x. (From Ref.[18])

is in particular strongly dependent on the polarization of the exciting laser light. Let us first concentrate on the p-polarized spectrum. Three transitions can be resolved. Two transitions occur in the wavelength regime of the excitonic hh- and lh-ground-states of the reference sample. Note the shift of these transitions of about 0.2 nm (0.4 meV) to smaller wavelengths with respect to the corresponding reference transitions. A third additional transition can be observed, which is 1.3 nm away on the smaller wavelength side of the hh-ground-state at about 807 nm.

We will demonstrate that these transitions represent quasi 1D exciton transitions. We have performed calculations to estimate the confinement energies using a 2D finite potential-well model. Taking the values for the exciton binding energies into account, which we deduce from magneto-optical experiments (see below), we can indeed ascribe the observed transitions to excitonic transitions related to quantum confined 1D subbands. In particular, the third new transition, which will be labeled hh_{12} in the following, is due to the second 1D subbands in the conduction- and valence-band. The indices of hh_{ij} denote the quantum numbers for the z- and

x-direction, respectively.

As another direct manifestation of the quantum confinement one expects that the energy separation between the hh_{11}- and hh_{12}-transition should increase with decreasing wire width. We indeed find this wire-width dependence if we compare in Fig. 3 the PLE spectra of QW wire samples with different values of L_x (L_x is here the geometrical wire width as determined with an SEM). The upper curve is the corresponding reference spectrum of a QW sample from the same wafer as presented in Fig. 2. The two peaks in the reference spectrum at 808.93 nm and 805.48 nm are the e_1-hh_1 and e_1-lh_1 transitions, respectively. Due to slight changes of the QW parameters across the wafer no splitting related to monolayer islands is observed in this case. All QW wire spectra exhibit three transitions. The two ground-state transitions, hh_{11} and lh_{11}, are blue-shifted with respect to the corresponding reference transitions. For the case of L_x = 100 nm, the third transition, hh_{12}, is only separated by 0.75 nm (1.5meV) from the hh_{11} transition. With decreasing geometrical wire width this separation increases to 2.6 meV for L_x = 70 nm and to 3.3 meV for L_x = 60 nm. In addition, also the energy separation between the hh_{11} and lh_{11} excitons increases. These effects are rather small, nevertheless they seem to indicate a systematic wire-width dependence of the confinement for the hh_{11} and lh_{11} states and of the binding energies of the corresponding excitons. The intrinsic nature of the hh_{12}-transition has further been confirmed by temperature-dependent PLE spectroscopy and a series of PLE measurements, where we set the detection wavelength to the different resonance positions. However, in all these measurements we were not able to resolve the lh_{12}-transition, which should be located for L_x = 70 nm at about 803 nm. The weakness of transitions between higher 1D electron- and lh-subbands has also been observed for InP/InGaAs QW wire samples.[8]

A very interesting observation for the QW wire samples is a characteristic polarization dependence. As can be seen in Fig. 2, the lh_{11} and hh_{11} transitions are dominant when we excite the QW wire with p-polarized laser light. However, the shape of the spectra drastically changes, when we switch to s-polarized excitation. In this case the spectrum is dominated by the hh_{12} transition such that the hh_{11}-transition can only be observed as a shoulder. The lh_{11}-transition is nearly absent in the spectrum. At least two effects can be responsible for this polarization dependence. First the symmetry of the exciton wavefunctions, in particular due to the inherent heavy hole - light hole mixing in quantum wires.[22,29,34,35] (Strictly speaking there is no pure light hole or heavy hole state in a quantum wire even at k = 0.) Secondly, one has to keep in mind also electrodynamic effects. The electromagnetic fields of the exciting or emitted photons are modified by the laterally structured surface and adjacent wires are coupled via electromagnetic fields. These effects depend on the polarization and we expect that they are strongly pronounced in particular in our deeply etched grating structures. Some aspects of these grating coupler effects, the so-called local field corrections, are discussed in Refs.[32,33].

For a further confirmation of the 1D character and for additional quantitative information we have performed magnetooptical PLE experiments in magnetic fields B oriented perpendicularly to the layers.[13,14] This method is well established for 2D QW excitons.[39–41] The exciton energies in a magnetic field are determined by an interplay between the excitonic Coulomb interaction, governed by the effective Rydberg energy and by the magnetic energy which is determined by the cyclotron resonance energy.[42] For low exciton states, in particular for the ground state and small magnetic fields, the Coulomb interaction dominates and the exciton energy is only affected by the diamagnetic shift. This diamagnetic shift is proportional to the square of the transversal extent of the exciton wave function and can such be used to study the expected shrinkage of the 1D exciton wave function. For higher exciton states and larger B the magnetic energy dominates and the transitions obtain the character of Inter-Landau-Level transitions with a linear B-dependence. These Inter-Landau-Level transitions can be easily observed with increasing magnetic field. Extrapolation to B=0 allows then a determination of the exciton binding energy.

The energy positions in the PLE-spectra of a QW wire with L_x=70 nm and of the corresponding reference-QW in a magnetic field are summarized in Fig. 4. For clarity, we did not

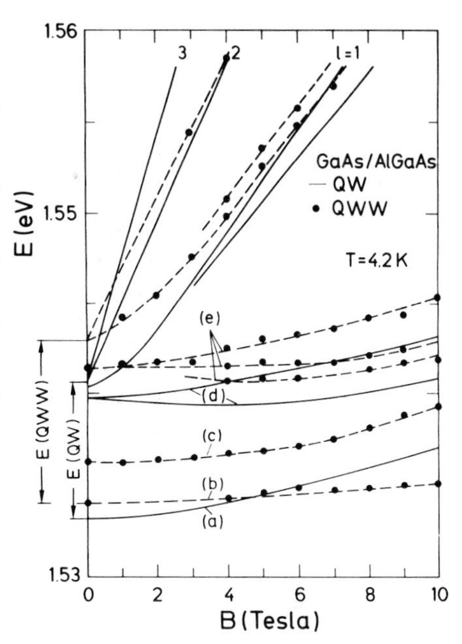

Fig. 4: Energy positions of the maxima in the magnetic-field dependent PLE-spectra of a $L_x \approx 70$ nm QW wire (QWW). The solid lines indicate the magnetic field dispersion for the transitions of the corresponding reference QW. (a),(b),(c),(d) and (e) denote the hh_1, hh_{11}, hh_{12}, lh_1 and lh_{11} transitions, respectively; the index l labels the order of the Inter-Landau Level transitions. In addition, the exciton binding energies for the QW and the QW wire, E(QW) and E(QWW), respectively, which were deduced from zero-field extrapolations, are indicated. They show directly the increased binding energies of the 1D excitons. The reduced dimensionality is also reflected in the smaller diamagnetic shift of the QW wire sample. (From Ref.[13]) .

plot the magnetic field dispersion of the QW in discrete data points but as solid lines. The magneto dispersion of the reference sample is very similar to earlier measurements on QW systems, e.g. Refs.[39–41]. The hh_1 and lh_1 transitions show a diamagnetic shift, which is stronger for the hh_1 transition reflecting the weaker Coulomb interaction of the hh_1 exciton. The magnetic field dependence of the QW wires is strikingly different. The hh_{11} transition of the QW wire shows a smaller diamagnetic shift as compared to the reference sample, indicating a stronger exciton binding energy. The diamagnetic shift of the hh_{12} transition demonstrates its excitonic character, which is less pronounced than that of the ground-state exciton. These features are qualitatively expected for 1D magneto-excitons and confirm our interpretation of the hh_{11} and hh_{12} transitions.

We analyzed the magnetic field dependence following similar evaluations for 2D systems, e.g. Refs.[40,41]. The analysis was performed at sufficient low magnetic fields (below 3 Tesla), where we observed no significant deviation from a quadratic energy-shift. The diamagnetic shift ΔE of the hh-ground state exciton of the reference sample was 0.51 meV at 3 Tesla, which is consistent with published values.[40,41] The corresponding exciton binding energy is $E_B = 7.5$ meV, which was determined by extrapolating the Inter-Landau-Level transitions to B=0. The value of 7.5 meV agrees very well with reported binding energies of comparable QW samples.[43,44] For the hh_{11} transition of the QW wire we could determine $E_B = 8.7 meV$, which indicates an enhancement of the ground-state excitonic interaction in QW wire by about 15% with respect to the 2D QW reference sample. From the diamagnetic shift of the excitons we evaluate that the transverse extension of the hh_1 exciton of the reference QW was 1.09 a_B (a_B is the effective Bohr radius), if we follow the analysis of Ossau et al.[41] This value is in excellent agreement with the calculated value of Bastard[45] for a 14 nm QW. The corresponding value of the QW wire hh_{11} exciton (0.97 a_B) is by about 11% reduced, which directly indicates the shrinkage of the excitonic wavefunction due to the additional lateral confinement.

The experiments described above demonstrate clearly quantum confinement effects on excitons. However, they also leave a lot of open questions, which is generally true for all current optical experiments on quantum wires and dots also in other groups[25]. One question is, for example, why in most of the structures the PL intensity is so low, in spite of the expected increased oscillator strength? Does, for instance, a chemically clean and perfect crystalline etched *GaAs* surface give a perfect confinement or does it 'intrinsically' lead to undesired nonradiative decay? Overgrowth of etched structures (e.g. Ref.[24]) or other passivation mechanisms, which perhaps occur due to redeposition also for deep-mesa-etching techniques, seems to be helpful to increase the luminescence efficiency. However also here the exact boundary conditions for the confinement of excitons, i.e., the 'active' wire width, and possible nonradiative decay channels are not really known. We think that in progress on the work on quantum wires and dots one should try to perform on one and the same quantum wire sample as many different experimental and analytical methods as possible to achieve a consistent picture not only of the physics of low-dimensional systems but also of the real structure of the sample. In this spirit we have recently prepared quantum wire sample and performed investigations with SEM and also x-ray diffraction[47] which gave us access to the crystalline structure in the wire. We have further performed linear PL excitation[19], Raman spectroscopy and, in time resolved experiments, high-density excitation which revealed a quantum confined 1D electron-hole plasma[20]. In the following we would like to review PL excitation spectroscopy on these relatively wide QW wires which gives us insight into the transition from 2D excitons in QW to 1D excitons in narrow QW wires such as discussed above.[19]

4. CENTER-OF-MASS QUANTIZATION OF EXCITONS

Quantum wire structures have been fabricated from a multiple quantum well (MQW), consisting of 25 GaAs quantum wells of L_z = 10.6 nm sandwiched between 15.3nm wide barriers, which were grown by MBE on top of a 1 μm thick $Al_{0.36}Ga_{0.64}As$ optical confinement layer.[46] By holographic lithography we prepared a mask of photoresist stripes oriented along the [110]-direction with a 280 nm periodicity. With three reactive ion etching steps of decreasing depth we etched in a $SiCl_4$-plasma rectangular grooves through all the MQW layers. The overall etching depth was 760 nm, whereby in the second step the topmost quantum wells were removed. The resulting QW wire-structures contain 7 of the original quantum wells. SEM shows that the geometrical width of the wires amounts to $L_x = 140 \pm 10$ nm. We have also performed x-ray diffraction measurements on the same sample, which directly probes the crystalline part of the etched wires.[47,20] We find that the crystalline part of the wire is $L_x^a = 60 \pm 10$ nm. From this we conclude that there is on eather side of the wire a region of about $L_d = (L_x - L_x^a)/2 = 40$ nm, which is either sligthly damaged, penetrated by channelled species during the etching process or shows a small intermixing at the interfaces of the QW's. We note that the wires discussed here have been exposed to several and relatively long etching processes. So the width L_d will be smaller on other wires with shorter etching times. For the sample here the optical measurements demonstrate that the crystalline part of the wire determined in the x-ray diffraction is the 'active' wire width L_x^a which determines the confinement effects.

In PL measurements on the 2D reference part of the sample we observe heavy and light hole (lh) excitonic transitions at 1.5539eV and 1.5666eV, respectively, and the peaks of the hh-exciton coincide in PL and PLE, i.e., there is no detectable Stokes-shift. The PL spectrum of the microstructured part of the sample is, compared to the reference, red-shifted and shows a broad low-energy tail. Therefore we conclude that mainly localized states contribute to the PL signal. The absolute intensity has dropped by a factor of 50. This means, taking into account the reduced number of quantum wells and their reduced area, that the emission efficiency per quantum well has only dropped by factor of 4. The polarization dependent PLE spectra in Fig. 5 reveal that the splitting of the hh-exciton is threefold. The ground state (1) can only be observed in s-polarized PLE, i.e. the electric field vector of the exciting and detected light is perpendicular to the plane of incidence and parallel to the wires. The highest energy state (3) is observed in both polarizations, while the intermediate transition (2) is only observable in p-polarization. Similarly as in the case of the narrow quantum wires we have also performed magnetic field dependent measurements up to 7T with the field oriented along the growth direction. We find that the lowest excitonic state in the QW wire shows essentially the same diamagnetic shift as the corresponding hh transition in the 2D reference. Since, as we have discussed above, the

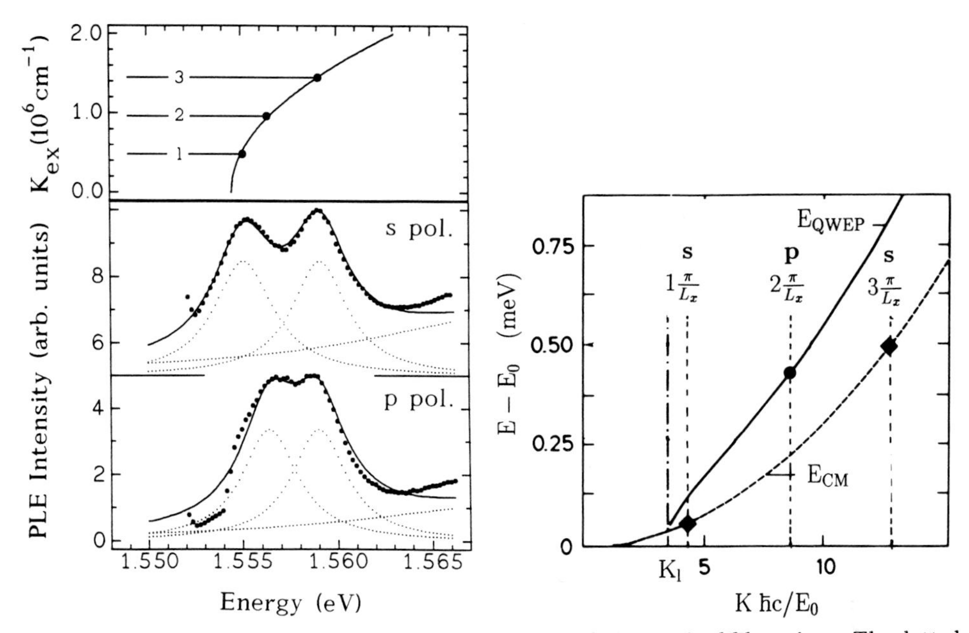

Fig. 5 (left): Polarization dependent PLE spectra of the CM-quantized hh exciton. The dotted curves display the contribution of the different Lorentzian broadened transitions. The top part shows the reconstruction of the CM-dispersion ($M^*_{ex} = 0.178 m_0$) for L^a_x= 65 nm. Fig. 6 (right): Schematic diagram of the free CM exciton dispersion E_{CM} (dashed line), of the X-mode-QW exciton polariton mode, E_{QWEP} (solid line) and of photons in GaAs (dashed-dotted line). The energies of the CM quantization can be constructed from the dispersion at $K = n\pi/L^a_x$. The parameters of the dispersion are chosen for GaAs with E_0 =1.54 eV for the exciton energy at K=0.

diamagnetic shift is a measure for the extent of the excitonic wavefunction, this indicates that the relative motion of the electron and hole is not affected by the lateral confinement in these wider wires.

We therefore ascribe the energy splittings to a center-of-mass (CM) quantization of the excitons in the wide quantum wires. This behaviour is expected for $L^a_x \gg a_B$, where the lateral wire confinement does not quantize the electronic wavefunctions directly. However, due to the wire structure the motion of the exciton as a whole is not free in the plane of the original well, since it is 'reflected' by the lateral boundaries. In terms of quantum mechanics, the reflection leads to the formation of quantized standing waves perpendicular to the direction of the wires. Consequently, the original continuous 2D energy dispersion of the exciton's motion becomes now a set of discrete points, at least in the direction of the lateral confinement.

The kinetic energy for the CM motion is given by $E^{kin}_{CM} = \frac{\hbar^2 K^2_{ex}}{2M^*_{ex}}$ where $M^*_{ex} = m^*_e + m^*_{hh}$ is the total mass of the exciton. In the direction perpendicular to the wires K_{ex} can have only quantized values:

$$K^n_{ex} = \frac{\pi}{L^a_x} n \quad , n = 1, 2, 3, ... \tag{1}$$

This quantization is sketched in Fig. 6. We determine the exact positions of the transitions by fitting the PLE spectra with Lorentzian curves, as is shown by the dotted curves in Fig. 5. Within 5% accuracy the spectral positions E^n_{CM} obey the following empirical relation:

$$E^n_{CM} = E_0 + C_1 * n^2 \ [meV], \quad E_0 = 1554.5meV, \ C_1 = 0.5 \pm 0.02, \tag{2}$$

and show a characteristic quadratic n-dependence of an 'infinite' square well confinement. The second term therefore yields the experimental quantization energies of the confined CM motion. To evaluate the quantization length, the active wire width, L^a_x we use $M^*_{ex} = m^*_e + (\gamma_1 + \gamma_2)^{-1} = 0.178m_0$, where the in-plane mass of the heavy hole is taken from the diagonal elements of the Luttinger-Kohn matrix. Under this assumption L^a_x is evaluated to be 65nm, and, as can be seen in the upper part of Fig. 5, the fit of the experimental data to the dispersion of the exciton's center-of-mass motion is reasonable. This value of the 'active' wire width L^a_x is in agreement with the value for the crystalline part of the wire determined from the x-ray diffraction measurements and also with a value deduced from the confinement length in high-excitation 1D electron-hole plasma on the same sample[20].

An interesting feature is the strong polarization dependence of the spectra, in particular the appearance of the peak n=2 only in p-polarized PLE. At present we are not sure whether this polarization dependence is an intrinsic property or whether grating coupler effects induce the emission of X-mode quantum well excitonic polaritons.[18] Excitons, also in QW, are automatically coupled to photons and thus have the character of polaritons.[48,49] The polariton spectrum consist of several modes. The lowest mode, E_{CM} in our sketch in Fig. 6, sometimes called the Y-mode[48,49], is, in first approximation, not affected by coupling to photons and thus directly reflects the CM-motion and is sometimes called the instantaneous exciton. The X-mode QW exciton polariton, E_{QWEP}, has an increased energy with respect to the E_{CM} mode. In a bulk material this increase corresponds to the longitudinal- transverse (LT) splitting of excitons. Andreani and Bassani have shown that the corresponding splitting for QW excitons, a manifestation of dynamic exchange effects, is small.[31] QW exciton polaritons also have been observed in QW systems.[50,51] They also play a dominant role in microstructured QW sytems and wide QW wires.[10,11] These aspects have recently been reviewed in more details.[18,52]

D'Andrea and Del Sole[32] have recently evaluated reflectance spectra for QW wires which structural properties were not too different from the samples in our experiments discussed above (Fig. 5). They found that in s-polarized reflectance spectra only CM-quantized states with *odd* n are observable. This agrees well with our observation. Unfortunately, it is not possible to analyze their p-polarized spectrum in such a straightforward manner, especially for small n. We expect, as indicated in Fig. 6, that (neglecting coupling to higher 1D subband states) PL emission of the *odd* states is excited in s-polarization, whereas *even* states are excited in p-polarization. Unfortunately the energy separation between the exciton and the X-mode exciton-polariton is small[31,48,49] and is exceeded by the linewidth by about one order of magnitude. So we are not sensitive to it concerning the small difference in transition energy and cannot distiguish polariton aspects from the energetic position. (We like to note that the highest CM state shows a more complex behavior in a magnetic field and refer here to the original paper[19]) Similarly as demonstrated here for the transition from 2D excitons via a CM quantization to 1D excitons Tuffigo et al have recently also demonstrated the transition from 3D bulk excitons to 2D QW excitons in the $CdTe$ system[53]. Interference of exciton polariton waves in thin GaAs layers has been demonstrated by Kusano et al[54] and Chen et al[55]. These experiments also show different excitation strengths for odd and even CM modes.

5. SUMMARY

We have demonstrated the transition from 2D QW excitons to 1D excitons. In wider QW wires we find a quantization of the CM motion of the hh quantum well exciton. The energetic positions of the threefold splitted hh excitonic state well reconstruct its original continuous CM-dispersion, when a quantization length L^a_x of the active wire width is used. The polarization effects demonstrate the influence of polariton aspects. One-dimensional quantum confined excitons are observed in quantum wires with geometrical width as narrow as 60nm. The 1D character of these excitons is manifestated in the observation of several exciton resonances which are related to different, energetically separated quantum confined 1D subbands, by a reduced diamagnetic shift in a magnetic field, and by a 15% increased binding energy with respect to 2D reference samples.

ACKNOWLEDGEMENTS: We want to thank A. Fischer for the MBE-growth of the excellent MQW-structure, and C. Lange for the high quality scanning electron micrographs. This work was supported by the Bundesministerium für Forschung und Technologie (BMFT).

PRESENT ADDRESSES: * Dipartemento di Scienze dei Materiali, Universita di Lecce, Lecce, Italy, ** Department of Material Science, Technical University, Darmstadt, Germany.

REFERENCES

[1] R. Dingle, W. Wiegmann, and C.H. Henry, Phys. Rev. Lett. 33, 827 (1974).
[2] C. Weisbuch in "Physics and Application of Quantum Wells and Superlattices", Eds. E.E. Mendez and K. von Klitzing, Plenum Press, New York, p. 261 (1987).
[3] K. Kash, A. Scherer, J.M. Worlock, H.G. Craighead, and M.C. Tamargo, Appl. Phys. Lett. 49, 1043 (1986).
[4] J. Cibert, P.M. Petroff, G.J. Dolan, S.J. Pearton, A.C. Gossard, and J.H. English, Appl. Phys. Lett. 49, 1275 (1986).
[5] M.A. Reed, R.T. Bate, K. Bradshaw, W.M. Duncan, W.R. Frensley, J.W. Lee, and M.D. Shih, J. Vac. Sci. Technol. B4, 358 (1986).
[6] H. Temkin, G.J. Dolan, M.B. Panish, and S.N.G. Chu, Appl. Phys. Lett. 50, 413 (1987).
[7] Y. Hirayama, S. Tarucha, Y. Suzuki, and H. Okamoto, Phys. Rev. B. 37, 2774 (1988).
[8] D. Gershoni, H. Temkin, G.J. Dolan, J. Dunsmuir, S.N.G. Chu, and M.B. Panish, Appl. Phys. Lett. 53, 995 (1988).
[9] H.E.G. Arnot, M. Watt, C.M. Sotomayor-Torres, R. Glew, R. Cusco, J. Bates, and S.P. Beaumont, Superlattices and Microstr. 5, 459 (1989).
[10] M. Kohl, D. Heitmann, P. Grambow, and K. Ploog, Phys. Rev. B37, 10927 (1988).
[11] M. Kohl, D. Heitmann, P. Grambow, and K. Ploog, Superlattices and Microstr. 5, 235 (1989).
[12] A. Forchel, H. Leier, B.E. Maile, and R. German, Advances in Solid State Physics, ed. U. Rössler, Vieweg, Braunschweig (1988).
[13] M. Kohl, D. Heitmann, P. Grambow, and K. Ploog, Phys. Rev. Lett. 63, 2124 (1989).
[14] M. Kohl, D. Heitmann, P. Grambow, and K. Ploog, Surf. Sci. 229, 248 (1990).
[15] K. Kash, J.M. Worlock, M.D. Sturge, P. Grabbe, J.P. Harbison, A. Scherer, and P.S.D. Lin, Appl. Phys. Lett. 53, 782 (1988).
[16] K. Kash, J.M. Worlock, A.S. Gozdz, B.P. Van der Gaag, J.P. Harbison, P.S.D. Lin, and L.T. Florez, Surf. Sci. 229, 245 (1990)
[17] M. Kohl, D. Heitmann, P. Grambow, and K.Ploog, Phys. Rev. B41, 12338 (1990).
[18] M. Kohl, D. Heitmann, P. Grambow, and K.Ploog, Phys. Rev. B42, 2941 (1990).
[19] H. Lage, D. Heitmann, R. Cingolani, P. Grambow, and K.Ploog, Phys. Rev. B44, 6550 (1991).
[20] R. Cingolani, H. Lage, L. Tapfer, H. Kalt, D. Heitmann, and K.Ploog, Phys. Rev. Lett. 67, 891 (1991).
[21] K. Kash, B.P. Van der Gaag, D.D. Mahoney, A.S. Godz, L.T. Florez, J.P. Harbison, and M.D. Sturge, Phys. Rev. Lett. 67, 1326 (1991).
[22] M. Tsuchiya, J.M. Gaines, R.H. Yan, R.J. Simes, P.O. Holtz, L.A. Coldren, and P.M. Petroff, Phys. Rev. Lett. 62, 466 (1989).
[23] M. Tanaka and H. Sakaki, Appl. Phys. Letts. 54, 1326 (1989) (1982).
[24] E Kapon, D.M. Hwang, and R. Bhat, Phys. Rev. Letts. 63, 430 (1989).
[25] K. Kash, J. Luminescene 46, 69 (1990)
[26] G. Bastard in "Physics and Application of Quantum Wells and Superlattices", eds. E.E. Mendez and K. von Klitzing, Plenum Press, New York, p. 21 (1987).
[27] J.W. Brown and H.N. Spector, Phys. Rev. B35, 3009 (1987).
[28] G.W. Bryant, Phys. Rev. B37, 8763 (1988).
[29] U. Bockelmann and G. Bastard, Europhys. Lett. 15, 215 (1991).
[30] R. Del Sole and A. D'Andrea, in *Optical Switching in Low-Dimensional Systems* (1989), ed. H. Haug and L. Bányai, Plenum Press N.Y., p. 289.

[31]L.C. Andreani and F. Bassani, Phys. Rev. B41, 7536 (1990).
[32]A. D'Andrea and R. Del Sole, Superl. Microstr. 8, 425 (1990).
[33]U. Bockelmann and G. Bastard, to be published.
[34]G. E. Bauer and H. Sakaki, Phys. Rev. B44, 5562 (1991).
[35]P.C. Sercel and K.J. Vahala, Phys. Rev. B44, 5681 (1991).
[36]P. Grambow, T. Demel, D. Heitmann, M. Kohl, R. Schüle, and K. Ploog Microelectronic Engineering 9, 357, (1989).
[37]D. Heitmann in ' Electronic Properties of Multilayers and Low-Dimensional Semiconductor Structures' Eds. J. M. Chamberlain, L. Eaves and J.-C. Portal, Nato ASI Series B: Physics, Vol. 231, Plenum Press, New York, 1990, p. 151.
[38]M. Kohl, D. Heitmann, S. Tarucha, K. Leo, and K. Ploog, Phys. Rev. B39, 7736 (1989).
[39]J.C.Maan, G.Belle, A.Fasolino, M.Altarelli, and K.Ploog, Phys. Rev. B30, 2253 (1984).
[40]D.C. Rogers, J. Singleton, R.J. Nicholas, C.T. Foxon, and K. Woodbridge, Phys. Rev. B34, 4002 (1986).
[41]W. Ossau, B. Jäkel, E. Bangert, G. Landwehr, and G. Weimann, Surf.Sci. 174, 188 (1986).
[42]O. Akimoto and H. Hasegawa, J. Phys. Soc. Japan 22, 181 (1967).
[43]R.C. Miller, D.A. Kleinman, W.T. Tsang, and A.C. Gossard, Phys. Rev. B24, 1134 (1981).
[44]U. Ekenberg and M. Altarelli, Phys. Rev. B35, 7585 (1987).
[45]G. Bastard, E.E. Mendez, L.L. Chang, and L. Esaki, Phys. Rev. B26, 1974 (1982).
[46]R. Cingolani, K. Ploog, A. Cingolani, C. Moro and M. Ferrara, Phys. Rev. B42, 2893 (1990).
[47]L. Tapfer, G. C. LaRocca, H. Lage, R. Cingolani, P. Grambow, A. Fischer, D. Heitmann, and K. Ploog, Workbook of the 5th International Conference on Modulated Semiconductor Structures, p. 201, July 8-12 (1991), Nara (Japan) and Surf. Science, in press.
[48]M. Nakayama, Solid State Commun. 55, 1053 (1985).
[49]M. Nakayama and M. Matsuura, Surf. Sci. 170, 641 (1986).
[50]K. Ogawa, T. Katsumura, and H. Nakamura, Appl. Phys. Lett. 53, 1077 (1988).
[51]K. Ogawa, T. Katsumura, and H. Nakamura,. Phys. Rev. Letts. 64, 796 (1990).
[52]M. Kohl, D. Heitmann, P. Grambow, and K. Ploog in 'Condensed Systems of Low Dimensionality' Ed. J. L. Beeby et al., Nato ASI Series B: Physics, Vol. 253, Plenum Press, New York, 1990, p. 123.
[53]H. Tuffigo, R.T. Cox, F. Dal'Bo, G. Lentz, N. Magnea, H. Mariette, and C. Grattepain, Superlattices and Microstr. 5, 83 (1989), H. Tuffigo, B. Lavigne, R. T. Cox, G. Lentz, N. Magnea and H. Mariette, Surf. Science 229, 480 (1990).
[54]J. Kusano, Y. Segawa, M. Mihara, Y. Aoyagi and S. Namba, Sol. State Comm. 72, 215 (1989).
[55]Y. Chen, F. Bassani, J. Massies, C. Depairis, and G. Neu, Europhys. Lett. 14, 483 (1991).

Inst. Phys. Conf. Ser. No 123
Paper presented at the International Meeting on Optics of Excitons in Confined Systems, Giardini Naxos, Italy, 1991

Exciton quantization and optical properties in semiconductor quantum gratings

A. D'Andrea (a) and R. Del Sole (b)

(a) Istituto di Metodologie Avanzate Inorganiche del CNR,C.P.10, I-00016 Monterotondo Scalo (Roma),Italy
(b) Dipartimento di Fisica, II Universita'di Roma "Tor Vergata", v.E. Carnevale, I-00173 Roma, Italy

ABSTRACT: A variational wave function for excitons in quantum well wires is obtained. The reflectivity of a grating is computed by taking into account the spatial dispersion of excitons and local-field effects.

1. INTRODUCTION

Exciton dynamics in different confined systems, as quantum wells (QWs), quantum well wires (QWWs) and quantum dots (QDs) has been the object of extensive investigation in recent years. The improving ability in microstructure fabrication allows one to study new phenomena such as those due to exciton confinement in a grating of parallel QWWs (quantum grating, QG) (Tsuchiya et al 1989, Kohl et al 1989). In this paper we address the problem of the correct exciton quantization in a QWW, determine exciton energies and wave functions, and use them to study the linear optical response of a grating of parallel QWWs.

2. EXCITON WAVE FUNCTIONS IN QUANTUM WIRES

Let us consider a QWW parallel to the y-direction, of lateral dimensions L_x and L_z. We treat the case of direct excitons arising from two parabolic nondegenerate bands for electrons and holes. An appropriate variational wave function for this case is

$$\psi_{nm}(\mathbf{r},\mathbf{R})=N_{nm}P_n(x,X)Q_m(z,Z)\exp(iKY)\exp(-r/a), \tag{1}$$

where $\mathbf{R} = (X,Y,Z)$ is the center-of-mass position (X and Z are relative to the center of the wire), $\mathbf{r} = \mathbf{r}_e - \mathbf{r}_h$ is the relative electron-hole coordinate, K is the center-of-mass momentum along y, and N_{nm} is a normalization constant. The "recycling functions" P_n and Q_m enforce the fulfillment of the no-escape boundary conditions at x and z boundaries respectively. They have different expressions, according to whether the relevant dimension is smaller or larger than $3a_B$, where a_B is the exciton Bohr radius (D'Andrea and Del Sole 1990a). Since in real QWWs one dimension is usually of the order of some hundreds of Angstroms, we assume L_x to be always larger than $3a_B$. In this case we have:

$$P_n(x,X) = \cos(K_nX) - F_e(x)\cosh(P_xX) + F_o(x)\sinh(P_xX) \;, \qquad (2)$$

where K_n is the exciton momentum along X. P_x, the inverse dead-layer depth, is a variational parameter, and $F_e(x)$ and $F_o(x)$ are general even and odd functions, to be determined by fulfilling the boundary conditions, given by D'Andrea and Del Sole (1990b). The requirement of a continous derivative with respect to x, at x=0, leads to the quantization condition:

$$K_n tg(K_nL_x/2)+P_x tgh(P_xL_x/2)=0, \qquad (3)$$

which determines K_n. These equations hold for exciton states which are even for reflection with respect to the yz plane, that is for n = 1,3,5..... For the other values of n, associated with odd wave functions, a similar formula holds (D'Andrea and Del Sole 1990b).

If the other lateral dimension of the wire, L_z, is larger than $3a_B$, the recycling function Q_m is given by an expression analogous to (2). For $L_z<3a_B$, we choose the product of electron and hole subband wave functions (Bastard et al 1982):

$$Q_m(z,Z) = \cos(m\pi z_e/L_z)\,\cos(m\pi z_h/L_z) \qquad (4)$$

Electron and hole subbands with the same quantum number m = 1,2,3...have been considered, since only these can be excited optically.

The exciton energy must be minimized with respect to a and P_x for $L_z < 3a_B$, and with respect to P_x and P_z (with $a = a_B$) for $L_z > 3a_B$. We perform the calculation for GaAs wires with L_x=1500 Å, L_z=100 Å, using the parameters: ε_0 = 12.6, M = 0.3 m, and the effective Rydberg R*= 4.2 meV. We find $P_x = 0.06$ Å $^{-1}$.

3. OPTICAL RESPONSE OF EXCITONS IN QUANTUM GRATINGS

We consider a grating of parallel QWWs with a distance d between their axes. In the mixed representation (K_x,Z) the polarization induced by an electric field $E(K_x,Z)$ is:

$$P(K_x,Z)=(4\pi d)^{-1}\Sigma_G\{(\varepsilon_0-1)E(K_x+G,Z)+S_0\Sigma_{nm}P_n^*(0,K_x)P_n(0,K_x+G)$$
$$Q_m^*(0,Z)\int Q_m(0,Z')E(K_x+G,Z')dZ'/(E_{nm}+\hbar\omega)\} \quad , \qquad (5)$$

where $G = 2\pi l/d$ (l=0, ±1, ±2 ..) is a vector of the reciprocal grating, Z is within 0 and L_z, and S_0 is proportional to the strength of the transition (D'Andrea and Del Sole 1990b). We are considering QWWs embodied in a semi-infinite solid of dielectric constant ε_0 occupying the half-space z>0.

We solve Maxwell's equations for light incident in the xz plane, for both s polarization (electric field along y) and p polarization (electric field in the xz-plane). We use the method of Green's functions (Bagchi et al. 1979), determining self-consistently the fields in the QWWs. The local-field effect, namely the electromagnetic coupling at wave vectors G between the wires is fully taken into account. The details of the derivation will be given elsewhere (D'Andrea and Del Sole 1992). We start by neglecting local-field effects, that is by retaining only the macroscopic component of the field (G=0).

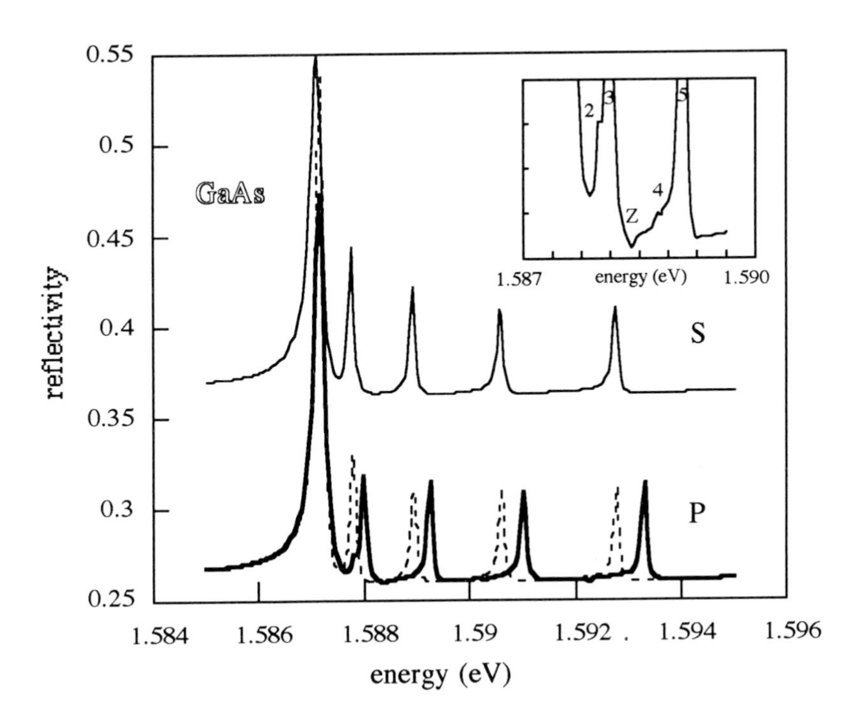

Fig. 1 Reflectivity of a GaAs quantum grating of depth L_z=100 Å and lateral dimension L_x=1500 Å, with a periodicity d=2500 Å. For p-light, the dashed line has been calculated considering only the macroscopic field (G=0), while the heavy line includes local-field effects. For s-light, the two calculations cannot be distinguished. An enlarged view of the heavy-line p-reflectivity is shown in the inset.

Let us consider first the s-light reflectivity, shown in Fig. 1; this curve can be well understood in terms of the first order expansion in L_z/λ (λ is the light wave lenght), which is routinely used in the case of the optical properties of the surfaces (McInthire and Aspnes 1971): The s-wave reflectivity is

$$R_s/R_s{}^0=1+4(\omega L_z/c)\cos\theta\,\mathrm{Im}[(\varepsilon_s-\varepsilon_b)/(\varepsilon_b-1)] \quad , \qquad (6)$$

where ε_b and $R_s{}^0$ are the dielectric constant and the reflectivity of the substrate, respectively, and θ is the angle of incidence. Since ε_b is real and constant in the energy range of interest, the reflectivity is proportional to the QG dielectric constant ε_s, and therefore has peaks at the quantized exciton energies. Only even excitons can be seen in the spectrum, while the odd ones give a negligible contribution, because of the small x-component of the light wave vector (odd excitons are in fact forbidden if the electric field is constant in the surface plane).

In the case of p-light (dashed curve in Fig. 1), there should be extremely weak structures in between the peaks, described by the first-order expansion (McInthire and Aspnes 1971):

$$R_p/R_p^0 = 1 + 4(\omega L_z/c)\cos\theta\, \mathrm{Im}\{[(\varepsilon_b - \sin^2\theta)(\varepsilon_s - \varepsilon_b) + \varepsilon_b^2 \sin^2\theta(\varepsilon_s^{-1} - \varepsilon_b^{-1})]/[(\varepsilon_b - 1)(\varepsilon_b\cos^2\theta - \sin^2\theta)]\}\ , \quad (7)$$

as arising from the term proportional to ε_s^{-1}. These structures are due to the excitation of the quantum well Z-polaritons predicted by Tassone at al. (1990). (The presence of the lateral confinement leads to a quantization of these modes, too.) The small oscillator strength of GaAs excitons makes such structures not visible in this curve.

The inclusion of local fields, namely that of a number (≈50) of G-vectors -until convergence is reached-, has little effect on the s-wave reflectivity, but shifts the p-wave structures by about 0.3 meV, as can be seen in Fig. 1 (D'Andrea and Del Sole 1990c). These results can be understood in terms of a simplified analytical treatment, that will be described elesewhere (D'Andrea and Del Sole 1992), and originate from the different properties of the Green's functions appropriate to s and p light. From a physical point of view, the transverse microscopic ($G \neq 0$) dipoles excited by the s-light interact less strongly than those in the x and z directions excited by the p-light (Tassone et al.1990). The relatively large wave vectors $K_x + G$ induced by the grating lead to increased contributions of the Z-polaritons and of the odd excitons: they are both visible in the heavy-line curve in Fig. 1, or, more clearly, in its expanded view shown in the inset. Here the numbers n = 2,3,4,5 identify the quantized exciton states, while the position of the Z-polariton is indicated by the symbol Z.

These results, although not comparable with experiments (Kohl 1989), because of the different geometry involved, indicate the extreme importance of the electromagnetic interaction between different wires in determining the optical properties of a grating.

4. REFERENCES

Bagchi A, Barrera R G, and Rajagopal A K 1979 Phys. Rev. B20 4824

Bastard G, Mendez E E, Chang L L and Esaki L 1982 Phys. Rev. B28 1974

D'Andrea A and Del Sole R 1990a Solid State Communications 74 1121

D'Andrea A and Del Sole R 1990b Phys. Rev. B41 1413

D'Andrea A and Del Sole R 1990c Superlattices and Microstructures 8 425 reported a wrong plot of the p-light reflectivity, because of a mistake in the computer program

D'Andrea A and Del Sole R 1992 to be published

Kohl M, Heitmann D, Grabow P and Ploog K 1989 Phys. Rev. Lett. 63 2124

McInthire J D E and Aspnes D E 1971 Surface Sci. 24 417

Tassone F,Bassani G F, and Andreani L C 1990 Il Nuovo Cimento D12 167

Tsuchiya M, Gaines J M, Yan R H, Simes R J, Holtz P O, Coldren L A and Petroff P M 1989 Phys. Rev. Lett. 62 466

Inst. Phys. Conf. Ser. No 123
Paper presented at the International Meeting on Optics of Excitons in Confined Systems, Giardini Naxos, Italy, 1991

Quasi one-dimensional excitons at Si(111)2 $\times$ 1

Lucia Reining[1],[2] and R. Del Sole[1]

[1] Dipartimento di Fisica, II Università di Roma, via E. Carnevale, 00173 Roma, Italy
[2] Centre Européen de Calcul Atomique et Moléculaire, Bât. 506, Université Paris-Sud, 91405 Orsay Cedex, France

ABSTRACT: We present theoretical evidence for the existence of strongly bound, quasi-one-dimensional excitons at Si(111)2$\times$1. On the basis of a self-consistent calculation of electronic states, excitons and optical properties, which involves for the first time a realistic treatment of the screened electron-hole interaction, we find a gap between surface states of 0.75 eV, and an optical spectrum in quantitative agreement with experiment. The exciton binding energy turns out to be E_b=0.3 eV. Higher excitonic states are not visible, due to the quasi one-dimensionality of the surface states.

1. INTRODUCTION

For many years, the excitonic- or single-particle- origin of the main peak in the optical spectrum of the Si(111)2$\times$1 surface, at 0.47 eV (Chiarotti et al 1971) has been object of discussion (Pandey and Phillips 1975, Del Sole and Tosatti 1977, Del Sole and Selloni 1984, Ciccacci et al 1986). On one hand, theoretical calculations predicted strongly bound ($\sim$0.3 eV) surface-state excitons. On the other hand, this result was contradicted by measurements of photoemission from n-doped samples, and STM, which both yielded a gap of about 0.5 eV, coincident with the optical peak and hence leaving no space for excitonic effects. Moreover, the experimental optical spectrum, which showed no secondary peaks typical for excitonic effects, could very well be fitted by a one-electron calculation.

A first hint to the solution of the puzzle has been given by a recent inverse photoemission experiment (Perfetti et al 1987), which found a gap between surface states of 0.75 eV (or 0.6 eV, according to a more critical interpretation (Cricenti et al 1990)). Moreover, a similar value of the gap has been suggested by a model calculation of the self-energy correction to the LDA eigenvalues (Reining and Del Sole 1990), and has also been confirmed independently by an ab-initio many-body perturbative GW calculation, which yielded 0.62 eV (Northrup et al 1991). On the other side, the use of the classical image potential result for the screening of the electron-hole pair ($(\epsilon_b + 1)/2 = 6.5$ for Silicon) in the calculations seemed to be doubtable, since the small optical gap between surface states, and the large oscillator strength of transitions across it (Chiaradia et al 1984, Cricenti et al 1988), implied that surface states might give a big contribution to the screening (Del Sole and Selloni 1984), leading to a reduction of the electron-hole (e-h) interaction and to smaller exciton binding energies. An accurate study of surface screening (Reining and Del Sole 1988, and to be published), however, has led us to discard this possibility, as a consequence of quantum effects, which strongly reduce screening at short distances with respect to classical predictions. Moreover, the one-dimensional character of the chains of atoms at Si(111)2×1 (Pandey 1981, 1982) leads to an antiscreening contribution of surface states at intermediate distances. The resulting position-dependent screening of the e-h interaction ranges between 4 and 10, and is hence even smaller than in the bulk. Instead, in this work we present a self-consistent calculation of the bandgap, screening and excitonic effects in the optical properties, which confirms the higher value for the gap (E_g=0.75 eV) and yields optical properties in excellent agreement with experiment, with an exciton binding energy of 0.3 eV.

2. CALCULATION AND RESULTS

We determine the dielectric response of the surface within a model, that -although simplified- embodies the basic features of surface states. We expand these states in the Dangling Bond (DB) orbitals and perform tight-binding calculations including first (t) and second (V) neighbor π-interactions. In this model, which has worked well in the interpretation of the anisotropy of optical and electron energy loss spectra (Selloni and Del Sole 1986), electrons belonging to different chains do not interact, so that the electron dynamics is essentially one-dimensional. Fitting the dispersion of the lower band and the Local Density Approximation (LDA) gap (Northrup et al 1991) yields t=-1.1 eV, V=0.35 eV, $E_{22} - E_{11} = 0.27eV$, where $E_{ii} = \langle i|H|i\rangle$, i=1,2 labeling the two DB orbitals in the surface 2×1 unit cell. By the way of contrast, we treat the substrate as a semi-infinite classical dielectrics. The surface part of the inverse dielectric function is essentially determined by the matrix M, which obeys the matrix equation (Hanke and Sham 1980)

$$M_{ss'}(\vec{q};\omega) = S_{ss'}(-\vec{q};\omega) + \sum_{tt'} M_{st}(\vec{q};\omega)[V_{tt'}(\vec{q}) - \frac{1}{2}V^{ex}_{tt'}(\vec{q})]S_{t's'}(-\vec{q};\omega). \qquad (1)$$

$\vec{q}$ is a vector in the first two-dimensional (2D) Brillouin zone, S is the Random Phase Approximation (RPA) polarizability matrix (Del Sole and Fiorino 1984), and $V_{ss'}$ is the lattice Fourier transform of the substrate-screened Coulomb interaction between two pairs (s and s') of DB orbitals. This matrix V contains the information about local field effects. V^{ex} is the lattice Fourier transform of the exchange interaction between pairs of DB orbitals. The exchange interaction is screened self-consistently by both the substrate and the surface states.

In a previous work (Reining and Del Sole 1988, and to be published) we have evaluated the static screening, which follows from (1) by setting ω=0. Self-consistency in the screened exchange was obtained by consecutive iteration of the calculation. The results serve here as an input for the calculation of the quasi-particle (QP) gap, according to the approaches of Bechstedt and Del Sole (1988) and Gygi and Baldereschi (1989): the difference between the self-energy $\Sigma(r, r', E)$ and the exchange-correlation potential V_{xc}(r) of the LDA is written in terms of the difference between the screening functions of the real system and of the homogeneous electron gas with the local density, neglecting dynamical-screening effects. Since large gap-corrections are expected, we avoid the use of perturbation theory, and determine the self-consistent GW wave functions $|\psi_{n\vec{k}}\rangle$ in the DB basis. The details of the calculation will be given elsewhere. The resulting self-consistent gap is 0.66 eV, in very good agreement with the ab initio result (0.62 eV).

These calculations confirm that a gap larger than the optical one should be expected. However, the precision in determining electron levels, even using ab initio methods, cannot be better than 0.1 or 0.2 eV. This is not sufficient, when one is looking at excitonic shifts, which are of the same order of magnitude, or smaller. In the present case, it is even difficult to determine the gap on the basis of the experimental results (Feenstra et al 1986, Martensson et al 1985, Uhrberg et al 1982, Perfetti et al 1987), since they are contradictory. In order to obtain a well-defined value of the gap, we determine the static dielectric constant $\epsilon^{opt,s}$ of the surface. This quantity has the dimension of length: roughly speaking, it is the dielectric constant of the surface layer times its thickness (Del Sole et al 1984). $\epsilon^{opt,s}$ can be obtained from the measured optical spectrum (Cricenti et al 1988) via a Kramers-Kronig transform. If only transitions involving DB orbitals (i.e. up to 1 eV) are considered in the experimental spectrum, a value of $\bar{\epsilon}^{opt,s} = 102$ Å is obtained for the average over the directions parallel to the surface. The calculated static dielectric constant is a very sensible function of the gap: it decreases by about a factor of 3 when going from E_g=0.45 eV to E_g=0.9 eV. Agreement with experiment is obtained for a gap of E_g=0.75 ± 0.06 eV, where the indetermination of 0.06 eV arises from a supposed uncertainty of 20% in the optical-peak height.

We can now compute the optical response, including excitonic and local-field effects, by assuming that the gap is 0.75 eV. We calculate the differential reflectivity, on the basis of the ω-dependent solution of (1). In order to include self-consistently the excitonic effects, the screened interactions obtained in the previous calculation of static screening (Reining and Del Sole to be published) are used.

The resulting differential reflectivity is shown in Fig. 1, together with the one-electron calculation.

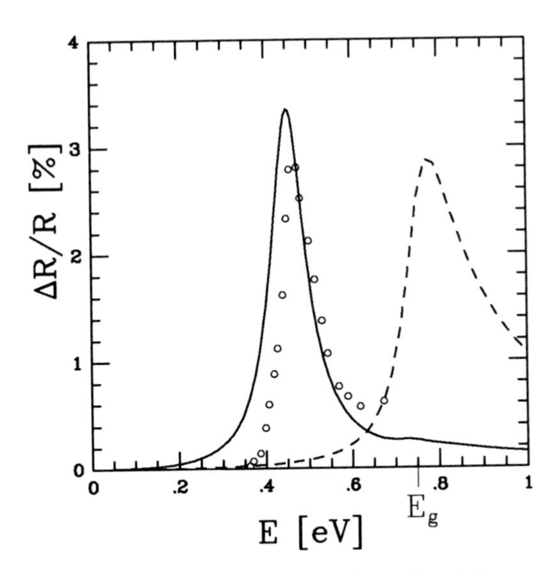

Fig. 1. Contribution of the surface states to the reflectivity, averaged over the directions parallel to the surface. Circles: experimental result (Ciccacci et al 1986). Dashed curve: one-electron calculation. Continuous curve: local field and excitonic effects are included.

It is evident that the latter (dashed curve) is quite different from experiment (Ciccacci et al 1986), since the peak value coincides with the assumed gap, 0.75 eV. On the other hand, the continuous curve, which includes the e-h interaction, has a peak at 0.45 eV, in very good agreement with experiment. This peak corresponds to the lowest exciton state, with a binding energy of 0.3 eV. The quasi one-dimensional character of the exciton shows up in the large binding energy and oscillator strength of the lowest-energy state, and in the vanishing oscillator strength of higher states. In fact, the lowest state of the one-dimensional hydrogen atom (Loudon 1959) is a delta-function localized at the nucleus, with infinite binding energy, while the wave functions of the excited states vanish there. In our case, the central-cell corrections, wich depend on the DB extension, yield a finite value of the binding energy of the lowest state, but only weakly affect the more extended higher lying states, which maintain their one-dimensional character, with vanishing wave functions for zero e-h separation.

In conclusion, we have given a coherent account of the screened e-h interaction, of the gap correction to LDA, of excitons and of the optical properties at Si(111)2×1, in terms of the dielectric response. In particular, we have presented the first calculation of surface excitons carried out on the basis of a realistic shape of the e-h interaction. We have shown that quasi-one-dimensional excitons can exist at clean surfaces.

3. REFERENCES

Bechstedt F and Del Sole R 1988 Phys. Rev. B **38**, 7710

Chiaradia P, Cricenti A, Selci S and Chiarotti G 1984 Phys. Rev. Lett. **52**, 1145

Chiarotti G, Nannarone S, Pastore R and Chiaradia P 1971 Phys. Rev. B **4**, 3398

Ciccacci F, Selci S, Chiarotti G, and Chiaradia P 1986 Phys. Rev. Lett. **56**, 2411

Cricenti A, Selci S, Ciccacci F, Felici A C, Goletti C, Zhu Yong, and Chiarotti G 1988 Phys. Script. **38**, 199

Cricenti A, Selci S, Magnusson K O and Reihl B 1990 Phys. Rev. B **41**, 12908

Del Sole R and Fiorino E 1984 Phys. Rev. B **29**, 4631

Del Sole R and Selloni A 1984 Phys. Rev. B **30**, 883

Del Sole R and Tosatti E 1977 Solid State Commun. **22**, 307

Feenstra R M, Thompson W A and Fein A P 1986 Phys. Rev. Lett. **56**, 608

Gygi F and Baldereschi A 1989 Phys. Rev. Lett. **62**, 2160

Hanke W and Sham L J 1980 Phys. Rev. B **21**, 4656

Loudon R 1959 Am. J. of Physics **27**, 649

Martensson P, Cricenti A and Hansson G V 1985 Phys. Rev. B **32**, 6959

Northrup J E, Hybertsen M S and Louie S G 1991 Phys. Rev. Lett. **66**, 500

Pandey K C and Phillips J C 1975 Phys. Rev. Lett. **34**, 1450

Pandey K C 1981 Phys. Rev. Lett. **47**, 1913; 1982 Phys. Rev. Lett. **49**, 223

Perfetti P, Nicholls J. M. and Reihl B 1987 Phys. Rev. B **36**, 6160

Reining L and Del Sole R 1988 Phys. Rev. B **38**, 12768

Reining L and Del Sole R 1990 *Proc. 20th Int. Conf. on the Physics of Semiconductors*, E. M. Anastassakis and J. D. Joannopoulod editors, World Scientific; p. 187

Reining L and Del Sole R to be published on Phys. Rev. B

Selloni A and Del Sole R 1986 Surf. Sci. **168**, 35

Uhrberg R I G, Hansson G V, Nicholls J M and Flodstrom S A 1982 Phys. Rev. Lett. **48**, 1032

Inst. Phys. Conf. Ser. No 123
Paper presented at the International Meeting on Optics of Excitons in Confined Systems, Giardini Naxos, Italy, 1991

Radiative recombination in GaAs-GaAlAs wires and dots

C M Sotomayor Torres [a] *, P D Wang [a], W E Leitch [a], H Benisty [b] and C Weisbuch [b]

(a) Nanoelectronics Research Centre, Department of Electronics and Electrical Engineering, University of Glasgow, Glasgow G12 8QQ, GB
(b)Thomson-CSF, Laboratoire Central de Recherche, B P 10, F-91404 Orsay Cedex, France

ABSTRACT: We present experimental and theoretical results on the low temperature luminescence intensity of dry etched dots and wires of GaAs-GaAlAs. Our model of the emission yield invokes slower momentum and energy relaxation mechanisms which become stronger as the lateral dimensions decrease. The dependence of the emission strength for wires upon wire length is postulated to be related to exciton diffusion and trapping mechanism.

1. INTRODUCTION

Strong luminescence intensity from nanostructures has been expected based on arguments of concentrated oscillator strength and superradience (see, for example, Hanamura 1988). The possibility of realising such predictions was significantly enhanced with the advent of semiconductors nanostructures such as wires and dots. Early work on GaAs-GaAlAs dots and wires by, for example, Kash et al (1986) gave impetus to research in this field. Many of the subsequent papers on etched structures (Clausen et al 1989, Forchel et al 1988, Sotomayor Torres et al 1991) did not report a luminescence enhancement, but on the contrary, a luminescence intensity decrease with decreasing lateral dimensions of semiconductors wires and dots. We reported the recovery of the luminescence intensity by overgrowth of etched dots (Arnot et al 1989), although uncontrolled blue shifts and linewidth broadening complicated qualitative analysis. To understand the luminescence behaviour of dots and wires variations of a model based on extrinsic phenomena have been put forward (Clausen et al 1989 and Forchel et al 1988). These models necessitate fitting parameters which are, at best, inaccurate since they may include as-grown defects, native surface traps, fabrication-induced traps and include uncertain parameters such as exciton drift velocity. Recently, Benisty et al (1991), based on the theory of Bockelmann and Bastard (1990) concerning slower relaxation of electrons via LA phonons in wires and dots, presented an "intrinsic" approach to understand the poor emission from dots and wires.

In this paper we report theoretical and experimental results concerning the luminescence efficiency of GaAs-GaAlAs nanostructures and show that the luminescence yield is limited by intrinsic phenomena leading to the electron and hole having different quantum numbers in the radiative recombination time scales.

2. EXPERIMENTAL DETAILS

The nanostructures were fabricated in MBE grown quantum well layers, written with an electron beam and $SiCl_4$ reactive ion etched (RIE). The starting material consists of 4nm, 6nm and 8nm quantum wells, separated by 20nm $Al_{0.3}Ga_{0.7}As$ barriers. A 10nm GaAs cap layer was grown at the top. The etching mask is laterally developed SrF_2/AlF_3 so that smaller dots (<60nm) can be readily fabricated. The etching depth is typically 140nm. The filling factor for our dots is 4% giving a typical number of approximately $8x10^4$ 60nm dots being exposed to a 40μm laser spot. The luminescence emission from the structures was characterised by 5K photoluminescence using standard equipment. Low excitation powers were used to avoid exciting luminescence from areas away from the pattern under study. The emission observed was only obtained from the patterns and mesas; no emission was observed from etched-back areas, and no stray emission from patterns other than those under observation was seen. The change in luminescence efficiency of the unpatterned starting material across the wafer was less than ±50% over 5mm. Overall, the measurements of luminescence intensity were reproducible to within a factor of 2; stray points result solely from imperfections in the fabrication technique. CL monochromatic images recorded at 15K of 15 free standing 1μm dots on a similar sample have evidenced dot-to-dot non-uniformity (Williams et al 1991) for any particular quantum well emission.This is particularly important since the emission from dots is normalized to that of a control area (mesa) for intensity comparisons. If a significant number of dots do not emit, then such comparison is related to the worst possible case.

3. EMISSION FROM WIRES AND DOTS

Fig. 1 shows one of the typical PL spectra of 60nm dots and mesa. It shows that the signal can be still detected from dots down to 60nm, although the intensity is only about 1% of the mesa.The emission from the three quantum wells is clearly observed with negligible linewidth broadening. Relative changes in the intensity from the wells in the mesa and the dot array are within a factor of two. Fig. 2 shows the normalised PL intensity as a function of dot size. It demonstrates that the PL intensity starts to decrease rapidly at around 100-200nm diameter for dots and at around 40-50nm for wires. Figure 3 shows the emission intensity of wires normalised to that of a control mesa area as a function of wire width. It can be seen that down to 100nm width/diameter all the points fall within two orders of magnitude from the mesa emission intensity.

We interpret our data using a model of luminescence yield of wires and dots which is based only on intrinsic phenomena (Benisty et al 1991). We have calculated the transition probability between dot (wire) electronic states taking into account both the phonon and electron wavevectors. We have then iterated a master equation on the occupation probabilities to obtain a steady state which allows the computation of fluxes and yields separately radiative and non-radiative recombination channels. The expected decrease in the recombination has an onset

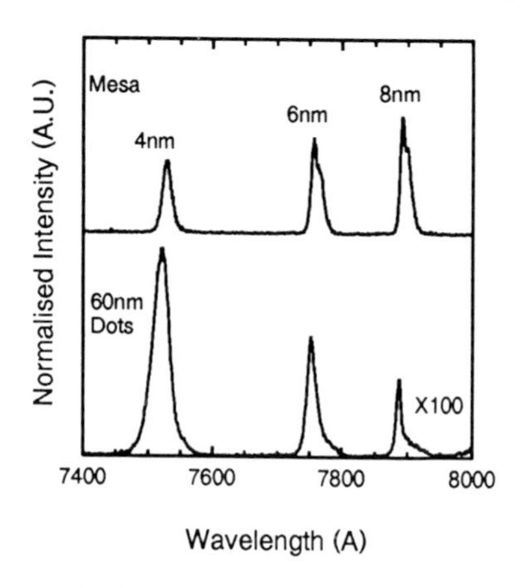

Fig. 1 4K luminescence spectra of GaAs-GaAlAs quantum wells. The three peaks correspond to the 4,6 and 8nm wells.

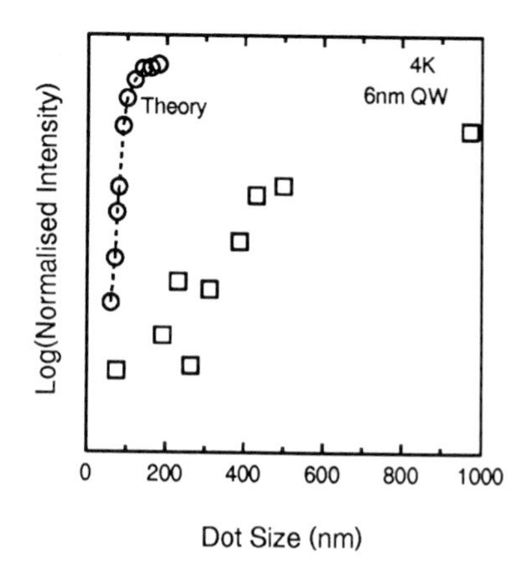

Fig.2 4K luminescence intensity normalised to the control mesa. Open circles are theoretical points using the theory of Benisty et al (1991).

around 150-200nm which agrees clearly with our experimental data. The discrepancies between the theory and the experiments can be accounted for by the dry etching damage which creates defects serving as the nonradiative recombination centre (Clausen et al 1990).

4. EXCITON DIFFUSION IN WIRES

For wires another mechanism needs to be taken into account. For a given wire width the intensity increases as the wire length decreases, suggesting a transport (exciton diffusion?) related phenomenon which is modified by the distribution of traps. Obviously, the shorter the wire, the smaller the probability of finding a trap and the larger the probability of radiative recombination (Leitch et al 1991). A model is currently being devised in which the emission efficiency in the high quality material of our structures is governed by the trap density. The carrier localisation due to well interface roughness is reduced in material with smooth interfaces. Thus the removal of 96% of the well area, together with a scaling of the remaining emission, could lead to an increase of the effective emission efficiency if nonradiative areas within a diffusion length of a trap are removed. This could lead even to an effective emission efficiency above that of the mesa. The GaAs layers grown just prior to the series of MBE quantum well layers used in this work exhibited record free carrier densities and mobilities (Stanley et al 1991). We expect our layers to have 4 x $10^{13}cm^{-3}$ free carriers. This carrier concentration is equivalent to a sheet donor density of 10 ionised donors per μm^2. The implication is that in wires of 1μm long by 100nm width there would be only one ionised donor per wire. For these and smaller dimensions the statistical appearance of defect-free wires is significant, and thus the possibility exists that this might give rise to more efficient emission for shorter wires and dots compared with the emission from longer wires of a given width. Fig. 4 shows the dependence of the emission intensity upon wire length for the 100nm wide wires. The solid lines shown are a fit using the inverse of the square root of the wire length dependence.

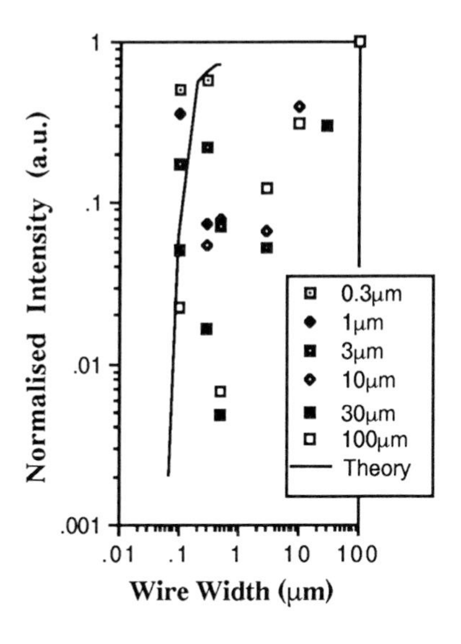

Fig. 3 6K luminescence intensity normalised to that of the control mesa from arrays of wires of various width and lengths. The solid line is the prediction of the theory of Benisty et al (1991).

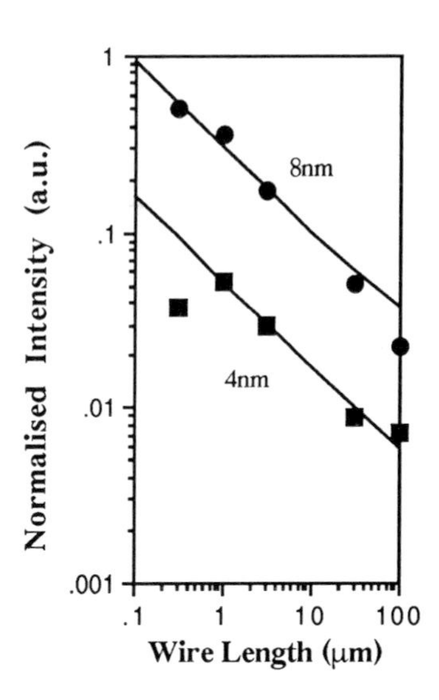

Fig.4 6K luminescence intensity, corrected for area coverage, as a funtion of wire length (L) of the 0.1μmwide wires. The solid lines are a fit using the inverse of the square root of L dependence.

5. DISCUSSION AND CONCLUSIONS

It is clear that intrinsic processes are limiting the emission intensity from dots and wires as demonstrated by the results shown above. One way to study this phenomenon is the use of picosecond luminescence techniques similar to previous time-resolved emission studies (Kash et al 1988 and Izrael et al 1990). This test is currently in progress. However, the effect of extrinsic (as-grown , fabrication-induced) processes, non-reproducible, etc cannot be neglected as indicated by the irregularities of the intensity shown by our cathodoluminescence work. The prospects of using such etched nanostructures for devices based on across-the-gap emission seem rather small. The slower carrier relaxation is, however, expected to increase the strength of intraband processes absorbing and emitting low energy photons and opening other range of possible devices. On the scientific part these results suggest that relaxation phenomena (electron-hole pair, phonons, etc) remain to be further explored theoretically and experimentally since it is clear that limiting energy and momentum spaces dramatically affects the optical properties of nanostructures.

ACKNOWLEDGEMENTS

This work is supported by the UK Science and Engineering Research Council and by the European Community ESPRIT BRA 3133 NANSDEV. The samples were grown by Dr C R Stanley and fabricated by Dr S Thoms and Ms H Wallace. We are grateful to the technical staff of the Nanoelectronics Research Centre for their expert assistance.

6. REFERENCES

* Nuffield Foundation Science Research Fellow

Arnot H E G, Watt M, Sotomayor Torres C M, Glew R, Cusco R, Bates J and Beaumont S P (1989), Superlattices and Microstructures **5**, 459.

Benisty H, Sotomayor Torres C M and Weisbuch C, to appear in Phys. Rev. B

Bockelmann U and Bastard G (1990) Phys Rev B **42**, 8947

Clausen Jr. E, Craighead H G, Worlock J M, Harbison J P, Schiavone L M, Florez L and Van der Gaag B P, Appl. Phys. Lett. **55**, 1427 (1989)

Forchel A, Leier H, Maile B E and German R, Festkörperprobleme **28**, 99 (1988).

Hanamura E (1988) Phys. Rev. B 37 1273

Izrael A, Sermage B, Marzin J-Y, Ougazzaden A, Azoulay R, Etrillard J, Thierry-Mieg V and Henry L, Appl. Phys. Lett. **56**, 830 (1990).

Kash K, Scherer A, Worlock J M, Craighead H C and Tamargo M C (1986) Appl. Phys. Lett. **49**, 1043

Kash K, Grabbe P, Nahory R E, Scherer a, Weaver A and Caneua C (1988) Appl. Phys. Lett. **53**, 2214

Leitch W E, Sotomayor Torres C M, Lootens D, Thoms S, Vand Daele P, Stanley C R, Demeester P and Beaumont S P (1991) to appear in Surf. Science

Sotomayor Torres C M, Leitch W E, Lootens D, Wang P D, Williams G M, Thoms S, Wallace H, Van Daele P, Cullis A G, Stanley C R, Demeester P and Beaumont S P, in: **Nanostructures and Mesoscopic Systems**, Eds W P Kirk and M A Reed, Academic Press (in press)

Stanley C R, Holland M C and Kean A H (1991) Appl. Phys. Lett. 58 478

Williams G M, Cullis A G, Sotomayor Torres C M, Thoms S, Lootens D, Van Daele P, Beaumont S P, Stanley C R and Demeester P, Institute of Physics Conference Series **117**, to be published.

Inst. Phys. Conf. Ser. No 123
Paper presented at the International Meeting on Optics of Excitons in Confined Systems, Giardini Naxos, Italy, 1991

Excitons in zero-dimensional nanostructures

Garnett W Bryant

Applied Physics Branch, Harry Diamond Laboratories, Adelphi, Maryland 20783-1197 USA

ABSTRACT: Exciton states in zero-dimensional quantum dots are determined by the competing effects of the electron-hole attraction and the single particle confinement. We present calculations of exciton and biexciton states in quantum dots, exciton-phonon relaxation rates, and exciton states in coupled quantum dots. The results reveal how this competition determines the properties of excitons confined in quantum dots.

1. INTRODUCTION

Confinement of excitons in zero-dimensional quantum dot nanostructures concentrates the optical oscillator strength into discrete transitions and provides the alluring promise of enhanced optical properties for these structures. Understanding exciton states in quantum dots is necessary to exploit this promise. Understanding confinement effects in dots presents challenges not addressed for wells and wires, because the single-particle spectrum for a quantum dot is discrete rather than a set of subbands, each with a continuum of states.

The properties of excitons confined in isolated quantum dots are determined by the competing effects of the electron-hole attraction and the three-dimensional size quantization. In large dots, the coulomb interaction, which scales as $1/L$ where L is the dot size, dominates and the excitons are bulk-like. In small dots, quantum confinement , which scales as $1/L^2$, is the critical effect. As the confinement in a quantum dot increases, coulomb-induced mixing of states becomes more difficult, even though the coulomb energies increase, and the excitons become uncorrelated electron-hole pairs frozen into the lowest-energy confinement states. Freeze-out of motion in the confined dimensions also occurs in quantum wells and wires when the confinement is increased. However coulomb-induced correlation of the unconfined motion in a well or wire is enhanced rather than frozen out by the confinement. A unique challenge to understanding excitons confined in dots is to correctly describe the transition from the correlated exciton in the bulk limit to the uncorrelated pair in the limit of complete confinement. Determining the length scale for this transition is necessary to accurately model the onset of the laser characteristics (Asada et al 1986), nonlinear optical response (Schmitt-Rink et al 1987), electroabsorption (Miller et al 1988) carrier induced bleaching (Sakaki et al 1990) and phonon scattering rates (Bockelmann and Bastard 1990) expected for quantum dots in the limit of complete confinement.The transition between these two regimes is different for different exciton properties. We present calculations of exciton and biexciton states in quantum dots, exciton-phonon relaxation rates, and exciton states in coupled quantum dots to show how this competition determines the properties of confined excitons.

The effects of three-dimensional confinement have been observed optically in both spherical microcrystallites and in quantum dots fabricated by confining lateral motion in two-dimensional quantum wells. In this paper we focus our attention on excitons confined in quantum dots fabricated from quantum wells. Original experiments (see for example Kash et al 1986, Reed et al 1986, Cibert et al 1986, and Temkin et al 1987) were not able to provide conclusive evidence for three-dimensional confinement in these structures, in part because of low photoluminescence efficiency due to surface recombination at exposed etched dot surfaces and in part because the dots were large and confinement effects were weak relative to the electron-hole attraction. Fabrication of quantum dots without exposed surfaces either by regrowth on the exposed surface (Arnot et al 1989), or by use of strain-induced lateral confinement (Kash et al 1989) improves photoluminescence efficiency. Confinement effects can be seen by use of strain-induced lateral confinement both in dot arrays and in single dots.

2. THEORY

Exciton states of electron-hole pairs confined in quantum dots have been determined by solving the electron-hole effective mass Schrodinger equation by use of a variational approach (Bryant 1988 and Tran Thoai et al 1990), as has been done for excitons in bulk and in quantum wells, by a configuration-interaction approach (Bryant 1988 and Chakraborty 1991), as was used to study multielectron systems in quantum dots (Bryant 1987), by perturbation theory for small dots where the electron-hole attraction is a small perturbation to the confinement energies (Banyai 1989 and Bryant 1990) and by a path integral approach (Pollock and Koch 1991).The electron-hole interaction is the coulomb interaction screened by the background dielectric constant. Image charge effects are unimportant for quantum dots made from lateral confinement of GaAs/AlGaAs quantum wells. The lateral confinement potential is typically treated as an infinite barrier hard wall potential although a parabolic confinement potential that accounts for charge depletion at the sidewalls or the strain-induced band deformation has been used (Chakraborty 1991) and is more realistic. Both models for the confinement give qualitatively the same results.

In this paper we model excitons confined in quantum dots which are constructed from narrow two-dimensional quantum wells by laterally confining the two-dimensional motion. Typically the width w of the two-dimensional dot is an order of magnitude less that the length L of the side of the dot. We model the dots as square thin plates as shown in Fig. 1. The lateral confinement is determined by infinite barrier potentials. Confinement in the well is also modeled with an infinite barrier potential. In our variational approach for an exciton in an isolated dot, the variational wave function f is chosen to be a product of the electron and hole lowest-energy single particle states and a linear combination s of Gaussian functions of the electron-hole separation to account for correlation in the exciton (see Bryant 1988)

$$f(\mathbf{r}_e,\mathbf{r}_h) = c(\mathbf{r}_e)c(\mathbf{r}_h)s(\mathbf{r}_e,\mathbf{r}_h),$$

$$c(\mathbf{r}_i) = \left(\frac{cos(kx_i)}{\sqrt{L/2}}\right)\left(\frac{cos(ky_i)}{\sqrt{L/2}}\right)\left(\frac{cos(qz_i)}{\sqrt{w/2}}\right), \tag{1}$$

$$s(\mathbf{r}_e,\mathbf{r}_h) = \sum_n c_n \, exp\{-a_n[(x_e - x_h)^2+(y_e - y_h)^2]\}.$$

with $k = \pi/L$ and $q = \pi/w$. Accurate energies and wave functions are obtained by use of 5 to 10 Gaussians. No correlation of z motion has been included because single particle levels in the well are sufficiently separated. With this wave function, electron-hole separations and oscillator strengths can also be calculated (Bryant 1988).

3. EXCITONS IN QUANTUM DOTS

The ground state energy E_{GS} of a confined heavy-hole exciton in a square, two-dimensional (w=0) dot, calculated by use of the variational approach, is shown in Fig. 1. Results are qualitatively the same for dots with a small finite thickness and for light-hole excitons. Effective masses and dielectric constant for GaAs have been used. The energies are scaled by the exciton effective Rydberg R_μ = 3.03 meV.The energies clearly reveal the transition from exciton to uncorrelated pair as the confinement increases. The shift of E_{GS} from the energy of an unconfined two-dimensional exciton, -4R_μ, is less than five per cent for L > 100 nm. For

$L < 10$ nm, the confinement energy dominates and the ground-state energy varies as $1/L^2$. At much smaller L this scaling will break down in real dots with finite barriers which cannot confine the exciton. The exciton kinetic energy E_{KE} and coulomb energy E_C ($E_{GS} = E_{KE} + E_C$) also show only a weak dependence on L for large dots. In contrast, the confinement energy E_{NI} of a noninteracting electron-hole pair displays a rapid variation as the confinement increases. In large dots the exciton kinetic energy is much larger than the confinement energy, indicating that the kinetic energy is due to correlation as an exciton. As confinement increases, the exciton kinetic energy changes more rapidly than the coulomb energy because the kinetic energy includes the contribution from the confinement energy. As L decreases, E_{KE} approaches E_{NI} and correlation becomes unimportant.

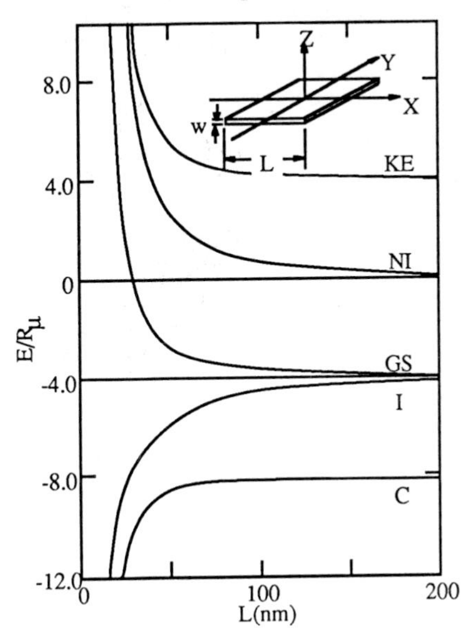

Fig. 1. Ground state (GS) energy of a heavy hole exciton confined in a square dot ($w = 0$). Exciton kinetic (K) and coulomb (C) energy, confinement energy (NI) and interaction energy (I) are shown. The inset shows the dot configuration.

The transition from exciton to uncorrelated electron-hole pair occurs in dots which are still bigger than the free exciton. The crossover occurs when the single particle level splittings become bigger than the interaction energies and the levels cannot be mixed by the electron-hole attraction. The electron-hole separation R of a heavy-hole exciton, calculated including the confinement and/or correlation, is shown in Fig. 2. The free-exciton radius (10.9 nm) is comparable to the size of the dot where confinement is complete and an order-of-magnitude smaller than the L where confinement effects begin. The electron-hole separation, as determined with a wave function which contains the correlation but not the confinement, increases as L decreases showing that dot confinement determines the exciton size as L decreases.

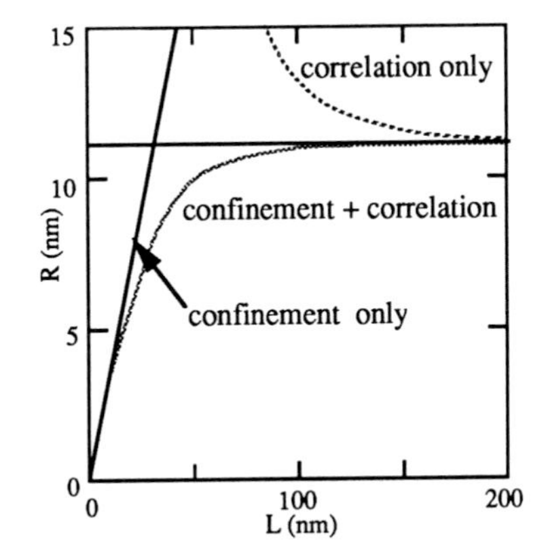

Fig. 2. Radius R of a heavy-hole exciton in a square dot ($w = 0$).

4. BIEXCITONS IN QUANTUM DOTS

The stability of the confined biexciton can be investigated by low order perturbation theory (Banyai 1989, Bryant 1990, and Pollock and Koch 1991) because, as shown for excitons, the coulomb interaction is a small perturbation to the confinement energy in small dots. In small dots the exciton energy has a contribution from zeroth- (kinetic and confinement energy) and first-order (coulomb energy of the electron-hole pair frozen in the unperturbed ground state). Higher order contributions to the exciton energy, due to electron-hole correlation, are unimportant for small dots. However the zeroth- and first-order contributions to the biexciton energy are exactly twice the zeroth- and first-order exciton energies. The stability of the biexciton is determined by the first finite contribution from perturbation theory, which is the second order term that includes the lowest order correlation effects. Correlation effects between interacting excitons stabilize the biexciton, relative to two independent excitons, and remain significant even at dot sizes where intraexciton correlation becomes unimportant.

The biexciton-exciton stabilization energy for a biexciton confined in a thin square GaAs dot, determined by third order perturbation theory (Bryant 1990) $\Delta = \Delta_2 + \Delta_3$, is shown in Fig. 3. The $L = 0$ limit is the second order term Δ_2 and the linear change is the third order term Δ_3. The stabilization energy is enhanced by a factor of 5 by confinement in a small dot. The following qualitative model can be deduced. The confined biexciton binding is strongly enhanced in small structures. The binding is greater for biexcitons formed from heavy holes because coulomb and hole-hole correlation are more important for heavier, more localized holes. The reduction in binding as L increases occurs more rapidly for small m_e/m_h. For small m_e/m_h, the biexciton acts as two excitons, each tightly bound so that exciton-exciton interactions are weak unless dot confinement forces the two excitons to overlap. However, for small m_e/m_h the coulomb energies are large and enhanced binding occurs when the excitons are forced by confinement to overlap. The exciton is less tightly bound when m_e and m_h are similar, so enhancements in binding energy in small dots are smaller but the exciton-exciton interaction has a longer range and some enhancement persists in larger dots.

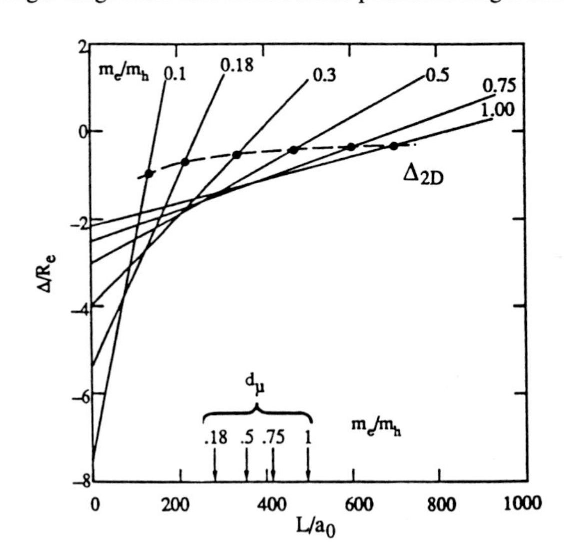

Fig. 3. Biexciton-exciton energy difference $\Delta = \Delta_2 + \Delta_3$ for the indicated electron-hole mass ratios. Energies are scaled by R_e = 5.4 meV, lengths by the Bohr radius. Δ_{2D} is the binding energy of a free two-dimensional biexciton (Kleinman 1983). Diameters of free two-dimensional excitons are indicated.

5. EXCITON-PHONON SCATTERING RATES

Recently Weisbuch et al (1991) suggested that the poor photoluminescence efficiency in quantum dots need not result from extrinsic damage due to processing but can also be an intrinsic property of quantum dots. They attribute this poor luminescence to the suppression of excited electron and hole energy relaxation by acoustical phonon scattering that occurs when electrons and holes are confined in a quantum dot (Bockelmann and Bastard 1990). Suppression of inter-hole relaxation was actually observed by Reed et al (1986).

Bockelmann and Bastard (1990) calculate electron-acoustical phonon scattering rates for single electrons confined in a quantum dot. Electron states are discrete. Energy conservation allows only a discrete spectrum of the acoustical phonons to contribute to the scattering. Conservation of electron-phonon momentum during scattering is relaxed when the electron is confined in a dot and the dot walls can share the transferred momentum. Bockelmann and Bastard find for small dots that energy conservation requires the emission of phonons with wave vectors too large to be accommodated by the dot even when momentum conservation is relaxed.

When the electron is correlated with a hole, the hole can also accommodate momentum not conserved during the electron-phonon energy relaxation. We have calculated scattering rates due to acoustical phonon emission when a confined electron-hole pair with an electron in an excited state relaxes by phonon emission to the correlated electron-hole ground state. To show how correlation can affect the relaxation rates we show in Fig. 4 the matrix element, $|\langle f_i | exp(iqx) | f_{gs} \rangle|^2$, that determines the phonon scattering rates. We assume that the initial excited exciton state f_i is a state of the form given in Eq. (1) but with the electron in its first excited state. Eq. (1) is used for the ground state f_{gs}. The actual matrix element is the product

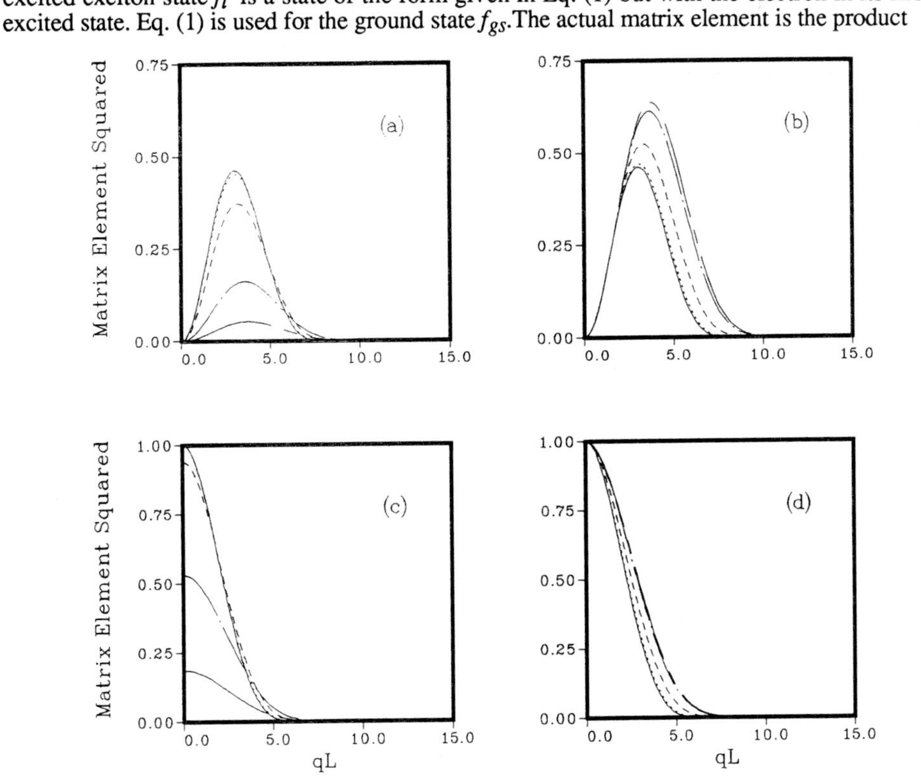

Fig. 4. Exciton-phonon scattering matrix element squared.

of matrix elements for lateral (x- and y-), and z-motion. We show the one-dimensional x-matrix elements: in Figs. (4a) and (4b) the electron is initially excited into the first excited x-state, and in Figs. (4c) and (4d) the electron is initially in the x-ground state and the first y-excited state. Two extremes are considered: no correlation in the initial state (Figs.(4a) and (4c)), and the same correlation in the initial state as in the ground state (Figs. (4b) and (4d)).In each figure the solid curve shows the matrix element when both initial and final states are uncorrelated. The one-dimensional matrix element for an free particle is a δ-function at $q = 0$ when the initial and final state are the same and a δ-function at finite q when the initial state is an excited state. The single particle scattering matrix element broadens when dot confinement relaxes the momentum conservation. The width varies as $1/L$. To illustrate the effect of correlation, we use a correlation factor with just one term. The four dashed curves show the effect of correlation when the decay coefficient in the exponent of the correlation factor is $a_1 = \alpha/L^2$ with $\alpha = 0.1$, 1, 10, and 100. Larger correlation does not change the matrix elements further. The scattering is suppressed if the initial state is uncorrelated and does not overlap well with the correlated ground state. When both states are correlated, there is a small enhancement of the matrix element and a small additional broadening of the q-dependence. However, correlation does not accommodate enough additional momentum transferred during phonon scattering to qualitatively change single particle scattering rates in dots.

6. EXCITONS IN COUPLED QUANTUM DOTS

Coupled double wells have attracted attention because their optical properties can be controlled by interwell charge transfer. Similar promise should exist for coupled quantum dots formed by confining lateral motion in coupled double wells. In this section we consider how interwell coupling modifies excitons confined in coupled quantum dots. We use a variational approach, based on a generalization of the wave function in Eq. (1), that is similar to the approach used by Dignam and Sipe (1991) for coupled double wells. We assume that the two wells are coupled through a finite barrier while lateral motion and motion away from the two wells is confined by infinite barriers. The lowest energy states localized in each well are used as a basis set of well states. Interwell tunneling is included to account for the mixing of the localized well states. The exciton wave function has a contribution for each configuration of the electron and hole in the two dots. Each contribution has the form of Eq. (1) with the appropriate electron and hole well states. A different correlation factor is used for each term because the correlation will depend on which wells the electron and hole occupy.

The energies of the ground state and the first excited state with the symmetry of Eq. (1) for a heavy-hole exciton in a double dot are shown in Fig. 5. The energy zero is the ground state energy of an unconfined, noninteracting pair in the double well. The double dots are laterally confined 5 nm GaAs wells separated by a 2 nm barrier (250 meV for electrons and 100 meV for holes). For thicker barriers, the two dots decouple and the exciton is localized in one well. For narrow barriers and large L, the electron-hole correlation is suppressed so that the electron and hole can independently occupy both dots to reduce the localization energy. Thus, for example, the exciton kinetic energy and the lateral confinement energy are nearly the same in large dots (in contrast to Fig. 1 for single dots). The reduction of coulomb effects in large coupled dots should make confinement effects easier to observe. Increasing the confinement in the double dot enhances the tendency of the electron-hole attraction to localize the pair in the same dot. Confinement enhances the correlation effects in double dots. This is shown dramatically for the excited state. In large dots E_{KE} and E_{NI} are nearly the same. However the exciton kinetic energy increases rapidly due to the increased correlation as L decreases. The size of the first excited state also shrinks rapidly when the correlation is enhanced.

The energies of the ground state and the first excited state of the heavy-hole exciton in a biased double dot are shown in Fig. 6. The dots are 5 and 4 nm GaAs wells separated by the same barrier as used for Fig. 5. An applied bias of 20 meV switches the ordering of the hole states. For weak confinement, the electron is in the wide well, the hole is in the narrow well, and correlation is suppressed. As L decreases, the exciton energy is minimized by localizing the hole to the wide well, despite the applied bias, to take advantage of increased coulomb energy. The exciton kinetic and coulomb energies increase rapidly as L decreases. When the exciton ground state localizes to the same well, the first excited state also changes dramatically. For large dots, the excited state is the first excited state for the configuration with the electron and hole in different wells and, as a result, has large but nearly canceling kinetic and coulomb energies. In small dots the first excited state remains delocalized but is now the lowest energy state with the electron and hole in different wells. The kinetic and coulomb energies are lower for that reason. Dramatic changes in double dot optical properties are expected when increasing the lateral confinement localizes the electron and hole to the same dot.

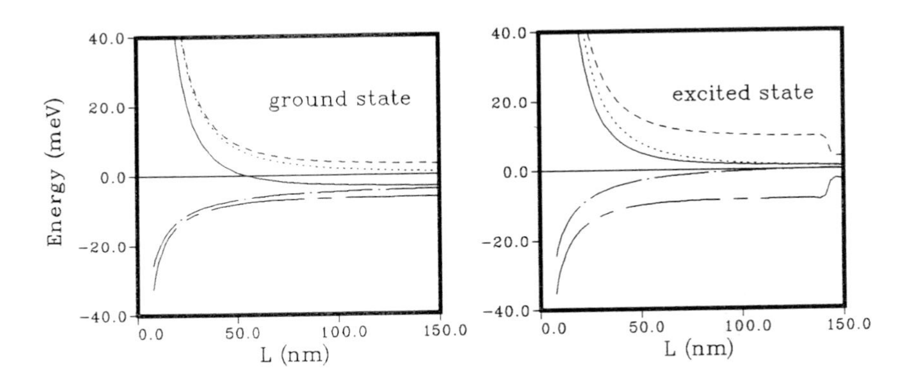

Fig. 5. Total energies of the exciton ground and first excited state (solid curves) in a double dot made from two 5 nm wells coupled by a 2 nm barrier. Exciton kinetic (dash) and coulomb energy (long-short dash), lateral confinement energy (dot) and interaction energy (dash-dot) are also shown.

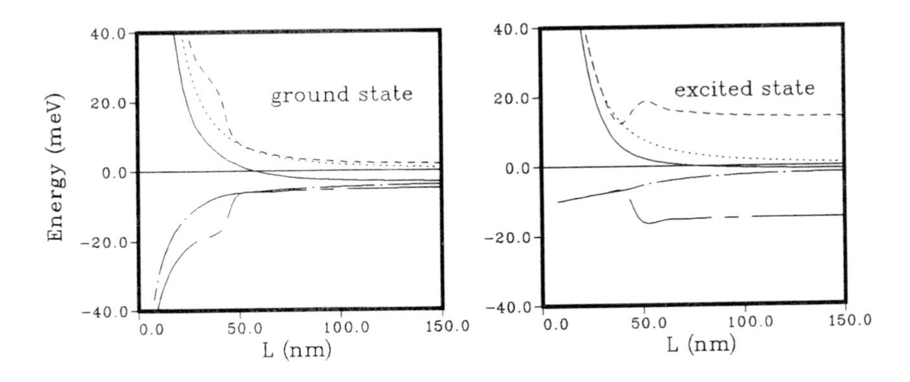

Fig. 6. Energies of the exciton ground and first excited state in a double dot made from 5 nm and 4 nm wells coupled by a 2 nm barrier. An applied bias of 20 meV reverses the ordering of the hole states. Curves are labeled as in Fig. 5.

7. SUMMARY

In the simple picture for excitons confined in quantum dots, the electron-hole attraction is the dominant effect in large dots while confinement is dominant in small dots. Increasing confinement weakens correlation in an exciton confined in an isolated dot. However, the picture is not as simple for interacting excitons and for excitons in coupled dots. Biexciton stabilization energies due to correlations between interacting excitons are enhanced by increasing the confinement. Coulomb effects can be weak for excitons in large coupled dots with the electron and hole delocalized. Increasing the confinement in double dots increases the correlation by localizing of the electron and hole to the same dot. The optical properties of double dots are sensitive to the electron-hole localization. Confinement-induced level shifts and changes in optical properties may be easier to observe for excitons in coupled dots than in single dots.

REFERENCES

Arnot H E G, Watt M, Sotomayor-Torres C M, Glew R, Cusco R, Bates J and Beaumont S P 1989 Superlattices and Microstructures **5** 459
Asada M, Miyamoto Y and Suematsu Y 1986 IEEE J. Quantum Electron. **QE-22** 1915
Banyai L 1989 Phys. Rev. B **39** 8022
Bockelmann U and Bastard G 1990 Phys. Rev. B **42** 8947
Bryant G W 1987 Phys. Rev. Lett. **59** 1140
Bryant G W 1988 Phys. Rev. B **37** 8763
Bryant G W 1990 Phys. Rev. B **41** 1243
Chakraborty T 1991 Proc. International Symposium on Nanostructures and Mesoscopic Systems (in press)
Cibert J, Petroff P M, Dolan G J, Pearton S J, Gossard A C and English J H 1986 Appl. Phys. Lett. **49** 1275
Dignam M M and Sipe J E 1991 Phys. Rev. B **43** 4084
Kash K, Scherer A, Worlock J M, Craighead H G and Tamargo M C 1986 Appl. Phys. Lett. **49** 1043
Kash K, Bhat R, Mahoney D D, Lin P S D, Scherer A, Worlock J M, Van der Gaag B P, Koza M and Grabbe P 1989 Appl. Phys. Lett. **55** 681
Kleinman D A 1983 Phys. Rev. B **28** 871
Miller D A B, Chemla D S and Schmitt-Rink S 1988 Appl. Phys. Lett. **52** 2154
Pollock E L and Koch S W 1991 J. Chem. Phys. **94** 6778
Reed M A, Bate R T, Bradshaw K, Duncan W M, Frensley W R, Lee J W and Shih H D 1986 J. Vac. Sci. Technol. **B4** 358
Schmitt-Rink S, Miller D A B and Chemla D S 1987 Phys. Rev. B **35** 8113
Temkin H, Dolan G J, Panish M B and Chu S N G 1987 Appl. Phys. Lett. **50** 413
Tran Thoai D B, Hu Y Z and Koch S W 1990 Phys. Rev. B **42** 11261
Weisbuch C, Sottomayor-Torres C M and Benistry H 1991 Proc. International Symposium on Nanostructures and Mesoscopic Systems (in press)

Inst. Phys. Conf. Ser. No 123
Paper presented at the International Meeting on Optics of Excitons in Confined Systems, Giardini Naxos, Italy, 1991

Excitons, biexcitons and optical nonlinearities in semiconductor quantum dots

S.W. Koch, Y. Z. Hu, B. Fluegel, and N. Peyghambarian

Optical Sciences Center and Physics Department, University of Arizona, Tucson, AZ 85721

ABSTRACT: Recent theoretical and experimental investigations of the linear and nonlinear optical properties of semiconductor quantum dots are summarized. Exciton and biexciton energies and wavefunctions are computed using a numerical matrix diagonalization scheme that includes the Coulomb interaction between the optically excited carriers, quantum confinement effects, and the confinement induced valence band mixing. Calculations of the nonlinear optical response indicate that the excitation of biexciton states leads to increasing absorption features on the high energy side of the bleached one-pair resonances. Calculated spectra are compared to results of pump-probe experiments using CdS and CdSe quantum dots in a glass matrix.

1. INTRODUCTION

In this paper we summarize some of our recent theoretical and experimental results on the linear and nonlinear optical properties of semiconductor quantum dots. Here, we use the term *quantum dot* to denote small clusters of semiconductor material (semiconductor microcrystallites) which have a size of the order of the exciton Bohr radius in the corresponding bulk semiconductor. Typically, these microcrystallites are embedded in some host material, like glass, certain liquids or organic solvents, which provides a confinement potential for the electronic excitations. The resulting three-dimensional quantum confinement leads to characteristic modifications of the linear and nonlinear optical spectra. Ideally, one expects a linear quantum dot absorption spectrum which consists of isolated absorption lines, in contrast to the excitons and continuum absorption in bulk materials. However, due to the large broadening and the size distribution of the crystallites in presently available samples these resonances have been experimentally observed only as relatively broad peaks (for a few examples see, e.g, Brus, 1984 and 1986; Ekimov *et al.*, 1984 and 1985; Borrelli *et al.* , 1987; Peyghambarian *et al.* , 1989*a* and 1989b; Roussignol *et al*, 1989; Wang *et al.* 1989).

It is well known that a theoretical treatment of the linear and nonlinear optical properties of bulk semiconductors requires the analysis of the optically generated electron-hole excitations (see e.g. Haug and Koch, 1990). Usually, one needs quantum mechanical many-body theory to compute the optical nonlinearities resulting from resonantly excited high density electron-hole systems. In contrast, however, in sufficiently small quantum dots an analysis of the optical properties with the standard nonlinear optics approach of expanding the material response in powers of the field, is reasonable because of the large separation between the energy levels of the excited electron-hole pairs. The linear absorption spectra are computed from the imaginary part of the linear susceptibility, $\chi^{(1)}$, and the lowest order nonlinearities are obtained from the third-order susceptibility, $\chi^{(3)}$, respectively.

To evaluate the expressions for $\chi^{(1)}$ and $\chi^{(3)}$, one needs the energies and wavefunctions of the relevant one- and two-electron-hole-pair states, i.e. of the exciton and biexciton states. In Sec. 2 we outline our calculation of exciton and biexciton energies and wavefunctions using a numerical matrix diagonalization method. Using the energies and eigenvectors for a sufficiently large set of states we then evaluate the various dipole matrix elements for transitions between ground-state, one- and two-electron-hole-pair states. In Sec. 3 we briefly discuss the confinement induced valence band mixing in quantum dots, and in Sec. 4 we compare computed spectra with experimental results. Sec. 5 contains a brief summary of our approach.

2. ELECTRON-HOLE-PAIR EXCITATIONS IN IDEAL QUANTUM DOTS

As direct consequence of the *mesoscopic* size of the semiconductor microcrystallites, the boundary conditions have an important influence on the physical properties. The finite dot size leads to the quantum confinement effects which manifest themselves in quantized energy eigenstates and modified electron and hole wavefunctions. In this section we present a brief summary of the theoretical approaches needed to analyze the optical properties of the quantum confined electron-hole excitations.

In order to simplify the calculations we idealize the shape of a quantum dot as simple sphere. For infinite confinement potential and without Coulomb interaction, one can solve the electron and hole problem in the spherical quantum dot analytically (Efros and Efros, 1982; Brus, 1984 and 1986). The single-particle wavefunctions $\phi(r,s)$ for this case are

$$\phi_N^i(r,s) = 0 \quad \text{for } |r| \geq R$$

and for $|r| < R$

$$\phi_N^i(r,s) = \sqrt{\frac{2}{R^3}}\,\frac{j_\ell\left(\alpha_{n\ell}\frac{|r|}{R}\right)}{j_{\ell+1}(\alpha_{n\ell})}\,Y_{\ell m}(\theta,\Phi)\,\delta_{\sigma s} \tag{1}$$

with

$$i = e, h;\; N = \{n,\ell,m\};\; n = 1, 2, 3, \ldots;\; \ell = 0, 1, 2, \ldots$$
$$m = -\ell, -\ell+1, \ldots, \ell-1, \ell;\; \sigma = -\frac{1}{2}, \frac{1}{2};$$

and the corresponding energy eigenvalues are

$$E_N^i = \frac{\hbar^2}{2m}\left(\frac{\alpha_{n\ell}}{R}\right)^2 = \left(\frac{a_{iB}}{R}\right)^2 E_{iR}\,\alpha_{n\ell}^2 \,. \tag{2}$$

In Eqs. (1) and (2) R denotes the radius of the quantum dot, j_ℓ is the ℓth order spherical Bessel function, $\alpha_{n\ell}$ is its nth root, $Y_{\ell m}$ is a spherical harmonic function, and the variables $|r|$, θ, Φ are the spherical coordinates of the vector r. E_{iR} and a_{iB} are Rydberg energy and Bohr radius evaluated using the mass of the electron (hole) for $i = e$ (h). Eq. (1) shows that, up to trivial scaling factors, the single-particle wavefunctions and energies for electrons and holes are equal. The single-particle energies, Eq. (2), in quantum dots form discrete levels, in contrast to the well-known case of bulk semiconductors, where one has a continuum. The spacing between the quantum dot levels increases like R^{-2} with decreasing radius.

As long as we ignore the Coulomb interaction, the wavefunction for an electron-hole pair, i.e.

an exciton, is simply the product of the corresponding electron and hole wavefunctions

$$\phi_N^{ex}(r,s) = \phi_N^e(r,s)\,\phi_N^h(r,s)\,, \tag{3}$$

and the exciton energy is the sum of the electron and hole energies, respectively. This so-called strong confinement approximation has often been used as first approximation to estimate the pair energies in quantum dots with a radius clearly smaller than the Bohr radius of the exciton. However, this approximation leads to wrong conclusions, especially, where energy differences or details of the electron-hole wavefunctions are involved. Furthermore, the strong-confinement approximation yields wrong selection rules for the optical transitions and, of course, all binding energies are exactly zero.

To obtain reliable results for the electron-hole excitations in quantum dots we used a numerical scheme (Hu *et al.* , 1990a and 1990b) which allows to treat the Coulomb interaction exactly for the relevant states. For this purpose, we expand the one- and two-electron-hole pair wavefunctions for the system with Coulomb interaction into the complete set of basis functions defined by the system without Coulomb interaction. The expansion for the one-pair-states is

$$\Phi^{ex}_{n_1 n_2 \ell_1 \ell_2; LM}(r_e, r_h) = \sum_{m_1 m_2} \langle \ell_1 m_1 \ell_2 m_2 | LM \rangle\ \phi^e_{N_1}(r_e)\,\phi^h_{N_2}(r_h), \tag{4}$$

where $\langle \ell_1 m_1 \ell_2 m_2 | LM \rangle$ is the Clebsh-Gordon coefficient and L and M label the angular momentum eigenstates of the pair. The corresponding expansion for the two-pair states involves the product of four single particle functions as has discussed in detail by Hu *et al.* (1990b).

The expansion in Eq. (4), and correspondingly for the two-pair states, is truncated at some finite number and the resulting Hamiltonian matrices are diagonalized numerically. For the regime $R \simeq a_B$ or smaller, the kinetic energy contributions which give the diagonal elements dominate over the Coulomb contributions which give the off-diagonal elements. Therefore, since we only want to compute the eigenfunctions and energies of the few lowest states, it is a rapidly converging procedure to take a finite matrix to represent the total, infinite matrix. The obtained results are then checked by increasing the size of the finite matrix.

3. VALENCE BAND-MIXING IN QUANTUM DOTS

To improve the idealized description of electron-hole-pair eigenstates in semiconductor quantum dots, we have recently extended our numerical matrix diagonalization scheme to include confinement induced valence band mixing (Xia, 1989; Vahala and Sercel, 1990; Koch *et al.*, 1991). Generally, the top two valence bands in most II-VI and III-V semiconductors are almost or completely degenerate at $k = 0$ (Γ point). The effective masses corresponding to the top valence bands are usually quite different, leading to the so-called heavy- and light-hole bands. These bands are strongly influenced by the quantum confinement in semiconductor quantum dots. To deal with this situation we adopt the so-called envelope function approximation to separate the Bloch functions of the valence band electrons into the lattice periodic parts and the envelope functions. In bulk materials the envelope functions are plane waves, but in quantum confined systems they are modified to satisfy the material boundary conditions. The lattice periodic part of the wavefunctions is assumed to be unchanged under quantum confinements conditions, which is a reasonable assumption as long as the microcrystallites are sufficiently large in comparison with the lattice unit cells.

For the envelope functions we use a method based on the well known $\mathbf{k}\cdot\mathbf{p}$ theory. The empirical Hamiltonian introduced by Luttinger (1956) to describe the semiconductor valence band structure in the limit of strong spin-orbit interaction has been transformed by Baldereschi and Lipari (1973) in terms of irreducible spherical tensor operators:

$$H_{BL} = \frac{\gamma_1}{2m_0}\left[p^2 - \frac{\mu}{9}\left(\mathbf{p}^{(2)} \cdot \mathbf{J}^{(2)}\right)\right]. \tag{5}$$

Here

$$\mu = \frac{6\gamma_3 + 4\gamma_2}{5\gamma_1},$$

m_0 is the free electron mass and γ_1, γ_2, and γ_3 are the Luttinger effective mass parameters. In Eq. (5) **p** is the linear momentum operator and **J** is the angular momentum corresponding to spin 3/2. $\mathbf{p}^{(2)}$ and $\mathbf{J}^{(2)}$ are spherical tensor operators of rank 2 which are discussed in detail by Baldereschi and Lipari (1973).

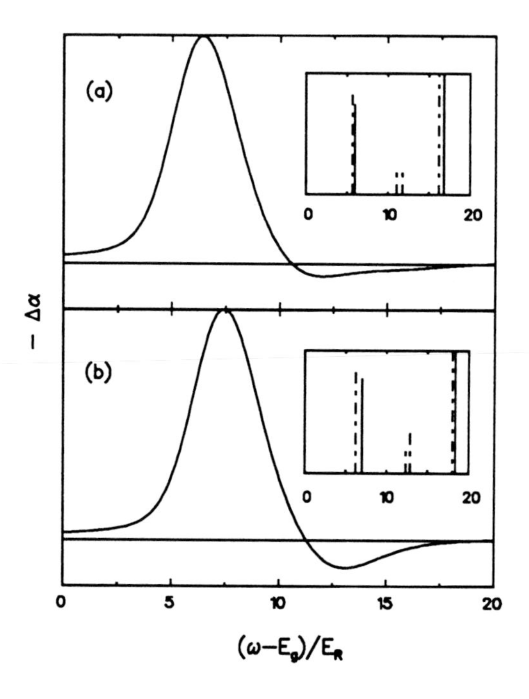

Fig. 1: Change in optical transmission, $-\Delta\alpha$, computed from the imaginary part of χ_3, as function of probe-energy detuning from the bulk semiconductor band-gap E_g (in units of the bulk-exciton Rydberg energy E_R). The shown results are for $R/a_0 = 1$, $\epsilon_2/\epsilon_1=1$ (a) and $\epsilon_2/\epsilon_1=10$ (*b*). The insets show the energetic position and the normalized oscillator strengths of the one- and two-pair transitions as dashed and full lines, respectively.

To compute the hole energy states in the presence of quantum confinement, we replace the parabolic kinetic energy term for the holes by the Hamiltonian (5). Then we repeat our numerical matrix diagonalization analysis outlined in the previous section (Koch *et al.*, 1991).

4. RESULTS AND COMPARISON WITH EXPERIMENTS

Using the results discussed in the previous sections, we evaluate the linear and nonlinear optical properties of the quantum dots. We restrict ourselves to χ^1 and χ^3 for which we need the energies of the eigenstates and the dipole elements between the eigenstates. Phenomenologically we also include relaxation between the states and, in order to compare with experimental results, we also include the size distribution of quantum dots in real samples.

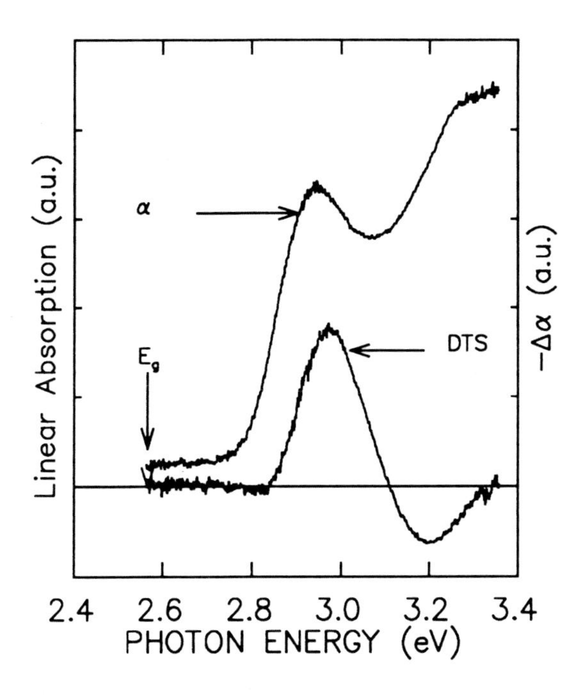

Fig. 2: Experimental results for CdS quantum dots in glass. The linear absorption spectrum is plotted together with the absorption changes and the spectral position of the pump.

Using parameters characteristic for CdS quantum dots in glass we show in Fig. 1 the change in optical transmission, $-\Delta\alpha$, which is proportional to the imaginary part of the third-order susceptibility. The insets show the one and two-pair dipole transition matrix elements as dashed and full lines, respectively. One clearly sees the ground-state biexciton on the low-energy side of the lowest one-pair state, as well as the higher biexciton states energetically between the two lowest one-pair resonances. The assumed relatively large homogeneous broadening leads to the suppression of the increasing absorption for the ground-state biexciton. However, the increasing absorption due to the excited-state biexcitons is clearly visible. In fact, this induced absorption feature is even enhanced through the surface polarization effects, which are present in Fig. 1*b*. These additional Coulomb terms increase the differences between our results and those of the strong-confinement approximation.

For comparison with the theory, we have performed femtosecond and nanosecond pump-probe experiments on CdS and CdSe quantum dots in glass matrices. As representative example we show in Fig. 2 our measurements for CdS quantum dots at 10 *K*. The pump pulse was tuned inside the energetically lowest exciton resonance and a broad-band cross-polarized probe pulse measured the absorption changes induced by the pump. In addition to the absorption changes, we also show the linear absorption spectrum and the energetic position of the pump. The absorption changes clearly show the induced absorption feature on the high-energy side of the exciton transition, in good qualitative agreement with the theoretical predictions.

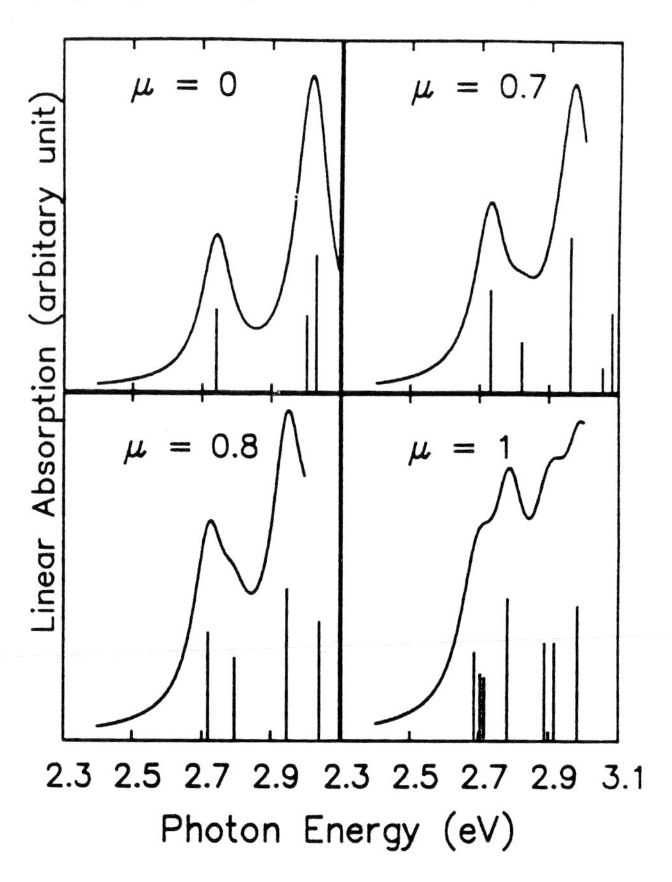

Fig. 3: Computed linear absorption spectra for semiconductor quantum dot and various heavy- and light-hole coupling constants μ. The dot radius $R = a_B$ and the mass ratio $m_e \gamma_1 / m_0 = 0.5$. The broadening of the spectra has been taken as $\gamma = 2E_R$. The inserted lines indicate the oscillator strengths for the dipole transitions.

In Fig. 3 we show a series of computed linear absorption spectra where we include the valence band mixing effects. The noticeable feature of these spectra is that with increasing μ, Eq. (5), a shoulder appears on the high energy side of the energetically lowest transition. This is due to the fact that the coupling between light hole and heavy hole opens new channels for dipole transitions. In the parabolic band approximation, the ground state exciton contains mainly the electron-hole-pair wavefunction $(n_e, n_h) = (1s, 1s)$, and the transition to the state $(n_e, n_h) = (1s, 1d)$ is dipole forbidden. However, when the coupling strength μ is large, the coupling between holes with $1d$ orbit and $1s$ orbit becomes very important. In contrast to the parabolic band approximation, the dipole transition between the $1s$-electron and the $1d$-hole becomes therefore dipole allowed.

In Fig. 4 we show an example of an experimentally obtained linear absorption spectrum for a CdSe quantum dot sample. The experimental spectrum reveals striking similarities with the calculated absorption spectrum for $\mu = 0.8$ shown in Fig. 3. The value $\mu = 0.8$ is roughly the correct combination of effective mass parameters for CdSe. Comparing theory and experiment one can assign the shoulder, which appears in Fig. 4 on the high energy side of the lowest quantum confined transition, to electron-hole transitions involving hole states with mixed *s* and *d* symmetry.

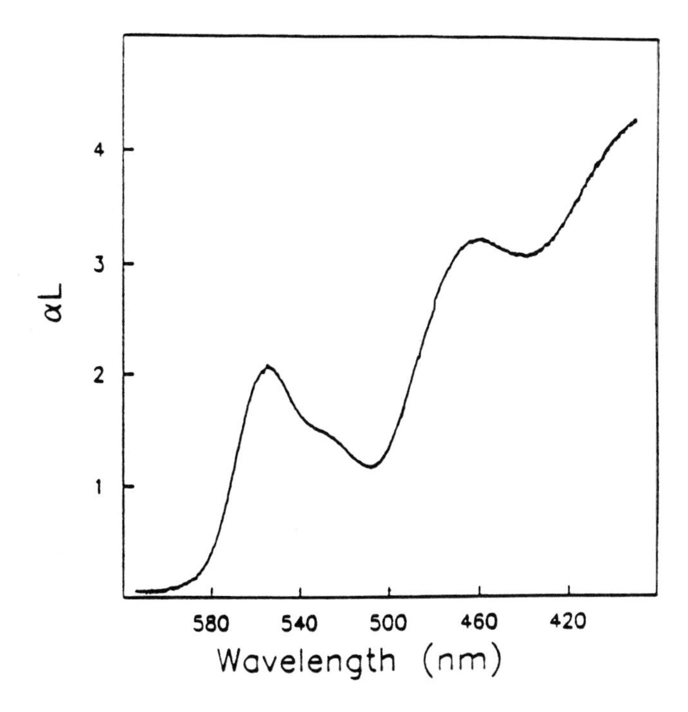

Fig. 4: Measured linear absorption spectrum of a CdSe quantum dot sample at $T = 10\ K$.

5. SUMMARY AND CONCLUSIONS

In summary, we have presented a numerical scheme to compute the one- and two-electron-hole-pair states with the accuracy limited only by available compute time or storage capacity. We show that Coulomb effects are responsible for modifications of the optical dipole transitions, causing a pronounced induced absorption feature on the high-energy side of the saturating one-pair resonance.

We then extended our approach to include the effects of quantum confinement induced valence band mixing. These effects lead to modifications of the hole energies and of the electron-hole pair wavefunctions which manifest themselves as shifts and substructures in the optical absorption spectra.

6. ACKNOWLEDGEMENTS

This work has been supported by grants from NSF, ARO/AFOSR (JSOP), ONR/SDI, the Optical Circuitry Cooperative at the University of Arizona, a NATO travel grant, and grants for CPU time at the Pittsburgh Supercomputer Center.

REFERENCES

Baldereschi *A* and Lipari *N U* 1973, Phys. Rev. **B8** 2697
Borrelli *N F*, Hall *D W*, Holland *H J*, and Smith *D W* 1987, J. Appl. Phys. **61** 5399
Brus *L E* 1984, J. Chem. Phys. **80** 4403
Brus *L E* 1986, IEEE J. Quant. Elect. **QE-22** 1909
Efros *Al L* and Efros *A L* 1982, Sov. Phys. Semicond. **16** 772
Ekimov *A* I and Onushchenko *A A* 1984, JETP Lett. **40** 1137
Ekimov *A I*, Efros *A L*, and Onushchenko *A A* 1985, Solid State Comm. **56** 921
Haug *H* and Koch *S W* 1990, Quantum Theory of the Optical and Electronic Properties of Semiconductors (Singapore: World Scientific)
Hu *YZ*, Koch *S W*, Lindberg *M*, Peyghambarian *N*, Pollock *E L*, and Abraham *F F* 1990a, Phys. Rev. Lett., **64** 1805
Hu *YZ*, Lindberg *M*, and Koch *S W* 1990b, Phys. Rev. **B42**, 1713
Luttinger *J M* 1956, Phys. Rev. **102** 1030
Koch *S W*, *Hu YZ*, and Peyghambarian *N* 1991, J. Cryst. Growth (to be published)
Peyghambarian *N*, Park *S H*, Morgan *R A*, Fluegel *B*, Hu *YZ*, Lindberg *M*, Koch *S W*, Hulin *D*, Migus *A*, Etchepare *J*, Grillon *G*, Hall *D W*, and Borrelli *N F* 1989a, in *Optical Switching in Low-Dimensional Systems*, ed. by Haug H and Banyai L (New York: Academic Press)
Peyghambarian *N*, Fluegel *B*, Hulin *D*, Migus *A*, Joffre *M*, Antonetti *A*, Koch *S W*, and Lindberg *M* 1989b, IEEE. J. Quant. Elect. **25** 2516
Roussignol *P*, Richard *D*, Flytzanis *C*, and Neuroth *N* 1989, Phys. Rev. Lett. **62** 312
Vahala *K J* and Sercel *P C* 1990, Phys. Rev. Lett. **65** 239
Wang *Y*, Herron *N*, Mahler *W*, and Sune *A* 1989, J. Opt. Soc. Am. **6** 808
Xia *J B* 1989, Phys. Rev. **B40** 8500

Inst. Phys. Conf. Ser. No 123
Paper presented at the International Meeting on Optics of Excitons in Confined Systems, Giardini Naxos, Italy, 1991

Effects of the valence band degeneracy on the optical properties of semiconductor microcrystals

Al.L.Efros

Physics Department E 16, Tech.Univ. Munich, 8046 Garching, Germany

ABSTRACT: We have investigated the effects of the degenerate valence band on the structure of absorption and luminescence spectra , and on the electron-hole pair - polar phonon coupling in small size microcrystals. It is shown that in small size CdS microcrystals the lowest quantum size level (QSL) of holes has p-type symmetry and optical transitions between this level and lowest QSL of electrons are forbidden. The polaron binding energy has been obtained analytically to show that its value increases reciprocally to the microcrystal radius.

1. INTRODUCTION

It is well known that quantitative description of the optical properties of semiconductors needs to take into account the real band structures. In bulk semiconductors and 2D semiconductor structures such considerations lead to the shifts of the optical transitions energies and/or to the appearing of some additional transitions which are forbidden in the framework of a simple parabolic band model. However, the real band structure in semiconductor microcrystals gives some qualitatively new effects, two of them we will discuss here.

2. ENERGY SPECTRUM AND WAVE FUNCTIONS OF HOLES

We have found the QSL's of holes in the framework of a six-band Hamiltonian describing the energy spectrum of cubic semiconductors with a finite value of the spin-orbit splitting Δ. If the holes move in a spherical potential it is possible to separate the variables in this Hamiltonian. Grigoryan et al (1990) obtained the system of equations for the radial components of the hole wave functions after separation of the angle dependences . Using their systems of equations we have determined the QSL's of holes. These radial functions in a spherical microcrystal can be generally represented by a sum of three spherical Bessel functions $j_l(k_r r)$. The arguments k_r correspond to the wave vector moduli dependences (k_l, k_h, k_s) on the energy eigenvalue E (three branches in the dispersion law for the valence band):

$$k_{l,s}^2 = m_o\left[2E(\gamma_1+\gamma)-\Delta(\gamma_1+2\gamma) \mp \{[2E(\gamma_1+\gamma)-\Delta(\gamma_1+2\gamma)]^2-4E(E-\Delta)(\gamma_1-2\gamma)(\gamma_1+4\gamma)\}^{1/2}\right]/$$

$$[\hbar^2(\gamma_1-2\dot{\gamma})(\gamma_1+4\gamma)]; \qquad k_h^2 = 2m\, E/[\hbar^2(\gamma_1-2\gamma)] \qquad (1)$$

where γ_1 and $\gamma_2=\gamma_3=\gamma$ are the Luttinger parameters which determine the effective masses of heavy and light holes $m_l = m_0/(\gamma_1+2\gamma)$, $m_h = m_0/(\gamma_1-2\gamma)$, where m_0 is the free electron mass.

The requirement that the radial functions should vanish at the surface of the microcrystal, which corresponds to an infinity high potential barrier, leads to the dispersion equations for the QSL's. The equation for the odd states takes the form (Grigoryan et al 1990):

$$(2j-1)2\varepsilon^{-}(1/k_s^2-1/k_l^2)j_{j-1/2}(k_l a)j_{j+3/2}(k_h a)j_{j-1/2}(k_s a)/j_{j-1/2}(k_h a)+$$
$$+3(2j+3)\Big([(\gamma_1+2\gamma)-2\varepsilon^{-}/k_l^2]j_{j+3/2}(k_s a)j_{j-1/2}(k_l a) - \qquad (2)$$
$$-[(\gamma_1+2\gamma)-2\varepsilon^{-}/k_s^2]j_{j+3/2}(k_l a)j_{j-1/2}(k_s a)\Big) = 0.$$

where $j = 1/2,\ 3/2,\ \ldots;\ \varepsilon = E/(\hbar^2/2m_h a^2)$ and $\delta = \Delta/(\hbar^2/2m_h a^2)$ are dimensionless energies; a is the microcrystal radius. For the even states we obtained:

$$(2j+3)2\varepsilon^{+}(1/k_s^2-1/k_l^2)j_{j+1/2}(k_l a)j_{j-3/2}(k_h a)j_{j+1/2}(k_s a) -$$
$$-3(2j-1)\Big([(\gamma_1+2\gamma)-2\varepsilon^{+}/k_s^2]j_{j+1/2}(k_s a)j_{j-3/2}(k_l a)j_{j+1/2}(k_h a) - \qquad (3)$$
$$-[(\gamma_1+2\gamma)-2\varepsilon^{+}/k_l^2]j_{j+1/2}(k_l a)j_{j-3/2}(k_s a)j_{j+1/2}(k_h a)\Big) = 0$$

where j=3/2, 5/2,.., and for j=1/2 the dispersion equation takes the form:

$$j_1(k_l a)j_1(k_s a) = 0 \qquad (4)$$

For $\varepsilon < \delta$, when $k_s^2 < 0$ the spherical Bessel functions have to be considered as functions of complex arguments.

2.1 VALENCE BAND WITH SMALL SPIN-ORBIT SPLITTING Δ

In the limit $m_l \ll m_h$ and for Δ=0 analysis of Eqs.2,3 shows that the ground state of holes is the even state with j=1/2. Hole energy and wave functions are described by the following expression:

$$E_{1/2} = \hbar^2\varphi_1^2/2m_h a^2,\quad \Psi_M = C^{-}j_1(\varphi_1 r/a)\sum_{m+\mu=M}\begin{pmatrix}3/2 & 1 & 1/2\\ \mu & m & -M\end{pmatrix}Y_{1,m}u_\mu \qquad (5)$$

where $\varphi_1 \approx 4.49$ is the first root of the Bessel function $j_1(x)$, $M=\mp 1/2$, u_μ are the Bloch functions of the four-fold degenerate valence band Γ_8 ($\mu=\pm 1/2,\pm 3/2$), $Y_{l,m}$ are the normalized spherical functions, $\begin{pmatrix}i & k & l\\ m & n & p\end{pmatrix}$ are the 3j Wigner symbols, $C^{-} \approx 6.51/a^{3/2}$. The wave functions Ψ_M are formed only by the spherical functions $Y_{1,m}$ and have p-type symmetry.

The relative intensities of the dipole-allowed optical transitions are determined by the square of the overlap integral of the electron Ψ_e and hole Ψ_h wave functions $K = |\int d^3r\Psi_e(\mathbf{r})\Psi_h(\mathbf{r})|^2$. This means that in the case Δ=0 transitions between the lowest state of holes and the lowest state of electrons, which has s-type symmetry, are forbidden (K=0). This case is realized in CdS nanocrystals where Δ =62 meV. Using Eqs.1,2 Grigoryan et al.(1990) have shown that the hole state with p-symmetry becomes lowest QSL in the microcrystals with radii smaller than 40 Å. It explains why in the small size CdS microcrystals only impurity transitions were observed in luminescence (Ekimov et al (1990)).

2.2 FOUR-FOLD DEGENERATE VALENCE BAND.

For $\Delta = \infty$ the ground state of holes is the odd state with j=3/2. This state is four-fold degenerate with respect to the momentum projection $M = \mp 3/2, \mp 1/2$ and corresponding wave functions have the form:

$$\Psi_M = 2 \sum_{l=0,2} R_l(r) \sum_{m+\mu=M} \begin{pmatrix} 3/2 & l & 3/2 \\ \mu & m & -M \end{pmatrix} Y_{l,m} u_\mu \tag{6}$$

In the limit $m_l/m_h \ll 1$ the energy and the wave functions $R_{0,2}$ were obtained by Efros and Rodina (1989):

$$E_{3/2} = \hbar^2\varphi_2^2/(2m_h a^2); \quad R_0 = C[j_0(\varphi_2 r/a) - j_0(\varphi_2)]; \; R_2 = Cj_2(\varphi_2 r/a) \tag{7}$$

where $\varphi_2 \approx 5.76$ is the first root of $j_2(x)$ and $C \approx 6.044/a^{3/2}$. The wave functions (6) differ significantly from the wave function of the electron ground QSL :

$$\Psi_e(r) = (1/\sqrt{2\pi a}) \sin(\pi r/a)/r \tag{8}$$

3. ELECTRON-HOLE PAIR - POLAR PHONONS COUPLING

In microcrystals containing one electron-hole pair their different wave functions lead to the space charge regions described by the charge distribution function $\rho(r)$:

$$\rho(r) = e[R_0^2(r) + R_2^2(r) - 4\pi\Psi_e^2(r)]r^2 \tag{9}$$

Polar phonons interact with this distribution of charge.
In spherical shaped microcrystal the orbital momentum l characterizes every phonon state. There are two types of polar phonons, longitudinal optical (LOP's) and surface phonons (SP's). The energies of SP's ω_l are determined by

$$[(l+1)/l]\, \varepsilon_d = - \varepsilon(\omega_l) = -\varepsilon_\infty(\omega_l^2 - \omega_{LO}^2)/(\omega_l^2 - \omega_{TO}^2) \tag{10}$$

where ε_d is the dielectric constant of the glassy matrix and $\varepsilon(\omega)$ is the dielectric function of the semiconductor. Transverse and longitudinal optical phonon modes (ω_{TO} and ω_{LO}) are connected by the ratio $(\omega_{LO}/\omega_{TO})^2 = \varepsilon_0/\varepsilon_\infty$ with high and low frequency dielectric constants (ε_∞ and ε_0).
The Hamiltonian of electron - LOP's interaction in the secondary quantization representation was found by Klein et al (1990):

$$U_{LOP} = f_{p-e} \sum_{l=0,1,..} \sum_{l \le m \le l, n} \Big[j_l(\alpha_{ln} r/a)/[\alpha_{ln} j_{l+1}(\alpha_{ln})] \Big] \Big(b_{lmn} Y_{lm} + b^+_{lmn} Y^*_{lm} \Big) \tag{11}$$

where α_{ln} are the serial numbers of the roots of the spherical Bessel functions $j_l(\alpha_{ln})=0$ (n=1,2,...); b_{lmn} and b^+_{lmn} are the operators of creation and annihilation of the spherical LOP's;

$$f_{p-e} = \Big((4\pi e^2/a)[1/\varepsilon_\infty - 1/\varepsilon_0]\hbar\omega_{LO} \Big)^{1/2} \tag{12}$$

For SP's this Hamiltonian has the form:

$$U_{SP} = f_{p-e} \sum_{l=1,2,..} \sum_{l \le m \le l} \sqrt{l\omega_{LO}/\omega_l}\; [\varepsilon_\infty/(l\varepsilon_\infty + (l+1)\varepsilon_d)](r/a)^l \Big(b^s_{lm} Y_{lm} + b^{s+}_{lm} Y^*_{lm} \Big) \tag{13}$$

where b^{s}_{lm} and b^{s+}_{lm} are the operators of annihilation and creation of SP's. Because the SP's can not have orbital momentum l=0 the sum in Eq.13 starts from l=1.

The strength of the electron-hole pair interaction with the phonons is determined by the square of the matrix elements of the transitions from a state without phonon to a state with phonon. Efros et al (1991) have shown that holes in the four-fold degenerate valence band interact with SP's with l=2 and with LOP's with l=2 and l=0. Electrons interact only with LOP's with l=0. This interaction has an opposite sign for electrons and holes and therefore it is partly compensated in neutral microcrystals containing one electron-hole pair. As a result the coupling of the electron hole pair with the LOP's with l=0 is determined by the radial charge distribution $\rho(r)$. In semiconductors with simple parabolic band $\rho(r)\equiv 0$ because the wave functions of electrons and holes are identical and there are no interaction with the phonons.

Interaction of the electron-hole pair with polar phonons leads to a polaron shift of their energy. This shift (polaron binding energy) has been determined by second order perturbation theory:

$$\Delta\varepsilon_M=\sum_{n}\frac{|\langle\Psi_M b^{+}_{00n}|U_{LOP}|\Psi_M\rangle-\langle\Psi_e b^{+}_{00n}|U_{LOP}|\Psi_e\rangle|^2}{\hbar\omega_{LO}}+\sum_{n,m,M'}\frac{|\langle\Psi_{M'}b^{+}_{2mn}|U_{LOP}|\Psi_M\rangle|^2}{\hbar\omega_{LO}}$$

$$+\sum_{m,M'}\frac{|\langle\Psi_{M'}b^{s+}_{2m}|U_{SP}|\Psi_M\rangle|^2}{\hbar\omega_2}\quad;\qquad(14)$$

This shift does not depend on the momentum projection M. A calculation using the wave functions of Eqs.6,7 gives the polaron shift in neutral microcrystal:

$$\Delta\varepsilon\ \approx 0.02\ (e^2/a)(1/\varepsilon_\infty-1/\varepsilon_0)\qquad(15)$$

This value is reciprocal to the microcrystal radius a and increases in small microcrystals. In ionized microcrystals containing one additional electron or hole the numerical coefficient in Eq.14 is 20 times larger because there is no charge compensation.

4. CONCLUSION

We have shown that the degeneracy of the valence band leads to some qualitatively new effects in the optical spectra of semiconductor microcrystals and it should be taken into account for the description of experimental data. Considering the valence band degeneracy we explained the absence of band edge luminescence in small size CdS microcrystals and the existence of the nonzero electron-hole pair - polar phonon interaction.

5. REFERENCES

Efros Al L and Rodina A V 1989 Solid Stat.Commun. 72 645

Efros Al L, Ekimov A I, Kozlowski F, Petrova-Koch V, Schmidbaur H and Shumilov S 1991 Sol. State Commun. 78 853.

Ekimov A I, Kudriavtsev I A, Ivanov M G and Efros Al L 1990 Journal of Luminescence 46 83

Grigoryan G B, Kazaryan E M, Efros Al L and Yazeva T V 1990 Sov.Phys.Solid State 32 1772

Klein M C,Hache F, Ricard D and Flytzanis C 1990 Phys.Rev.B 42 11123

Inst. Phys. Conf. Ser. No 123
Paper presented at the International Meeting on Optics of Excitons in Confined Systems, Giardini Naxos, Italy, 1991

Photoluminescence from GaAs microcrystallites

S Juen, J Baldauf, K Überbacher, R Tessadri[(a)], KF Lamprecht, RA Höpfel

Institute of Experimental Physics
University of Innsbruck, A-6020 Innsbruck, AUSTRIA

ABSTRACT: GaAs microcrystallites were produced by pulverization of monocrystalline bulk material and size-dispersed by means of gradient sedimentation. The samples were characterized by scanning and transmission electron microscopy as well as x-ray diffraction. The cw-photoluminescence spectra of GaAs crystallites show an increasing blue-shift of up to 24 meV with decreasing particle size. We discuss the observed shift on the basis of carrier quantum confinement.

1. INTRODUCTION

Several technologies have been developed in recent years to realize one- and zero-dimensional semiconductor structures. Most of them are based on lateral structuring of quantum wells and heterostructures, in which carriers are already confined in one dimension. Additional confinement is attained by lithographically patterned etching, gate controlled electrostatic potentials, or ion implantation. Recent developments are surveyed by Reed (1991) and Beaumont (1990). A different technique based on colloidal particle growth in solvents and glasses has been applied to produce II-VI semiconductor nanocrystallites (Bawendi 1990). The change of the electronic structure due to quantum confinement appears in a drastical modification of optical and electronic properties. In three-dimensionally confined systems especially the excitonic enhancement of third order optical nonlinearities has been studied theoretically and experimentally in detail (Schmitt-Rink 1987, Hu 1990). Information about the electronic structure is also contained in the spectral and temporal behaviour of the photoluminescence.

In the present paper we report on cw- and time resolved photoluminescence measurements from GaAs microcrystallites produced by a new method developed to study directly the transition from three-dimensional to zero-dimensional behaviour. Monocrystalline bulk material is fine-pulverized and subsequently size-dispersed in suspensions by the standard separation techniques sedimentation and centrifugation. The crystallites are deposited on substrates and characterized by scanning and transmission electron microscopy as well as x-ray diffraction. For photoluminescence measurements the crystallites are deposited on Si substrates. A strong blue-shift of the luminescence peaks of up to 24 meV is observed and discussed on the basis of carrier quantum confinement. Carrier lifetimes down to 2 ps are estimated from femtosecond laser spectroscopy for the microcrystallites.

2. SAMPLE PREPARATION

Semiisolating liquid-encapsulated Czochralsky (LEC) grown GaAs was pulverized in a micro-rapid-mill (Retsch) in an agate mortar filled with water for 20 minutes. The particle size distribution was determined with an optical sedimentometer and yielded an average diameter of less than 3 μm and a volume fraction of 10% for particles with a diameter below 0.8 μm. The subsequently performed size dispersion is based on the same principle as the functioning of the sedimentometer.

Gradient sedimentation was performed in a methanol/water solution. 0.5 ml of the suspension were mixed with 0.5 ml methanol and deposited on the 9 ml liquid column (diameter 1.1 cm) in the tube with a linear concentration gradient from 30% CH_3OH at the top decreasing to zero at the bottom. After 6 hours ten fractions of 1.0 ml were pipetted out one by one from the top of the suspension and refered to as F1 (top) to F10 (bottom). The viscosity gradient prevents intermixing during the extraction of fractions. A 1.0 μl drop from each of the fractions was put on a Si substrate, and after the evaporation of the solution the photoluminescence measurements were performed. After this, the samples were studied by a digital scanning electron microscope (Zeiss DSM 950). Additionally, transmission electron micrographs (Zeiss EM 902) were obtained from drops dried on standard electron microscope films. The remaining parts of the fractions were used for ion-coupled plasma (ICP) emission and x-ray diffraction analysis.

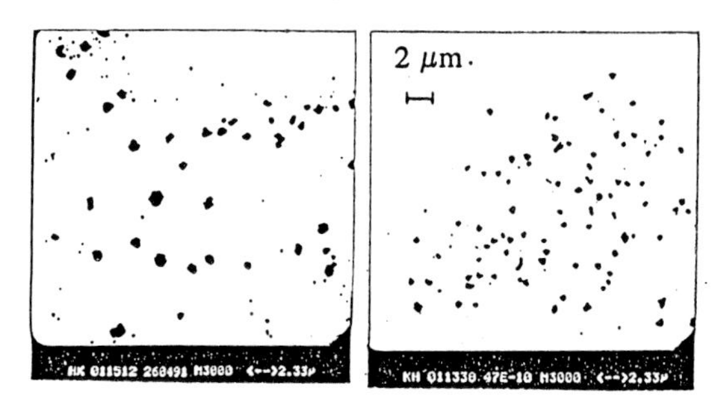

Fig. 1: Transmission electron micrographs of pulverized GaAs crystallites size-dispersed by gradient sedimentation.

Fig. 1 shows transmission electron micrographs of crystallites size-dispersed by gradient sedimentation. The average particle sizes of the samples F3, F7, and F9 were estimated from scanning and transmission electron micrographs. We obtained the values 0.84 μm (0.28 μm), 1.13 μm (0.49 μm), and 1.35 μm (0.42 μm) for the average radius of the samples F3, F7, and F9, respectively (standard deviation given in brackets). These values are by a factor of two larger than the calculated radii. In the calculation spherical GaAs particles were assumed and Stokes' equation was used for the friction force, since the Reynolds number is much smaller than one in the present particle size regime.

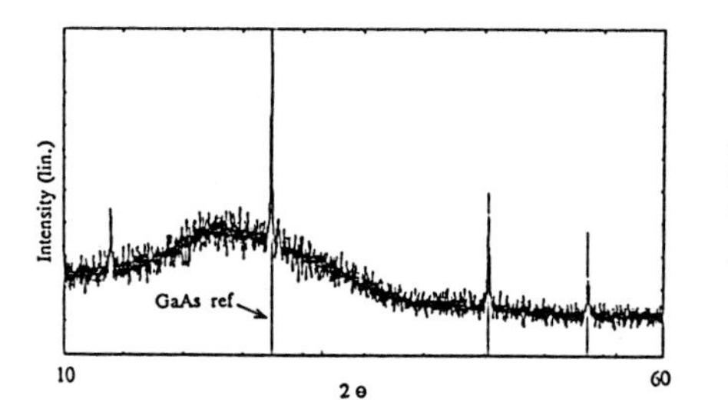

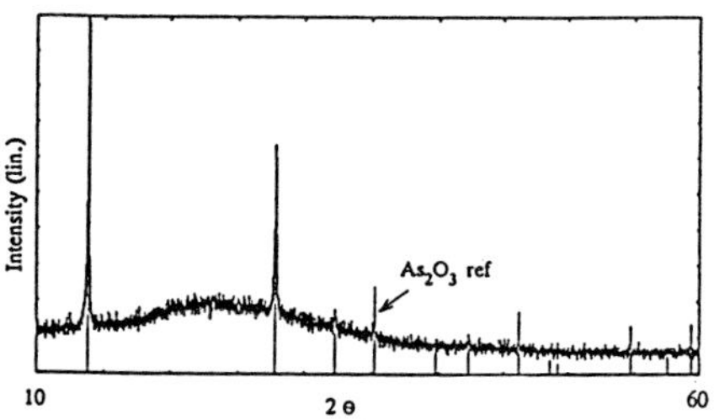

Fig. 2: X-ray powder spectra from samples F10 (left) and F1 (right). The vertical lines indicate positions and amplitudes of GaAs and cubic As_2O_3 reference data. Radiation source: Cu- KL_2- and KL_3.

Fig. 2 shows x-ray powder patterns from the samples F10 and F1 obtained with Cu- KL_2- and KL_3- radiation. The x-ray spectrum from sample F10 shows additionally to GaAs the presence of cubic As_2O_3 (arsenolite). The pattern from sample F1 reveals As_2O_3 only. The vertical lines in Fig. 3 indicate the positions and amplitudes of GaAs and arsenolite reference spectra. In the x-ray spectra of the intermediate samples only arsenolite could be detected. Additional information was obtained from ion-coupled plasma (ICP) emission analysis, which yielded a decrease of the Ga:As ratio (in atomic percent) with decreasing fraction number: The ratios 1 : 1 (F10), 1 : 1.8 (F8), 1 : 4 (F6), 1 : 26 (F4) and 1 : 38 (F2) were obtained (fraction number given in brackets). A comparison of these ratios with the corresponding x-ray spectra shows that a portion of Ga can be present only in an amorphous form.

3. PHOTOLUMINESCENCE

For cw-photoluminescence measurements the Si substrates with the adhered crystallites were immersed in liquid nitrogen and excited with a He-Ne laser (632.8 nm, 20 mW). The power density was 200 W/cm^2 at a focus diameter of 50 μm. The excited area was imaged on the entrance slit of a 0.275 m spectrograph with a liquid nitrogen cooled CCD-detector (348 x 576 pixels) in the exit focal plane. The spectra were recorded by integrating simultaneously a bandwidth of 309 nm with a resolution of 0.8 nm.

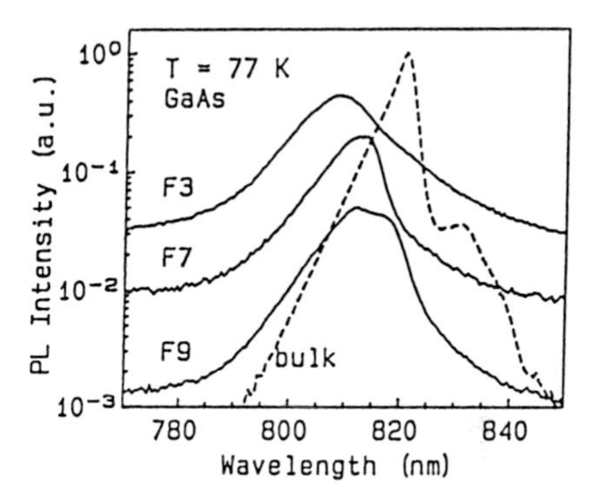

Fig. 3: Cw- photoluminescence spectra of GaAs microcrystallites on Si (full lines: samples F3, F7, F9) and bulk GaAs (dashed line) at T = 77 K. Excitation: He-Ne laser (632.8 nm, 200 W/cm^2).

Fig. 3 shows photoluminescence spectra from bulk GaAs and pulverized GaAs crystallites at a temperature of 77 K. The full lines were recorded from the samples F3, F7, and F9. The luminescence spectra represent an average of 20 to 30 particles. The remarkable features in these spectra are the shift of the luminescence maximum to a shorter wavelength, the spectral broadening and the nonexponential shape of the high and low energy tail. The shift of the luminescence maximum corresponds to a blue shift of 6 meV, 16 meV, and 24 meV for the samples F9, F7, and F3, respectively, relative to bulk GaAs. The luminescence intensity at its spectral peak position increased proportional to the square of the excitation power density and showed no shift of its spectral position up to a power density of 200 W/cm^2.

Time resolved photoluminescence measurements were performed with a colliding pulse modelocked (CPM) laser using the population correlation technique in a set-up as described by Lamprecht (1991) and Höpfel (1989). At room temperature, a single exponential carrier decay time of 130 ps was evaluated for the LEC-grown bulk GaAs. The analysis of the correlation signal of the microcrystallites revealed an exponential carrier decay with two time constants. The values 2 ps and 5 ps were estimated for sample F3.

4. DISCUSSION

The production of GaAs microcrystallites by pulverization of LEC-grown bulk material leads also to the formation of cubic As_2O_3 (Fig. 2), whose bandgap energy amounts nearly 5 eV. This is attributed to the transient temperature increase during fraction along the created surfaces, that enables the formation of As_2 and As_4 (Rosenblatt 1976). The results of the ICP emission and the x-ray diffraction analysis show that the crystallites contain a diminishing portion of crystalline and amorphous GaAs and an increasing portion of cubic As_2O_3 with decreasing crystallite size. Thus the actual sizes of the GaAs crystals are much smaller than the estimated particle sizes and the x-ray spectra are dominated by arsenolite if the oxide thickness reaches the dimension of GaAs.

In the following we discuss the spectral features of the crystallite cw-photoluminescence shown in Fig. 3. In principle, a blue-shift of the luminescence maximum of GaAs is caused by a) compressive stress, b) band-filling, and c) quantum confinement:

a) A hydrostatic pressure of 0.71 kbar, 1.88 kbar and 2.82 kbar is necessary to produce the observed energy shifts of 6 meV, 16 meV, and 24 meV, respectively, in bulk GaAs. No mechanism is seen which could lead to these large values moreover size-dependent.

b) No change of the luminescence shift was observed as a function of the excitation power density up to a value of 200 W/cm^2. Therefore we conclude that band-filling effects can give only a negligible contribution to the observed luminescence blue shifts.

c) Quantum size effects: In a spherical infinite potential well the lowest electronic transition energy is increased relative to the bulk GaAs band gap by 6 meV, 16 meV, and 24 meV for a sphere radius of 32 nm, 20 nm, and 16 nm, respectively. These values are obtained if the bulk GaAs effective carrier masses are used and the Coulomb attraction as well as the dielectric confinement is neglected (Brus 1984). The values for the radii become smaller if the Coulomb interaction is taken into account. Since the actual size of the GaAs crystals is only a diminishing fraction of the estimated particle sizes, and the luminescence blue-shift increases with decreasing particle size, we suggest that a major contribution to the observed blue-shifts originates from carrier quantum confinement. The removal of the fabrication-inherent As_2O_3 is necessary to exactly estimate the GaAs crystal size and is in progress by a special etching technique.

The nonexponential decay of the high energy cw-luminescence tail is attributed to the particle size and shape inhomogeneity. The fast luminescence decay is explained by an increase of the recombination center density due to dislocations and surface states.

5. CONCLUSION

Photoluminescence spectroscopy was performed on GaAs crystallites produced by a new method. A luminescence blue-shift of up to 24 meV was observed as a function of the particle size, and carrier lifetimes of a few picoseconds were measured. We suggest that the main contribution to the observed shifts in the luminescence spectra results from quantum confinement of carriers in GaAs surrounded by As_2O_3.

ACKNOWLEDGEMENT

This work has been supported by the Fonds zur Förderung der Wissenschaftlichen Forschung (FWF), Austria, under project number P7558.

[(a)]*Institute of Mineralogy and Petrography, University of Innsbruck.*

REFERENCES

Bawendi MG, Wilson WL, Rothberg L, Carroll PJ, Jedju TM, Steigerwald ML, and Brus LE, *Phys. Rev. Lett.* **65**, 1623 (1990), and references therein.

Beaumont SP and Sotomayor Torres CM, ed. of *Science and Engineering of One- and Zero-Dimensional Semiconductors*, (NATO ASI Series B Physics Vol. 214, Plenum Publishing, New York, 1990).

Brus LE, *J. Chem. Phys.* **80**, 4403 (1984).

Höpfel RA, Sawaki N, Wintner E, *Appl. Phys. Lett.* **55**, 460 (1989).

Hu YZ, Koch SW, Lindberg M, Peyghambarian N, Pollock EL, and Abraham FF, *Phys. Rev. Lett.* **64**, 1805 (1990).

Lamprecht KF, Juen S, Palmetshofer L, Höpfel RA, *Appl. Phys. Lett.* **59**, 926 (1991).

Reed MA and Kirk WP, ed. of *Nanostructures and Mesoscopic Systems*, Proceedings of the International Symposium, Santa Fe, May 1991 (Academic Press, San Diego, to appear).

Rosenblatt GM, in *Treatise on Solid State Chemistry*, Vol. 6A, edited by Hannay NB (Plenum Press, New York, 1976).

Schmitt-Rink S, Miller DAB, and Chemla DS, *Phys. Rev.* **B35**, 8113 (1987).

Inst. Phys. Conf. Ser. No 123
Paper presented at the International Meeting on Optics of Excitons in Confined Systems, Giardini Naxos, Italy, 1991

Surface polarization instabilities of electron–hole states in quantum dots

L Bányai, P Gilliot

Institut für Theoretische Physik der Universität, Frankfurt am Main

and

Y Z Hu and S W Koch

Optical Sciences Center and Department of Physics, University of Arizona, Tucson, Arizona 85721

ABSTRACT: The role of the dielectric surface polarization on the ground state of a Coulomb interacting electron-hole pair in a spherical semiconductor quantum dot embedded in a dielectric medium is studied through numerical solution of the Schrödinger equation. If the confining potential barrier is not high enough the particles may be selftrapped on the surface of the dot.

1. INTRODUCTION

The influence of the surface dielectric polarization on the electron-hole states in semiconductor quantum dots was first considered theoreticaly by Brus 1984. Recently Tran 1990 brought to attention in the context of quantum wells, that finite barriers give rise to principial difficulties in the treatment of the dielectric surface polarization due to the singularities of the selfenergy. However, with a phenomenological cut-off distance (of the same order of magnitude as the interatomic distance) it was shown, that by GaAs-GaAlAs quantum wells, although the potential barriers are low, the corrections due to the dielectric surface polarization are not too important due to the small difference of the dielectric constants.

In this paper we undertake an analysis of the interacting electron -hole pair ground state ("exciton") in a spherical quantum dot within the frame of a finite potential barrier and a cutt-off dielectric selfenergy, under the simplifying assumption of the same effective masses inside and outside the dot.

2. THE EFFECTIVE COULOMB ENERGY OF CHARGED PARTICLES INSIDE AND OUTSIDE A DIELECTRIC SPHERE.

A charged particle in the neighbourhood of a dielectric interface induces a surface polarization charge and correspondingly its potential energy depends on the distance to this interface. This induced surface charge acts also on other particles and therefore renormalizes the (Coulomb) interaction energy of these. The total effective Coulomb energy for two oppositely charged particles of absolute charge e in a dielectric sphere of radius R and dielectric constant ϵ_1 embedded in an infinite dielectric medium of dielectric constant ϵ_2 may be written as

$$W = \frac{1}{2}\Big(V(r_1/R,\, r_2/R) + V(r_2/R,\, r_1/R)\Big) + \Sigma(r_1/R) + \Sigma(r_2/R).$$

Here the selfenergy is given by

$$\Sigma(\rho) = E_R \frac{a_B}{R} \frac{\epsilon - 1}{\epsilon + 1} \left[\frac{1}{1 - \rho^2} - \frac{1}{\rho^2} \frac{\epsilon}{\epsilon + 1} \ln(1 - \rho^2) + \frac{\epsilon^2}{\epsilon + 1} \sum_{\ell=0}^{\infty} \frac{\rho^{2\ell}}{(\ell + 1)(1 + \ell\,(\epsilon + 1))} \right]$$

for $\rho < 1$

and

$$\Sigma(\rho) = E_R \frac{a_B}{R} \frac{\epsilon - 1}{\epsilon + 1}\, \epsilon \left[\frac{1}{1 - \rho^2} - \frac{1}{\epsilon + 1} \ln(1 - \rho^{-2}) + \frac{1}{\rho^2} \frac{\epsilon}{\epsilon + 1} \sum_{\ell=0}^{\infty} \frac{\rho^{-2\ell}}{(\ell + 1)(1 + \ell\,(\epsilon + 1))} \right]$$

for $\rho > 1$,

with $\epsilon \equiv \epsilon_1/\epsilon_2$ being the relative dielectric constant while a_B and E_R the Bohr radius and binding energy (Rydberg) of the exciton in the bulk semiconductor material of the dot.

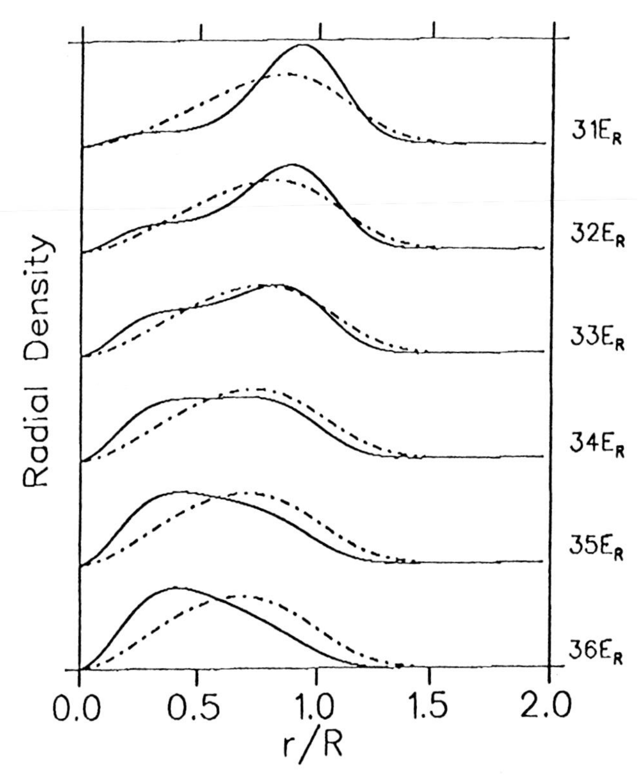

Figure 1. Radial distributions of electrons and holes by different potential barrier heights U_0 (in units of E_R) with $\epsilon = 10$, $\delta = 0.08$ and $m_h/m_e = 0.1$.

In what follows we shall use a regularized selfenergy $\Sigma^r(\rho)$, which is finite everywhere. It coincides with $\Sigma(\rho)$ for $|\rho - 1| > \delta$ and in the intervall $|\rho - 1| < \delta$ is just a linear interpolation between $\Sigma(1-\delta)$ and $\Sigma(1+\delta)$. The cut off parameter δ represents the exclusion of a layer whose thickness δ R should be of the order of the lattice constant, a domain in which the electrodynamics of the continous media breaks down.

The interaction energy between the particles is given by

$$V(\rho_1,\rho_2) = -2 E_R \frac{a_B}{R} \left\{ \frac{C(\rho_2)}{|\rho_1 - \rho_2|} + (\epsilon - 1) \sum_{\ell=0}^{\infty} \frac{1}{1 + \ell(\epsilon + 1)} P_\ell \left(\frac{\boldsymbol{\rho}_1 \boldsymbol{\rho}_2}{\rho_1 \rho_2} \right) d_\ell(\rho_1,\rho_2) \right\}$$

where

$$C(\rho) = \begin{cases} 1 \text{ for } \rho < 1 \\ \epsilon \text{ for } \rho > 1 \end{cases}$$

$$d_\ell(\rho_1,\rho_2) = \begin{cases} (\ell + 1)\, \rho_1^{\ell}\, \rho_2^{\ell} & \text{for } \rho_1 < 1 \,,\ \rho_2 < 1 \\ -\epsilon\, \ell\, \rho_1^{\ell}\, \rho_2^{-\ell-1} & \text{for } \rho_1 < 1 \,,\ \rho_2 > 1 \\ (\ell + 1)\, \rho_1^{-\ell-1}\, \rho_2^{\ell} & \text{for } \rho_1 > 1 \,,\ \rho_2 < 1 \\ -\epsilon\, \ell\, \rho_1^{-\ell-1}\, \rho_2^{-\ell-1} & \text{for } \rho_1 > 1 \,,\ \rho_2 > 1 \end{cases}$$

The series defining the interaction energy are also diverging for $r_1 \rightarrow R$ and simultaneously $r_2 \rightarrow R$, but the divergence is integrable and therefore allows physically meaningful solutions of the Schrödinger equation.

3. NUMERICAL SOLUTION OF THE SCHRÖDINGER PROBLEM OF AN ELECTRON-HOLE PAIR CONFINED TO A SEMICONDUCTOR SPHERE.

Besides their Coulomb interactions, the electron and the hole feel a confinement potential, chosen as a finite potential barrier of height U_0, which we take for simplicity to be the same for both particles. Thus we are interested to find the lowest-lying eigenenergies and wavefunctions of the hamiltonian operator

$$H = -\frac{\hbar^2}{2m_e} \nabla_e^2 - \frac{\hbar^2}{2m_h} \nabla_h^2 + U_0 [\, \theta(r_e - R) + \theta(r_h - R) \,] + W^r(\mathbf{r}_e, \mathbf{r}_h)$$

for an electron-hole pair with effective electron and hole masses m_e and m_h respectively. The superscript by the effective Coulomb energy means, that the selfenergy was replaced by the regularised one.

We numerically solved the Schrödinger equation (for the ground state) with the two-particle hamiltonian (2.1) using two different approximations. The simplest one was the Hartree (or selfconsistent potential) approximation. Another approach was a suitable modification of the method of Hu 1990, which enables to achieve a high degree of precision. It consists in developing the two-body wave-function in sums of products of a complete system of one-particle eigenfunctions and diagonalizing the hamiltonian matrix obtained with a finite but great number of basis functions.

We did perform numerical calculations with both described methods using identical physical parameters. The values used in these calculations were: an effective electron-hole effective mass ratio $m_e/m_h = 0.1$, a dielectric constant ratio $\epsilon = 10$, a dot radius identical with the exciton Bohr radius $R = a_B$ and a cut-off parameter $\delta = 0.08$. The height of the potential barrier U_0 was varied in discrete steps in a wide domain to catch the whole scenario.

In Fig. 1 the radial distribution of electrons and holes for different potential barrier heights U_0 obtained by the second method are shown. At $U_0 = 40\ E_R$ the barrier is equivalent still to an infinite one, but in the following interval up to 30 E_R both particles change gradually their position to the surface of the sphere, while the electron still remains wider spread than the heavy hole. This picture survives after the complete elimination of the confinement barrier ($U_0 = 0$). The particles remain confined in the minimum of the potential energy given by the selfenergy.
Therefore, according to our results, where large relative dielectric constants (typical for quantum dots embedded in glasses) are used, a reasonable cut-off distance for selfenergy of .08 a_B is enough to create a strong potential minimum, that is able to bring the particles on the surface of the dot already at potential heights of order of 30 E_R. For CdS values of the exciton parameters (a_B = 30 Å and E_R = 30 meV) this means for example, that we did admit the validity of the electrostatics of continuous media outside a layer of about 5 Å and the dangerous potential barriers are about .9 eV. All these values are quite close to numbers obtained from reasonable estimates for the conditions realized by semiconductor quantum dots in glasses or liquids. Hence, the effects discussed in this paper may very likely play a role in explaining experimentally observed features.

Essentially the same scenario takes place also in the Hartree approximation, but there are some quantitative differences.

Acknowledgements.

One of the authors (P. G.) acknowledges for a Lavoisier Grant of the french Ministère des Affaires Etrangères during his stay in Frankfurt.

The work of the Tucson group has been supported by grants from NSF, ARO, AFOSR (JSOP), and the Pittsburgh Supercomputer Center. Furthermore, we acknowledge a NATO travel grant which supports the Tucson - Frankfurt collaboration.

References:

Brus L E 1984 J. Chem. Phys. 80, 4403
Hu Y Z, Lindberg M, and Koch S W 1990 Phys. Rev. B42, 1713
Tran Thoai D B, Zimmermann R, Grundmann R and Bimberg D 1990 Phys. Rev. B42, 5906

Inst. Phys. Conf. Ser. No 123
Paper presented at the International Meeting on Optics of Excitons in Confined Systems, Giardini Naxos, Italy, 1991

Theory of exciton population induced nonlinear absorption in large microcrystallites

L. Belleguie and L. Bányai[†]
Institut de Physique et Chimie des Matériaux de Strasbourg,
Unite Mixte 380046 CNRS - ULP - EHICS
Groupe d'optique nonlinéaire et d'optoélectronique
5, rue de l'Université - 67084 Strasbourg CEDEX
†Permanent address: Institut für Theoretische Physik der Universität Frankfurt, Robert-Mayer Str. 8, D-6000 Frankfurt am Main, Germany.

ABSTRACT: A model of excitons in large semiconductor quantum dots ($R/a_o \gg 1$, along the ideas of Efros-Efros (1982) about the exciton mass center quantization, is used to describe many exciton states in materials with large exciton binding energy. In the light of this theory the recent pump-probe experiments on CuCl quantum dots are interpreted as nonlinear absorption induced by an average population of a few excitons per dot.

1. INTRODUCTION

Exciton optics of semiconductor microscrystallites is a subject of many recent experimental and theoretical studies. Efros and Efros (1982) have developed a theory of the lowest electron-hole pair state (exciton) in large microcrystallites based on the quantization of the center of mass motion. Recent pump-probe experiments on large CuCl microcrystallites (Gilliot 1989 and 1990) (on the scale of the exciton Bohr radius a_o) motivate our interest in a theoretical investigation of the optics of a many-exciton system confined in a microsphere. Here the typical feature is the importance of hundreds of quantum states, whose energetical separation is comparable to their line-width.

Starting from the ideas of Efros-Efros, we elaborate a tractable model of interacting, confined excitons in the asymptotic domain of large microcrystalls. We show that in the presence of an (induced) equilibrium exciton concentration the linear absorption shows a blue shift,saturation and blue tailing of the main excitonic absorption peak compared to the one in the absence of such a population.

The pump induced population dependent linear absorption coefficient of this model is investigated separately for very small exciton populations per dot, (including only the vacuum and the one exciton states) and for more excitons in the dot within a boson model without spin (in an assumed macrocanonical distribution) with the help of the mean field approximation in the many-body linear response formula.

2. LINEAR RESPONSE WITH SPIN

In a microcrystallite whose radius R is greater than the excitonic radius a_o ($\lambda \equiv \frac{R}{a_o} \gg 1$), the excitonic eigenenergies are given by:

$$\varepsilon_{ln} = E_G - E_R + \frac{\hbar^2 \alpha_{nl}}{2(m_e + m_h)\, R^2}, \tag{1}$$

where α_{nl} are determined by $j_l(\alpha_{nl}) = 0$; j_l are Bessel functions of the first kind. E_G and E_R are respectively the band gap and the excitonic Rydberg energy in the Bulk material.

For the energies and polarization matrix elements only the leading terms in $\frac{1}{\lambda^3}$ are retained. The two-pair states are treated in the Heitler-London approximation as properly-normalised anti-symmetrized products of one-pair functions. These are approximate solutions of the four particle hamiltonian (two electrons and two holes). The matrix elements of the polarization operator are calculated under the same model. Then the corresponding eigenenergies are given by:

$$E_{\nu_1\nu_2} = \varepsilon_{\nu_1} + \varepsilon_{\nu_2} + E^x_{\nu_1\nu_2}, \qquad (\nu = n, l, m) \tag{2}$$

where $E^x_{\nu_1\nu_2}$ is the exchange energy.
The polarization matrix elements are given by

$$< 0 \left| P^{(-)} \right| \nu >= p_{cv} g^d_\nu \tag{3}$$

$$< \nu \left| P^{(-)} \right| \nu_1\nu_2 >= \frac{p_{cv}}{\sqrt{1 + \delta_{\nu_1\nu_2}}} \left\{ \delta_{\nu\nu_2} g^d_{\nu_1} + \delta_{\nu\nu_1} g^d_{\nu_2} + g^x_{\nu\nu_1\nu_2} \right\}, \tag{4}$$

The matrix elements, in the $\lambda \to \infty$ asymptotic limit are given in Belleguie 1991. A two electron-hole pair state besides the quantum numbers ν_1, ν_2 actually has to be charaterized also by the total spin state of the two electrons s^e, s^e_z and holes s^h, s^h_z. We are interested to study the modification of the optical absorption spectra in the neighbourhood of the bulk exciton due to the existence of an induced pair population. This system is then optically tested by a weak field whose frequency lies in the vicinity of the bulk exciton line. According to linear response theory, the absorption (imaginary part of the susceptibility) of the test-field is given by:

$$\alpha(\omega) \propto \frac{1}{R^3} \sum_{\alpha,\beta} \rho^0_{\alpha\beta} \left\{ \mid P^{(-)}_{\alpha\beta} \mid^2 \; \delta(E_\alpha - E_\beta - \hbar\omega) - \mid P^{(-)}_{\beta\alpha} \mid^2 \; \delta(E_\alpha - E_\beta + \hbar\omega) \right\} \tag{5}$$

Applying this formula to the case of a canonical distribution of excitons on the levels (1) and (2) with spin included, we have calculated the linear absorption of a weak test field when one exciton is present in the sphere. Figure 1 shows the calculated curves. (a) is the linear spectrum without induced exciton, (b) with one exciton present in the sphere and assumed to be canonically distributed over levels (1) and (c-e) mean sum of (a) and (b) as: p_0 (a) + p_1 (b) with $p_0 + p_1 = 1$ and $p_0 = 3/4$, 1/2 1/3. The temperature is T=10°K, and the homogeneous width is $\Gamma = 0.005\, E_R$.

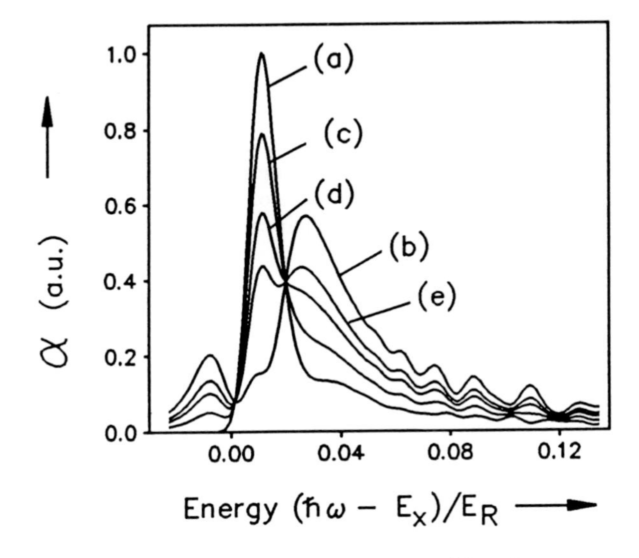

Figure 1. linear response with spin
(a) linear, (b) with one exciton in the sphere
(c-e) average of (a) and (b), see text.

3. The exciton-Boson model.

The previous model is in no way easy to generalize when more than two particles are involved. Instead of that, we have introduced a spinless simplified boson picture. The Hamiltonian of the Bose system will be chosen in such a way to have the same one- and two-boson spectrum as the one- and two-pair electron-hole states of our asymptotic approximation Eq.(1-4).

$$\mathbf{H} = \sum_{\nu} \epsilon_\nu b_\nu^+ b_\nu + \frac{1}{2} \sum_{\nu\nu'} E^x_{\nu\nu'} b_\nu^+ b_{\nu'}^+ b_{\nu'} b_\nu, \tag{6}$$

Similarly, we shall define the boson operator structure of the polarization operator as

$$\mathbf{P}^{(-)} = p_{cv} \left\{ \sum_{\nu} g^d_\nu b_\nu + \sum_{\nu\nu'\nu''} g^x_{\nu\nu'\nu''} b_\nu^+ b_{\nu'} b_{\nu''} \right\} \tag{7}$$

in order to get the same matrix elements as in Eqs. 3 - 4. The characteristic features of this bosonic model are the existence of an exchange interaction between excitons and of an exchange polarization giving rise to density-dependent effects.
With the aid of the formula 5, we have evaluated the linear response of a sphere filled with excitons assumed to be in macrocanonical distribution within a mean field approximation [see Belleguie 1991]. The resulting absorption curves (in arbirary units) in an R = 10 a_o microsphere, corresponding to an average number of bosons in the

initial state <N> = 0, 5 and 10 are given on Fig.2 for the respective temperatures of T = 0 (a), 100 (b) and 200 (c) ^{0}K . No inhomogenous width was introduced, since we have seen already, that the one given by the Lifshitz Slezov distribution (Lifshitz 1959) of radii is negligible and the true distribution is unknown. The homogenous widths Γ taken for the three curves are 0.03, 0.045 and 0.09 E_R respectively. The temperatures and homogenous widths we considered are heuristical, but according to the plausible scenario, that with the increase of the pump intensity not only the number of excited bosons increases, but also their temperature and consequently their energetical uncertainty.

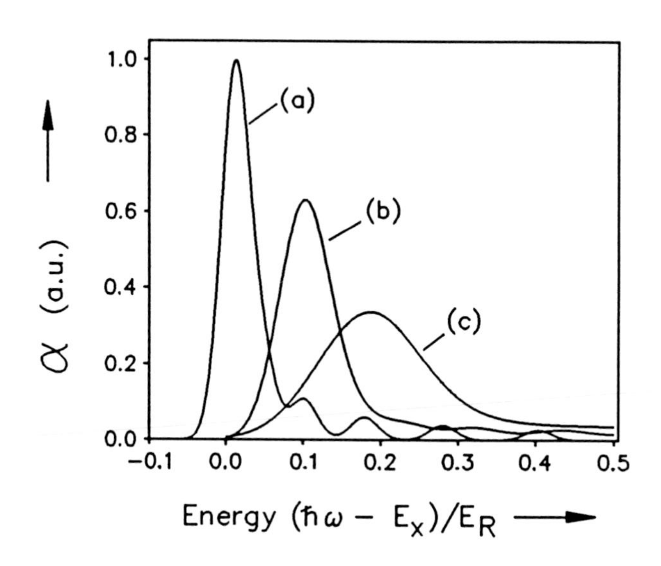

Figure 2
Mean-field many-body linear response theory

The general features are in a good qualitative agreement with the above mentioned experimental pump-probe results (Gilliot 1989).

References

Efros Al. L., Efros A. L., Fiz. Tekh. Poluprovodn., 16, 1209 (1982)
Belleguie L. and Bányai L. to be published in Phys. Rev. B (1991).
Gilliot P., Merle J.C., Levy R. , Robino M. and Hönerlage B., Phys. Stat. Sol. **(b)153**, 403 (1989)
Gilliot P., Hönerlage B., Levy R. and Grun J.B., Phys. Stat. Sol. **(b)159**, 259 (1990)
Lifshitz I. M. and V. V. Slezov, Soviet Physics JETP, **35(8)**, 331(1959)

Inst. Phys. Conf. Ser. No 123
Paper presented at the International Meeting on Optics of Excitons in Confined Systems, Giardini Naxos, Italy, 1991

Absorption dynamics in photoexcited CuBr microcrystals due to exciton–exciton interaction

R Baltramiejūnas, G Tamulaitis, S Pakalnis

Institute of Physics, Lithuanian Academy of Sciences, Goštauto 12, 2600 Vilnius, Lithuania

ABSTRACT: The spectral and temporal behaviour of transient absorption of CuBr-doped glass in $Z_{1,2}$ exciton region was investigated at band-to-band excitation by using excite-and-probe method. The exciton-band damping due to exciton-exciton collisions was observed. Radiative exciton-exciton collisions are shown to have a dominant role in exciton decay in a wide range of exciton density, and the reaction constant of such an interaction was evaluated to be equal to $5 \cdot 10^{-8} cm^3 s^{-1}$.

1. INTRODUCTION

Exciton-exciton interactions play the important role in the dynamics of dense exciton gas causing the exciton dephasing (elastic collisions) as well as their recombination (radiative collisions). The latter effect has been interpreted by Benoit a la Guillaume et al (1967) as the collision when one of the interacting excitons annihilates emitting a photon while the other changes its energy and quasimomentum in the excitonic state according to conservation laws. The radiative exciton-exciton interaction has been claimed as a prevailing recombination mechanism at a high density of photoexcited excitons in many of semiconductor crystals. Like in a bulk crystal, dense exciton gas can be generated in a semiconductor-doped glass, when the average radius of microcrystals significantly exceeds the exciton Bohr radius. In the present work, the mechanisms of nonlinear excitonic absorption in CuBr-doped glass were investigated, and the reaction constant of radiative exciton-exciton cillision was evaluated.

2. EXPERIMENTAL

The excitons were generated in CuBr microcrystals with the average radius of 140 Å, embedded in a glass matrix. So the microcrystal radii significantly exceeded the exciton Bohr radius in CuBr (12.5 Å). Therefore, quantum confinement effects (Ekimov et al 1985) were negligible, and we were able to treate the microcrystal excitation as a system of bulk-like excitons. Due to a large binding energy of $Z_{1,2}$ exciton in CuBr (107 meV), the exciton is stable even at room temperature

manifesting itself in the linear absorption spectrum as a clearly resolved peak. Excite-and-probe method was used to study the light-induced changes of this absorption band.
The third harmonic of $KGd(WO_4)_2:Nd^{3+}$ laser radiation (Stokes component with 5 ps FWHM pulse duration and 3.05 eV quantum energy) was used for excitation. The excitation intensity was sufficiently high to excite the initial concentration corresponding to at least some tens of excitons per microcrystal. The sample was probed by the tunable radiation of parametric generator pumped by the second harmonic of the same laser.

3. DAMPING OF EXCITON ABSORPTION BAND

The excitation-induced modification of the optical density of the sample in the $Z_{1,2}$ exciton region is illustrated in Fig.1. The most characteristic features of $Z_{1,2}$ absorption band transformation are the bleaching in the vicinity of the band peak and the concurrent increase of the absorption in remote

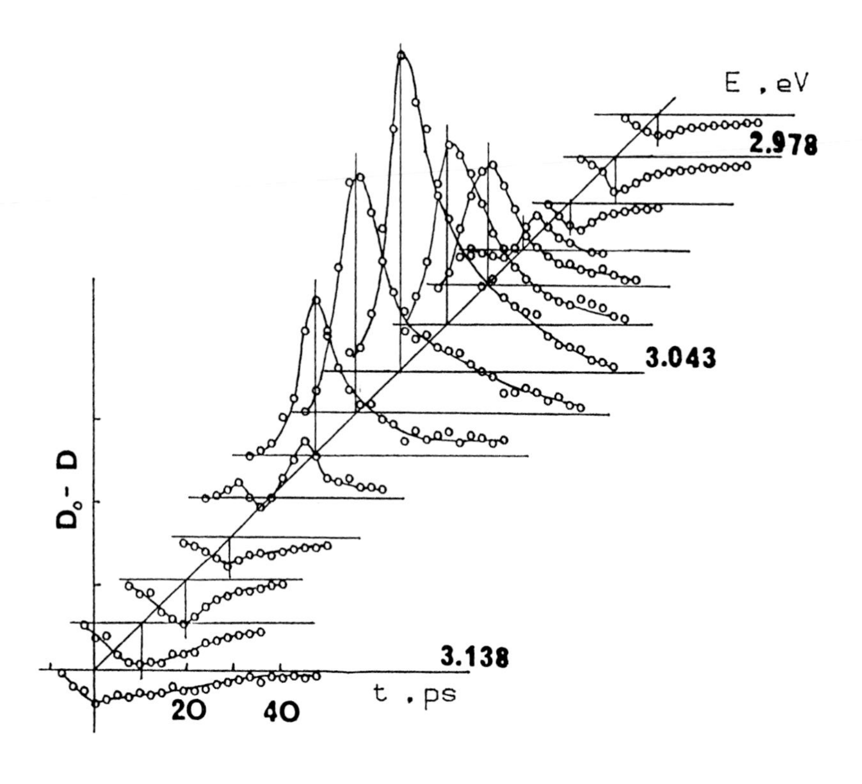

Fig. 1. Time evolution of transient optical density at different probe quantum energies.

spectral regions. In order to describe the spectral changes of the exciton-absorption band, a damping model was employed. The exciton-band having a Lorentzian shape, the relationship between transient absorption coefficient $\Delta\alpha$ and induced damping $\Delta\Gamma$ is simply expressed as

$$\Delta\alpha = \text{const}\ \frac{(\omega_0-\omega)^2 - \Gamma^2 - \Gamma\cdot\Delta\Gamma}{((\omega_0 - \omega)^2 + (\Gamma + \Delta\Gamma)^2)((\omega_0 - \omega)^2 + \Gamma^2)}\ \Delta\Gamma \qquad (1)$$

The proportionality constant as well as exciton energy ω_0 and damping parameter Γ were obtained by fitting the Lorentzian shape to the $Z_{1,2}$ exciton band in the linear-absorption spectrum. The results of the calculation according to (1) with $\Delta\Gamma$ = 20 meV are presented in Fig.2. A fair coincidence of the calculated spectrum with experimental results proves the damping model to be appropriate.

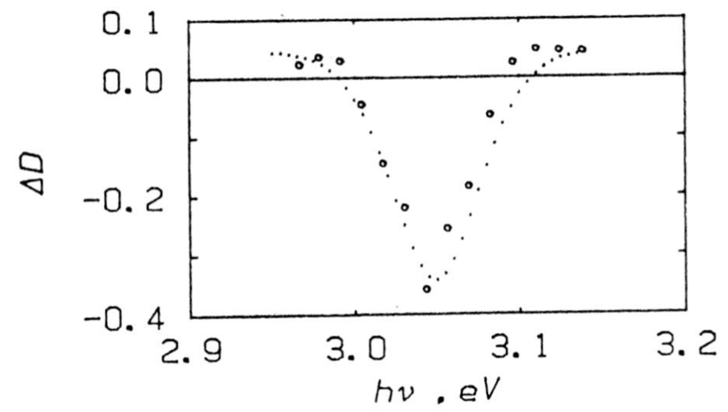

Fig. 2. Experimental (circles) and calculated (dot line) spectra of transient absorption coefficient

The exciton-exciton collisions causing the damping in our experimental conditions were treated in terms of the classical collision theory, since the criterion for the classical approximation (Thomas et al 1976) is satisfied. According to the classical theory of gasses the induced damping is proportional to exciton density n_x and temperature T:

$$\Delta\Gamma \approx 2\cdot h\cdot a_B^2 (k_B T/M_x)^{1/2} n_x . \qquad (2)$$

Here a_B and M_x are Bohr radius and translational mass of the exciton, correspondingly. Although such a classical approach is insufficient for the determination of n_x value, the direct proportionality $\Delta\Gamma$ versus n_x was employed when describing the time behaviour of exciton concentration.

4. TIME BEHAVIOUR OF EXCITON DENSITY

In order to exclude the possible free-electron influence on the exciton decay, the sample was cooled from room temperature, where electron concentration was equal to one tenth of the exciton concentration, down to 90 K where electron density was negligable. Despite the different free-electron concentrations, the spectral features and time evolution of exciton band transformation are very similar at both temperatures. So the exciton dynamics can be described by a single equation

$$\frac{dn_x}{dt} = G - \frac{n_x}{\tau} - C \cdot n_x^2 \qquad (3)$$

During a certain period of time, when the generation term G turns to zero, but the concentration is sufficiently high, the time interval when generation term G turns to zero but the exciton density is sufficiently high , equation (3) reduces to a simple form with a single term in the right, expressing the contribution of radiative exciton-exciton interaction with reaction constant C. According to the reduced rate equation, constant C becomes the only fitting parameter when describing the exciton decay during appropriate time interval. In order to evaluate constant C we elucidated the exciton density relaxation from the time evolution of optical density according to (1) and (2), following the procedure used by Ugumori et al (1982). The value of $5 \cdot 10^{-8} cm^3 s^{-1}$ was received for constant C. It is worth noting that the decay time of nonlinear absorption in CuBr-doped glass is signifficantly shorter than in a bulk CuBr. On one hand, it is caused by high exciton concentration in microcrystal. Spatial expansion of the excitation, characteristic of the bulk matherial, is excluded by natural boundaries of the microcrystal. So the exciton density in the microcrystals is higher than that in the bulk matherial at the same excitation intensity. The intensive surface recombination, on the other hand, effectively destroys the excitons when their concentration drops to the level corresponding to a few excitons per microcrystal. So the optical nonlinearities of CuBr-doped glasses relax to zero during few tens of picoseconds without any accumulation of permanent component.

A la Guillaume Cl B, Debever J-M and Salvan F 1967 Phys. Rev. 177 567

Ekimov A I, Efros Al L and Onushchenko A A 1985 Solid State Commun. 56 921

Thomas G A et al 1976 Phys. Rev. B 13 1692

Ugumori T et al 1982 Jpn.J.Appl.Phys. 21 1588

Inst. Phys. Conf. Ser. No 123
Paper presented at the International Meeting on Optics of Excitons in Confined Systems, Giardini Naxos, Italy, 1991

Polarized lumiscence in CdS_xSe_{1-x} doped glasses

M.A. Chamarro*, C. Gourdon and P. Lavallard

Groupe de Physique des Solides, CNRS UA 17, Universités Paris 6 et 7, 2 place Jussieu, Tour 23, F-75005 Paris, France

ABSTRACT: Low-temperature absorption and photoluminescence spectra of 30 nm mean size $CdS_{0.4}Se_{0.6}$ nanocrystals grown in a glass matrix have been studied. Under quasi-resonant excitation with circularly polarized light a degree of polarization of 16% due to optical pumping has been observed. Analysis of the effect of an applied magnetic field on the degree of polarization shows that the nanocrystals have wurtzite structure. Luminescence excited with resonant and linearly polarized light, is also partially polarized (12%) because of the selective absorption of light by the nanocrystals.

1. INTRODUCTION

Linear and non-linear optical properties of semiconductor nanocrystals grown in a transparent oxide glass matrix have been extensively studied. If crystallites are so small that electron-hole paires are confined in a reduced volume, it has been demonstrated that optical properties are modified by size effects (Ekimov 1991). On the other hand, according to their size, nanocrystals may have hexagonal or cubic structure. Nanocrystals structure is reflected in the electronic and optical properties of these materials. We show that the polarization properties of luminescence reveal the lattice symmetry of nanocrystals. Information about the origin of the luminescence bands is also obtained.

2. EXPERIMENTAL

The glasses were prepared at the Schott Glasswerke Center. They have nearly the same composition as the commercial filters Schott RG 630. The nanocrystals are obtained by heating the samples at 700°C during two days. Vibrational Raman scattering and X-ray diffraction experiments confirm that the mole fraction of selenium is close to 0.6. The crystallites are elongated. By using high resolution transmission electronic microscopy the mean length and width are evaluated to be 37 and 26 nm respectively (Allais and Gandais 1990).

Luminescence was excited by a dye laser (DCM) pumped by an Argon laser. The time resolved luminescence spectra were analysed by a double spectrometer followed by a synchroscan streak camera. For polarization studies an optoelastic modulator was used.

3. RESULTS

Absorption and luminescence spectra at 2K of a sample containing 30 nm mean size nanocrystals of $CdS_{0.4}Se_{0.6}$ are shown in Fig.1. A broad and featureless absorption

---

* On leave of absence from Instituto de Ciencias de Materiales, Universidad de Zaragoza-CSIC, 50009 Zaragoza, Spain

spectrum is observed which corresponds to a negligible confinement effect. The long tail on the low-energy side can be attributed to two different types of transitions: valence-band-to-donor-level and acceptor-level-to-conduction-band transitions. The luminescence spectrum begins at an energy somewhat lower than the onset of the band-to-band transition. In addition to a broad band due to donor-acceptor recombination (D-A), two other bands at 1.969 and 1.995 eV are observed in the photoluminescence spectrum. From studies of temperature behaviour, excitation intensity dependence and time resolved luminescence these two lines have been attributed to donor-hole recombination (D-h) and to the annihilation of an exciton bound to a neutral donor (I_2), respectively. At low temperature, a non-linear increase of the higher-energy line intensity with the exciting light intensity has been observed. This indicates that the emitting state is created in a bimolecular process as expected for an exciton. On the other hand, the high-energy lines disappear at about 60 K and a new band appears at slightly higher photon energy. The energy position of this band moves towards lower energy with increasing temperature. This is the behaviour expected for an exciton bound at low temperature with a binding energy of a few meV. Furthermore, the decay time ($\tau \sim 500$ ps) of the 1.995 eV line is very close to the lifetime of the I_2 line in CdSe (Gourdon et al 1988). We conclude that the complex emitting state is an exciton bound to a neutral donor D^0X. As will be shown below, this assumption is confirmed by the analysis of the polarization of luminescence under quasi-resonant excitation. The other line at 1.969 eV is very likely due to a donor-valence band transition as found by Ekimov in CdS nanocrystals (Ekimov et al 1990).

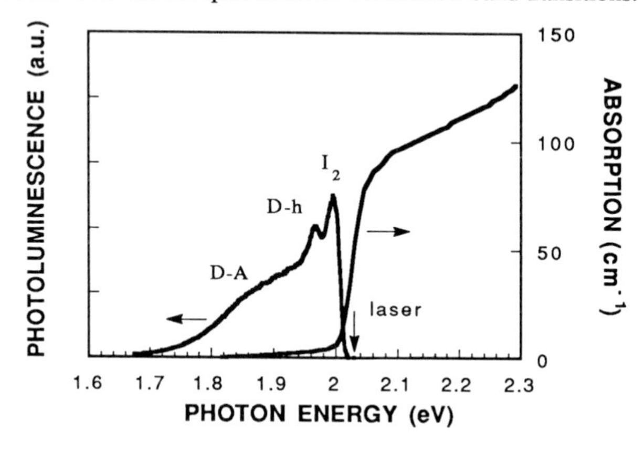

FIG.1 Absorption and luminescence spectra at 2K

In bulk semiconductors, interband absorption of circularly polarized light produces optical orientation of magnetic moments of carriers relative to the direction of exciting light. If the spin relaxation time is not too short as compared with the carrier lifetime, recombination radiation is also partially circularly polarized. The sum and the difference of the two luminescence spectra with right- and left-circularly polarizations, are shown in Fig. 2, for a left-circularly polarized excitation close to the I_2 line. The degree of circular polarization defined as DCP = $(I_{\sigma+}-I_{\sigma-})/(I_{\sigma+}+I_{\sigma-})$ is also displayed in this figure. The maximum DCP is equal to 0.16 on the high energy side of the I_2 line.

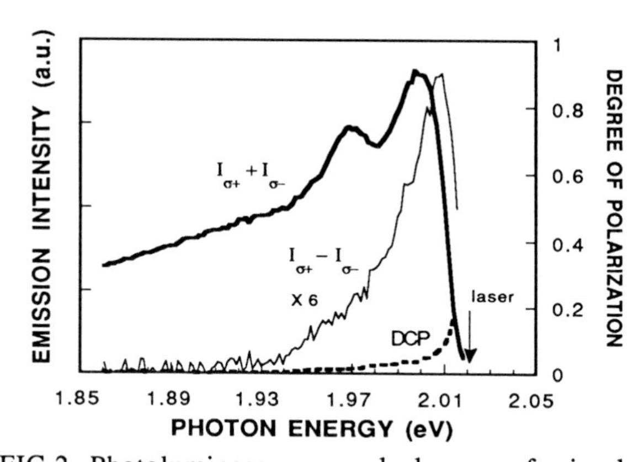

FIG.2 Photoluminescence and degree of circular polarization at 2K. The exciting laser beam is σ_+ polarized. The degree of circular polarization is defined as: DCP = $(I_{\sigma+}-I_{\sigma-})/(I_{\sigma+}+I_{\sigma-})$.

In cubic crystals, owing to the degeneracy of the valence band at k=0, there is a strong spin relaxation of the holes. It is therefore possible to observe only the orientation of the electrons. On the other hand, the application of a magnetic field perpendicular to the direction of exciting light wavevector leads to depolarization of luminescence (Hanle effect). In hexagonal crystals, the degeneracy of the valence band is lifted and both electron and hole orientation can be achieved. However the g-factor of the hole, and consequently also of the exciton, is equal to zero in the direction perpendicular to the crystal axes. As a consequence, application of a transverse magnetic field has no effect on the orientation of these particles and the Hanle effect is absent (Thomas et al 1962). Optical pumping on the exciton bound to a neutral donor was indeed observed in CdS (Gross et al 1971). In such a complex the two electrons have antiparallel spins and only the hole is oriented.

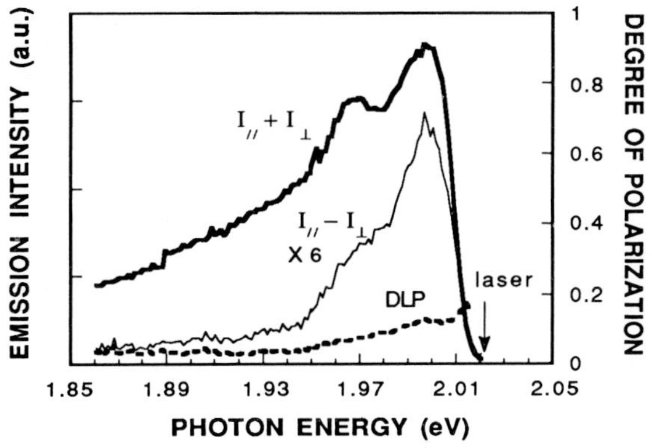

FIG.3 Photoluminescence and degree of linear polarization at 2K. The exciting laser beam is linearly polarized. The degree of linearpolarization is defined as: $DLP=(I_{//}-I_{\perp})/(I_{//}+I_{\perp})$

In our samples, application of a transverse magnetic field up to 1.2 T does not vary the DCP. On the contrary, a longitudinal magnetic field of 1.2 T enhances the DCP up to 0.21. From these results it can be concluded that the g-factor is extremely anisotropic, and almost zero, for a magnetic field transverse to the propagation direction of light. The anisotropy of the g-factor indicates that the involved luminescent state corresponds to a hole or an exciton in a hexagonal structure. The c-axes of the wurtzite-type nanocrystals are randomly oriented in glass, but optical pumping is most efficient for those nanocrystals which have their c-axis along the propagation direction of light. A transverse magnetic field is then normal to the c-axes of the crystals, but has no effect, since the $g_{\perp}$ factor of a hole or an exciton is zero. The maximum theoretical value of DCP, calculated for uniaxial crystals with a random distribution of c-axes in glass matrix, is equal to 0.71 (Chamarro et al). Spin relaxation occurs during the lifetime of luminescent states and decreases the observed DCP. As a result, the maximum experimental DCP (0.21) is lower than the theoretical value (0.71).

If the higher-energy luminescent state were formed with only one electron and one hole, under resonant excitation with linearly polarized light, it would be possible to align the pair and to observe linearly polarized luminescence. Lines are indeed partially polarized under linearly polarized excitation (Fig. 3). The degree of linear polarization (DLP) displayed in these figure has a maximum close to the laser line. Its spectrum is not the same as the DCP spectrum. It extends to low energy and even some DLP is observed on the donor-acceptor line. If optical alignment existed, a decrease of DLP would be observed when a longitudinal magnetic field is applied (resonant Faraday rotation (Bonnot et al 1974)). In our experiments, the application of a longitudinal magnetic field up to 2 T does not decrease the DLP. Then it can be concluded that the involved state is not an exciton. Taking into account results obtained under circularly polarized excitation, we conclude that the luminescent state corresponding to the higher-energy line is a D^0X complex, which has the same symmetry as a hole.

The observed DLP is due to the selective absorption of light by nanocrystals. According to theoretical optical selection rules, the valence-band-to-conduction-band transition in wurtzite-type crystals is allowed only for polarization of the light perpendicular to the c-axis. Then, among the randomly oriented nanocrystals, those with their c-axis normal to the polarization-vector of light are preferentially excited. Recombination radiation is emitted with the same selection rules. A good agreement is expected between the theoretical and the experimental DLP values, since the DLP is not affected by relaxation processes. The maximum experimental DLP, 0.12 is very close to the theoretical value, 0.15 calculated for nanocrystals with randomly oriented c-axes (Chamarro et al).

The observed variation of the DLP with wavelength originates from the overlap of lines which have different DLPs. The D-h line, in the conditions of fig. 3, is mainly excited not by band-to-band or excitonic transitions as the I_2 line but very likely by valence band-to-donor transitions. These transitions occur at $k \neq 0$ and selection rules are not as strict as for $k=0$. As a result the nanocrystals selection by polarized light is not very effective and the DLP of the D-h line is less than the DLP of the I_2 line. By tuning the laser wavelength closer to the D-h line a larger DLP, up to 0.1, has been observed.

4. CONCLUSION

We have observed optical pumping of the exciton bound to a neutral donor in $CdS_{0.4}Se_{0.6}$ nanocrystals. The analysis of the effect of an applied magnetic field on the DCP shows that in our samples nanocrystals have wurtzite structure. Under resonant linearly polarized excitation the luminescence is polarized because of the selective absorption of light by nanocrystals.

5. REFERENCES

Allais M and Gandais M 1990, J. of Appl. Crystall. 23 418.
Bonnot A, Planel R and Benoit à la Guillaume C 1974, Phys. Rev. 9 690.
Chamarro M A, Gourdon C and Lavallard P, to be published.
Ekimov A I, Kudyavtsev I A, Ivanov M G and Efros Al L 1990, J. Lum. 46 83.
Ekimov A I 1991, Conference of the European Physical Society, Exeter, April 1991, to appear in Physica Scripta
Gourdon C, Lavallard P and Dagenais M 1988, Phys. Rev. B 37 2589
Gross E F, Ekimov A I, Razbirin B S and Safarov V I 1971, JEPT Lett. 14 70.
Thomas D G and Hopfield J J 1962, Phys. Rev. 128 2134

Inst. Phys. Conf. Ser. No 123
Paper presented at the International Meeting on Optics of Excitons in Confined Systems, Giardini Naxos, Italy, 1991

Picosecond studies of confined excitons and biexcitons in CuCl doped glasses

B. Kippelen, L. Belleguie, P. Faller, P. Gilliot, J.B. Grun, B. Hönerlage, R. Levy

Institut de Physique et Chimie des Matériaux de Strasbourg
Unité Mixte 380046 CNRS - ULP - EHICS
Groupe d'Optique Nonlinéaire et d'Optoélectronique
5, rue de l'Université - 67084 Strasbourg Cedex, France.

ABSTRACT : We report picosecond excite and probe transmission measurements as well as luminescence experiments performed on CuCl quantum dots embedded in a glass matrix. The nonlinear optical properties due to the creation of confined excitons and biexcitons in these samples are investigated. An optical gain has been evidenced.

1. INTRODUCTION

Materials in which quantum confinement effects lead to new nonlinear optical properties have attracted much attention recently. Particularly, glasses doped with different semiconductors have been grown by several techniques. In such materials composed of semiconductor quantum dots dispersed in a transparent glass matrix, the elementary excitations are confined in all three dimensions. Apart commercial glasses doped with II-VI semiconductors, samples containing CuCl have been fabricated. In the latter bulk material, the exciton binding energy is high (190 meV) which results in a small excitonic Bohr radius of 7 Å. Therefore, since the mean radius of the semiconductor microspheres is always larger than the excitonic Bohr radius, the confinement can be treated in the weak limit where only the center of mass motion is quantum confined (Efros 1982). The optical properties are then strongly governed by interacting excitons and by their possibility to form biexcitons. The existence of biexcitons and their properties in semiconductors of low dimensionality are the subject of many theoretical and experimental investigations (Banyai 1989, Hönerlage 1991, Hu 1990, Itoh 1990, Levy 1991).

In section 2.1, transmission experiments in a test-pump configuration in the picosecond range are presented. From the nonlinear absorption data, the nonlinear refractive index has been calculated by Kramers-Kronig transformations. The magnitude of the nonlinearities is discussed in terms of a figure of merit. Section 2.2 is devoted to high excitation luminescence which has been measured for various confinement conditions in samples containing microcrystallites whose radii are ranging from 110 Å to 30 Å. The study of this luminescence has been concentrated on the radiative recombination of biexcitons into excitons. Finally, we report the experimental observation of an important optical gain in quantum dots. Its origin can be attributed to biexciton-exciton transitions.

2. EXPERIMENTAL PROCEDURE AND RESULTS

The light source used for our experiments was a dye laser synchronously pumped by 35 ps pulses from a mode-locked, frequency tripled Nd:YAG laser. Using Exalite 389 dye in dioxane, it produced, after amplification, individual pulses of 20 ps duration at a repetition rate of 5 Hz. For the test-pump experiments, the broad-band probe consisted in the amplified spontaneous emission from BiBuQ dye in toluene, transversally pumped by a part of the Nd:YAG UV pulses. For the luminescence experiments, the samples were excited by the latter UV pulses (354 nm). The transmitted probe output or the luminescence was directed to a spectrometer with a gated multichannel read-out. The samples were cooled down in a continuous flow cryostat to liquid helium temperature.

2.1 Nonlinear absorption

In our experiments, the nonlinear absorption was measured by the changes of the transmitted broad-band probe pulse covering the spectral region of the Z_3 exciton, when the sample is under the excitation of a spectrally narrow laser pulse in resonance with the exciton. From these absorption measurements nonlinear refractive index changes were derived via Kramers-Kronig relation. Our experiments showed that the excitonic absorption line shifts to the blue, undergoes a saturation, and that induced absorption shows-up on its high energy side (Gilliot 1989, Kippelen 1991). This behaviour has been shown to be explained, in the case of large exciton densities in big quantum dots, by a boson model (Belleguie 1991). It is due to important exchange effects in the exciton-exciton interactions. Moreover, as shown in Fig.1.a the nonlinear absorption shows an asymetry which can be ascribed to size distribution effects.

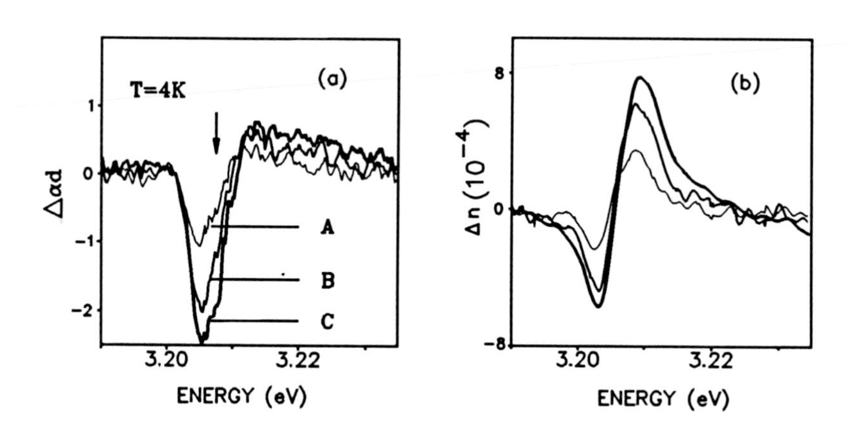

Figure 1. (a) Nonlinear absorption changes in a sample containing dots with 110 Å radius and (b) corresponding nonlinear refractive index changes for different pump intensities (A : $I = 760\ kW/cm^2$, B : $I = 1.76\ MW/cm^2$ C : $I = 4\ MW/cm^2$)

In order to characterize the magnitude of the nonlinearity, we introduce the nonlinear change in refractive index σ_{ex} induced per exciton and per cubic centimer of CuCl. We determine σ_{ex} from the calculated index change $\Delta n = \sigma_{ex} N_{sc}$ where N_{sc} is the density of excitons created per cubic centimeter of the semicondcutor. It is given by :

$$N_{sc} = \frac{I\alpha_{eff}\tau}{f\hbar\omega} \quad (1)$$

where I is the excitation power density, τ the duration of the laser pulses, $\hbar\omega$ the laser photon energy and f is the fraction in volume of the semiconductor in the transparent matrix. In our samples, f is of the order of 0.5 %. The value of σ_{ex} has been calculated for samples containing quantum dots of different sizes. We could show that the efficiency of the nonlinearity slightly increases when the mean radius of the microspheres decreases from 110 Å to 70 Å. However, its highest value has been found to be $\sigma_{ex} = -2.\ 10^{-22}\ cm^3$ which is three orders of magnitude lower than for III-V multiple quantum wells. This value of the nonlinearity could be increased if the sample preparation is improved, especially if the problems due to the size dispersion of the microcrystallites can be resolved.

2.2 Luminescence and Optical Gain

When a band-to band excitation is applied using the Nd:YAG UV pulses to the CuCl quantum dots, luminescence bands appear in the spectral region below the Z_3 exciton (Hönerlage 1991) . As shown in Fig.2.a, two bands can be observed whose spectral locations are at $\hbar\omega = 3.164\ eV$ and $\hbar\omega = 3.173\ eV$. In this spectral region, in bulk material, the luminescence corresponds to the radiative decay of biexcitons, leaving in the crystal a longitudinal or a transverse exciton, which give rise to the two well-known M bands. In the microcrystallites, the situation is different because of the lack of translational invariance. Therefore, the origin of the two bands observed in the biexciton luminescence of our samples is different and not well understood at present. We must notice that these observations are different from the one made by Itoh (1991) in CuCl microcrystallites inside a NaCl matrix where one luminescence band is observed under a band-to-band excitation. Luminescence spectra have been measured for different confinement conditions in samples containing quantum dots whose radii are ranging from 110 Å to 25 Å. The position of the two peaks shows a slight low energy shift when the radius of the dots is decreased but this shift is much smaller than the one of the exciton. Following Efros' theory, the blue biexciton-level shift due to the confinement should be half the one of the exciton level since the biexciton effective mass is nearly twice the exciton mass. These observations could indicate that the binding energy of the biexciton increases with the confinement as predicted theoretically (Hu 1990 , Takagahara 1989).

In a last set of experiments, an optical gain has been evidenced in the spectral region of the luminescence discussed above. The experimental configuration is the same as for the nonlinear absorption measurements but the experiments were carried out in thicker samples (1 mm). When the samples are under a strong transient band-to-band excitation or excited in resonance with the exciton a strong amplification of the transmitted test pulse is observed as shown in Fig.2.b. The luminescence due to the pump pulse is recorded and substracted afterwards. This strong optical gain appears already at pump levels of a few MW/cm^2. Its spectral shape shows that the two transitions observed in the luminescence experiments are involved in the process. Therefore, we believe that the gain is due to a transient population inversion between the biexciton and exciton levels. The dynamics of the gain is complex since a biexciton recombines into an exciton and a photon, but two excitons can couple again to form a biexciton.

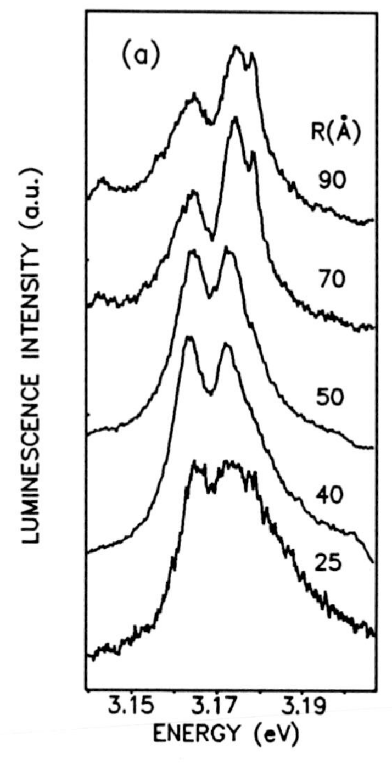

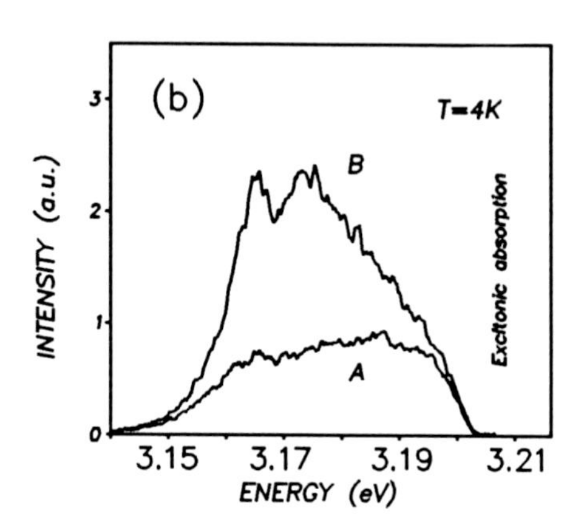

Figure 2. (a) Luminescence spectra for different samples
(b) Transmitted test pulse A) without pump pulse B) when a 25 MW/cm^2 pump pulse is applied (R = 50 Å)

The temporal evolution of the gain could be measured for different delays between the pump and the test pulses. Its decay can be approximated by a double exponential function containing a fast (160 ps) and a slow component (6 ns) for both transitions.

References

Banyai L., Phys. Rev. B, 39, 8022, (1989)
Belleguie L., Banyai L., to be published in Phys. Rev. B
Efros Al., Efros A., Fiz. Tekh. Poluprodovn., 16, 1209, (1982)
Gilliot P., Merle J.C., Levy R., Robino M., Hönerlage B., Phys. Stat. Sol. (b) 153, 403, (1989)
Hönerlage B., Mager L., Gilliot P., Levy R., Annales de Phys., 16, 49, (1991)
Hu Y.Z., Koch S.W., Lindberg M., Peyghambarian N., Pollock E.L., Abraham F.F., Phys. Rev. Lett., 64, 1805, (1990)
Itoh T., Nonlinear Optics, 1, 61, (1991)
Kippelen B, Levy R, Faller P., Gilliot P., Belleguie L., to be published in Appl. Phys. Lett. (1991)
Levy R, Gilliot P., Mager L., Hönerlage B., to be published in Phys. Rev. B (1991)
Takagahara T., Phys. Rev. B, 39, 10206, (1989)

Inst. Phys. Conf. Ser. No 123
Paper presented at the International Meeting on Optics of Excitons in Confined Systems, Giardini Naxos, Italy, 1991

Exciton transfer and magneto-excitons in biased double quantum wells and superlattices

R. Ferreira and G. Bastard
Laboratoire de Physique de la Matière Condensée, Ecole Normale Supérieure
24 rue Lhomond F75005 Paris (France)

ABSTRACT: We present theoretical results bearing on the exciton eigenenergies and assisted transfer in biased double quantum wells and on magneto - excitons in biased superlattices.

1. INTRODUCTION

Double quantum wells (DQW) are among the simplest systems exibiting the quantum coupling effects which in superlattices lead to the formation of the minibands. Various theoretical and experimental works have been reported on the excitonic states in symmetric (SDQW) and asymmetric (ADQW) structures. In the latter structures combined action of a narrow intermediate barrier and of the long range coulombic interaction gives rise to two types of excitons: **i**) the direct excitons, formed of electron and hole states essentially localized in the same well; **ii**) the crossed excitons, formed between an electron and a hole mainly localized in different wells. In addition, important admixture effects are expected to occur each time two band-to-band transitions which involve a common conduction or valence state anticross.
Relaxation of excited excitonic states would determine the low temperature quantum vertical transport of excitation in DQW systems. We have discussed (Bastard et al 1989) the assisted scatterings of independent electrons and holes in single and multiple quantum wells. More recently excitonic effects have been included and results for exciton assisted scattering times have been reported (Ferreira et al 1990 b). In sections 2 to 4 we review our results on the energy states and vertical transport of excitons in double quantum wells.
The electronic states of a superlattice under parallel electric and magnetic fields show interesting dimensionality properties (Claro et al 1990; Ferreira et al 1990 a). The suppresion of the delocalized motion along the growth axis by the electric field and both the intraband and exciton ladders of states have attracted much attention in the last few years (Mendez and Agulló - Rueda 1989; Voisin et al 1988, Schneider et al 1989). In particular direct and crossed excitons have been observed, associated with electron and hole states centered in well separated periods of the biased structure (up to $|\Delta p|=5$: Mendez and Agulló - Rueda 1989). More recently we have considered the evolution with an additional magnetic field of the superlattice absorption (Ferreira et al 1990). The excitonic interaction is enhanced by the magnetic field. This strongly affects the absorption spectra at high fields. We review briefly in section 5 our results on magnetoexcitons in biased superlattices.

2. EXCITON STATES IN DOUBLE QUANTUM WELLS

In the parabolic effective mass approximation the exciton Hamiltonian reads

$$H = H_{C.M.} + H_{rel} + H_e + H_h - (e^2/\kappa)/r \tag{1}$$

where the four first terms account for the in plane center of mass, in plane relative electron-hole, independent electron and independent hole z motions, respectively. The last term is the coulombic interaction with r denoting the electron-hole distance and κ the relative dielectric constant of the heterostructure. The H_h Hamiltonian is taken as the heavy or light diagonal terms of the Luttinger matrix. In this diagonal approximation the center of mass motion is completly decoupled from the relative motion and thus contributes to the exciton energy only by an additive term ($\hbar^2 K_{C.M.}{}^2/2M$, where M is the sum of the electron and hole in-plane masses). H_e and H_h hamiltonians are diagonalized in the standard way, and their eigenvalues and eigenfunctions are denoted by E_n, $| E_n >$, for H_e , and H_m, $| H_m >$ for H_h .
The exciton DQW wavefunction Ψ has been the subject of a number of theoretical studies (Galbraith and Duggan 1989, Golub et al 1990, Ferreira et al 1990 b, Dignam and Sipe 1991, Zrenner et al 1991). We construct Ψ as a linear combination of all the 1S-like exciton states that can be generated in a decoupled way. That is, to each couple of states ($|E_n>$, $|H_m>$) we can associate an 1S-like exciton state, with wavefunction given by

$$\Psi_{n,m}(z_e,z_h,\rho) = (2/\pi\lambda_{n,m}{}^2)^{1/2} < z_e | E_n > < z_h | H_m > \exp\{ -\rho/\lambda_{n,m} \} \tag{2}$$

where $\lambda_{n,m}$ are variational parameters which are determined by minimizing H for a given pair (n,m) of electron and hole states. In a second step Ψ is written as

$$\Psi = \Sigma_{n,m}\, \alpha_{n,m}\, \Psi_{n,m} \tag{3}$$

The coefficients $\alpha_{n,m}$ and the exciton energies (ε) are then determined after solving the $(NxM)^2$ system $< \Psi | H - \varepsilon | \Psi >=0$, where N (M) is the number of electron (hole) states that generate the basis $\Psi_{n,m}$. The oscillator strengths (O.S.) are given by

$$O.S. = (2/\pi) |\Sigma_{n,m}\alpha_{n,m}(a_0{}^*/\lambda_{n,m}) < E_n | H_m > |^2 \tag{4}$$

where $a_0{}^*$ is the bulk effective Bohr radius. In Fig.1 we show the dependence with the barrier thickness (B) of the exciton binding energies ($\Delta\varepsilon$) and O.S. for an unbiased 80Å/B/40Å GaAs - $Ga_{1-x}Al_xAs$ ADQW. Parameters values are x=0.3, electron (hole) barrier heights equal to 213 meV (140 meV), and κ=12.5. The results for the direct exciton (associated mainly with the E_1-HH_1 levels) depend only slightly on the barrier thickness, the (dashed) curves connecting monotonically the not very different 120Å (B=0) and 80Å ($B\rightarrow\infty$) isolated wells' $\Delta\varepsilon$ and O.S. values (7.7 meV, 0.32 and 8.1 meV, 0.40 respectively). On the other hand, the crossed exciton results (full curves) show a strong dependence on B. This state is principally related to the E_2-HH_1 pair, where E_2 (HH_1) is well centered in the

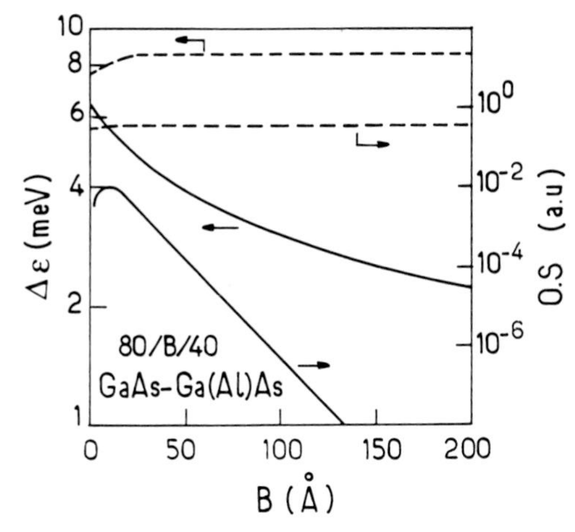

Fig. 1 -Intermediate barrier thickness dependence of the binding energies and O.S. of the direct (dashed lines) and crossed (full lines) excitons for a 80Å/B/40Å unbiased ADQW.

narrow (wide) well for large B. For B=0 it gives an estimate of the "ground" exciton state attached to E_2 for an isolated single quantum well of thickness 120Å ($\Delta\varepsilon$=6.7 meV ; O.S.=0). In the finite B case the overlap integral of the E_2 and HH_1 levels is not zero, and the state becomes optically observable. O.S. decreases exponentially with B for thick enough barriers, and both the binding energy and the oscillator strength vanish when $B \rightarrow \infty$ (decoupled DQW). The maximun O.S. value occurs near B=10Å in the present calculation, and corresponds to an overlap integral $\langle E_2 | HH_1 \rangle \approx 0.22$. The crossed exciton binding energy decreases with B, but is still of the order of 2.5 meV for barriers as large as 150Å, as a consequence of the long range coulombic interaction.

Optical studies have been recently performed on a biased ADQW structure (Ferreira et al 1990 b). Direct and crossed excitons have been shown to be important to understand the low temperature vertical transport of excitation in such systems. In Fig. 2 we show the electric field dependence of the band-to-band (dashed curves) and excitonic transitions for a 78Å/55Å/35Å ADQW. The origin of the electrostatic potential is taken at the center of the middle barrier and we focus on an electric field region around the $F=F_0 \approx -54$ kV/cm value, where the E_2 and E_1 levels anticross. Two independent excitons are obtained in the simplest approximation explicited in eq.(2). They correspond either to coupling the electronic state E_1 or E_2 to the hole state HH_1 (the ground hole level if F<0). The $\varepsilon_{n,m}$ (n=1,2;m=1) exciton energies at this approximation are shown as solid curves in Fig. 2. The presence of a direct and a crossed exciton with very different binding energies (≈ 8.1 and 3.5 meV, respectively) accounts for the shift in the electric field posi tion of the energy anticrossing of the two exci tonic levels (at $F_1 \approx -57$ kV/cm $< F_0$). In this

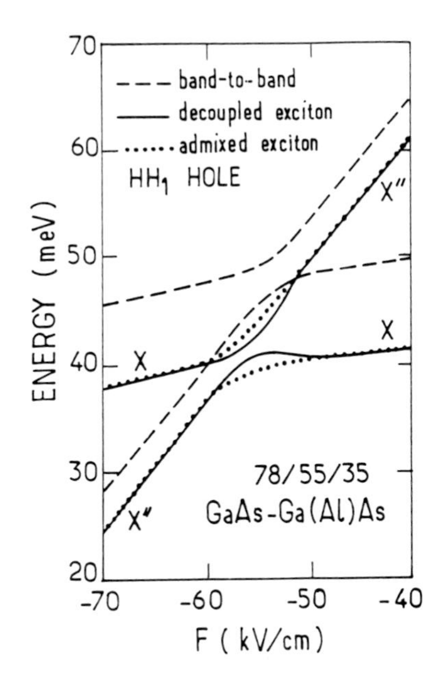

Fig. 2 Electric field dependence of the various transition energies for the 78Å/55Å/35Å GaAs-$Ga_{0.7}Al_{0.3}As$ DQW. Dashed: band-to-band. Full: decoupled excitons. Dotted: admixed excitons.

decoupled approximation the excitonic minigap (≈1.6meV at F_1) is smaller than the band-to-band one (≈3.5meV at F_0). However, strong coupling effects between $\Psi_{2,1}$ and $\Psi_{1,1}$ are expected if $|\varepsilon_{2,1} - \varepsilon_{1,1}| \leq$ Ry* (the bulk Rydberg: ≈ 5.5 meV). The results of the calculations for admixed excitons are plotted as dotted curves in Fig. 2. The interaction between the two $\Psi_{n,m}$ states (non diagonal terms in $<\Psi|H-\varepsilon|\Psi>$) displaces the field where the anticrossing takes place (to F_2≈-58 kV/cm $< F_1 < F_0$) and widens the excitonic minigap (≈ 3.5meV at F_2). Note finally that $\Psi_{2,1}$ and $\Psi_{1,1}$ represent respectively the direct and the crossed states if $F<<F_0$ (the crossed and the direct states if $F_0<<F<0$). O.S. calculations for the ADQW excitonic levels considered in Fig. 2 are presented in Fig.(3). In Figs. 2 and 3 direct (crossed) excitons are labeled X (X"), and dotted (full) curves refer to the admixed (decoupled) exciton calculations. In both types of calculations we find that the O.S. for X is only weakly dependent on F (≈ 0.3), but strongly varies with F in the X" case, decreasing rapidly away from the anticrossing region. We also want to stress the strengthening (weakening) of the crossed admixed transition, as compared to the crossed decoupled one, when $F<F_1$ ($F>F_1$). This feature is consistent with optical studies performed in a 78Å/55Å/35Å biased ADQW where the luminescence of the crossed exciton X" only appears below a threshold reverse voltage (-3.5 Volts, corresponding approximatively to $F=F_1$), and which loses rapidly intensity.

The z_e dependence of the exciton wavefunctions (after integrating $|\Psi|^2$ over the other variables) in the decoupled (full line) and admixed (dotted) approximations are shown in Fig. 4 (for F=-58kV/cm). Labels 1 or 2 in this figure are such that energy(1) < energy(2).

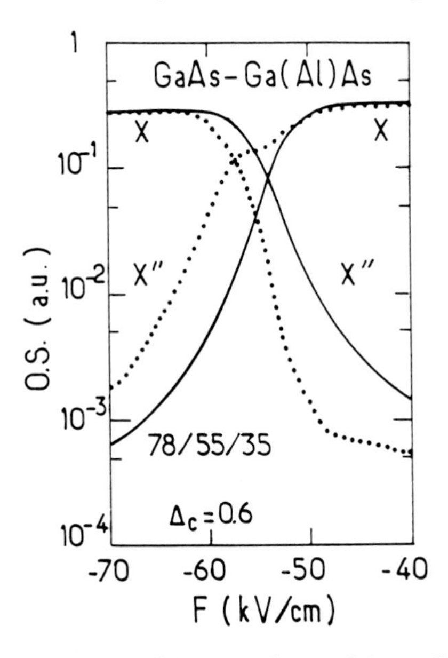

Fig. 3 - Electric field dependence of the oscillator strenghts of the X and X" excitonic levels in the decoupled (full lines) and admixed (dotted lines) approximations. ADQW in Fig. 2.

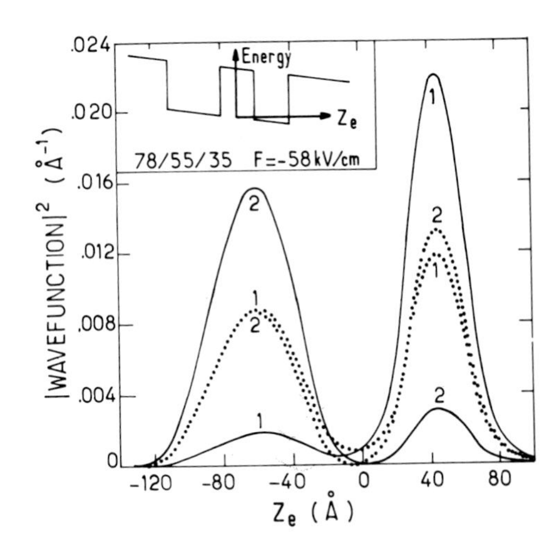

Fig. 4 - z_e dependence of the decoupled (full lines) and admixed (dotted lines) excitonic wavefunctions for the ADQW in Fig. 2. F= -58 kV/cm. See the text for the labels.

3. EXCITONIC TRANSFER TIMES

The Fermi golden rule is used to calculate the level width $h/2\tau_i$ of an initial excitonic state Ψ_i due to transitions to all final states Ψ_f induced by a static potential $V(\mathbf{r}_e,\mathbf{r}_h)$. $K_{C.M.}=0$ in the initial state. As scattering potential we consider unscreened ionized impurities sitting on two inverted interfaces and randomly distributed in the layer plane. In this case we have $V(\mathbf{r}_e,\mathbf{r}_h,\mathbf{R}_i) = V_e(\mathbf{r}_e,\mathbf{R}_i) + V_h(\mathbf{r}_h,\mathbf{R}_i)$. In Fig. 5 we show the dependence with F (around F_0) of the excitonic transfer times for the 78Å/55Å/35Å ADQW. Dashed dotted (full) curves refer to the admixed (decoupled) approximation. The areal impurity density is $10^{10}cm^{-2}$. In addition to the X and X" excitons discussed previously we have also considered the X' and X''' excitons, obtained by pairing the two E_1 and E_2 electron states with the first hole state mainly centered in the narrow well. For comparison we present also the results for the band-to-band scattering time ($E_2 \rightarrow E_1$; dotted line). Pronounced interference effects are easily seen in Fig. 5. They transform the single minimun τ(F) curve (decoupled excitons) into curves with two minima (admixed excitons); raise by about one decade the transfer times, as compared to the decoupled results; and shift negatively (positively) the field for the energy anticrossing of the X-X" states (of the X'-X''' states) as compared to F_0. We can show, for instance for the X↔X" transfer, that in this electric field range there is:

$$\tau^{adm}/\tau^{dec} \approx (\alpha_{2,1}^2 - \alpha_{1,1}^2)^{-2} \langle E_2|F(Q_\perp^{dec})|E_1\rangle^2 / \langle E_2|F(Q_\perp^{adm})|E_1\rangle^2$$

$$F(Q_\perp) = \exp\{-Q_\perp |z_e - z_{imp}|\} \qquad (5)$$

where $Q_\perp^2 = (2M/\hbar^2)(\varepsilon_i - \varepsilon_f)$, and $\alpha_{2,1}^2 + \alpha_{1,1}^2 = 1$, where the α's are the coefficients of Ψ_i in eq.(3) and z_{imp} the impurity position. In the decoupled approximation $|\alpha_{2,1}|=1$ since the initial state is formed exclusively by the second conduction state (E_2) and the ground valence state (H_1). The mixing between the decoupled excitons makes the α's to deviate from their decoupled 0 or 1 values (e.g., $|\alpha_{2,1}| \approx 0.84$ at F=-56kV/cm), thereby increasing τ^{adm}. Another feature leading to an increase of τ^{adm} is the widening of the excitonic minigap near the anticrossing. This leads to a larger in-plane momentum transfer ($Q_\perp^{adm} > Q_\perp^{dec}$) and thus to a decrease of the scattering efficiency (since M is large and the Fourier components of the scattering potential decay faster than $Q_\perp^{-1}$).

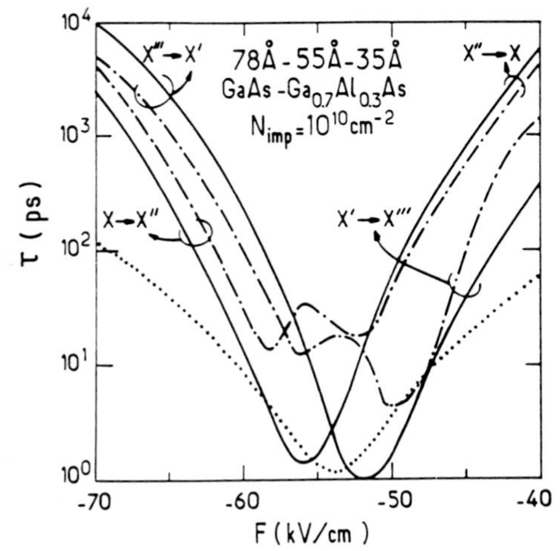

Fig. 5 - Dependence with F of the elastic (ionized impurities) scattering times of exciton in the decoupled (full lines) and admixed (dashed-dotted lines) approximations for both the X↔X" and X'↔X''' processes. Dotted line: $E_2 \rightarrow E_1$.

4. COMPARISON WITH EXPERIMENTAL RESULTS

We show in Fig. 6 the results (Ferreira et al 1990 b) of optical c.w. photoluminescence (PL) experiments performed at T = 2K on the 78Å/55Å/35Å ADQW (as symbols). Namely we show a plot of the bias dependence of the integrated PL intensities of the various PL lines. We have solved rate equations for the direct and crossed excitons populations, which incorporate direct↔crossed scattering times, pumping rates for the direct excitons and radiative recombinations (non radiative processes are neglected). In the results of section 2 only the electrons effectively transfer, the hole acting mostly as a spectator. In the rate equations we also account for the reverse processes (with hole transfer). Inter-well hole transfer times have been estimated from previous independent hole results which account for the complications of the valence band in-plane dispersions (Bastard et al 1989). The numerical results for the variation with F of the various PL intensities (population over the radiative recombination time, both F dependent) are plotted as lines in Fig. 6. Two features are noticeable in these results: **i)** the appearence of a crossed PL line below ≈-3.5 Volts: the crossed exciton is observable only when it is the ground exciton state of the ADQW; **ii)** the existence of two distinct resonances (around -2.2 and -3.5 Volts) in the I and I' intensities which are symmetrically located around the bare electronic resonance $E_1 - E_2$: this is the unambiguous signature of the excitonic nature of the states involved in the tunnelling processes at low temperatures.

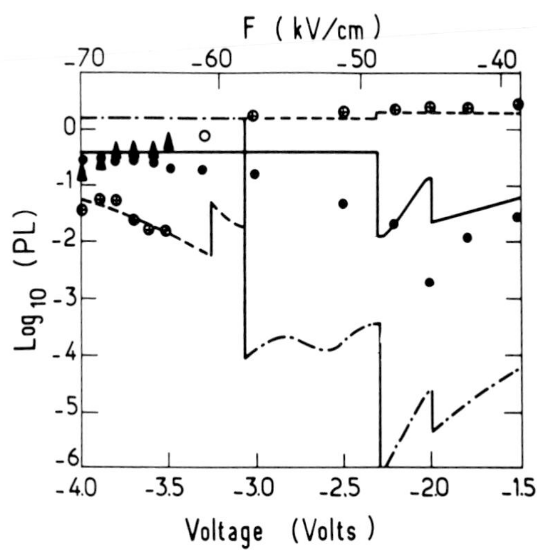

Fig. 6 - Experimental (symbols) and calculated (lines) PL intensities for the 78Å/55Å/35Å biased ADQW.

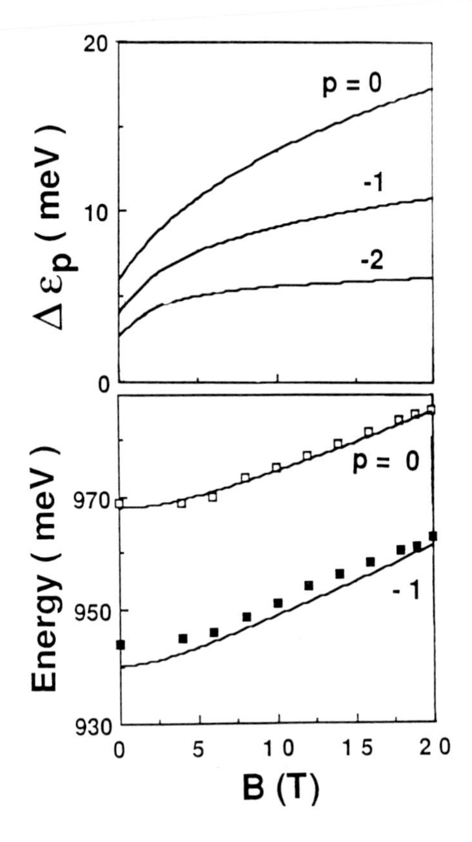

Fig. 7- Upper part: calculated B dependences of the p=0,-1,-2 exciton binding energies. Lower part: calculated (lines) and measured (symbols) B dependences of the p= 0 and p= -1 excitonic transitions. F=35 kV/cm.

5. MAGNETOEXCITONS IN BIASED SUPERLATTICES

We have proposed a variational approach in order to treat both direct and crossed magnetoexciton states in biased superlattices. Let d be the superlattice period, the magnetic field B=0 and F strong enough such that eFd > $\Delta\varepsilon_{p,q}$ where $\Delta\varepsilon_{p,q}$ is the binding energy of the decoupled exciton $\Psi_{p,q}$ for the electron/hole pairs separated by $|p-q|$ periods of the superlattice, with $p,q=0,\pm1,\pm2,...$ $\Delta\varepsilon_{p,q}=\Delta\varepsilon_{p-q}$ is the larger for p=q and decreases rapidly with increasing electron-hole separation $|p-q|d$. Mixings of direct (p=q) and crossed ($p\neq q$) excitons should then be small. Also the mixing of states with same p-q value but centered in different periods are expected to be weak. In that approximation for each hole state there corresponds a ladder of exciton levels. By symmetry the q'-th ladder (associated to the q'-th hole level) matches exactly the q-th one by a rigid translation of the first by (q-q')eFd. Let us focus on the q=0 ladder and ommit in the following the hole label for the decoupled excitons (for a comprehensive survey of Wannier - Stark excitons see Dignam and Sipe 1991).Under the previous approximations a magnetic field applied along the growth axis modifies only the in-plane reduced motion of the excitons. Direct and crossed excitons will be differently affected by the same magnetic field, since they have very different in-plane extensions (λ_p increases rapidly with increasing $|p)|$. Let us focus first on the direct exciton. For small B values such that $\lambda_0<<\lambda_c$ (λ_c=cyclotron radius) ε_0 displays only a small diamagnetic shift with B. When $\lambda_0\approx\lambda_c$ the coulombic and magnetic energies are of the same order and at very high fields ($\lambda_0>>\lambda_c$) the reduced motion is completly determined by the magnetic field (Landau orbits). In that limit $\Delta\varepsilon_0$ increases roughly as $\sqrt{B}$ whereas the energy of the free Landau levels increase linearly with B. So the transition energy ε_0 growths roughly linearly with B for strong fields. Let us consider now the crossed excitons. As stated before $\lambda_p>>\lambda_0$ if $|p|>>0$. For a given B value all crossed states with $|p|>pp$ ($0\leq|p|<pp$) will be in a regime of strong (weak) magnetic field, where pp is defined such that $\lambda_{pp}\approx\lambda_c$. $\Delta\varepsilon_{p\neq0}(B)$ increases initially with B and saturates roughly at $e^2/(\kappa|p|d)$ as $B\rightarrow\infty$: the divergence found for $\Delta\varepsilon_{p=0}(B)$ is lifted since the carriers confined into orbits of radius $\lambda_p\approx\lambda_c\rightarrow0$ are located in different wells of the superlattice. So crossed exciton states enter more rapidly in the regime where their transition energies ε_p increase linearly with B. Consequently it would be difficult to observe the diamagnetic regime of very crossed states (high $|p|$). We show in fig. 7 the B dependence of the transition energies (p=0,-1) and of the binding energies $\Delta\varepsilon_{p=0,-1,-2}$ for a superlattice of 40Å wells/46Å barriers widths under a bias corresponding to F≈35kV/cm. Note that the good agreement between the B dependence of the p = 0, -1 transition energies has been reached using the measured superlattice parameters.

ACKNOWLEDGEMENTS

We are pleased to thank Drs C Delalande, H W Liu, P Rolland, B Soucail and P Voisin for their collabarotion. This work has been supported by the Centre National de la Recherche

Scientifique and by a CNET contract (906B067 007909245). One of us (R F) thanks the CAPES (Brazil) for financial support.

REFERENCES

Bastard G, Delalande C, Ferreira R and Liu HW 1989 J. of Lumines.**44** 247

Chen YJ, Koteles ES, Elman BS and Armiento CA 1987 Phys. Rev. **B36** 4562

Claro F, Pacheco M and Barticevic Z 1990 Phys. Rev. Lett. **64** 3058

Dignam M M and Sipe J E 1991 Phys. Rev. **B43** 4097

Ferreira R, Soucail B, Voisin P and Bastard G 1990 a Phys. Rev. **B42** 11904

Ferreira R, Delalande C, Liu HW, Bastard G, Etienne B and Palmier JF 1990 b Phys. Rev. **B42** 9170

Ferreira R 1991 Phys. Rev. **B43**, 9336

Fukuzawa T, Mendez EE, Hong JM 1990 Phys. Rev. Lett.**64** 3066

Galbraith J and Duggan G 1989 Phys. Rev. **B40** 5515

Golub J. E. , Kash K. , Harbison J. P. and Florez L. T. 1990 Phys. Rev. **B41** 8564

E E Mendez and F Agullô - Rueda 1989 J. of Lumines.**44** 223

Meynadier M H, Nahory R E, Worlock J M , Tamargo M C , Miguel J L and Sturge M D 1988 Phys. Rev. Lett. **60** 1338

Sauer R, Thonke K and Tsang WT 1988 Phys. Rev. Lett. **61** 609

Schneider H, von Klitzing K and Ploog K 1989 Superlatt. Microstruct.**5** 305

Voisin P, Bleuse J, Bouche C, Gaillard S, Alibert C and Regrény A 1988 Phys. Rev. Lett.**61** 1639

Zrenner A, Leeb P, Schäfer J, Böhm G, Weimann G, Worlock JM, Florez LT and Harbison JM 1991to be published in Surf. Sci.

Inst. Phys. Conf. Ser. No 123
Paper presented at the International Meeting on Optics of Excitons in Confined Systems, Giardini Naxos, Italy, 1991

Quantitative analysis of transmission spectra of GaAs/(Ga,Al)As multiple quantum wells

S Frisk (a), JL Staehli (a), LC Andreani (b), A Bosacchi (c) and S Franchi (c)

(a) Institut de Physique Appliquée, Ecole Polytechnique Fédérale, PH-Ecublens, CH-1015 Lausanne, Switzerland; (b) Institut Romand de Recherche Numérique en Physique des Matériaux, PH-Ecublens, CH-1015 Lausanne, Switzerland; and (c) CNR-Istituto MASPEC, Via Chiavari 18/A, I-43100 Parma, Italy

ABSTRACT. We measured the optical absorption spectra of a series of GaAs/(Ga,Al)As multiple quantum wells. We interpret these spectra by making use of recently developed theories. We compare quantitatively spectral positions, oscillator strengths and absorption coefficients of the observed excitonic and subband-to-subband transitions with the available theoretical results. The overall agreement between experiment and theory is quite good, and most of the observed peaks can readily be identified.

1. INTRODUCTION

The optical properties of semiconductor quantum wells (QW) have been investigated for several years by now (see eg the book by Bastard (1988)). It turned out that for a thorough understanding of the observed optical spectra accurate theoretical calculations were needed. In particular, the fourfold degenerate valence band states at the Γ point of the Brillouin zone of bulk compounds having diamond or zincblende structure split and strongly mix in quantum structures, making theory more involved (Sanders and Chang 1987, Andreani 1989).

Using one of the most elaborate theories of the electronic states in QW (Andreani and Pasquarello 1990), we present a detailed analysis of experimental transmission spectra. We find a quite good agreement between theory and experiment, in particular the energies of the lowest four exciton states agree within a few meV with the calculations. However, we still observe some systematic discrepancies.

2. EXPERIMENTAL

The optical transmission measurements were performed on a series of multiple QW grown by molecular beam epitaxy on a GaAs substrate. Their width is ranging from 2.0 to 18.0 nm, and the Al content of the barriers varies between 20 and 40 atomic %. Between the wells and the substrate there is a several tenths of μm thick (Ga,Al)As layer, in this way the GaAs substrate could be removed by using a selective etch (LePore 1980). The samples were anodized to avoid interference due to multiple reflections of light between the front and back surfaces.

The transmission measurements were made using a incandescent lamp, and the samples were immersed in superfluid He. The light propagated in a direction perpendicular to the layers and thus its polarization was parallel to the well planes. The transmitted light was analysed with a grating monochromator and detected by a photo multiplier tube having a GaAs cathode. The spectra were recorded using standard lock-in techniques. Examples of measured spectra of the absorption coefficient $\alpha(h\nu)$ (ie the probability per well that a photon is absorbed) are shown in Fig. 1. For the determination of $\alpha(h\nu)$, the transmission

$T(h\nu)$ of the N wells present in a sample (typically 25) was approximated by

$$T(h\nu) = (1 - \alpha(h\nu))^N \approx exp(-N\alpha(h\nu)), \tag{1}$$

ie with an absorption coefficient of 0.1 this approximation causes an error of about 5 % which is comparable to the experimental accuracy.

3. DISCUSSION

The measured spectra were compared with accurate theoretical calculations performed as discussed by

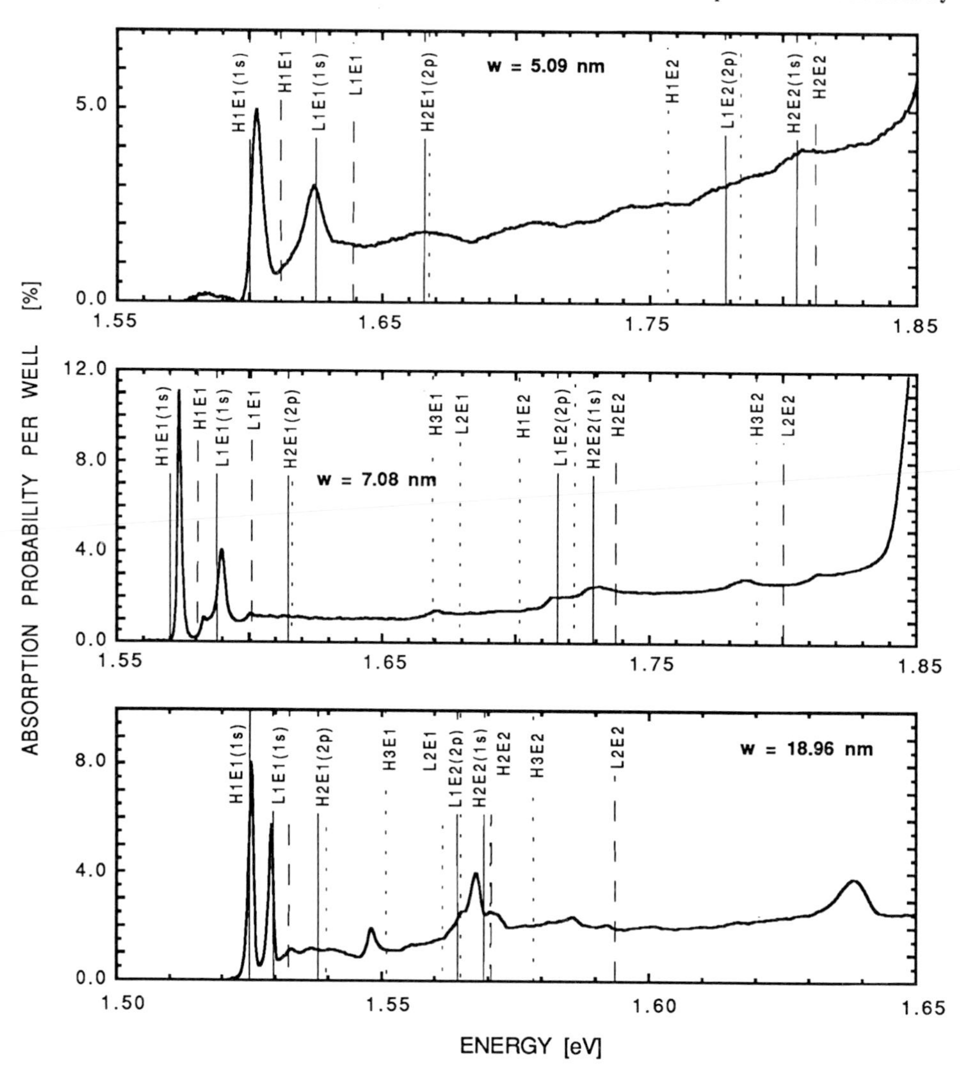

Figure 1. Absorption spectra of quantum wells of different widths w for light propagating perpendicularly to the well planes. The vertical lines denote the calculated spectral positions of different transitions: solid lines correspond to excitonic states (labelled 1s and 2p), while dashed and dotted lines mark the positions of transitions between the subband edges. Hi, Li and Ei label subband nr i of heavy holes, light holes and electrons, respectively.

Andreani and Pasquarello (1990), ie accounting for valence band mixing, nonparabolicity of the bulk conduction band, coupling between excitons belonging to different subbands and the difference of the dielectric constants between well and barrier materials. The energies of the different electronic levels were calculated with a precision of about 2 meV.

As usual, from the parameters used for the epitaxial deposition of the layers, the widths w of the QW and the Al content x of the $Ga_{1-x}Al_xAs$ barriers can be determined only approximately. While x was determined from the luminescence or transmission spectra of the barrier material, w was determined by fitting the measured spectral positions of the strong H1E1(1s) (Hi, Li and Ei denote subband nr i of heavy holes, light holes and electrons, respectively), L1E1(1s), L1E2(2p) and H2E2(1s) exciton peaks. We obtained a good overall agreement of the measured with the calculated transition energies. In particular, it turns out that also the energy of the L1E2(2p) exciton is well accounted for by theory (Andreani and Pasquarello 1990). The remaining deviations between theory and experiment which are a few meV could in part be due to strain in the layers. However, we found that the calculated splitting between the H1E1(1s) and L1E1(1s) excitons is systematically higher than what has been measured, in the narrowest wells this discrepancy can be as high as 10 meV.

In Fig. 1, for the 5.09 nm and the 7.08 nm wells, the linewidth of the L1E1(1s) exciton is larger than that of the H1E1(1s) exciton, which for $w < 13$ nm is degenerate with the continuum of the H1E1(1s) exciton (Pasquarello and Andreani 1991). The additional broadening of the L1E1(1s) exciton is about 1.2 meV, in agreement with theory.

The oscillator strength (OS) of the fundamental heavy and light hole excitons can be determined quite easily (for the latter the strength of the transitions to the H1E1 continuum has to be accounted for). The measured values are shown in Fig. 2, together with the results of theoretical calculations (Andreani and Pasquarello 1990). Here also overall agreement is found, however as already observed earlier (Masselink et al 1985) the light hole excitons in wide QW seem to have an OS which is systematically higher than predicted by theory. This is probably due to lack of convergence with the number of subbands used in the theoretical calculation, since the OS is very sensitive to coupling to higher subbands for wide wells. The OS of the higher excitons are more difficult to extract from the spectra, because they overlap with

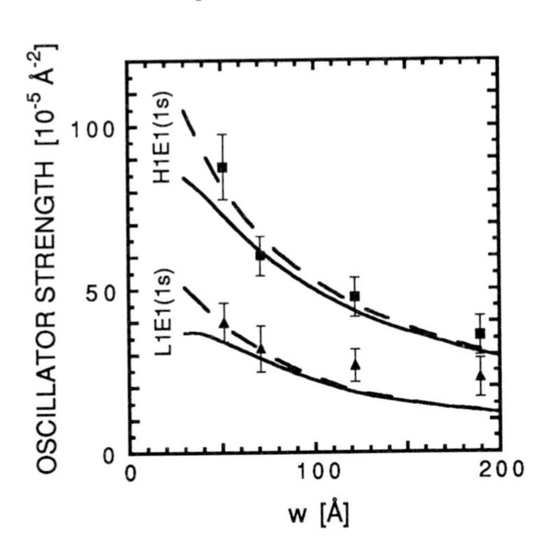

Figure 2. Oscillator strengths of of the two lowest 1s exciton peaks (H1E1(1s) and L1E1(1s)) versus well width w. Points: measured; curves: theory (Andreani and Pasquarello 1990); dashed lines are for an Al concentration in the barriers of 25 %, solid lines for one of 40 %.

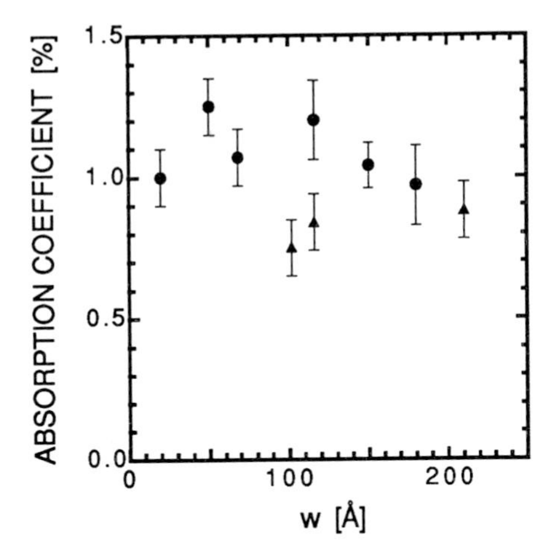

Figure 3. Absorption coefficient of the lowest subband to subband transitions, H1E1 + L1E1 + H2E1, versus well width w. Circles: measured, triangles: calculated (Sanders and Chang 1987)

transitions to continuum states of lower subbands, and since their peaks are often wider and thus less pronounced. In any case, we note that the transition to the L1E2(2p) exciton becomes stronger than that to the H2E2(1s) exciton once the well is wider than 8 nm, as expected from theory (Andreani and Pasquarello 1989).

The absorption coefficient (per well) for the transitions between continuum states of the lowest hole and electron subbands does not depend on the width of the wells (see Fig. 3) (Andreani 1989). Further, as shown by Sanders and Chang (1986), the sum of the three subband to subband transitions H1Ej, L1Ej and H2Ej (j = 1, 2) is close to the staircase function predicted by the simple model of decoupled subbands. The measured magnitudes of the first step (H1E1 + L1E1 + H2E1) are shown in Fig. 3, they are somewhat higher than predicted by theory.

4. CONCLUSION.

We quantitatively compared the optical transmission spectra of multiple QW with theoretical expectations. We find a good general agreement between experiment and theory. However, the observed energetic splitting of the L1E1(1s) and H1E1(1s) exciton levels is systematically smaller than theoretically expected.

Acknowledgements. This work has in part be supported by the *Fonds National Suisse de la Recherche Scientifique*. We thank WT Wu for her valuable help during the experiments.

5. REFERENCES

Andreani LC, 1989: *Theory of Excitons and Polaritons in Semiconductor Quantum Wells*, thesis, Scuola Normale Superiore, Pisa

Andreani LC and Pasquarello A, 1989: *Superlattices and Microstructures* **5**, 59

Andreani LC and Pasquarello A, 1990: *Phys Rev B* **42**, 8928

Bastard G (1988): *wave mechanics applied to semiconductor heterostructures*, les editions de physique, Les Ulis

LePore JJ, 1980: *J Appl Phys* **51**, 6441

Masselink WT et al, 1985: *Phys Rev B* **32**, 8027

Pasquarello A and Andreani LC, 1991: *Phys Rev B* **44**, 3162

Sanders GD and Chang YC, 1987: *Phys Rev B* **35**, 1300

Inst. Phys. Conf. Ser. No 123
Paper presented at the International Meeting on Optics of Excitons in Confined Systems, Giardini Naxos, Italy, 1991

Excitons and minibands in (InGa)As-GaAs strained layer superlattices

Karen J. Moore[+]

Manchester Polytechnic, Department of Mathematics and Physics, John Dalton Building, Chester Street, Manchester M1 5GD, UK.

ABSTRACT: Photoluminescence excitation measurements on (InGa)As-GaAs strained layer superlattices reveal both parity allowed and forbidden transitions at the mini-Brillouin zone centre and edge. This allows the evolution of the superlattice minibands to be mapped out, with decreasing barrier thickness. The spectra also demonstrate the significant effects of the electron-hole Coulomb interaction within the miniband on the continuum absorption shape. These effects are carefully considered in the design of a sample which clearly reveals the $\Delta n=0$ exciton resonance below the saddle point.

1. INTRODUCTION

This report describes low temperature (4K) photoluminescence (PL) and photoluminescence excitation (PLE) investigations of excitons associated with the formation of minibands in strained layer (InGa)As-GaAs superlattices (SLs). The compressive strain in the alloy layers, which produces a significant increase in the splitting of the light- and heavy-hole exciton states, allows a much clearer identification of features in the region of the heavy-hole continuum (including the saddle point exciton) in this material system compared to previous observations in lattice -matched structures (Deveaud et al 1989, Song et al 1989).

The (InGa)As-GaAs SL samples used in these investigations were all grown by molecular beam epitaxy at Philips Research Laboratories, Redhill. A total of five samples were studied, comprising a set of four structures with a nominal indium fraction in the alloy layers of 0.06 and a single sample with a nominal fraction of 0.12. The four samples have 20 periods of 5 nm (InGa)As wells and GaAs barriers of either 20, 15, 10 or 5 nm. The fifth sample is also a 20 repeat structure, but has 2.5 nm (InGa)As wells separated by 10 nm GaAs barriers. Details of the well and barrier widths and indium concentrations derived from x-ray diffraction (XRD) measurements have been discussed previously (Moore et al 1990). In this report each sample is simply identified by its nominal (well)-(barrier) thickness.

2. CALCULATIONS OF MINIBAND DISPERSION

The predicted evolution of all the expected transitions for these samples, both at the centre and at the edge of the minizone, is best displayed as shown in Fig. 1. Each transition has been calculated as a function of GaAs barrier thickness, for a fixed value of (InGa)As well thickness and composition. (We have chosen average values of both parameters, obtained by taking the mean of the XRD results.) Included in the figure are both the allowed $\Delta n=0$ transitions and the parity forbidden e1-hh2 transitions. All details of the calculation can be found in Pan et al 1988. Studying Fig. 1 provides a clear picture of the expected evolution of the minibands with decreasing barrier thickness. For example, if we consider the e1-hh2 transition, then for the (5 nm)-(20 nm) sample we expect to see only one peak in the PLE spectrum. However, as the barrier thickness is decreased this feature should split into a doublet, corresponding to the centre and edge transitions.

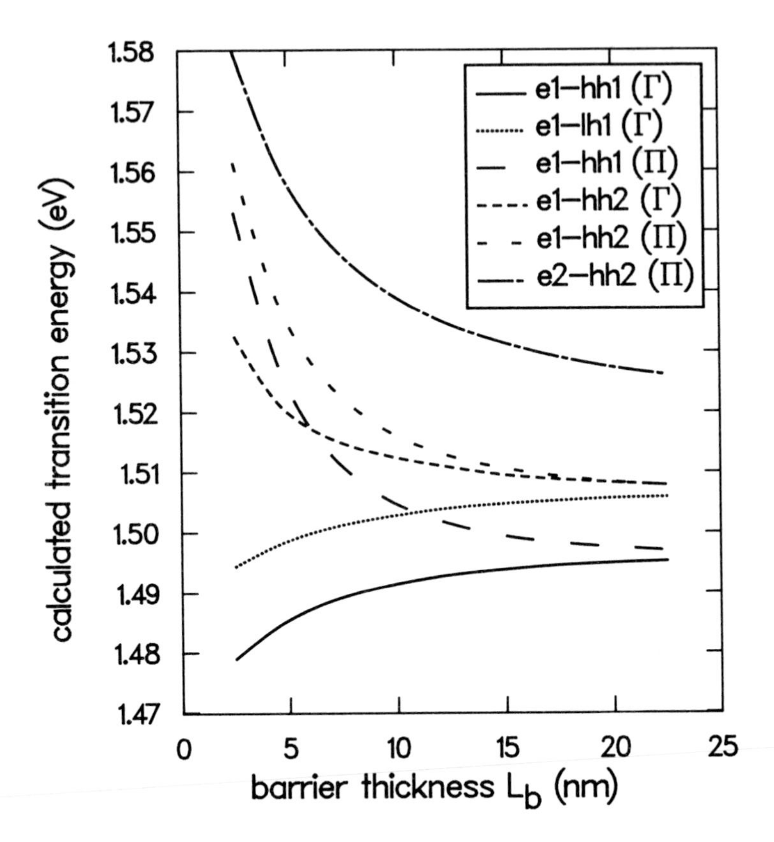

Fig. 1 Calculated band-to-band transitions as a function of GaAs barrier thickness, for a fixed (InGa)As well width of 6.4 nm and indium fraction of 0.045. Transitions at the minizone centre are labelled as Γ and those at the zone edge as π.

3. EXPERIMENTAL RESULTS AND DISCUSSION

The PL and PLE data of the four samples are illustrated in Fig. 2. The position of the light-hole (lh) exciton has been confirmed by circularly polarised PLE measurements on all the samples. All other features have been identified as having heavy-hole character. The peak at about 1.515 eV is the bulk GaAs exciton arising from absorption in the buffer and substrate material. Comparison of the measured energy positions of peaks 1 and 2 with our envelope calculations leads us to assign these features to the symmetry-forbidden e1-hh2 exciton transitions. Peak 1 corresponds to the transition at q=0, ie. at the centre of the mini-Brillouin zone, and is referred to as the e1-hh2(Γ) exciton, while peak 2 is a direct transition at the 1D Brillouin zone edge, ie. at $q=\pi/d$, which we call e1-hh2(π). The evolution of the e1-hh2 transitions proceeds as anticipated from the calculations illustrated in Fig. 1. The splitting between peaks 1 and 2 slowly increases, allowing us to map out the dispersion of this band and to measure directly the combined miniband width of the e1 and hh2 subbands. A further prominent feature appears in the spectra of Fig. 2, above the GaAs band edge, labelled as peak 3. This we assign to the allowed e2-hh2(π) minizone edge exciton state. As predicted in Fig. 1 this feature moves rapidly to higher energy with decreasing barrier thickness. A detailed comparison of the measured exciton positions with our calculations is given in Moore et al 1990.

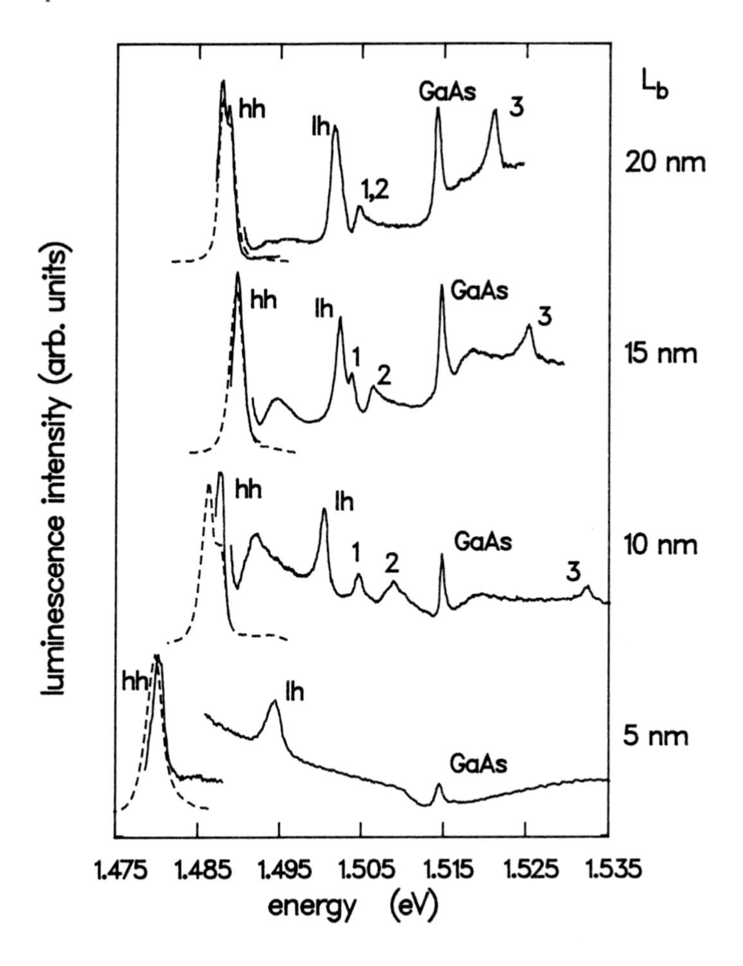

Fig. 2. PL and PLE spectra, recorded at 4 K, of the four (InGa)As-GaAs superlattices. The nominal barrier thickness, L_b , is given for each sample.

In addition to the evolution of the exciton states with decreasing barrier thickness, there is a dramatic change in the shape of the PLE spectra, in the region of the heavy-hole continuum states. For the sample with the thickest barriers, the steplike continuum is entirely consistent with the two dimensional density of states we would anticipate. However, as the barrier thickness is reduced the spectral shape in this region develops into a broad peak; the greatest enhancement occuring when the width of the electron miniband is comparable to the exciton binding energy. The evolution of this feature can be correlated with an increase in the dispersion of the SL bandstructure in the growth direction and compares most favorably with the theoretically generated absorption curves of Chu and Chang 1987. Unfortunately, this masks the e1-hh1(π) transition in these samples. In order to reveal its position we must design a structure with a large miniband width while at the same time ensuring that other excitonic states, such as the light-hole, are not expected to appear in the same energy range of the spectrum. The fifth sample in these investigations forefils these criteria. The PLE spectrum of the (2.5 nm)-(10 nm) sample is shown in Fig. 3.

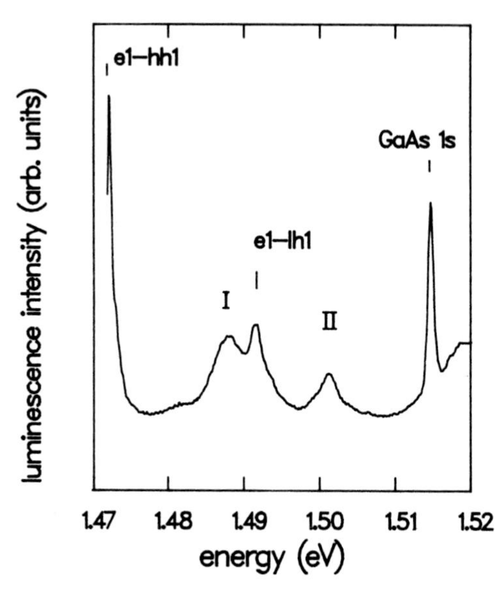

Fig. 3. PLE spectrum, from the nominally (2.5 nm)-(10 nm) (InGa)As-GaAs SL sample.

The e1-hh1(Γ), e1-lh1(Γ) and bulk GaAs excitons are clearly identified. By comparison with our calculation we ascribe peak II to the parity forbidden e1-hh2 (Γ) exciton transition. Peak I is a new feature which we assign as the exciton resonance associated with the n=1 electron and heavy-hole minibands at q=π/d, denoted e1-hh1(π). Unlike the e1-hh2 transitions, both the electron and hole states are derived from M_1 critical points in the bandstructure, and we refer to this peak as a saddle-point resonance.

4. SUMMARY

We have used PLE spectroscopy to study the evolution of superlattice minibands in (InGa)As-GaAs structures. The observed exciton peaks are consistent with theoretical predictions of Δn=0 and forbidden intersubband transitions at the centre and edge of the superlattice minizone. In addition, as the miniband dispersion becomes comparable to exciton binding energy a significant enhancement of the heavy-hole continuum absorption is noted. Finally, the Δn=0 e1-hh1(π), saddle point resonance feature is clearly identified.

5. ACKNOWLEDGEMENT

Many thanks to Geoffrey Duggan for invaluable contributions to this work.

6. REFERENCES

+ Previous address: Philips Research Laboratories, Redhill, RH1 5HA, UK.

Chu H and Chang Y-C 1987 Phys. Rev. B36, 2946
Deveaud B, Chomette A, Clerot F, Regreny A, Maan JC, Romestain R, Bastard G, Chu H and Chang Y-C 1989 Supperlatt. Microstruct. 6, 183
Moore KJ, Duggan G, Raukema A and Woodbridge K 1990 Phys. Rev. B42, 1326
Pan SH, Shen H, Hang Z, Pollak FH, Zhuang W, Xu Q, Roth AP, Masut RA, Lacelle C and Morris D 1988 Phys. Rev. B38, 3375
Song JJ, Jung PS, Yoon YS, Chu H, Chang Y-C and Tu CW 1989 Phys. Rev. B39, 5562

Inst. Phys. Conf. Ser. No 123
Paper presented at the International Meeting on Optics of Excitons in Confined Systems, Giardini Naxos, Italy, 1991

Optical gap engineering of ZnSe-ZnTe strained layer superlattices

P. Boring, P. Lefebvre and H. Mathieu

Groupe d'Etudes des Semiconducteurs, Université de Montpellier II: Sciences et Techniques du Languedoc. Case Courrier 074, 34095 Montpellier Cedex 5, France.

ABSTRACT : Interesting candidates for the achievement of blue emission, via semiconductor based devices, might be ZnSe-ZnTe superlattices, which should allow to obtain efficient excitonic recombination lines, with short wavelengths. We propose an exploration of the possibilities of ZnSe-ZnTe superlattices, via envelope function calculations, including the strain-induced coupling of light hole states with spin-orbit split off ones, due to important internal biaxial strains. Quantitative arguments are proposed, concerning the optimization of the energies, oscillator strengths and lifetimes of confined excitons in these systems.

1. INTRODUCTON

The growing interest in the field of optoelectronic devices, such as semiconductor lasers, having their emitting band in the short-wavelength range of the visible spectrum (blue light), calls the attention of the researchers on zinc-based II-VI compounds, due to their wide direct optical gaps. This paper presents a prospective investigation of the optical excitonic properties of ZnSe-ZnTe superlattices (SL's), accounting for the relatively poor knowledge on the parameters of both binaries. In particular, our calculations show that these SL's should be type II, in most cases : we thus propose a tentative evaluation of the designs of the structures, which should be the more able to present "blue excitons" (around ~ 2.6 eV), with optimized oscillator strengths.

2. THEORY

Our theoretical approach was made within the framework of the Envelope Function Approximation (EFA), accounting for the internal strainfields induced by the strong lattice mismatch between the two materials, via the simple theory of elasticity (Mathieu et *al.* 1988). In a first step, varying the thicknesses of ZnSe and ZnTe layers, we computed the energies of the fundamental transitions e_1hh_1 and e_1lh_1, using a model including the coupling between the Γ_8 upper valence bands and the spin-orbit split-off band, of Γ_7 symmetry. In the present case of growth along the (001) direction, both kinetic and strain Hamiltonians describing the Γ_8 and Γ_7 valence bands contain a nonvanishing element of coupling between the light-hole (lh) and the split-off-holes (so), while the heavy-hole band remains decoupled, thus parabolic. Comparisons with more usual models, neglecting such a coupling, provide an important correction to the energies of light-holes (up to ~ 100 meV!). This was already confirmed by experiments, in the case of III-V microstructures (Gil et *al.* 1991) : the confinement energies of light holes are reduced by the coupling. The widths of the fundamental energy gaps of the materials, relatively to their spin-orbit splitting, allows to ignore the **k.p** coupling with the Γ_6 conduction band. Next, by reasonable estimations of excitonic binding energies and oscillator strengths, we have attempted to put together the conditions necessary to the constitution of wide and optically efficient excitonic gaps.

3. NUMERICAL RESULTS

In spite of the uncertainty of some parameters of the calculation like, in particular, the valence band offset (Ev(ZnTe)-Ev(ZnSe)), for which the literature gives values ranging from 0.29 eV (Harrison et al 1986) to more than 1 eV (Rajakarunanayake et al 1988, Hong Wu et al 1990), our calculations show that the SL is most probably type II : since the difference between the gaps of ZnSe and ZnTe is of the order of 400 meV, one can easily see, as in Figure 1, that the energy minima for the electrons lie in the ZnSe layers, inasmuch as the effects of strain favour this situation. The electrons are then confined in the ZnSe layers, while the heavy holes lie in the ZnTe layers. The location of the light holes depends on both the value of the offset and the nature of the buffer layer on which the SL is grown. Preliminary studies allow us to conclude that the highest energy gaps are obtained with a ZnSe buffer layer, which will be the case considered in the following. The light holes are then always confined in the ZnTe layers, in accordance with the more recently published values of the valence band offset (≥ 500 meV).

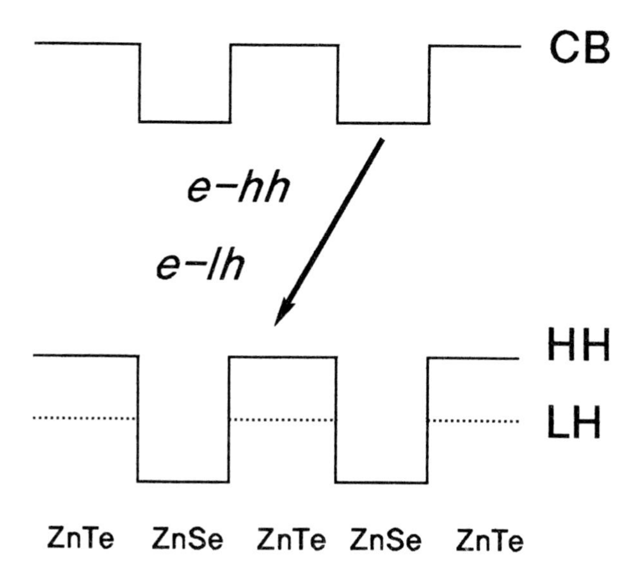

Figure 1. *Schematic view of the conduction and valence band profiles for a*

As a first result, we thus see that the fundamental excitonic gap in ZnSe-ZnTe SL's, grown on a ZnSe buffer layer, is type II. Consequently, the binding energies of the related excitons are expected to be lower or of the order of the Rydberg in ZnSe (17.35 meV, from Hölscher et *al.*1985) or in ZnTe (14.10 meV from Neumann et *al.*1988). Table 1 displays the calculated values for the band-to-band optical gap, for various designs of the SL : in spite of the reduction of the light-hole energies due to the coupling with split-off states, this gap always involve heavy-hole states. Two values are proposed, depending on whether a valence band offset of 0.5 eV or 1.0 eV is chosen. We have restricted ourselves to thicknesses lower than 5 monolayers for several reasons : *i*) large gaps are obtained only for thin ZnTe layers, *ii*) : the large lattice mismatch implies very small critical thicknesses for coherent growth, *iii*) : in order to obtain sufficient excitonic oscillator strengths, it is necessary to impose conditions of large electron-hole envelope function overlap integral. This last argument is crucial, within the present "type II" context, as shown by Table 2, since the overlap integral for the e_1hh_1 transition rapidly decreases when the period of the SL increases.

Number of ZnTe monolayers \ Number of ZnSe monolayers	1	2	3	4	5
5	2.55	2.55	2.55	2.55	2.50
4	2.55	2.55	2.55	2.55	2.55
3	2.60	2.60	2.60	2.60	2.55
2	2.60	2.65	2.65	2.65	2.65
1	2.65	2.70	2.70	2.70	2.75

Table 1(a) : *Excitonic transitions calculated with a valence band offset of 500 meV, for a SL lattice-matched onto a ZnSe buffer layer. The energies are given in eV.*

Number of ZnTe monolayers \ Number of ZnSe monolayers	1	2	3	4	5
5	2.50	2.45	2.35	2.30	2.25
4	2.55	2.45	2.40	2.30	2.25
3	2.55	2.50	2.40	2.35	2.30
2	2.55	2.55	2.50	2.45	2.40
1	2.65	2.65	2.65	2.60	2.60

Table 1(b) : *The same as in (a), but for a valence band offset of 1.0 eV.*

Table 2(a) : *Electron-to-heavy hole overlap integral calculated for a valence band offset of 500 meV.*

Number of ZnTe monolayers	1	2	3	4	5
5	0.98	0.92	0.82	0.72	0.62
4	0.98	0.93	0.84	0.74	0.65
3	0.99	0.95	0.87	0.78	0.69
2	0.99	0.97	0.91	0.84	0.76
1	1.00	0.99	0.97	0.94	0.90

Number of ZnSe monolayers

Table 2(b) : *The same as in (a), but for a valence band offset of 1.0 eV.*

Number of ZnTe monolayers	1	2	3	4	5
5	0.94	0.80	0.64	0.49	0.37
4	0.96	0.83	0.68	0.53	0.42
3	0.97	0.91	0.72	0.59	0.47
2	0.98	0.91	0.79	0.66	0.54
1	0.99	0.96	0.90	0.82	0.72

Number of ZnSe monolayers

CONCLUSIONS

From our prospective calculations, one can conclude that, if the valence band offset is of the order of 1.0 eV, it will be really difficult to build a ZnSe-ZnTe superlattice with a large gap (*e.g.* ~ 2.60 eV) and a sufficient oscillator strength, unless growth techniques become sharp enough, so as to produce one-monolayer-wide layers. On the other hand, if the offset is of ~ 0.5 eV, interesting characteristics could be reached with layer thicknesses between 2 and 4 monolayers for ZnSe, and between 2 and 3 for ZnTe, which is not so compulsive. Of course, experimental data are desirable to clarify this point.

REFERENCES

Gil B., Lefebvre P, Boring P., Moore K.J., Duggan G., and Woodbridge K. 1991, Phys. Rev.B **44**, 1942

Harrison W.A., Tersoff J; 1986 Journal Vacuum Sci. Techn. **61**, 275

Hölscher H.W., Nöthe A., and Uihlein Ch. 1985 Phys. Rev. B **31** 2379

Hong Wu Yi, Shizuo Fujita, Shigeo Fujita 1990 J. Appl. Phys. **67** 908

Mathieu H., Allegre J., Chatt A., Lefebvre P., and Faurie J.P. 1988 Phys. Rev.B **38**, 7740

Neumann Ch., Nöthe A., and Lipari N.O. 1988 Phys. Rev. B **37** 922

Rajakarunanayake Y., Miles R.H., Wu G.Y., and Mc Gill T.C. 1988 Phys. Rev B **37** 10212

Inst. Phys. Conf. Ser. No 123
Paper presented at the International Meeting on Optics of Excitons in Confined Systems, Giardini Naxos, Italy, 1991

Excitonic effects in the absorption spectra of GaAs/AlGaAs superlattices

[a]V Capozzi, [b]JL Staheli, [c]C Flesia, [d]D Martin, [a]V Augelli and [a]GF Lorusso

[a]*Dipartimento di Fisica, Universitá di Bari, Via Amendola 173, Bari, Italy*
[b]*Inst. de Physique Appliquée, Ecole Polytechn. Fédérale, Lausanne, Switzerland*
[c]*Départ. des Mathématiques, Ecole Polytechn. Fédérale, Lausanne, Switzerland*
[d]*Inst. de Micro-Optoéléctronique, Ecole Polytechn. Fédérale, Lausanne, Switzerland*

ABSTRACT: Transmission spectra of short period of GaAs/$Al_xGa_{1-x}As$ (x=0.3) superlattices have been measured at the temperature of 2 K. The calculated energy ranges of the first carrier minibands are in agreement with the experimental absorption spectra. At the upper energy edges of the minibands, two bumps are present. They have been attributed to saddle point excitons.

1 INTRODUCTION

Excitonic effects in semiconductor superlattices (SL) are less extensively studied than in quantum-wells. Coulomb interaction between electrons (e) and holes (h), strongly affects the absorption spectra of semiconductors, because of the formation of excitonic states (Bassani 1975). In SL the zone-folding effects of the band structure in the growth direction z, give rise to an M_0-type singularity at the centre of the Brillouin minizone (BZ) at k_z=0 (where k_z is the wave-vector along the z direction) and to an M_1-type critical point (saddle point) at the BZ edge ($k_z=\pi/d$, where d is the SL period), as far as the joint density of the states (JDOS) is concerned (Bassani 1975). In this contribution, we discuss absorption measurements at T=2 K of short period (25Å/25 Å) SL of GaAs/$Al_xGa_{1-x}As$ (x=0.3). Besides the exciton lines at k_z=0, new structure are observed near the M_1 singularity of JDOS, which can be attributed to saddle point excitons.

2 EXPERIMENTAL METHODS

The SL samples were grown by means of MBE technique. They consist of a GaAs substrate upon which a buffer layer of GaAs (0.3 μm thick) was grown; the SL consists of 200 periods of GaAs/$Al_xGa_{1-x}As$ (x=0.3) layers, cladded between two thick layers (0.05 μm) of $Al_{0.3}Ga_{0.7}As$. Each period of the SL nominally contains 9/9 layers of GaAs/$Al_{0.3}Ga_{0.7}As$. In order to perform transmission measurements, the GaAs substrate was etched by using standard etching solution (Le Pore 1980) on a region having a diameter of about 0.5 mm. The Al concentration was checked by performing photoluminescence (PL) measurements of the thick AlGaAs layers, at the temperature of 2 K. From the spectral position of the recombination line of the AlGaAs exciton ground-state, it has been obtained x=0.30 ± 0.02. Furthermore, the width of the n=1 heavy-hole exciton line in the PL spectra at 2 K is about 3 meV. This is an indication of the good quality of the samples if a comparison with other PL-linewidth is made (see e.g. PL-spectra at 2K of A. Chomette et al. (1989)). The samples were immersed in superfluid He at T=2 K and illuminated by a tungsten-filament lamp on a spot of about 50 μm. The transmitted light was analyzed by means of a double spectrometer (dispersion of 40Å/mm) and detected by an optical multichannel analyzer system. The spectral resolution is 4 Å.

3 RESULTS AND DISCUSSION

Fig. 1 shows the band structure of the investigated SL in the z direction. Due to the short period of SL structure, only the first carrier miniband are present in the wells. These calculations were performed following the envelope function model (see e.g. Bastard 1987). The carrier parameters used in the calculations are the following: 0.065 m_o (0.914 m_o) for the GaAs ($Al_{0.3}Ga_{0.7}As$) electron effective masses, 0.392 m_o (0.503 m_o) for the GaAs ($Al_{0.3}Ga_{0.7}As$) hh-effective masses; 0.089 m_o (0.233 m_o) for the GaAs ($Al_{0.3}Ga_{0.7}As$) lh-effective masses. The band-gaps of the GaAs and $Al_{0.3}Ga_{0.7}As$ are 1.519 eV and 1.962 eV, respectively. A band-offset ratio of 0.65/0.35 is used corresponding to a conduction (valence) band discontinuity of 288 meV (155 meV) between the two semiconductor materials. Fig. 2 shows an optical absorption spectrum of a SL sample of GaAs/$Al_{0.3}Ga_{0.7}As$ measured at T=2 K. The two resolved peaks X_1^{hh} at 730.7 nm and X_1^{lh} at 724.5 nm correspond to the ground state of the heavy-hole (hh) and light-hole (lh) excitonic transitions, respectively. Between these two peaks a weak shoulder at about 728.4 nm (which is more or less pronounced depending on the sample quality) is present. It can be attributed to the transition to the excited 2s state of the hh-exciton (X_2^{hh}) (Miller et al. 1985). The presence of the X_2^{hh} permits to determine the hh exciton binding energy. It results to be 5.4 meV, in agreement with the value obtained from PL-excitation measurement (Chomette 1987) and also from variational calculations performed by Chomette et al. (1987). This value is close to the exciton binding energy (4.2 meV) of bulk GaAs (Sell 1972).

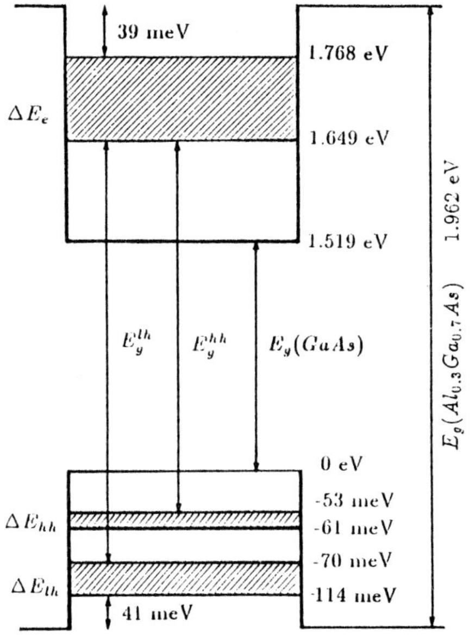

Fig.1: Calculated band structure (T=2 K) of the GaAs/$Al_{0.3}Ga_{0.7}As$ superlattice having 8 monolayers of GaAs and 9 monolayers of AlGaAs. ΔE_e, ΔE_{hh} and ΔE_{lh} are the minibands width for electrons, heavy-holes and light-holes, respectively.

In fact, because of the short period (d=50 Å) of the investigated SL, the exciton is not entirely confined in the well, but it is spread over a few layers (the exciton Bohr radius of bulk GaAs is about 100 Å) and displays a nearly 3D character. We mention that due to the experimental width of the hh-exciton line (about 3 meV) the 2s state of the hh-exciton can not be distinguished from the edge of the hh energy gap E_g^{hh}.

Since the binding energy of the hh- and lh-excitons are very similar (Miller and Kleinmann 1985), the energy difference between the two excitonic lines is caused by of the lh-hh splitting of the SL valence band. This splitting mainly depends on barrier height and well thickness of the semiconductor SL. In our case the lh-hh splitting results to be of 14.6 meV which is in a quite good agreement with the calculation performed by Chomette et al. (1987). We observe that the line width of the n=1 lh-exciton is broader than that of the ground-state hh-exciton. This is due to the Fano's resonance (Fano 1961) between the X_1^{lh} bound state and the continuum states of the hh1-e1 band to band transitions.

In fig. 2, we also report the calculated energy ranges of the optical transitions (hh1-e1 and lh1-e1) from the centre up to the edge of the BZ, as deduced from fig. 1.

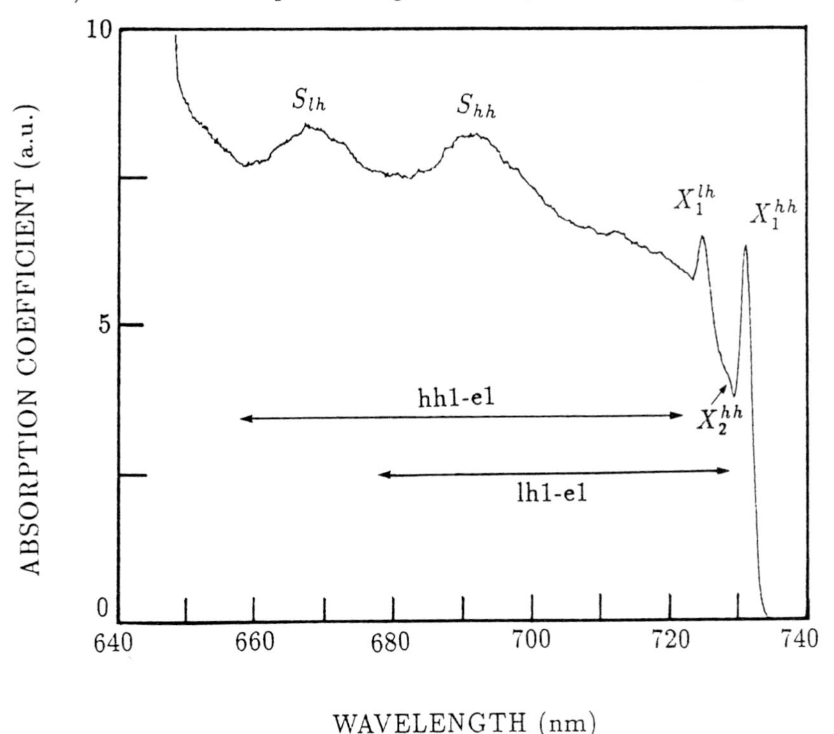

Fig.2: Absorption spectrum at T=2 K of a superlattice sample of GaAs/$Al_{0.3}Ga_{0.7}As$ formed by 200 periods (each period contains 8 monolayers (22 Å) of GaAs and 9 monolayers (25 Å) of AlGaAs, cladded between two layers of $Al_{0.3}Ga_{0.7}As$ (0.05 μm thick). The labels X_1^{hh} and X_1^{lh} mean: heavy-hole and light-hole ground-state excitons, respectively. S_{hh} and S_{lh} stand for heavy-hole and light-hole saddle point excitons. hh1-e1 and lh1-e1 are the calculated heavy-hole/electron and light-hole/electron optical transition energies corresponding to the centre and to the edge of the Brillouin minizone, as deduced from fig. 1.

The SL band structure has been calculated for different thicknesses of GaAs/AlGaAs layers, in order to account for the atomic layer fluctuations which can occur during the MBE growth process. The obtained values of the (E_g^{hh}, $E_g^{hh}+\Delta E_e+\Delta E_{hh}$) and ($E_g^{lh}$, $E_g^{lh}+\Delta E_e+\Delta E_{lh}$) were compared with the absorption spectrum of fig. 2. We found that the best sequence which matched the low energy part of the spectrum, corresponds to a sample with 8 monolayers of GaAs and 9 of AlGaAs. This calculation agrees with the uncertainty of the thickness calibration of the MBE machine, which for our sample is just one atomic layer.

In fig. 2 two bumps are present at about 691.4 nm (S_{hh}) and 667.4 nm (S_{lh}), respectively. Their intensity depends on the sample quality. These two structures can be attributed to hh- and lh-saddle point excitons associated with the high energy edge of the relative SL miniband JDOS which has an M_1 critical point. In fact, the e-h Coulomb interaction leads to excitonic resonances also below the energy of the saddle point M_1. These resonances are not bound states, as the excitonic levels associated to the M_0 critical point, but they are resonances due to the persistence of e-h correlation in the continuum of the miniband states. Moreover, we remark that the S_{lh} and S_{hh} structures are quite broad features, because of the Fano's resonance (Fano 1961) with the extended states of SL minibands. Similar structures, even if less pronounced, were found by Deveaud et al. (1989) in the short period SL by means of luminescence-excitation spectroscopy, and they were assigned to saddle point excitons too.

In summary, several structures appear in the low temperature absorption spectra of GaAs/AlGaAs superlattices, measured at T=2 K. They are attributed to optical transitions at the centre and at the edge of the Brillouin minizone in direction perpendicular to the SL layers.

AKNOWLEDGMENTS

We express our gratitude to L.C. Andreani and to K. Maschkle for useful discussions, and to A. Pasquarello for the collaboration in the calculations. We also tank P. Favia for her help in the measurements. This work was partially supported by "Fonds National Swiss pour la Recherche Scientifique".

REFERENCES

Bassani F and Pastori-Parravicini G 1975 "Electronic States and Optical Transitions in Solids" (Oxford:Pergamon) Chaps 5 and 6

Bastard G 1988 "Wave Mechanics Applied to Semiconductor Heterostructures" (Paris:Editions de Physique) Chap III

Chomette A et al 1987 Europhys. Lett. **4** 461

Chomette A et al 1989 J. Lumin. **44** 265

Deveaud B et al 1989 Phys. Rev. B **40** 5802

Fano U 1961 Phys. Rev. **124** 1866

Le Pore JJ 1980 Appl. Phys. **51** 6441

Miller RC and Kleinmann DA 1985 J. Lumin. **30** 520

Sell DD 1972 Phys. Rev. **136** 3750

Inst. Phys. Conf. Ser. No 123
Paper presented at the International Meeting on Optics of Excitons in Confined Systems, Giardini Naxos, Italy, 1991

Quantum beats between excitonic sublevels in GaAs/AlAs short period superlattices

C. Gourdon*, P. Lavallard* and R.Planel**

*Groupe de Physique des Solides - CNRS UA17- Universités Paris 6 et 7 - Tour 23
2, place Jussieu - F75251 Paris cedex 05 - FRANCE
**Laboratoire de Microstructures et Microélectronique - CNRS UP 20 - 196, avenue Henri Ravera - F92220 Bagneux - FRANCE

ABSTRACT: In short period GaAs/AlAs superlattices a splitting of the optically allowed lowest exciton states was recently evidenced. Upon excitation of a coherent superposition of the two sublevels, quantum beats appear in photoluminescence time decay. From the period of the oscillations, very small splittings in the range of 5 to 10 μeV are determined. Exciton lifetime and spin relaxation time are obtained from the stationary and time-resolved degree of polarization. Type I and type II superlattices are compared. An explanation is proposed for the origin of this splitting.

1. INTRODUCTION

Photoluminescence (PL) spectra of short period type II pseudo-direct GaAs/AlAs superlattices (SL) of standard quality exhibit a zero-phonon line and phonon replicas. This luminescence arises from the recombination of weakly allowed indirect excitons built from the heavy hole state and the X_z electron state. At low temperature the PL linewidth, Stokes shift and temperature behavior show that excitons are localized by layer thickness fluctuations. From the D_{2d} symmetry of the superlattice the optically allowed excitonic states are expected to be doubly degenerate. However, recently, a splitting of these states was independently evidenced by the study of the degree of linear polarization of luminescence on the one hand (Permogorov et al. 1990, Ivchenko et al. 1990), and by optically detected magnetic resonance (ODMR) results on the other hand (van Kesteren et al.1990). Moreover, it was shown that the two sublevels, hereafter named X and Y, correspond to excitonic dipoles aligned along the [110] and [$\bar{1}$10] crystallographic directions, respectively. From the observation of quantum beats in excitonic emission, first reported by van der Poel et al.(1990), we determine the splitting between the two excitonic sublevels. We obtain the spin relaxation time τ_s and the lifetime τ_0 from the analysis of the experimental results in the frame of the density-matrix formalism. Finally we discuss the possible mechanisms which this splitting may originate from.

2. EXPERIMENTAL DETAILS

The samples were grown by molecular beam epitaxy. The composition of type II samples BC09 and E219 is close to 17/11 (GaAs/AlAs layer thickness in Å) and 18/12, respectively. In order to get some insight in the PL behavior when crossing the transition from type II to type I SL, we studied a special sample (E913) grown with a composition gradient. The rotation of the substrate holder was stopped during the growth to induce spatial variation of the composition in the layer plane (x-y plane). This sample is 23.4 mm long. Along this direction (x direction), the SL composition varies from 20.8/12.2 (type II side) to 22.8/11.2 (type I side). The progressive disappearance of phonon replica, increase of the zero-phonon line intensity and shortening of PL decay time along x direction clearly evidence a smooth transition from type II to type I SL. A type I SL sample (C609) of composition 20/8.7 was also studied.

PL is selectively excited by a picosecond dye laser synchronously pumped by an argon laser. PL time decay is studied with a synchroscan streak camera with a resolution of 30 ps. The degree of linear polarization (DLP) is defined as usual as $\rho=(I_{//} - I_{\perp})/(I_{//} + I_{\perp})$, and the degree of circular polarization (DCP) as $\rho=(I_{\sigma+} - I_{\sigma-})/(I_{\sigma+} + I_{\sigma-})$. Unless specified, they are measured at the maximum of the PL band with the excitation photon energy set about 6.5 meV above. A photoelastic modulator is used to measure the stationary DLP and DCP.

3. SPLITTING, SPIN RELAXATION TIME AND LIFETIME

Since the splitting of the two excitonic sub-levels X and Y is much smaller than kT and than the laser spectral width, these two states are coherently excited by a laser pulse linearly polarized along [100] direction. Owing to their small energy difference ΔE, these states dephase while decaying. Fig. 1 shows the time dependence of $\rho_{<100>}$ in sample E913. From the period of the oscillations ($T=h/\Delta E$), the mean energy splitting ΔE is found to be equal to 6.3 μeV. Setting the detection photon energy closer to the laser only very slightly changes the decay and time period (one finds $\Delta E\approx5.7$ μeV). Similarly, almost no change of the splitting is observed when scanning the laser photon energy over 6 meV inside the PL line while keeping a fixed detuning. In samples BC09 and E219 the splitting is found equal to 9.3 ± 0.5 and 8.3 ± 0.5 μeV, respectively. For this last sample, an independent determination of the splitting in agreement with the present one, but less accurate, was obtained from the effect of a magnetic field on the DLP (Ivchenko et al. 1990).

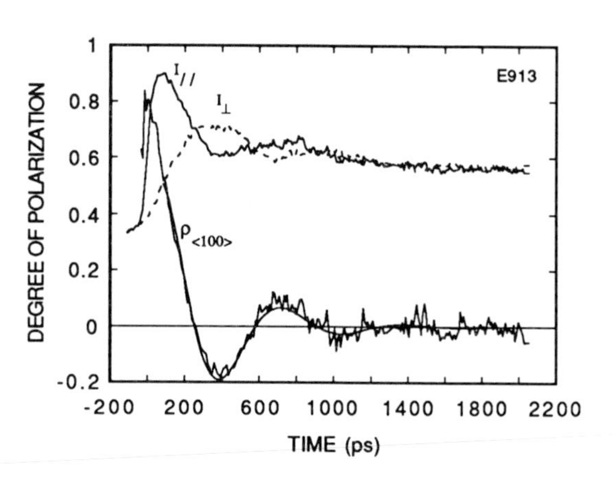

FIG.1 Degree of polarization $\rho_{<100>}$ for the composition-graded sample E913 in a type II region close to the type I-type II transition (x=15.7, composition 22/11.5). The exciting beam is linearly polarized along a ‹100› axis. PL intensities $I_{//}$ and $I_{\perp}$ are detected with polarization parallel and perpendicular to the excitation, respectively. $\rho_{<100>}$ is fitted with $\rho(0)\exp(-\delta t/\hbar)\cos(\Delta Et/\hbar)$ where $\Delta E=6.3$ μeV, $\delta=2$ μeV and $\rho(0)=0.5$. The time origin is taken at the maximum of $I_{//}$ signal.

The stationary DLP was calculated by Ivchenko (1990) in the frame of the density-matrix formalism. The four excitonic states built from the $J_z=\pm3/2$ hole and the $J_z=\pm1/2$ electron are taken into account, assuming that the splitting and the electron-hole exchange interaction are smaller than kT. Along the same lines, we derived the time-dependent DLP as a function of the total spin relaxation time τ_s $\left(\tau_s^{-1}=(\tau_s^e)^{-1} + (\tau_s^h)^{-1}\right)$, the lifetime τ_0, the transverse and longitudinal relaxation times T and T' defined as $T^{-1}=\tau_0^{-1}+\tau_s^{-1}$ and $T'^{-1}=\tau_0^{-1}+(2\tau_s)^{-1}$. $\rho_{<110>}(t)$ depends only on τ_s. The calculation gives $\rho_{<110>}(t) = \dfrac{2}{1+\exp(t/\tau_s)}$. The intensity difference $\left(I_{110}(t) - I_{\bar{1}10}(t)\right)$ is proportional to $\exp-(t/T)$. Although the PL decay time is longer than the 12 ns pulse period of our laser, we can estimate τ_s and T for sample E913 close to the transition region (composition ≈22/11.5). We find $\tau_s \approx 32$ ns, $T \approx 14$ ns and then $\tau_0 \approx 28$ ns. These values are consistent with the stationary DLP $\rho_{<110>}=T'/\tau_0$ found equal to 0.7. With τ_s of the same order of magnitude as τ_0 and $\Delta E\tau_s/\hbar >> 1$, one calculates that $\rho_{<100>}(t)$ decays and oscillates like $\frac{\exp(-t/4\tau_s)}{ch(t/2\tau_s)}\cos\left(\frac{\Delta E}{\hbar}t\right)$. From the large value of τ_s it is obvious that the fast damping of the oscillations of $\rho_{<100>}(t)$ (fig.1) is not due to spin relaxation but rather to the distribution of ΔE values. Averaging over the quantum beats, using a lorentzian distribution of ΔE values of FWHM 2δ brings about an additional damping term $\exp\left(-\frac{\delta}{\hbar}t\right)$. Although this averaging method is not fully justified if δ is not much smaller then the mean splitting ΔE, $\rho_{<100>}(t)$ was fitted in fig.1 with $\left[\rho(0)\exp\left(-\frac{\delta}{\hbar}t\right)\cos\left(\frac{\Delta E}{\hbar}t\right)\right]$. We find that the distribution of ΔE values is rather large: δ=2 µeV for ΔE=6.3 µeV.

4. FROM TYPE II TO TYPE I SUPERLATTICES

In the composition-graded sample E913, the splitting could be measured only in the transition region between type II and type I SL where a satisfactory PL signal-to-noise ratio can be achieved with resonant excitation and where the decay time is not too long. From type II to type I SL (x<20 mm) the quantum beats amplitude strongly decreases but the splitting remains rather constant. We think that, even in the type I side, these oscillations are due to localized HH_1-X_z states. Calculation shows that such states indeed exist at the same energy as HH_1-Γ localized states. They can give a weak contribution to PL at long times. Farther in the type I region as well as in the type I sample C609, no oscillations of $\rho_{<100>}$ are observed but the stationary DLPs $\rho_{<110>}$ and $\rho_{<100>}$, although approaching a common value, are still anisotropic and the stationary DCP remains very weak (fig.2). These results seem to indicate the existence of a splitting in these type I SLs. However, we cannot get a definite conclusion about its value since spin relaxation mainly occurs during the risetime of luminescence (within the first 100 ps). This is not taken into account in our model.

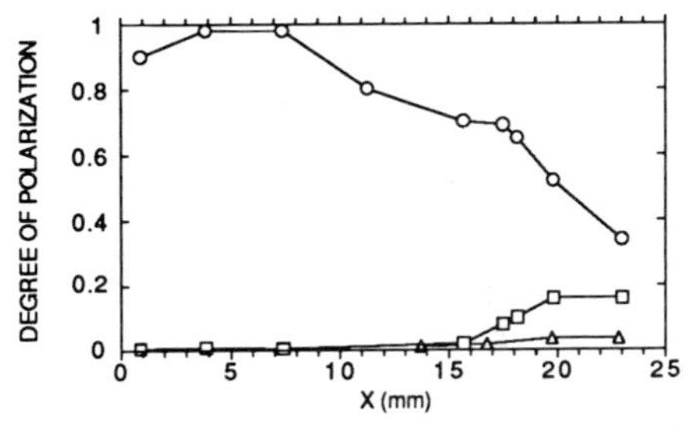

FIG.2 Stationary DLPs $\rho_{<110>}$(circles) and $\rho_{<100>}$ (squares) and DCP (triangles) are plotted as a function of the position of the exciting laser spot on the sample surface for sample E913 with composition gradient. The smallest x values correspond to type II SL and the largest to type I SL. The transition region between type II and type I determined from PL spectra is roughly located between between x=13 and x=20.

5. DISCUSSION

The origin of the splitting has to be related to the localization of excitons since the DLP disappears when excitons are thermally delocalized by raising the temperature up to 20 K. Let us assume that localized excitons are subjected to a local anisotropic perturbation directed along ‹110› axes. Interface defects which present rather well defined edges aligned along ‹110› crystallographic directions (Neave et al. 1983) may be the source of this perturbation. By analogy with excitons in hexagonal compounds (Akimoto et al. 1968), the anisotropic perturbation must lead, in the first approximation, to a splitting proportional to the electron-hole exchange energy, to the strength of the perturbation and inversely proportional to the energy separation between heavy and light hole levels $\Delta(LH_1\text{-}HH_1)$. Since the exchange energy is proportional to the overlap of HH_1 and X_z enveloppe functions, we plotted the energy splitting measured for various samples as a function of $\frac{|\langle HH_1|X_z\rangle|^2}{\Delta(LH_1\text{-}HH_1)}$ in figure 3. The almost linear behavior is in agreement with the proposed model and suggest that the strength of the anisotropic perturbation related to the mean splitting is rather constant at least in the range of SL compositions which has been investigated. For the graded sample E913 the rather constant mean value of the splitting found in the transition region is in agreement with calculations giving an almost constant overlap of HH_1 and X_z enveloppe functions. Although this calculation was achieved using the SL delocalized enveloppe functions and not localized ones, we believe that this calculation is meaningful as for the general trend.

Van der Poel (1990) pointed out that, contrary to type I SL, in type II SL an increase of AlAs thickness at the expense of GaAs thickness lowers the electron confinement energy but increases the hole confinement energy. Therefore, taking into account Coulombic interaction, electron and hole are expected to be localized on each side of the defect edge along a ‹110› direction. A well defined splitting means that these defects have a minimum size larger than the in-plane exciton Bohr radius. Contrary to the samples studied by van der Poel, our samples show a broad distribution of the splitting values around the mean value. We think that the spatial extension of localized wavefunctions along the growth direction can explain the dispersion of the splitting. Our samples are characterized by very thin AlAs layers. Calculation shows that the enveloppe wave function of the heavy-hole localized in an enlarged GaAs slab (+1 monolayer) inside the SL extends over the two adjacent GaAs layers in the case of our samples whereas it is mainly confined in the enlarged GaAs slab in the case of the samples studied by van der Poel.

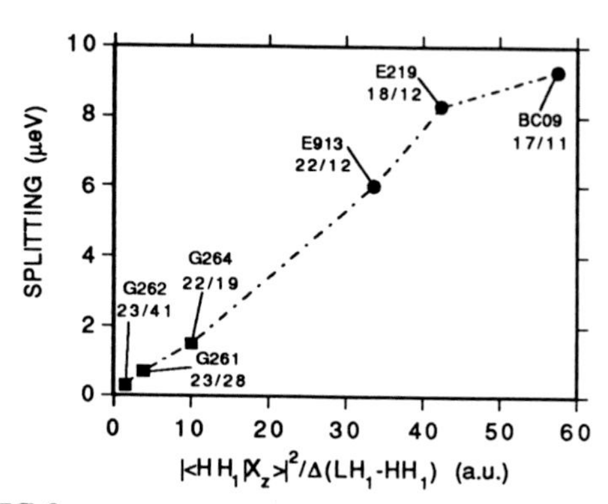

FIG.3 The excitonic splitting measured for various SL samples is plotted as a function of the overlap of HH_1 and X_z enveloppe functions divided by the difference of confinement energies for heavy and light holes. Each sample is labelled with its name and composition. Our samples are represented by dots and the samples studied by van Kesteren (1990)(see table I of this ref.) and van der Poel (1990) by squares.

It is worth noticing that this qualitative explanation does not hold anymore if the existence of a splitting in type I SLs is confirmed. To explain the splitting, another mechanism of intrinsic origin was considered: the additional term in the band dispersion, cubic in wave vector k, which arises from spin-orbit interaction (Pikus and Titkov 1984). This term would give the right order of magnitude of the splitting (1 to 20 μeV) for an electron localized in the x-y plane. However the splitting should depend very much on the spatial extension of the localized function and hence on the localization energy, which was not observed. Furthermore, no dependence on the SL composition is expected contrary to our results.

Obviously more work is needed to reach a definite conclusion about the origin of the splitting between the optically allowed excitonic states in short-period pseudo-direct superlattices. Polarization properties of photoluminescence of type I samples clearly deserves further investigation.

We acknowledge the help of Roland Tessier in the calculation of SL wavefunctions.

Akimoto O, Hasegawa H 1968, Phys. Rev. Lett. 20 916

Ivchenko E L, Kochereshko V P, Naumov A Yu, Uraltsev I N, Lavallard P 1990, Proceedings of the 5th Conference on Superlattices and Microstructures Berlin

Neave J H, Joyce B A, Dobson P J and Norton N 1983, Appl.Phys A 31 1

Permogorov S, Naumov A, Gourdon C, Lavallard P 1990, Solid State Comm. 74 1057

Pikus G E, Titkov A N 1984, in Optical orientation chap.3 p.80 edited by Meier F and Zakharchenya B P, Elsevier Science Publ. 1984

van der Poel W A J A, Severens A L G J, Foxon C T 1990, Optics Comm. 76 116

van Kesteren H W, Cosman E C, van der Poel W A J A, Foxon C T 1990, Phys. Rev. B 41 5283

Inst. Phys. Conf. Ser. No 123
Paper presented at the International Meeting on Optics of Excitons in Confined Systems, Giardini Naxos, Italy, 1991

Carrier-induced excitonic red shift in CdTe/CdZnTe superlattices

M. K. Jackson *, D. Hulin *, J.-P. Foing *, N. Magnea+ and H. Mariette+

*Laboratoire d'Optique Appliquée, Ecole Normale Supérieure de Techniques Avancées
Ecole Polytechnique, 91120 Palaiseau, France

+Laboratoire de Physique des Semiconducteurs, C.E.N.- Grenoble,
38041 Grenoble Cedex, France

+Laboratoire de Spectrométrie Physique, Université J. Fourier, Grenoble,
38402 St. Martin d'Hères Cedex, France.

ABSTRACT: We report observation of a red shift of excitonic absorption in type-I CdZnTe/CdTe superlattices when carriers are created by an above-gap pump pulse. The red shift lasts for several picoseconds after excitation, and is not observed when the pump is resonant with the excitonic absorption. An explanation for the red shift in terms of dynamic band bending caused by the spatial separation of electrons confined in the CdTe layers and energetic holes delocalized in both CdZnTe and CdTe layers is proposed.

1. INTRODUCTION

Recent progress in the epitaxial growth of wide bandgap II-VI semiconductors now makes possible the fabrication of heterostructures of quality rivalling that of III-V devices. These materials are of great practical and scientific interest for a variety of reasons. Practical interest is due to the direct bandgaps available from the near infrared into the blue regions of the spectrum. Recent progress towards doping II-VI semiconductors both *p*- and *n*-type, a long-standing and difficult problem, are very encouraging, making II-VI light emitters a real possibility. Scientific interest stems from the differences between II-VI heterostructures and the more-studied III-V ones. One intriguing difference is the availability of II-VI systems such as the CdTe/CdZnTe system, for example, in which the valence band offset between the two materials is very small. This allows the study of systems where electrons are confined to the CdTe layers, but holes can be confined in either layer, or not well confined at all.

Optical pump-probe experiments have proved to be an excellent way to study semiconductors in non-equilibrium conditions. The sample absorption can be probed following excitation by a short pump pulse with a temporal resolution better than 100fs. Measurements of the absorption as a function of time give direct information about the carrier densities and energy distributions in the sample in non-equilibrium conditions. The evolution of these quantities yields valuable information about a variety of relaxation processes.

In this paper we report ultrafast pump-probe measurements of absorption in CdTe/CdZnTe superlattices. Time-resolved measurements allow us to investigate the effects of the weak hole confinement on the excitonic optical response in the presence of free carriers.

2. EXPERIMENT

We have studied two CdTe/CdZnTe superlattices grown by molecular beam epitaxy on (100) $Cd_{0.96}Zn_{0.04}Te$ substrates (Ponchet 1990). The first, sample Z339, consists of 10 periods of 12.8nm CdTe wells and 12.6nm $Cd_{0.90}Zn_{0.10}Te$ barriers. The second, Z322, is similar: 36 periods of 14.0nm CdTe wells and 13.8nm $Cd_{0.91}Zn_{0.09}Te$ barriers. Pump-probe absorption experiments were performed at approximately 13K, using a colliding pulse mode-locked laser, amplified at 6.5kHz with a copper vapour laser (CVL). The amplified pulses are focused on an ethylene glycol jet to form a continuum, part of which is selected with interference filters, reamplified in a dye jet also pumped by the CVL, and used as a pump. The pump bandwidth is 8nm and the duration is approximately 150fs. A portion of the continuum is used as a test, and has a duration of less than 100fs.

In Fig.1 we show the time-resolved absorption spectra αl for sample Z339, for several pump-probe delays, as indicated. For comparison, the dotted line shows the spectrum for t=-1ps, corresponding to the unperturbed, or linear, absorption. The zero of delay corresponds to pump-probe coincidence, within approximately 250fs.

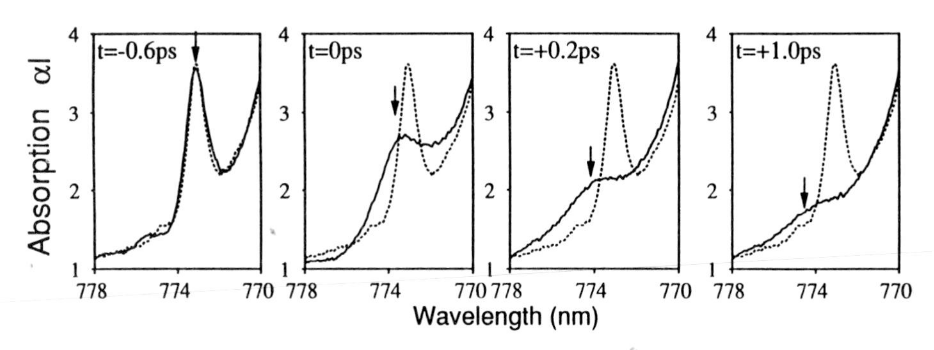

Fig. 1. Time-resolved absorption spectra αl for sample Z339, with a pump at 760nm, for several pump-probe delays, as indicated. The spectrum for t=-1ps is shown as a dotted line for comparison. The wavelength corresponding to peak excitonic absorption is indicated by an arrow in each spectrum.

From the spectrum for t=-1ps shown as a dotted line in Fig.1, we see that the linear absorption spectrum is dominated by the excitonic absorption at approximately 773nm, with substrate absorption and band-to-band absorption of the superlattice contributing to a rising absorption for short wavelengths. Note that although the substrate interferes with transmission of the probe, pump absorption is unaffected since the pump is incident from the superlattice side. The pump spectrum for the data shown in Fig.1 was centered on 760nm. In each spectrum, the wavelength of peak excitonic absorption is indicated by an arrow. The exciton undergoes a significant red shift, an absorption bleaching, and a broadening, due to the free carriers created by the pump. The red shift lasts for at least 1ps.

To investigate the effect of the pump wavelength on the excitonic absorption, we repeated the experiment for various pump wavelengths. These experiments were done with the thicker sample, Z322. In Fig. 2 we show the center wavelength of the exciton absorption, as a function of time, for two pump wavelengths. The solid line corresponds to the pump at 760nm. The red shift (positive peak shift) is seen to appear and then disappear, on a timescale of approximately 1ps; for longer times, a blue shift is observed, which eventually decreases for longer times. The dashed curve shown in Fig. 2 is for a pump centered at 772nm. The peak shift behavior with time is strikingly different in this case, which corresponds to direct creation of excitons. No red shift is observed; the spectra are dominated by a blue shift that quickly reaches a peak, and then decreases monotonically. Note that the intensities of the pump beams in the two experiments were chosen to give similar absorption bleaching at long times, corresponding to excitation of roughly the same density of carriers.

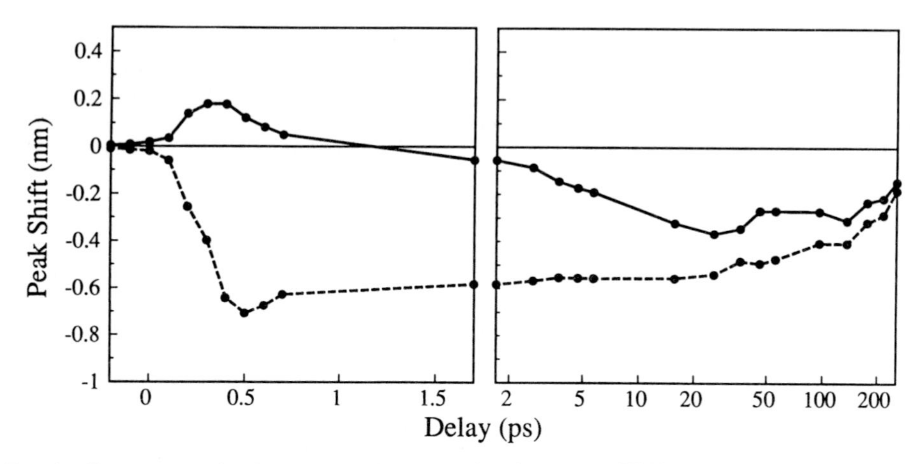

Fig. 2. The peak shift of the excitonic absorption in sample Z322 as a function of time, for two different pump wavelengths. The solid curve corresponds to a pump at 760nm, and the dashed curve to a pump at 772nm. The zero of delay has been arbitrarily set to coincide with the beginning of a change in absorption. Note the change of time scale at 1.7ps.

3. DISCUSSION

We now discuss the origin of the red shift described above. It has been established (Tuffigo 1991) that our samples have a type-I excitonic bandgap. Therefore, we compare the red shift we observe to previous studies of type-I excitons in III-V multiple quantum wells (MQWS). In GaAs/GaAlAs MQWS, either no shift (Knox 1985) or a blue shift (Peyghambarian 1984, Masumoto 1985, Hulin 1986) of the exciton was observed. This blue shift is due to many-body interactions among the excitons (Schmitt-Rink 1985). An excitonic red-shift has very recently been reported (Chemla 1991) in magnetic fields when pumping above the exciton energy. This was attributed to the optical Stark effect, an interaction between the intense electromagnetic field of the pump pulse and the excitonic state, and is unrelated to the presence of carriers created by the pump. In summary, an excitonic red-shift due to the presence of free carriers has not been observed in III-V MQWS of type I.

By examining the temporal behavior of the red shift shown in Fig. 1, we can immediately see that it is *not* due to the optical Stark effect. If it were, the shift would decrease on a timescale corresponding to the pump duration; in our experiments, that is 150fs. The zero of delay in Fig. 1 is estimated to be accurate to better than 250fs, so we see that the observed red shift lasts significantly after the pump pulse. Because the observed red shift lasts significantly longer than the pump, we conclude that it is caused by carriers created by the pump. However, referring to Fig. 2, we see that the red shift is not observed if excitons are resonantly excited by a pump at 772nm. It therefore appears that the red shift only occurs in the case when carriers are created with significant excess energy.

One possible reason for free carriers to cause a red shift of the exciton is band-gap renormalization, which causes a red shift of the free-carrier energies. However, in III-V materials an *excitonic* red shift due to this has never been observed. Zimmermann (1978) and Schmitt-Rink (1989) have shown that the band-gap renormalization is offset by a reduction in the exciton binding energy, with the result that the excitonic transition energy stays almost constant. Therefore, although we cannot exclude its contribution, we consider band-gap renormalization an unlikely explanation for the red shift we observe.

We propose an explanation for the red shift based on the delocalization of energetic holes. The holes can be expected to have a significant fraction of the 27meV excess energy when pumping at 760nm, either immediately, or through scattering by hot electrons. Both electrons and holes are created in the CdTe layers, but the weakly-confined holes are quickly able to delocalize throughout both layers. Strongly confined electrons stay in the CdTe layers. The electric field caused by such a charge separation causes a red shift of the excitonic absorption (Sasaki 1990). We have calculated the red shift by numerically solving for the lowest bound-state energy for electrons and heavy holes. We assumed a hole charge uniformly distributed throughout both CdTe and CdZnTe layers, and an electron charge uniformly distributed throughout the CdTe layers. For a density of $4x10^{11}cm^{-2}$, we calculate a red shift of 1meV (0.5nm). This red shift is approximately what is observed experimentally. There are several effects that have been left out of this very simple model, including screening, decreased exciton binding energy, and many-body effects. However, this model does provide a simple explanation for the red-shift, and suggests some similarity to type-II AlGaAs/AlAs superlattices.

4. CONCLUSIONS

In conclusion, we have performed time-resolved absorption measurements on CdTe/CdZnTe superlattices, and observed a red shift of excitonic absorption during early stages after photoexcitation. We attribute this red shift to delocalization of hot holes. We have proposed an explanation of the red shift in terms of the transient band bending caused by the separation of the hot holes from the electrons which remain confined in the CdTe layers.

ACKNOWLEDGMENTS

It is a pleasure to acknowledge helpful discussions with A. Mysyrowicz, R. Romestain, Y. Merle d'Aubigné, and C. Tanguy. One of us (M.K.J.) would like to acknowledge financial support in the form of a NATO Science Fellowship administered by the National Science and Engineering Research Council of Canada. The Laboratoire d'Optique Appliquée is CNRS URA 1406, and the Laboratoire de Spectrométrie Physique is CNRS URA 8.

REFERENCES

D.S. Chemla, this conference, in *Proceedings of the International Meeting on the Optics of Excitons in Confined Systems*, A. D'Andrea, R. Del Sole, R. Girlanda, and A. Quattropani, eds., (The Institute of Physics, 1991), in press.
D. Hulin, A. Mysyrowicz, A. Antonetti, A. Migus, W.T. Masselink, H. Morkoç, H.M. Gibbs, and N. Peyghambarian, Phys. Rev. B **33**, 4389(1986).
W.H. Knox, R.L. Fork, M.C. Downer, D.A.B. Miller, D.S. Chemla, C.V. Shank, A.C. Gossard and W. Wiegmann, Phys. Rev. Lett. **54**, 1306(1985).
Y. Masumoto, S. Shionoya, and H. Okamoto, Opt. Commun. **53**, 385(1985).
N. Peyghambarian, H.M Gibbs, J.L. Jewell, A. Antonetti, A. Migus, D. Hulin, and A. Mysyrowicz, Phys. Rev. Lett. **53**, 2433(1984).
A. Ponchet, G. Lentz, H. Tuffigo, N. Magnea, H. Mariette, and P. Gentile, J. Appl. Phys. **68**, 6229(1990).
F. Sasaki, T. Mishina, and Y. Masumoto, Phys. Rev. B. **42**, 11426(1990).
S. Schmitt-Rink, D.S. Chemla, and D.A.B. Miller, Phys. Rev. B. **32**, 6601(1985).
S. Schmitt-Rink, D.S. Chemla, and D.A.B. Miller, Advanc. in, Phys. **38**, 89(1989).
H. Tuffigo, N. Magnea, H. Mariette, A. Wasiela, and Y. Merle d'Aubigné, Phys. Rev. B. **43**, 14629(1991).
R. Zimmermann, Phys. Stat. Sol. B. **90**, 175(1978).

Inst. Phys. Conf. Ser. No 123
Paper presented at the International Meeting on Optics of Excitons in Confined Systems, Giardini Naxos, Italy, 1991

Absorption coefficient and refractive index of GaAs-AlAs superlattice optical waveguides

R.Grousson, P.Lavallard, V.Voliotis, M.A.Chamarro*, M.L.Roblin, R.Planel**

Groupe de Physique des Solides-CNRS UA 17- Universités Paris 6 et 7
2, place Jussieu, Tour 23- F75005 Paris - FRANCE
*Instituto de Ciencias de Materiales-Universidad de Zaragoza
**Laboratoire de Microstructures et de Microélectronique CNRS-196 Av.Henri Ravera, 92220 Bagneux - FRANCE

ABSTRACT: The optical properties of a planar guide made of a GaAs-AlAs short period superlattice are determined at ambient temperature in a broad spectral range. The refractive index is obtained from interference measurements. From the determination of the losses of the guide towards the GaAs substrate, we deduce the absorption coefficient of the material. The superlattice shows a long tail of absorption at photon energies lower than the band gap. For both polarizations parallel and perpendicular to the layers, we determine the absolute value of the excitonic absorption coefficient of a quantum well embedded in the structure.

1. INTRODUCTION

The study of optical guides of superlattices (SL) containing an enlarged quantum well (QW) is interesting for basic study as well as for applied research (Weiner et al 1985a, 1985b). For light propagating along the plane of the layers, the optical field can be either parallel or perpendicular to the layers. Furthermore, the long propagation distance allows to measure small absorption coefficients. Such structures are used to guide the stimulated emission from a QW or to modulate light with an applied electric field. To design real structures, one needs to know the optical constants of the guide (Sonek et al 1986). In this communication we present experimental results on a leaky guide. The guide transmission is measured in a broad spectral range extending from the SL band gap to the GaAs substrate band gap in both polarization directions, parallel and normal to the layers. An absolute value of the absorption coefficient is obtained by the study of the reabsorption of luminescence in the guide. On the other hand we directly determine the SL refractive index from interference measurements. This allows us to calculate the losses of the guide towards the GaAs substrate. The comparison between the two absorption coefficients gives us the residual absorption in the material in both polarizations.

2. EXPERIMENTAL

The sample was grown by MBE on a (001) oriented semi-insulating GaAs substrate at 630°C. The waveguide structure consists of 1.8 μm of GaAs-AlAs superlattice. The thickness of the GaAs and AlAs layers are 17.5 and 10.5 Å , respectively. A single 34 Å thick layer is deposited after 0.238 μm and forms a single QW not centered in the SL. A capping layer of 350 Å of GaAs ends the growth sequence. The sample was cleaved along (110) planes and glued on the cold finger of a cryostat across a 0.5 mm diameter hole and between two gold wires. The layer planes are set out vertically in order to minimize the effect of vertical contraction or dilation of the cryostat. Care was taken to avoid vibrations of the sample and temperature variation of the external vase of the cryostat. In these conditions we are able to study the properties of the guide at low as well as at ambient temperature. In this communication, the main results are presented at ambient temperature. Light from a high pressure mercury arc lamp was tightly focused on one cleaved face of the sample with a 5.6 mm focal length lens. The transmitted light was collected by a second microscope lens and focused on an intermediate slit in order to prevent spurious light to reach the detector. An image of the intermediate slit was made on the entry slit of a spectrometer. Finally, light was detected with a photomultiplier followed by a lock-in amplifier.

3. RESULTS

Figure 1 shows in logarithmic scale the transmitted light spectra taken at ambient temperature for polarizations parallel and perpendicular to the layer planes. The same spectra at low temperature are shown in figure 2. We checked that the spectrum of the transmitted light was the same for different sample lengths. The steep rise of absorption on the high energy side is due to the SL. The absorption tail increasing at small photon energies is mainly due to the losses towards the GaAs substrate. The structure at 1.42 eV at ambient temperature comes from the resonant variation of the GaAs refractive index at the energy gap. A more distinct structure is even clearly seen at low temperature at the exciton energy 1.515eV as shown in fig 2. The two peaks at 1.670 (1.765) and 1.715 eV (1.810) correspond to QW heavy and light exciton, respectively, at ambient and low temperature.

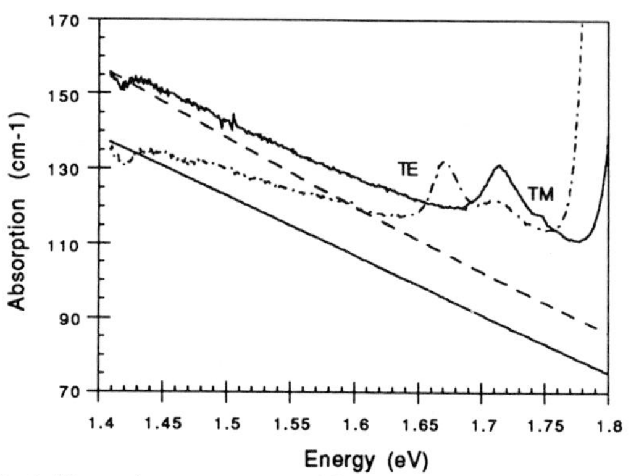

Fig 1.Absorption spectra at ambient temperature for parallel and perpendicular polarization to the layers plane. The solid and dotted curves represent the losses in the GaAs substrate for parallel and perpendicular polarization, respectively.

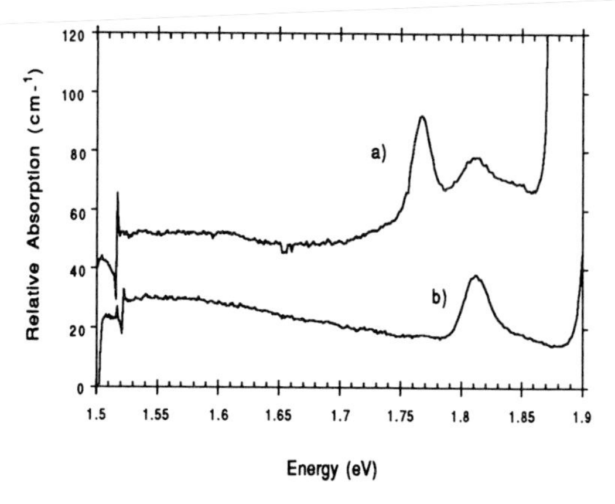

Fig 2.Absorption spectra at low temperature for a) parallel and b) perpendicular polarization to the layers plane.

In order to determine the absolute value of the absorption coefficient at ambient temperature, we measured the reabsorption of the QW luminescence which propagates into the guide. A He-Ne laser was focused at several positions on the (001) surface. We detected the intensities I_1 and I_2 of the luminescence light outgoing from both cleaved faces. In order to keep the output solid angle constant for all the measurements, a pin-hole was placed at the intermediate slit position. The logarithm of the ratio I_1/I_2 detected at the peak of the heavy exciton band (hv=1.675 eV) is plotted in figure 3 as a function of the laser spot position. Half the value of the slope is equal to the absorption coefficient, α= 130±5 cm^{-1} .Using this as a reference, we plotted in figure 1 the absorption coefficient in absolute values.

The SL refractive index for a polarization parallel to the layers, $n_{//}$, was determined by measuring the interferences of the light transmitted through the sample along the growth axis or reflected from the sample surface (Suzuki et al 1983). Figure 4 shows the variation of $n_{//}$ with the wavelength. The value found at 1.5 eV is in agreement with the theoretical value given by Kahen and Leburton (1985). It is slightly larger than the value of the equivalent alloy (Afromowitz 1974). From the very well known refractive index of the substrate (Sell et al 1974, Casey et al 1975, Apnes et al 1983) and the measured value of $n_{//}$, we calculate the losses from the guide towards the substrate in a three layers model, air-SL-GaAs (Yariv 1975). The losses are shown in figure 1 as a solid curve. An upper value of birefringence $\Delta n<1\%$ is obtained by measuring the interferences of TE and TM modes in the guide. Then for polarization normal to the layers, we use the same value of the refractive index as the one found for the parallel polarization to calculate the losses. This is shown in figure 1 as a dotted curve.

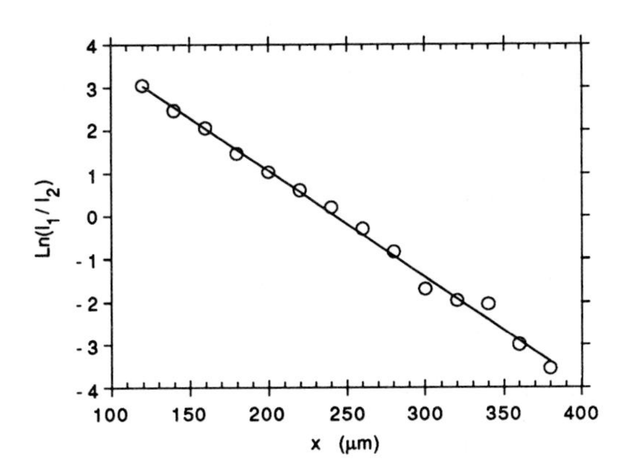

Fig 3 Ratio between the luminescence intensities going out of both sides of the sample as a function of the position of the laser spot on the sample (001) surface at photon energy 1.675 eV.

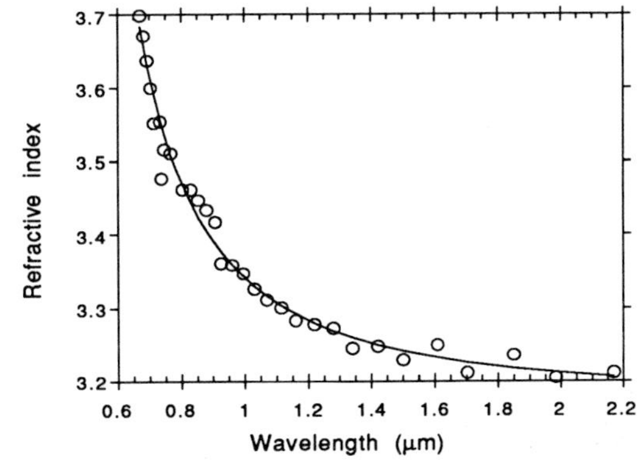

Fig 4. Plot of the refractive index versus the wavelength for a parallel polarization to the layers plane, at ambient temperature. The determination was done from interferences measurements.

Figure 5 shows the residual absorption of the SL and the QW when the losses towards the substrate are substracted from the total absorption.
The QW absorption is superimposed to a low energy tail which is very likely due to imperfections in the crystal growth. It is worthwhile noticing that in usual transmission experiments, such low absorption coefficients cannot be measured. The QW absorption spectra are consistent with the theoretical selection rules in polarized light when taking into account a broadened two-dimensional continuum (Chemla 1984). The absolute value of the exciton absorption can be calculated with

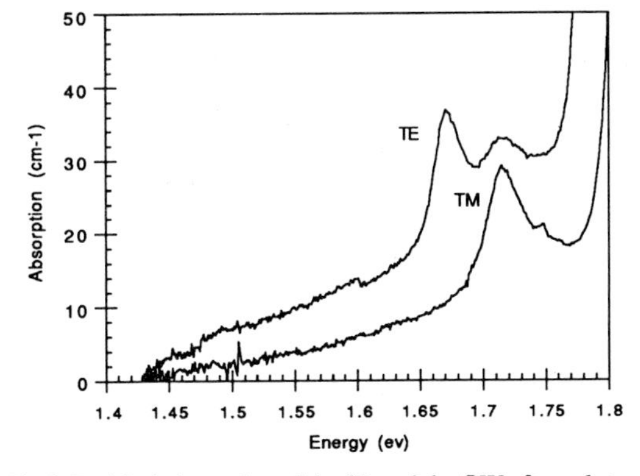

Fig 5. Residual absorption of the SL and the QW after substracting the losses from the total absorption.

the help of the filling factor f, which is defined as the ratio of the integrated amplitude of the mode at the QW position to the total integrated amplitude of the mode. For our structure f is equal to 10^{-3}. The QW heavy exciton absorption coefficient at the peak is then evaluated to be 17000 cm^{-1}. The full width of the line at 1/e intensity is 40 meV. The peak absorption coefficient is in agreement with theoretical calculations (Bastard 1988), if one assumes an exciton binding energy $E_b = 10$ meV.

4. CONCLUSION

We determined the optical constants of a SL guide. The study of the guide losses towards the substrate allowed us to determine the residual absorption in the SL. We showed that a weak absorption tail persists at energies lower than the SL band gap. We checked the selection rules of a single QW and measured the absolute value of the excitonic absorption. All the results were presented at ambient temperature. In fact, they could also be obtained at low temperature if the dielectric constant of GaAs were known. This method of investigation can be generalized to others structures. The large propagation distance allows to study materials where the absorption coefficient is weak as for example type II SL close to the band gap. This method can also be used to study important absorption coefficient of a material embedded in the guide like for the single QW excitonic absorption of our structure. Coupled QW for example might be studied in the same way.

ACKNOWLEDGMENTS

We are grateful to G.Parjadis de la Rivière and to the Laboratoire d'Optique des Solides-Université Pierre et Marie Curie Paris-France for help with the interference measurements.

REFERENCES

Afromowitz M A 1974 Solid State Commun.15 , 59
Apnes D E , Studna A A 1983 Phy. Rev. B 27 , 985
Bastard G 1988 Wave mechanics applied to semiconductor heterostructures (Editions de Physique - France) pp 262-272
Casey H C, Sell D D, Wecht K W 1975 J. Appl. Phys. 46 ,250
Chemla D S, Miller D A B, Smith P W, Gossard A C and Wiegmann W 1984 IEEE J. Quantum Electrn. QE-20 , 265
Kahen K B and Leburton J P 1985 Appl. Phys. Lett. 47 , 508
Sell D D, Casey H C, Wecht K W 1974 J. Appl. Phys. 45, 2650
Sonek G J, Ballantyne J M, Chen Y J, Carter G M, Brown S W, Koteles E S and Salerno J P 1986 IEEE J. Quantum Electron. QE-22 , 1015
Suzuki Y and Okamoto H 1983 J. Electron. Mater. 12, 397
Weiner J S, Chemla D S, Miller D A B, Haus H A, Gossard A C, Wiegmann W, and Burrus C A 1985a Appl. Phys. Lett. 47 , 664
Weiner J S, Miller D A B, Chemla D S and Damen J C 1985 Appl. Phys.Lett. 47 , 1148
Yariv A 1975 Quantum Electronics-2nd Edition (Wiley - New York) pp 539-541

Inst. Phys. Conf. Ser. No 123
Paper presented at the International Meeting on Optics of Excitons in Confined Systems, Giardini Naxos, Italy, 1991

Sum rules in nonlinear optics: prospects for nonlinear processes in quantum wells

F Bassani and S Scandolo

Scuola Normale Superiore, 56100 Pisa, Italy

ABSTRACT: It is shown that all sum rules known in linear optics depend only on the causality condition since the presence of an additional light beam does not modify them. Additional sum rules exist for nonlinear processes. The study of a three level system shows how the nonlinear sum rules relate to the relevant parameters (Rabi frequency, oscillator strength, electron density) and how they characterize the nonlinear absorption lineshapes. Applications to quantum well optics are discussed.

1. Introduction

Sum rules are general constraints on the frequency dependence of the optical functions due to the fact that their integrated values are determined by the physical properties of the medium. The most famous of them is the Thomas-Reiche-Kuhn (TRK) sum rule (see e.g. Bethe and Salpeter 1957) which states that the sum of the oscillator strengths of all optical transitions is equal to the number of electrons. Another important sum rule due to Altarelli et al (1972) is the vanishing of the integral of $[n(\omega) - 1]$, where n is the index of refraction of the medium. Such rules have been verified experimentally with the use of synchrotron radiation (Smith and Shiles 1978, Altarelli and Smith 1974), and have been found very useful in determining the frequency dependence of the optical constants outside the region of measurements and the spectral lineshapes under specific conditions (Stern 1963).

All sum rules are generally proved under two hypotheses, causality and linearity of the response to an electromagnetic field. After the advent of laser technology, however, a large number of nonlinear phenomena have become accessible to experimental investigation (Shen 1980), so that it becomes important to address the problem of the existence and of the role of sum rules in nonlinear optical phenomena. In the case of quantum wells and other mesoscopic structures, nonlinear effects may be of particular relevance, so that proving the existence of sum rules becomes very important.

The purpose of this article is to illustrate the basic sum rules of nonlinear optics, and to present their implications to a three level model, as a suitable scheme to interpret nonlinear properties of quantum wells. Though originally the sum rules were obtained directly from the quantum mechanical definition of the oscillator strengths and of the optical constants, a more general procedure is to use the Kramers-Kronig dispersion relations to derive the asymptotic behavior of the optical constants and to compare it with the asymptotic behavior independently computed (Landau and Lifshitz 1960, Bassani and Altarelli 1983).

We will show that this procedure can be extended to the nonlinear case and leads to the conclusion that all the known sum rules do not depend on linearity, and new sum rules can be obtained for nonlinear phenomena. Applications to a three level model for quantum wells will be briefly described.

In section 2 we give the nonlinear definitions of the optical constants and the conditions for the validity of dispersion relations. In section 3 we show how this leads to a number of sum rules and we describe their immediate implications. In section 4 we discuss the three level model and its applicability to quantum well nonlinearities. In section 5 we give some conclusions.

2. Nonlinear susceptibility and dispersion relations

The general relation between the polarizability and the applied electric fields, making use of time translation independence, can be written as:

$$\begin{aligned} P(t) = &\int_0^\infty dt_1\; G^{(1)}(t_1)E(t-t_1)+ \\ &+\int_0^\infty dt_1 \int_0^\infty dt_2\; G^{(2)}(t_1,t_2)E(t-t_1)E(t-t_2)+... \\ &+\int_0^\infty dt_1 ... \int_0^\infty dt_n\; G^{(n)}(t_1,...,t_n)E(t-t_1)...E(t-t_n)+... \end{aligned} \tag{1}$$

where $G^{(n)}(t_1,...,t_n)$ is a real symmetric function which verifies the causality condition

$$G^{(n)}(t_1,...,t_i,...,t_n)=0 \quad if \quad t_i<0\;.$$

The Fourier transform gives the relation between the polarization $P^{(n)}(\omega)$ and the susceptibility $g^{(n)}(\omega_1,...,\omega_n)$

$$\begin{aligned} P^{(n)}(\omega) = \frac{1}{(2\pi)^{n-1}} \int d\omega_1 ... \int d\omega_n\; g^{(n)}(\omega_1,...,\omega_n) \times \\ \times E(\omega_1)...E(\omega_n)\delta[\omega-(\omega_1+...+\omega_n)]\;. \end{aligned} \tag{2}$$

We notice that $g^{(n)}$ is a complex function and that

$$g^{(n)}(-\omega_1,...,-\omega_n) = g^{(n)*}(\omega_1,...,\omega_n). \tag{3}$$

In first order we obtain the usual linear properties. At higher orders the response function contains specific nonlinear effects. If we consider in particular two radiation beams we can observe that second order contains second harmonic generation, sum and difference frequency generation and optical rectification. The third order contains not only third harmonic generation, but also dissipative effects such as two-photon absorption and stimulated Raman scattering, whose magnitudes are proportional to the imaginary part of $g^{(3)}(\omega,\omega_2,-\omega_2)$, where ω is the probe light frequency on which the measurements are performed and ω_2 is the pump beam frequency responsible for the nonlinearities.

Extending previous results by Price (1963), Caspers (1964) and Ridener and Good (1974, 1975), we prove that dispersion relations are valid for the response functions to all orders, provided the phenomena we consider are represented by a response function which allows a straight line with positive slope in the n-dimensional space $\omega_1,\omega_2,...,\omega_n$ (Bassani and Scandolo 1991).

Then, if we define the straight line in the parametric form

$$\omega_k(s) = v_k s + w_k \qquad k = 1, 2, ..., n \tag{4}$$

the response function $g^{(n)}(s)$ satisfies the dispersion relations on the real s axis

$$Re[g^{(n)}(s)] = \frac{1}{\pi} P \int ds' \frac{Im[g^{(n)}(s')]}{s' - s}$$

$$Im[g^{(n)}(s)] = -\frac{1}{\pi} P \int ds' \frac{Re[g^{(n)}(s')]}{s' - s} \tag{5}$$

provided that all the v_i coefficients are positive. We can observe that not all the phenomena allow dispersion relations, because some are characterized by frequencies ω_i which cannot be represented by eq.(4) with positive v_i. Among the above mentioned processes for instance, optical rectification $(g^{(2)}(\omega, -\omega))$ and nonlinear absorption due to the photons of the same beam $(g^{(3)}(\omega, \omega, -\omega))$ do not satisfy dispersion relations of type (5). For the other above mentioned processes dispersion relations (5) hold, and consequently they also hold for the complex nonlinear dielectric function $\varepsilon^{NL}(\omega, \omega_2, \mathcal{E}_2)$ related to the pump beam intensity $|\mathcal{E}_2|^2$ by a power expansion whose coefficients are the $g^{(n)}(\omega_1, \omega_2, -\omega_2, ...)$. We can then write dispersion relations for the total dielectric function

$$\varepsilon(\omega, \omega_2, \mathcal{E}_2) = \varepsilon^L(\omega) + \varepsilon^{NL}(\omega, \omega_2, \mathcal{E}_2). \tag{6}$$

The real and imaginary parts are related by

$$\varepsilon_1(\omega, \omega_2, \mathcal{E}_2) - 1 = \frac{2}{\pi} P \int_0^\infty d\omega' \frac{\omega' \varepsilon_2(\omega', \omega_2, \mathcal{E}_2)}{\omega'^2 - \omega^2}$$

$$\varepsilon_2(\omega, \omega_2, \mathcal{E}_2) = -\frac{2\omega}{\pi} P \int_0^\infty d\omega' \frac{\varepsilon_1(\omega', \omega_2, \mathcal{E}_2)}{\omega'^2 - \omega^2} \tag{7}$$

Similar relations hold for all optical constants related to ε, including the index of refraction $n = \sqrt{\varepsilon}$ for which an "ad hoc" demonstration can be found (Nussenzweig 1972). Additional dispersion relations can be obtained for the nonlinear part of all optical constants. By considering the properties of the analytic functions $\omega^2 \varepsilon^{NL}(\omega, \omega_2, \mathcal{E}_2)$ (Bassani and Scandolo 1991) we find for instance

$$\omega^2 \varepsilon_1^{NL}(\omega, \omega_2, \mathcal{E}_2) = \frac{2}{\pi} P \int_0^\infty d\omega' \frac{\omega'^3 \varepsilon_2^{NL}(\omega', \omega_2, \mathcal{E}_2)}{\omega'^2 - \omega^2}$$

$$\omega \varepsilon_2^{NL}(\omega, \omega_2, \mathcal{E}_2) = -\frac{2}{\pi} P \int_0^\infty d\omega' \frac{\omega'^2 \varepsilon_1(\omega', \omega_2, \mathcal{E}_2)}{\omega'^2 - \omega^2} \tag{8}$$

All the above dispersion relations can be used directly to relate dispersive to dissipative phenomena to all orders, and can be of immediate use in the analysis of experimental data. They can also lead to a number of sum rules which we will describe in the next section.

3. Sum rules for the nonlinear dielectric function

Sum rules can be derived by considering the properties of the above dispersion relations. Besides the linear sum rules amply discussed in the literature and widely used (Bassani and Altarelli 1983), additional sum rules are obtained for the nonlinear functions. Some are contained in expressions (7) and (8) and can be found by setting $\omega = 0$. Others can be obtained by considering the asymptotic behavior of the optical functions as obtained from expression (8) with the use of the superconvergence theorem (Altarelli et al 1972) and comparing it with the asymptotic behavior independently found from their definition (Bassani and Scandolo 1991).

Among sum rules of the first type we wish to mention the following ones:

$$\varepsilon_1(0,\omega_2,\mathcal{E}_2) = 1 + \frac{2}{\pi}\int_0^\infty d\omega' \frac{\varepsilon_2(\omega',\omega_2,\mathcal{E}_2)}{\omega'} \tag{9}$$

$$\int_0^\infty \varepsilon_1^{NL}(\omega,\omega_2,\mathcal{E}_2)d\omega = 0 \tag{10}$$

$$\int_0^\infty \omega\varepsilon_2^{NL}(\omega,\omega_2,\mathcal{E}_2)d\omega = 0 \tag{11}$$

The sum rule (11) was previously derived by Peiponen (1987) for the case of the anharmonic oscillator. Here, in proving its general validity we also obtain the other sum rules. The sum rule (9) extends the analogous results of linear optics to the case of nonlinear optics and allows the calculation of the nonlinear contributions to the static dielectric function, which modifies the dielectric screening properties of impurity and electron-hole potentials. The sum rule (10), and the similar one obtained for the real part of the index of refraction, implies that sum rule $\int[n(\omega)-1]d\omega = 0$ does not depend on linearity since the nonlinear contribution integrates to zero. The most important of the above sum rules is the (11) because it is the nonlinear extension of the TRK sum rule of linear optics. It implies that the presence of an additional light beam preserves the integrated value of the absorption coefficient, though the form of the spectrum can be drastically modified. This also implies that $\varepsilon_2^{NL}(\omega,\omega_2,\mathcal{E}_2)$ must became negative at some frequency values so as to compensate the well known two-photon absorption and stimulated Raman scattering processes.

Additional sum rules can be obtained by considering the definition of the time dependent response function, as given by Kubo (1957), and using it to find its asymptotic behavior for $\omega \to \infty$. The expansion gives (Bassani and Scandolo 1991):

$$\varepsilon^{NL}(\omega,\omega_2,\mathcal{E}_2) = -\frac{c(\omega_2,\mathcal{E}_2)}{\omega^4} + o(\omega^{-4}), \tag{12}$$

where c is a real constant which depends on the properties of the medium and on the intensity and frequency of the additional light beam, and gives a measure of the nonlinearity. We do not report its general expression because it is rather complicated, but we will give its values case by case. Since the imaginary part of ε^{NL} vanishes for $\omega \to \infty$ much faster than the real part, whose asymptotic behavior is given by (12), we can use the superconvergence theorem in the dispersion relation (8) and, by comparing it with (12), we obtain the following relevant sum rules

$$\int_0^\infty \omega^2\varepsilon_1^{NL}(\omega,\omega_2,\mathcal{E}_2)d\omega = 0 \tag{13}$$

$$\int_0^\infty \omega^3\varepsilon_2^{NL}(\omega,\omega_2,\mathcal{E}_2)d\omega = \frac{\pi}{2}c(\omega_2,\mathcal{E}_2) \tag{14}$$

Equations (13) and (14) have no counterpart in linear optics and give a measure of the nonlinearity of the system under consideration. As the frequency spectrum of the optical functions is experimentally investigated, the criterion for the extension of all nonlinear effects is the saturation of the above sum rules.

4. Nonlinear properties of quantum wells: the three level model

Experimental evidence which is beginning to cumulate indicates that nonlinear effects are particularly enhanced in the case of quantum wells, with respect to those present in a homogeneous medium (Chemla et al 1988, Mysyrowicz et al 1986, Fröhlich et al 1990). Theoretical

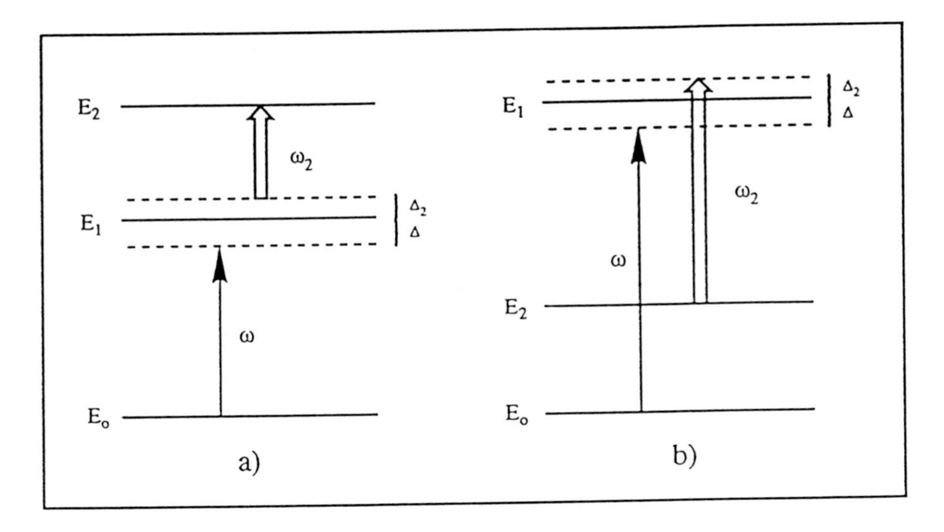

Fig. 1 Three level system resonantly interacting with two radiation beams of frequency ω and ω_2. a) Ladder configuration; b) Lambda configuration.

studies of the nonlinear properties of quantum wells and superlattices have been also carried out (Schmitt-Rink et al 1988, Chemla 1991) and prove that the nonlinear optical response strongly depends on dimensionality and on the internal structure of excitons.

In the case of multiple quantum wells sufficiently separated to neglect the miniband effect, the excitons formed from holes in the heavy hole (hh) and the light hole (lh) bands and from electrons in the first (e1) and second (e2) sublevels, hh-e1, hh-e2, lh-e1, lh-e2, form two approximately noninteracting three level systems. If we consider the lower hole states hh2, we also have a three level system, with the intermediate excited state higher than the final excited state. The two situations are schematically shown in Fig. 1.

The susceptibility in the three level model has been evaluated first by Hänsch (1971) in the approximation that ω is not too far from the resonance condition and in the rotating wave approximation. As a necessary improvement to insure that the sum rules are satisfied (Bigot and Hönerlage 1984), the antiresonance pole must also be considered so that the expression for the complex dielectric function becomes

$$\varepsilon(\omega,\omega_2,\mathcal{E}_2) = 1 + \frac{4\pi n|\langle 0|ex|1\rangle|^2}{E_{10} - \hbar\omega - i\Gamma_1 - \frac{\alpha_2^2}{E_{20} \pm \hbar\omega_2 - \hbar\omega - i\Gamma_2}} + \frac{4\pi n|\langle 0|ex|1\rangle|^2}{E_{10} + \hbar\omega + i\Gamma_1 - \frac{\alpha_2^2}{E_{20} \pm \hbar\omega_2 + \hbar\omega + i\Gamma_2}} \qquad (15)$$

where n is the electron density, $\alpha_2 = \langle 1|ex|2\rangle\mathcal{E}_2/2$ is the Rabi energy which gives a measure of the nonlinear strength, and the $+(-)$ sign refers to the lambda (ladder) configuration.

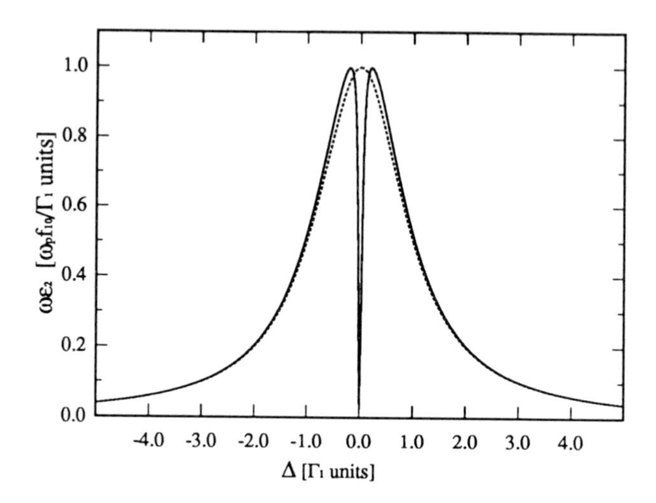

Fig. 2 First order absorption (dotted line) and total absorption (continuous line) in the lambda configuration, with $\Gamma_1 \simeq 10^3\ \Gamma_2$.

The computed value of the c constant of expression (14) near the resonance $\Delta_2 = 0$ is

$$c \simeq \frac{3}{2}\left(\frac{4\pi ne^2}{m}\right)\left(\frac{\alpha_2}{\hbar}\right)^2 f_{01}, \tag{16}$$

where f_{01} is the oscillator strength of the first order transition $0 \rightarrow 1$.

The consequence of the sum rules can be observed immediately in the case of the lambda configuration and with $\Delta_2 = 0$, shown in Fig. 2. The nonlinear contribution becomes negative at zero detuning ($\Delta = 0$) and nearly cancels the absorption. This effect has been first observed in atomic physics experiments (Alzetta et al 1976).

A similar effect occurs in the ladder configuration for zero detuning ($\Delta_2 = 0$), as shown in Fig. 3. The first order absorption splits into a doublet, but its integrated value remains constant. It is called the Autler-Townes effect in atomic physics.

When the pump beam is slightly off resonance, as in the experiments of Fröhlich et al (1990), besides the two-photon peak which appears at $\hbar(\omega + \omega_2) = E_{20}$, we have a blue shift and a decrease of the first order resonance. This decrease is required by the absorption sum rule (11) to compensate the appearance of the two-photon peak. This is shown in Fig. 4 for the case with $\hbar\omega_2 > E_{21}$. A red shift of the E_{10} occurs when $\hbar\omega_2 < E_{21}$.

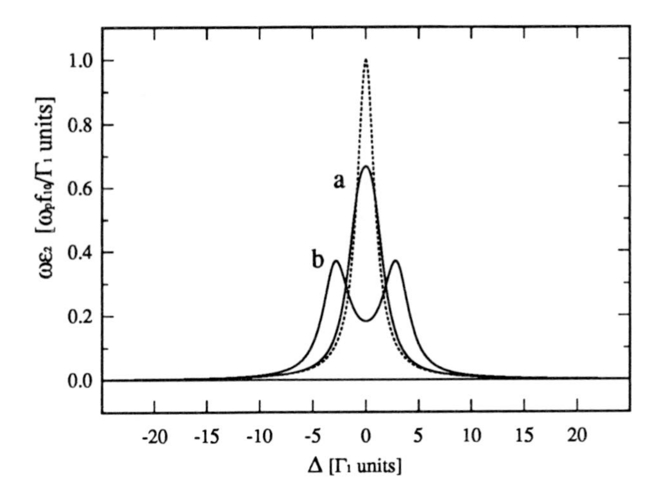

Fig. 3 First order absorption (dotted line) and total absorption (continuous line) for the ladder configuration. The curves a) and b) refer to $\alpha_2 = \Gamma_1$ and $\alpha_2 = 3\ \Gamma_1$ respectively.

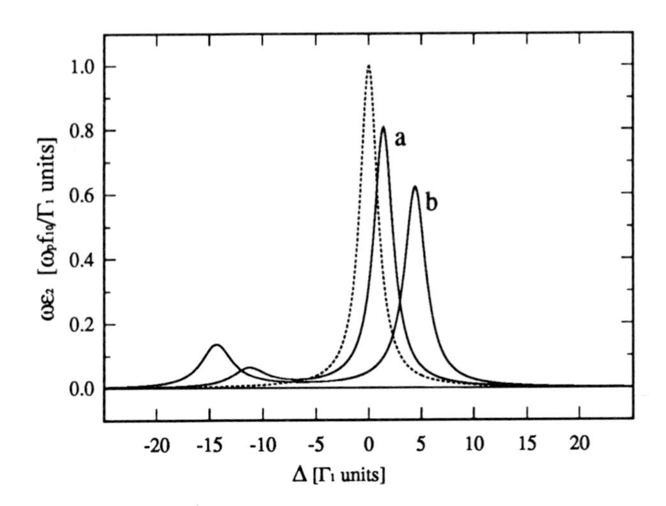

Fig. 4 First order absorption (dotted line) and total absorption (continuous line) for the ladder configuration when $\Delta_2 = 10\,\Gamma_1$. In case a) $\alpha_2 = 3\ \Gamma_1$ and in case b) $\alpha_2 = 8\ \Gamma_1$.

5. Conclusions

We have derived the dispersion relations for the nonlinear optical functions, and specified the appropriate conditions for their validity. Under the same conditions sum rules are obtained. Some of them establish that the known sum rules of linear optics depend only on causality. Others are specific to the nonlinear contributions and give a measure of the importance of nonlinearity. The three level model verifies the sum rules and seems to be appropriate for quantum well excitons, as experimentally observed by Fröhlich et al (1990).

This work was performed under a contract of C.N.R. (Consiglio Nazionale delle Ricerche). Important suggestions from D. Fröhlich are gratefully acknowledged.

References

Altarelli M, Dexter D L, Nussenzweig H M and Smith D Y 1972 Phys. Rev. B 6 4502
Altarelli M and Smith D Y 1974 Phys. Rev. B, 9 1290
Alzetta G, Gozzini A, Moi L and Orriols G 1976 Nuovo Cim. B36 5
Bassani F and Altarelli M 1983 Interaction of radiation with matter, in Handbook of Synchrotron Radiation, vol. 1a, ed E E Koch (Amsterdam: North Holland)
Bassani F and Scandolo S 1991 Phys. Rev. B (to appear)
Bethe H E and Salpeter E E 1957 Quantum mechanics of one and two electron atoms (Berlin: Springer-Verlag) sec. 61
Bigot J Y and Hönerlage B 1984 Phys. Stat. Sol. (b) 121 649
Caspers W J 1964 Phys. Rev. 133 A1249
Chemla D S, Miller D A B and Schmitt-Rink S 1988 in Optical nonlinearities and instabilities in semiconductors, ed Haug H (CA: Academic Press Inc)
Chemla D S 1991 High density excitations in confined systems, this volume
Fröhlich D, Neumann Ch, Uebbing B and Wille R 1990 Phys. Stat. Sol. (b) 159 297
Hänsch T W 1970 Z. Phys. 236 213
Kogan Sh M 1963 Sov. Phys. JETP 16 217
Kubo R 1957 J. Phys. Soc. Japan 12 570
Landau L D and Lifshitz E M 1960 Electrodynamics of continuous media (Oxford: Pergamon Press)
Mysyrowicz A, Hulin D, Antonetti A, Migus A, Masselink W T and Morkoc H 1986 Phys. Rev. Lett. 56 2748
Nussenzweig H M 1972 Causality and dispersion relations (New York: Academic Press)
Peiponen K-E 1987 J. Phys. C 20 2785
Price P J 1963 Phys. Rev. 130 1792
Ridener Jr. F L and Good Jr. R H 1974 Phys. Rev. B 10 4980
Ridener Jr. F L and Good Jr. R H 1975 Phys. Rev. B 11 2768
Schmitt-Rink S, Chemla D S and Haug H 1988 Phys. Rev. B 37 941
Shen Y R 1984 The principles of Nonlinear Optics (New York: J Wiley)
Smith D Y and Shiles E 1978 Phys. Rev. B 17 4689
Stern F 1963 Elementary theory of the optical properties of solids, in Solid State Physics, vol. 15, eds F Seitz and D Turnbull (New York: Academic Press)

Inst. Phys. Conf. Ser. No 123
Paper presented at the International Meeting on Optics of Excitons in Confined Systems, Giardini Naxos, Italy, 1991

Raman scattering in bulk semiconductors and semiconductor microstructures

Manuel Cardona

Max-Planck-Institut für Festkörperforschung, Heisenbergstr. 1, D-7000 Stuttgart 80, Federal Republic of Germany

ABSTRACT: The efficiencies (cross sections) for scattering by phonons in conventional semiconductor microstructures and their bulk constituents are discussed. Experimental results, in absolute units, are compared with microscopic calculations based on measured or calculated electron-phonon coupling constants. Two types of coupling mechanisms are considered: the standard, deformation potential-type and that induced by the electric polarization which accompanies the phonons. These two mechanisms and their quantum-mechanical interference lead to a rich phenomenology, involving detailed selection rules which can be experimentally verified. In the case of MQW's (or superlattices) the phenomenon of confined optical modes is observed. It can be used to map the bulk dispersion relations with a remarkable accuracy and even for their determination in cases where they are not known (e.g. AlAs). Multiphonon processes, and related double and triple resonance phenomena are also presented. Finally, recent data on spin-flip scattering at bound excitons and excitons localized at interface roughness are discussed.

1. INTRODUCTION

The simplest Raman scattering process observed in semiconductors is perhaps that of scattering by a single phonon whose wavevector must be very close to the center (Γ-point) of the Brillouin zone (BZ). The latter follows from the fact that the wavevector $\vec{k}_L$ of the laser light usually employed is very small ($k_L \leq 4\pi/\lambda$, λ = wavelength in the medium) compared with the relevant electronic lengths (lattice constant, exciton radius). This scattering process involves six different diagrams which can be obtained by permuting the phonon and the two photons in those of Fig. 1. These represent the most resonant processes when the two intermediate electronic states have nearly the same energies and the latter coincide with either the laser frequency ω_L or that of the scattered phonons ω_S or both (double resonance, Cerdeira et al., 1986, Alexandrou et al., 1988,1989). On the left of Fig. 1 the case of uncorrelated electron-hole $(e-h)$ pairs as intermediate states is shown. The phonon can couple to either the electron or the hole through electron-phonon interaction and both diagrams, representing scattering amplitudes, must be added algebraically and then squared. To the right of Fig. 1 the uncorrelated $e-h$ intermediate states have been replaced by correlated ones, i.e., excitons. The phonon now couples to them through exciton-phonon interaction which can be reduced to the effect on the e and h components and an effect on the excitonic envelope function. For deformation potential interaction, and taking into account the fact that $q \approx 0$, the phonon-exciton coupling constant reduces simply to the sum of the interactions between the corresponding uncorrelated electrons and holes. In the case of the strongly q-dependent Fröhlich interaction the leading terms in q^{-1} for electrons and holes fortunately cancel, thus removing the divergence in the scattering amplitude. The first non-vanishing term results from the exciton envelope function and is proportional to $q\left(\frac{1}{m_e} - \frac{1}{m_h}\right)$, where m_e and m_h are the effective masses of electrons and holes, respectively (M. Cardona, 1982).

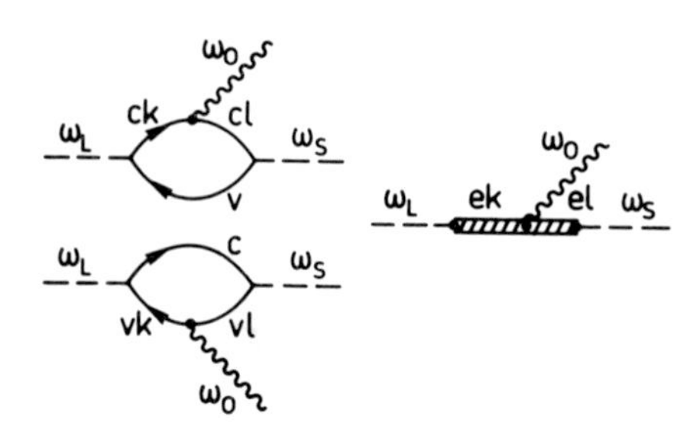

Fig. 1: Feynman diagrams for the most resonant contribution to the Raman scattering by one phonon. Left: uncorrelated pairs, right: correlated pairs.

The Stokes scattering efficiency (per unit solid angle Ω and unit pathlength) is usually written as (Cantarero et al., 1989)

$$\frac{dS}{d\Omega} = \frac{\omega_S^3 \omega_L}{c^4} \frac{\hbar}{2V_C M^* \omega_0} \frac{n_S}{n_L} \left| \hat{e}_S \cdot \overleftrightarrow{R} \cdot \hat{e}_L \right|^2 [n_B(\omega_0) + 1] \tag{1}$$

where ω_0 is the phonon frequency, V_C the primitive cell (PC) volume and M^* its reduced mass, n_S and n_L are refractive indices $\hat{e}_S$ and $\hat{e}_L$ polarization vectors and $\overleftrightarrow{R}$ the Raman tensor of the phonon under consideration in units of a polarizability per unit length of phonon displacement, i.e., dimensions of L^2, and $n_B(\omega_0)$ the Bose-Einstein factor.

In the case of zincblende-type materials the dipole-allowed Raman tensors for phonons polarized along $\hat{x}$, $\hat{y}$, and $\hat{z}$ are, respectively:

$$\overleftrightarrow{R}_x = \begin{pmatrix} 0 & 0 & 0 \\ 0 & 0 & a \\ 0 & a & 0 \end{pmatrix}; \quad \overleftrightarrow{R}_y = \begin{pmatrix} 0 & 0 & a \\ 0 & 0 & 0 \\ a & 0 & 0 \end{pmatrix}; \quad \overleftrightarrow{R}_z = \begin{pmatrix} 0 & a & 0 \\ a & 0 & 0 \\ 0 & 0 & 0 \end{pmatrix}. \tag{2}$$

For backscattering on a [001] surface $\overleftrightarrow{R}_z$ corresponds to LO, $\overleftrightarrow{R}_x$, $\overleftrightarrow{R}_y$ to TO phonons.

2. DEFORMATION POTENTIAL AND FRÖHLICH INTERACTION: INTERFERENCE EFFECTS

In the electron-phonon interaction Hamiltonian one usually distinguishes between deformation potential and Fröhlich interaction terms. The former are assumed to be q-independent, i.e. local in real space, while the latter are found only for the longitudinal components of ir-active phonons. They correspond to the electrostatic polarization which accompanies such phonons and the corresponding Hamiltonian has the behavior $\sim q^{-1}$, singular for $q \to 0$. Near the lowest direct gap of zincblende E_0 the deformation potential interaction acts only on the valence states and yields intra and interband terms (Fig. 2). The former lead to strong resonances near E_0 but vanish at $E_0 + \Delta_0$ which is only affected by the less strongly resonant interband terms. The Fröhlich Hamiltonian acts, as already mentioned, on all states in a band-index-diagonal way since it is basically a c-number. Thus it leads to diagonal Raman tensors with apparent Γ_1 symmetry which is incompatible with that of the phonons. This paradox results from the fact that the Fröhlich scattering is dipole forbidden. The corresponding Raman tensor is proportional to the phonon wavelength q and LO phonons with $q \neq 0$ have A_1 symmetry in the group of $\vec{q}$. We write the corresponding Raman tensor as

$$\overleftrightarrow{R}_F = \begin{pmatrix} 1 & 0 & 0 \\ 0 & 1 & 0 \\ 0 & 0 & 1 \end{pmatrix} a_F \, . \tag{3}$$

LO phonons usually scatter both via dp and Fröhlich. For $\hat{e}_L \parallel [100]$ and $\hat{e}_S \parallel [010]$ (or similar) only dp scattering occurs while $\hat{e}_L \parallel \hat{e}_S \parallel [100]$ the scattering only takes place via Fröhlich

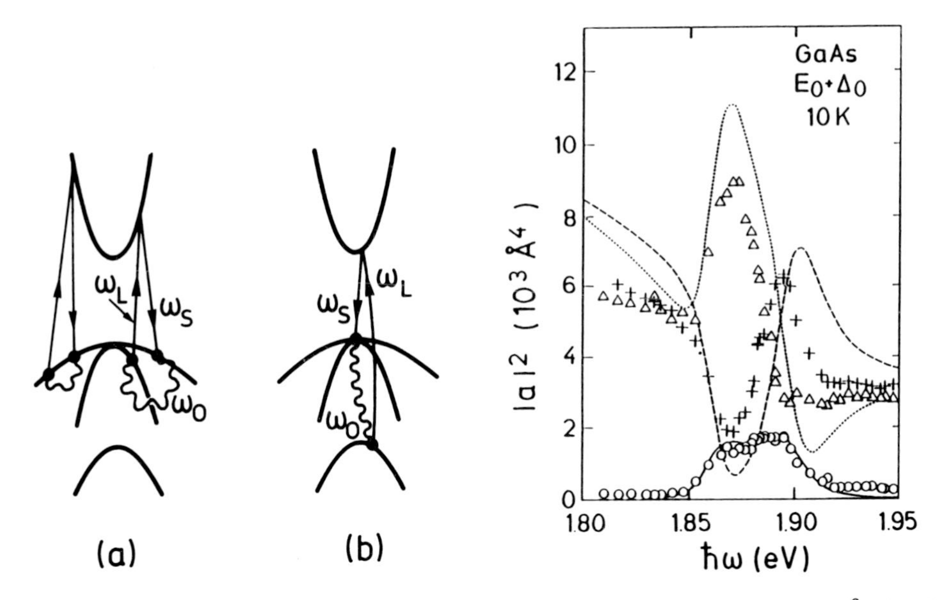

Fig. 2: Band diagrams for scattering by deformation potential near the direct gap of GaAs. (a) intraband terms (b) interband terms.

Fig. 3: Raman polarizability $|a_F|^2$ (solid curve) and interference terms $|a_F + a|^2$ (dashed curve) $|a_F - a|^2$ (dotted curve) near the $E_0 + \Delta_0$ gap for backscattering on a (001) GaAs face. The points are experimental (Cantarero et al., 1989b).

interaction (i.e. for LO phonons). Of particular interest are the cases $\hat{e}_L \parallel \hat{e}_S \parallel [110]$ and $\hat{e}_L \parallel \hat{e}_S \parallel [1\bar{1}0]$. The squared contracted Raman tensor of (1) (Raman polarizability) becomes:

$$\begin{aligned} \hat{e}_L \parallel \hat{e}_S \parallel [110] : &\quad [a + a_F|^2 \\ \hat{e}_L \parallel \hat{e}_S \parallel [1\bar{1}0] : &\quad |a - a_F|^2 \end{aligned} \tag{4}$$

Equation (4) offers a beautiful example of quantum mechanical interference of scattering amplitudes for the same final state and different scattering channels. These interferences have been observed near the $E_0 + \Delta_0$ edge of GaAs (Fig. 3). This figure also shows the observed pure Fröhlich scattering (dots) while the lines represent microscopic calculations of the absolute Raman polarizability compared with the experimental value.

The calculation were performed using excitonic functions as intermediate states and without adjustable parameters. They lead to excellent agreement with the experimental data. Note that resonable agreement had also been found earlier with a calculation involving uncorrelated pairs. Such calculations, if performed without adjustable parameters, lead to absolute values much smaller than those found experimentally, especially for the Fröhlich contribution. This was remedied by Menéndez and Cardona (1985) by enhancing the q-vector in an *ad hoc* manner through scattering by charged impurity centers.

Calculations of efficiencies for scattering mediated by the E_0 and $E_0 + \Delta_0$ excitons involve coupling by the phonons between discrete and discrete states (D), discrete-continuum (D-C) and continuum-continuum states. In the case of deformation potential interaction the calculations can be strongly simplified if one assumes that the reduced excitonic masses are the same for light and heavy hole excitons. This approximation, which is rather good since these masses are dominated by the electron mass, leads to equal excitonic radii for both exciton series and to excitonic wavefunctions which are orthogonal unless $n = n'$ regardless of whether the excitons are related to the light or heavy hole. Consequently, the phonon coupling can only take place between exciton states with the same n or, in the continuum, between states with the same quantum number. We thus only have D ($n = n'$) and C terms. Incoming and outgoing resonances result depending on whether the incoming $\hbar\omega_L$ or the outgoing $\hbar\omega_S$ equals the excitation energy for a given $n = n'$ state. Experimental and theoretical data for the incoming resonance of GaAs ($n = 1$ are shown in Fig. 4 in absolute Raman polarizability

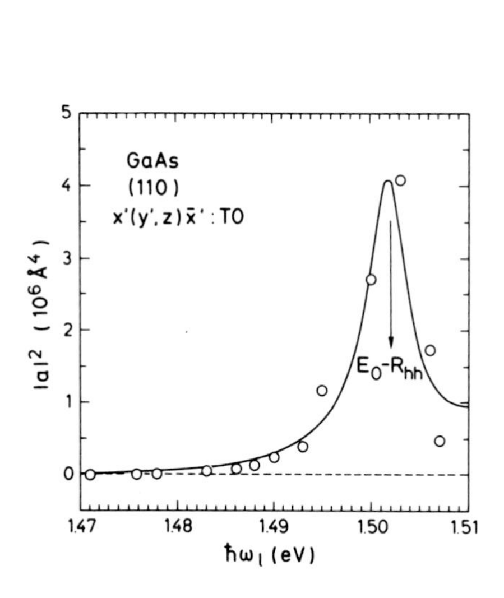

Fig. 4: Calculated (solid line) and measured (dots) $|a|^2$ for deformation potential scattering near the $E_0 + \Delta_0$ gap of GaAs (Cantarero et al, 1989a).

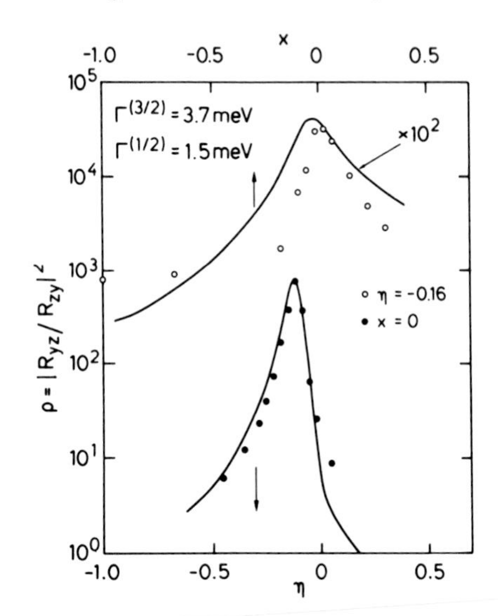

Fig. 5: Theoretical (solid lines) double resonant profile $|R_{yz}|^2$ normalized to the nonresonant $|R_{zy}|^2$ for GaAs under [001] stress vs. reduced laser freuqency η and reduced stress x. The points are experimental (Trallero-Giner et al., 1988).

units. The agreement is excellent. Note that the outgoing resonance cannot be observed in this case because it is covered by the luminscence background. This can be avoided by studying the direct gap resonance of an indirect gap material such as GaP of AlSb (Cantarero et al., 1989a; Gavrilenko et al., 1990). We conclude this section by recalling that excitonic interaction is of the essence for the interpretation of resonant profiles and absolute scattering efficiencies in bulk semiconductors. In the case of the Fröhlich mechanism the efficiencies calculated for uncorrelated pairs are several orders of magnitude smaller than the measured ones. This can be corrected by increasing the scattering wavevector (to which the scattering amplitude is proportional) in an *ad hoc* way (of dubious physical meaning) through the assumption of a sizeable amount of impurity scattering.

3. DOUBLE AND TRIPLE RESONANCES

They occur when both the incident and the scattered frequency coincide with electronic gaps. MQW's are ideal objects for such phenomena. The valence bands split into light and heavy hole components because of the wavefunction confinement. It is easy to choose the MQW parameters, especially the well width, such that the splitting equals the energy of the scattering particle (e.g. an optical phonon) (Kleinman et al., 1987). Similar effects can be achieved by splitting the valence bands through application of uniaxial stress in bulk samples (Cerdeira et al., 1986; Trallero-Giner et al., 1988). The stress-induced double resonances can also be quantitatively explained (including absolute efficiencies) on the basis of excitonic excitations (Trallero-Giner et al., 1988). We display in Fig. 5 the double resonance observed by application of stress for a fixed resonant laser frequency and also for a fixed stress (corresponding to double resonance) vs. laser frequency. We note that the stress along z splits the valence bands into the $(\frac{3}{2}, \pm\frac{1}{2})$ and the $(\frac{3}{2}, \pm\frac{3}{2})$ components. The latter do not contain the z-component of the valence wave functions and therefore the corresponding edge exciton cannot be excited by z-polarized light. Consequently the Raman tensor becomes asymmetric: The R_{yz} (or R_{xz}) component is double resonant while R_{zy} is not, an interesting and rare occurrence for phonon scattering (common, however, in scattering by magnetic excitations).

Doubly and even triply resonant phenomena appear in scattering by two phonons. Two mechanisms contribute to such scattering: the electron–two-phonon interaction taken to first order and the electron–one-phonon interaction taken to second order (i.e., iteration). The former is isomorphic (as vertex in a Feynman diagram) with the electron-phonon interaction and leads to similar double resonances both for stressed bulk samples and for MQW's (Alexandrou et al., 1989). The iterated first order electron-phonon interaction can lead to triple resonances (Fig. 6, right) in which not only the incoming and the outgoing beam are resonant with a critical point but also the intermediate state (after emission of one phonon) is resonant with the discrete of excitonic excitations. Such triple resonances can be observed for 2LO phonons

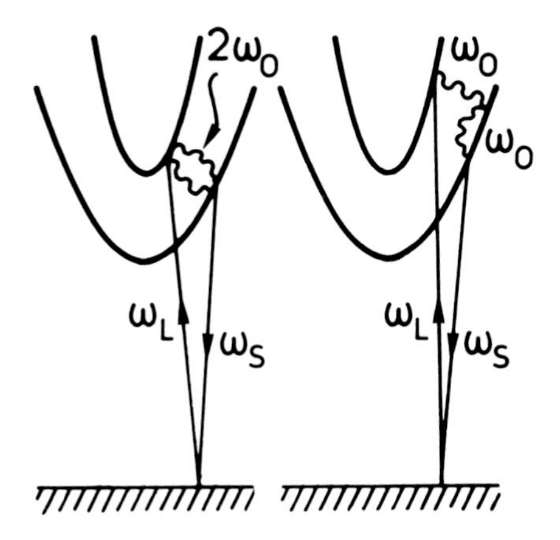

Fig. 6: Excitonic diagrams for doubly resonant scattering via electron-two-phonon interaction (left) and for triply resonant, iterated electron–one-phonon interaction (right).

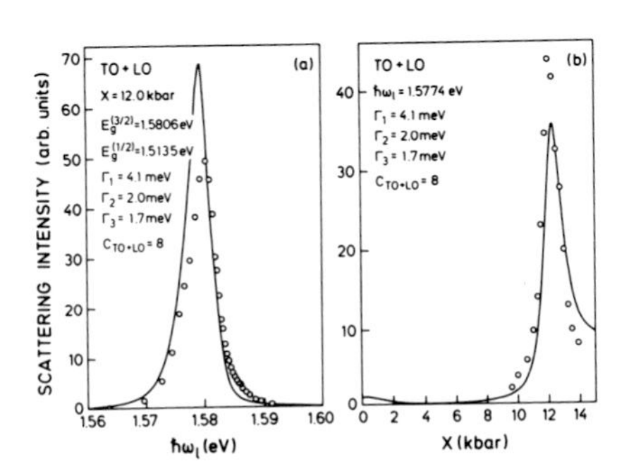

Fig. 7: Triply resonant profiles for scattering by LO+TO phonons in GaAs. The points are experimental. (Alexandrou et al., 1988).

(2 × Fröhlich), LO+TO (Fröhlich + deformation potential) and 2TO phonons (2 × deformation potential). In the 2LO and LO+TO cases they lead to the appearence of dominant triply resonant scattering for crossed polarizations involving Fröhlich interaction, a rather unusual effect (Alexandrou et al., 1988; Cardona and Trallero-Giner, 1991). We show in Fig. 7 the stress induced triple resonance for LO+TO phonons in GaAs.

Similar triple resonances have also been observed in MQW's. The data of Fig. 8 were obtained for circular incident and scattered polarizations, the normal modes for the direction of propagation along the quantization axis. Again here we note the dominance of scattering for *crossed polarizations* induced by *Fröhlich interaction.* This phenomenon is a consequence of the *lh-hh* mixing for $\vec{k} \neq 0$ at the point where the intermediate state of the triple resonance occurs (Cardona and Trallero-Giner, 1991).

We note that triple resonant scattering involving one phonon plus elastic scattering by defects (probably interface roughness) has been reported (Alexandrou et al., 1988). It also appears for crossed circular polarizations and near resonance it becomes stronger than the q-dependent first-order scattering by one phonon. Since it involves Fröhlich interaction (see Sect. 4), it only appears for confined phonons with quantization number m = even (Sood et al., 1985; Jusserand and Cardona, 1984). Very close to resonance it completely swamps the deformation potential scattering, which should be observed for crossed polarizations for phonons with m = odd. The theoretical selection rules, parallel polarizations (x, x) for m = even (Fröhlich) and

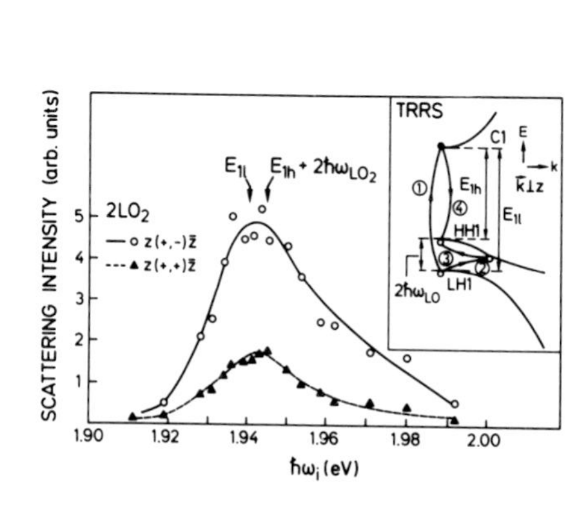

Fig. 8: Triply resonant profiles for scattering by 2LO phonons (see Fig. 9) in a GaAs/AlAs MQW. The points are experimental, the lines a guide to the eye (Alexandrou et al., 1988).

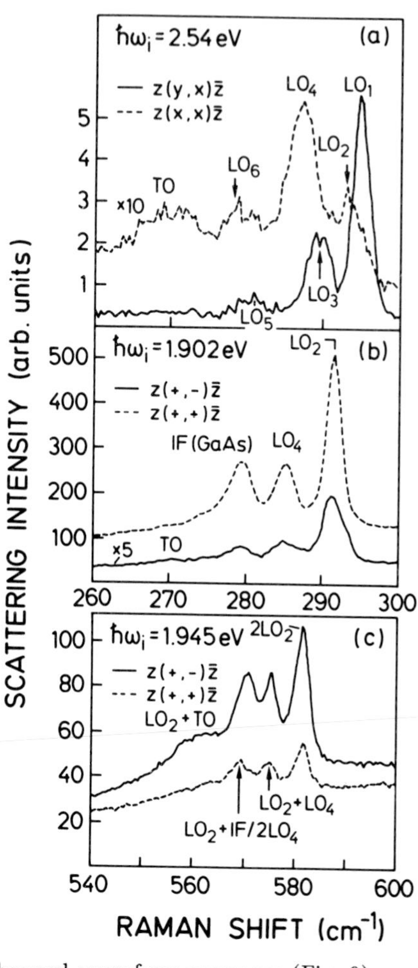

Fig. 9: Nonresonant (a) and resonant (b) scattering by GaAs-like LO phonons in a GaAs/AlAs MQW. (c) scattering by two phonons (see Fig. 8) (Alexandrou et al., (1988).

perpendicular (x, y) for m = odd are, however, observed away from resonance (Fig. 9).

4. RESONANT SCATTERING IN MQS'S AND SUPERLATTICES

No detailed calculations of resonant Raman profiles including excitonic interaction have appeared so far for MQW's. Such profiles have, however, been investigated rather profusely, usually in arbitrary units of scattering efficiency.

We show in Fig. 10 a resonant profile obtained for GaAs/GaAlAs MQW near the $n = 2$ $hh \rightarrow e$ transitions. The solid curve represents a two-dimensional calculation based on defect induced (*ad hoc*!) Fröhlich interaction and uncorrelated electron-hole pairs. As mentioned above, a rather good fit to the experimental data is obtained but its physical significance is dubious. The fit reproduces well the experimental finding of a dominant outgoing resonance, the canonical behavior. The opposite can be observed, however, under special conditions, e.g., through interference of two scattering channels such as $n = 1$ and $n = 2$ excitons: a stronger incoming than outgoing resonance may result (Zucker et al., 1988).

The selection rule of nonresonant LO scattering given above (dp: m = odd; Fröhlich: m = even) makes possible the assignment of the observed peaks to reliable values of m. The corresponding frequencies can be mapped rather accurately onto bulk dispersion relations with $k = (\pi/d + \delta)m$, where δ is a small length which takes into account residual penetration

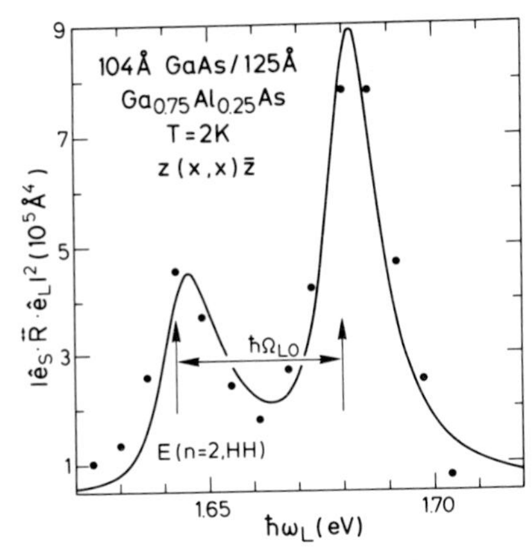

Fig. 10: Experimental resonant profile for scattering by LO phonons near the $n = 2$ hh exciton for a GaAs/GaAlAs MQW. The solid line has been calculated for uncorrelated pairs, Fröhlich interaction plus impurity scattering (Kauschke et al., 1987).

into the barriers (Jusserand and Paquet, 1986). Such results can be used to predict bulk dispersion relations in cases where they have not been measured such as for AlAs (Mowbray et al., 1991). Next we discuss the possible mechanisms for LO-phonon scattering by Fröhlich interaction in MQW's and superlattices. We notice that m = even LO phonons have A_1 symmetry which is Raman allowed in parallel polarizations for the D_{2d} point group of a [001] MQW. Since it is forbidden in the bulk (T_d), it is of interest to consider the mechanism that makes it allowed when the symmetry is lowered to D_{2d}. It is generally accepted that the main contribution arises from Fröhlich interaction but with the scattering wavevector replaced by a sample property, namely $m\pi/d$. If the barriers for electrons and holes had infinite heights the Fröhlich effect on the electrons would cancel that on the holes and no scattering would be observed. Since these heights are not infinite, the electron and hole wave functions penetrate different amounts into the barriers. The phonons are more strongly confined then the electrons, hence the (electron-hole)-phonon matrix element:

$$\int \left(\varphi_e^2(z) - \varphi_h^2(z)\right) \phi(z) dz, \qquad (5)$$

(where $\varphi_{e,h}$ are uncorrelated electron and hole wavefunctions and $\phi(z)$ is the electrostatic (Fröhlich) potential of the phonon) does not vanish. The same result is obtained for the exciton-phonon matrix element.

Besides the "allowed" Fröhlich mechanism just discussed, two other "forbidden" (i.e. q-dependent) mechanisms are possible although they have not been experimentally identified. One of them is the two-dimensional version of the bulk effects for q in the layer plane (q_x). The other is induced by q_z; it should only be important in superlattices (very short period MQW's) in which there are finite and different electron and hole masses along the growth direction. These "forbidden" Fröhlich mechanisms should only apply to ir-active phonons, i.e., to confined phonons for m = odd (contrary to the "allowed" Fröhlich coupling which obtains for m = even).

Near resonance multiphonon effects are also observed. For short period superlattices they may involve phonons of both well and barrier (Mowbray et al., 1991a). For not too short periods only multiphonon scattering by phonons with m = even is observed. For very short periods m = odd (corresponding to so-called interface modes for $k_x \gg k_z$) are also seen. An interpretation of these observations has been given by Mowbray et al. (1991b).

We show in Fig. 11 the resonance profiles of various multiphonon peaks observed in a $(GaAs)_6/(AlAs)_6$ superlattice. The inset is meant to prove that they peak under outgoing resonance.

5. RESONANT SPIN FLIP SCATTERING

Observations of scattering involving spin flip, resonant at bound excitons (BE) or at excitons localized at interface fluctuations in GaAs/AlGaAs MQW's have been recently reported by

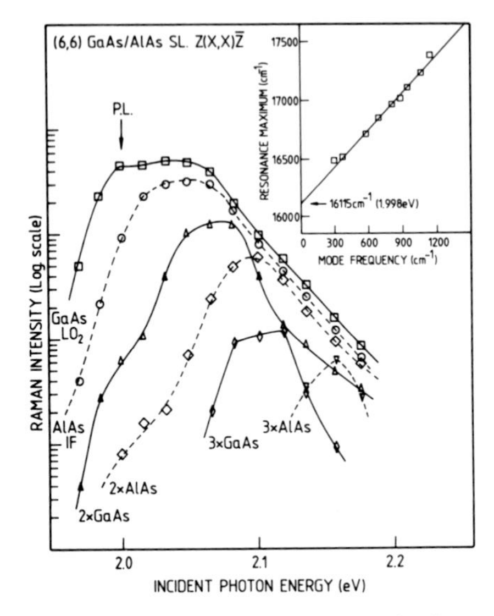

Fig. 11: Resonant profiles for multiphonon scattering in a $(GaAs)_6/(AlAs)_6$ MQW. The inset demonstrats the outgoing nature of the resonances (Mowbray et al., 1991b).

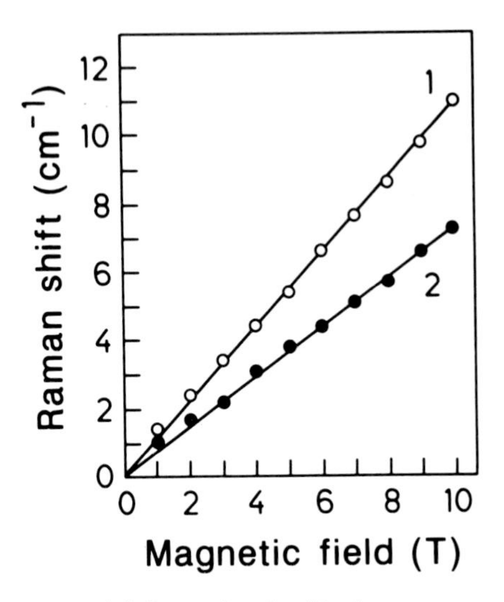

Fig. 12: (1) Scattering by flipping an acceptor bound hole in a Be-doped GaAs/GaAlAs MQW vs. magnetic field. (2) Similar observations for an undoped MQW; the shift corresponds to the Zeeman splitting of the exciton (Šapega et al., 1992).

Sapega et al. (1992). We show the dependence of the Stokes shift on magnetic field observed for a Be-doped (1) and an undoped (2) MQW in Fig. 12. The former is resonant at the bound exciton (BE) energy and thus it is believed to correspond to a spin filp of the hole bound to the acceptor (with an effective Landé factor $g^* = 2.3$). The latter resonates near the energies of an LE exciton. It is believed to be due to double resonant scattering with emission of one acoustic phonon, involving as intermediate states excitons Zeeman-split in a magnetic field ($g^* = 1.5$). These observations should be extended to the study of non-linear Zeeman splittings which have been predicted for larger period MQW's (Bauer, 1989).

REFERENCES

Alexandrou A, Cardona M and Ploog K 1988 Phys Rev B 38 2196
Alexandrou A, Trallero-Giner C, Kanellis G and Cardona M 1989 Phys. Rev. B 40 1013
Bauer G E W 1989 High Magnetic Fields in Semicond. Physics II (Berlin: Springer) p 240
Cantarero A, Trallero-Giner C and Cardona M 1989a Phys. Rev. B 39 8388
Cantarero A, Trallero-Giner C and Cardona M 1989b Phys. Rev. B 40 12290
Cardona M 1982 Light Scattering in Solids II (Heidelberg: Springer) p 19
Cerdeira F, Anastassakis E, Kauschke W and Cardona M 1986 Phys. Rev. Lett. 57 3209
Cardona M and Trallero-Giner C 1991 Phys. Rev. B 43 9959
Gavrilenko V I, Martínez D, Cantarero A, Cardona M and Trallero-Giner C 1990 Phys. Rev. B 42 11718
Jusserand B and Cardona M 1989, Light Scattering in Solids V (Heidelberg: Springer) p 49
Jusserand B and Paquet D 1986 Phys. Rev. Lett. 56 1751
Kauschke W, Sood A K, Cardona M and Ploog K 1987 Phys. Rev. B 36 1612
Kleinman D A, Miller R C and Gossard A C 1987 Phys. Rev. B 35 664
Menéndez J and Cardona M 1985 Phys. Rev. B 31 3696
Mowbray D J, Cardona M and Ploog K 1991a Phys. Rev. B 43 1598
Mowbray D J, Cardona M and Ploog K 1991b Phys. Rev. B 43 11815
Sapega V F, Cardona M, Ploog K, Ivchenko E L and Mirlin D N 1992 Phys. Rev. B, in press
Sood A K, Menéndez J, Cardona M and Ploog K 1985 Phys. Rev. Lett. 54 2111
Trallero-Giner C, Alexandrou A and Cardona M 1988 Phys. Rev. B 38 10744
Zucker J E, Pinczuk A and Chemla D S 1988 Phys. Rev. B 38 4287

Inst. Phys. Conf. Ser. No 123
Paper presented at the International Meeting on Optics of Excitons in Confined Systems, Giardini Naxos, Italy, 1991

Investigation of intraband optical Stark effect in multiple quantum wells

D. Fröhlich, Ch. Neumann, S. Spitzer, B. Uebbing and R. Zimmermann[1]

Institut für Physik, Universität Dortmund, 4600 Dortmund 50, FRG
[1]permanent adress: Zentralinstitut für Elektronenphysik, O-1086 Berlin, FRG

ABSTRACT: Experimental results for the intraband optical Stark effect in multiple quantum wells are reported for different polarizations of the pump laser with respect to the growth direction (z-direction). In the theoretical treatment it is shown that for polarization perpendicular to z (x-Stark effect) one gets a blue shift, whereas the z/y-Stark effect shows a resonance behaviour. In the analysis of the experimental results the inhomogeneous field distribution along the growth direction is taken into account.

1. INTRODUCTION

Nonlinear optical methods are extensively used for study of electronic properties of quantum-well structures. For recent reviews on linear and nonlinear optical properties of these man-made structures we refer to Schmitt-Rink, Chemla and Miller (1989) and Cingolani and Ploog (1991). One can distinguish two categories of nonlinear experiments. For illustration of this point we consider two-photon processes. In the first category we place the classical two-photon absorption (TPA) experiments, where the electronic system under investigation exhibits resonances only for the **sum** of the photon energies. Despite the fact that at least one of the laser beams has a high intensity which might exceed $10^8\,W/cm^2$, TPA can be considered as a weak probe of the electronic system. The two-photon transition rate is calculated by second-order perturbation theory, which leads to pronounced resonances only for the sum of the photons. Additional resonance enhancement in TPA can be achieved if one of the laser beams is close to a strong one-photon resonance which is the case if one uses a tunable dye laser ($\hbar\omega \sim 1-2\,eV$) and a CO_2 laser ($\hbar\omega \sim 0.1\,eV$). This "near resonance TPA" experiments simplify very often the detailed interpretation of the polarization dependence of the data, since one has to consider only the dominant term of the sum over virtual transitions to one-photon allowed intermediate states. Experiments of two-photon magneto-absorption in multiple quantum wells (MQW) by Fröhlich et al. (1988) demonstrate this point. In this case only the dye laser induces the interband transition, whereas the CO_2 laser induces the intraband Landau transitions as proven by the polarization dependence. There are many TPA experiments where a one beam pumping technique is used (Tai et al. 1989, Catalano et al. 1989, Nithisoontorn et al. 1989). In this case one does not expect a resonance enhancement since both laser photons are at half the resonance energy. Nevertheless one can study the polarization dependence which shows characteristic differences for polarization parallel or perpendicular to the growth direction. Catalano et al. (1989) show clearly resolved resonances for both configurations. By comparison of one-photon and two-photon data they gain a consistent assignment of P-excitons as well. Pasquarello and Quattropani (1990a, 1990b, 1991) have discussed in detail the two-photon transition rate. They show that for transitions with the polarization in the layer planes the contribution of continuum states is comparable to the contribution due to transitions between discrete states. For polarization of both beams parallel to the growth direction pronounced resonances to S-excitons of the light hole (LH) exciton are predicted which are due to the $n = 2$ subband either of the valence or the conduction band. With a two-laser setup, where both polarizations can be set independently it should be possible to detect these resonances also for the heavy hole (HH) exciton.

In the second category we place two-photon experiments where the electronic system cannot be considered as independent of the laser fields. In this case a high power laser is in resonance or close to resonance with two electronic levels which allow dipole transitions. This is the classical case of the optical Stark effect, which was first reported for molecular transitions by Autler and Townes (1955). The first experiments were done in the microwave and radio-frequency region. For a review of the observation of splittings and shifts of atomic levels in radiation fields at optical frequencies we refer to Bonch-Bruevich and Khodovoi (1968). The observation of such an effect in solids, however, seemed to be rather difficult, because the linewidths of electronic transitions in solids are in general much larger than in atoms and molecules. The first experiments of an optical Stark effect in a semiconductor were reported by Fröhlich et al. (1985) for exciton transitions in Cu_2O. The authors observed the dynamical coupling of the 1S and 2P excitons by an intense CO_2 laser beam. These experiments were later done on MQW (Fröhlich et al. 1987). In this case the authors mainly observed the near resonant coupling between the lowest conduction band sublevels by the intense CO_2 laser beam. In both cases (Cu_2O and MQW) the photon energy of the high intensity beam (pump beam: $\hbar\omega \sim 0.1$ eV) is small as compared to the lowest exciton transitions which are induced by the low intensity test beam ($\hbar\omega \sim 2$ eV and 1.5 eV for Cu_2O and MQW, respectively). It can thus be assumed that population effects of excitons or free carriers due to many photon transitions can be disregarded. In the cases discussed so far the high intensity laser couples excited states of the **same** band to band transition. Zimmermann (1990) has therefore introduced the term "intraband optical Stark effect" for this "three-level optical Stark effect" (Fröhlich et al. 1990a). Much more attention has been given to the "interband optical Stark effect" (Zimmermann 1990) where the photon energy of the pump laser is close to the first exciton transition, and consequently virtual exciton populations play a decisive role. First experiments on MQW were reported by Mysyrowicz et al. (1986) and von Lehmen et al. (1986). In this case the authors observed a characteristic blue-shift of the HH- and LH-exciton transitions which depends on the detuning of the pump beam from the first exciton resonance and its intensity. Knox et al. (1989) have investigated in detail the intensity dependence of this interband dynamical Stark effect in quantum wells with the use of femtosecond spectroscopy. They found out that at high intensities two-photon absorption leads to real carrier excitation, which then competes with virtual exciton effects. For literature on the theory of the interband optical Stark effect and further experiments on other semiconductors we refer to a review by Zimmermann (1990).

In this contribution we report new results on the intraband optical Stark effect of MQW. In the first experiment of Fröhlich et al. (1987) the emphasis was on the investigation of the inter-subband coupling by the intense CO_2 laser, whose polarization had a large component parallel to the growth direction (z-direction). The authors considered only two oscillators (HH and LH transition) for the derivation of the nonlinear susceptibility, which is then used to describe the differential transmission spectrum. First experiments with the polarization of the CO_2 laser in the plane of the MQW (in-plane Stark effect) were reported recently by Fröhlich et al. (1990b). The authors used a CO_2 laser of much higher intensity then in the earlier experiments (Fröhlich et al. 1987). The main features of the in-plane Stark effect were a blue shift of exciton lines, a reduction of oscillator strength, an increase of damping and the stability of the continuum absorption. In the theory of this effect it was shown that the rotating wave approximation is not applicable since no resonant transitions are involved. However a few-states expansion in the length gauge was applied which is not reliable, as shown in section 3.

After a short description of the experimental setup we will present the theory of the intraband optical Stark effect. The experimental results are then fitted by the theoretical expressions derived before.

2. EXPERIMENTAL DETAILS

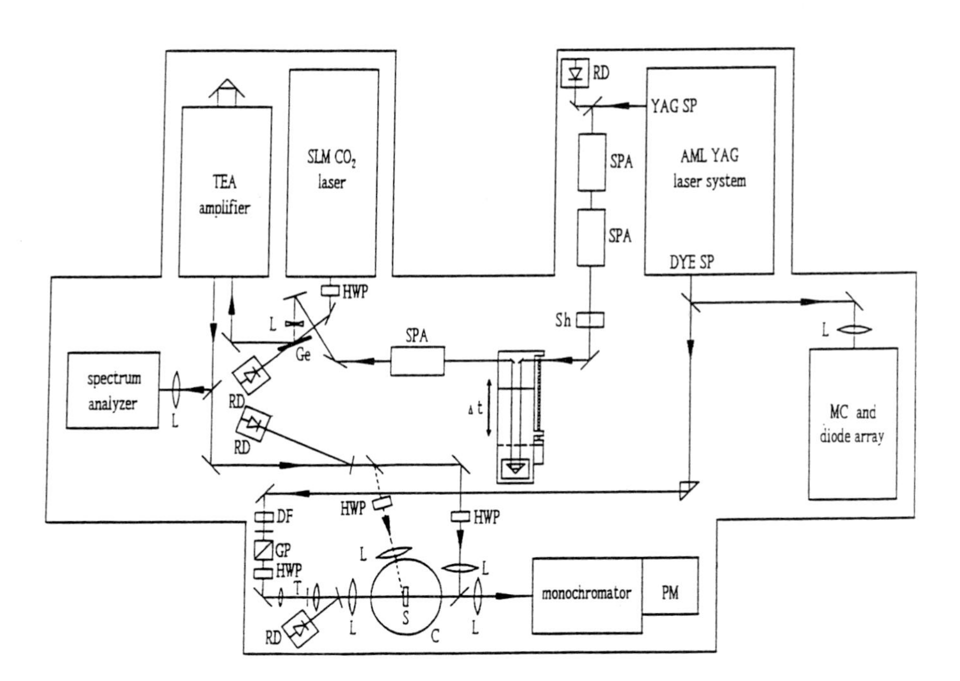

Fig. 1. Block diagram of the experimental setup. AML YAG laser system, active-mode-locked YAG pumped dye laser system; C, cryostat; DF neutral density filter; Δt, optical delay; Ge, Germanium plate; GP, Glan polarizer; HWP, half-wave plate; L, lens; MC, monochromator; PM, photomultiplier; RD, reference diode; S, sample; Sh, shutter; SLM CO_2 laser, single longitudinal mode CO_2 laser; SP, single pulse; SPA, single pulse amplifier; T, telescope; TEA amplifier, transversely excited atmospheric CO_2 amplifier.

There are two novel aspects to the experiments reported in this contribution. i) The CO_2 laser is coupled under an angle into the MQW which allows to measure simultaneously the in-plane and the mixed in-plane and intersubband Stark effect as a function of CO_2 laser intensity. ii) The photon energy of the CO_2 laser is tuned in discrete steps between 0.114 and 0.134 eV. This allows for the first time to measure the resonance behaviour of the inter-subband Stark effect. The theoretical analysis of the experiments clearly shows that it is not posssible to unfold the spectra for the mixed CO_2 laser polarization (z/y spectra) in order to get pure spectra for the in-plane (y-polarization) and inter-subband (z-polarization) Stark effect. In the theory we derive an expression for the susceptibility which is used to analyse the experimental data.

The experiments were performed with a picosecond-setup as shown in Fig. 1. The active-mode-locked YAG-dye laser system is described in detail by Fröhlich et al. (1990a). The CO_2 laser pulse from a single mode CO_2 laser (Laser Science PRF-150S) is shortened from about 150 ns to 250 ps by plasma switching (Fröhlich et al. 1990a). The CO_2 pulse is coupled through a ZnSe prism (n = 2.4) under an angle of incidence of 70^0 into the MQW in order to get a large electric field component parallel to the z-axis. Details of the coupling are shown in Fig. 2. The polarization of the CO_2 laser can be changed from horizontal (z/y-spectrum) to vertical (x-spectrum) by a half-wave plate.

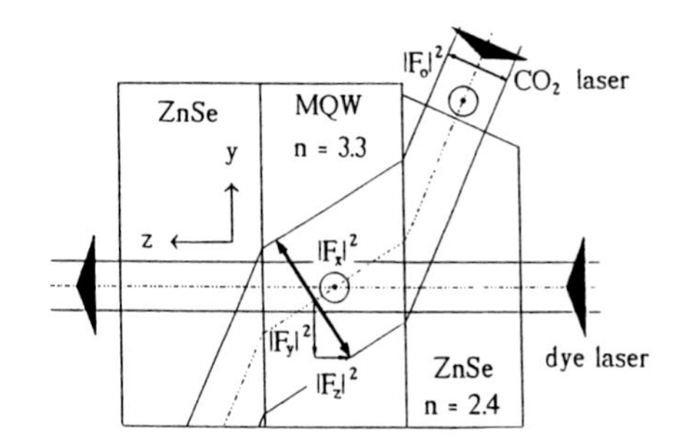

Fig. 2. Schematics of the optical coupling of CO_2 and dye laser pulses into the MQW.

3. THEORY OF THE INTRABAND OPTICAL STARK EFFECT

In any theoretical treatment of nonlinear optics, the invariance with respect to gauge transformations has to be checked carefully. We start here with the Coulomb gauge for the pump field introducing its vector potential **A** in the longwave limit. Even if the pump laser frequency is much below the energy gap, a small induced virtual density is excited which can be written as A^2/μ, well known for any static vector potential (diamagnetic shift). Other density effects as band filling and Coulomb correlation can be neglected. A derivation of this result will be given elsewhere. The interband polarization $\hat{P}$ due to the weak test field F^{test} obeys then

$$\left[-i\hbar\partial_t + E_g + \frac{1}{2m_e}(-i\hbar\partial_{\mathbf{r}_e} - e\mathbf{A}(t))^2 + \frac{1}{2m_h}(-i\hbar\partial_{\mathbf{r}_h} + e\mathbf{A}(t))^2 + V_e(z_e) + V_h(z_h) - \frac{e^2}{\varepsilon_0\,|\mathbf{r}_e-\mathbf{r}_h|}\right]\hat{P}(\mathbf{r}_e,\mathbf{r}_h,t) = \tag{1}$$

$$= r_{cv}\,\delta(\mathbf{r}_e - \mathbf{r}_h)\,F^{test}(t)$$

The electron-hole Coulomb interaction is screened by the dielectric constant ε_0, and r_{cv} is the interband dipole matrix element. A real space representation has been used which is a convenient starting point for the quantum well characterized by the confining potentials for electrons and holes, $V_e(z_e)$ and $V_h(z_h)$.

The transformation from the momentum gauge into the length gauge is mediated via

$$\hat{P}(\mathbf{r}_e, \mathbf{r}_h, t) = \exp\left[i\frac{e}{\hbar} \mathbf{A}(t)(\mathbf{r}_e - \mathbf{r}_h)\right] \tilde{P}(\mathbf{r}_e, \mathbf{r}_h, t) \tag{2}$$

giving

$$\left[-i\hbar\partial_t + E_g - \frac{\hbar^2}{2m_e}\partial^2_{r_e} - \frac{\hbar^2}{2m_h}\partial^2_{r_h} + V_e(z_e) + V_h(z_h) \right. \tag{3}$$

$$\left. - e\mathbf{F}(t)(\mathbf{r}_e - \mathbf{r}_h) - \frac{e^2}{\varepsilon_0 |\mathbf{r}_e - \mathbf{r}_h|} \right] \tilde{P}(\mathbf{r}_e, \mathbf{r}_h, t) =$$

$$= r_{cv}\ \delta(\mathbf{r}_e - \mathbf{r}_h)\ F^{test}(t)$$

with pump field $\mathbf{F}(t) = -\partial_t \mathbf{A}(t)$.

The nonlinear susceptibility is related to the polarization at zero electron-hole distance (valid for direct allowed interband transitions).

The polarization is expanded as

$$\hat{P}(\mathbf{r}_e, \mathbf{r}_h, t) = \sum_{ijn} u_{ei}(z_e)\ u_{hj}(z_h)\ \Phi_n(\vec{\rho}_e - \vec{\rho}_h)\ \hat{P}_{ijn}(t) \tag{4}$$

using as basis sublevel states $u_i(z)$ and exciton wave functions $\Phi_n(\vec{\rho})$. z is directed along the growth direction, and $\vec{\rho}$ is the two-dimensional vector in the quantum well plane. Since our basis is complete, the result for the susceptibility does not depend on the gauge used. But, for computational reasons, it would be desirable to truncate the basis to a few states. Then, the gauge has to be chosen carefully. Model calculations have shown that for the problem at hand best results are obtained with a mixed gauge: z-motion in length gauge and $\vec{\rho}$-motion in momentum gauge. We give a qualitative explanation in what follows. The non-diagonal elements in the length gauge are of type

$$e\mathbf{F}(t)\,(\alpha|\mathbf{r}|\beta) \qquad (\alpha = ijn) \tag{5}$$

whereas in the momentum gauge

$$\frac{e\hbar}{m^*}\mathbf{A}(t)(\alpha|\partial_{\mathbf{r}}|\beta) = e\,\mathbf{F}(t)\,\frac{E_\alpha - E_\beta}{\hbar\omega_p}\,(\alpha|\mathbf{r}|\beta) \tag{6}$$

enters. ω_p is the frequency of the pump laser, and a commutator relation has been used for the second form. If the energy differences between relevant states are small with respect to $\hbar\omega_p$ the momentum gauge gives thus smaller coupling elements. With $\hbar\omega_p \sim 120$ meV in the present case this happens for the in-plane motion since here excitonic transitions refering to the same sublevel pair are active (binding energy ~ 10 meV). These transitions are induced by in-plane polarized light (x-polarization). In contrast, for z-polarized light, inter-sublevel transitions are invoked which have a large spread in energies, among them one which is nearly resonant with the laser. Consequently, the length gauge is to be prefered. Treating optical transitions in the hydrogen atom, Bassani et al. (1977) found the length gauge to be superior for the same reason. The choice of a preferential gauge for two-photon interband transitions is not as clear (Pasquarello and Quattropani (1988) and references therein).

Thus, with the mixed gauge

$$\hat{P}(\mathbf{r}_e, \mathbf{r}_h, t) = \exp\left[i\frac{e}{\hbar} \mathbf{A}_z(t)(z_e - z_h) \right] P(\mathbf{r}_e, \mathbf{r}_h, t) \tag{7}$$

and the expansion (eq. 4) we get

$$\left[-i\hbar\partial_t + E_{ei} + E_{hj} + \mathcal{E}_n + \frac{e^2}{2\mu} A_x^2(t)\right] P_{ijn}(t) + \frac{i\hbar e}{\mu} A_x(t) \sum_m (n|\partial_x|m)\, P_{ijm}(t)$$

$$- eF_z(t) \left[\sum_k (ei|z|ek)\, P_{kjn}(t) - \sum_k (hj|z|hk)\, P_{ikn}(t) \right] \tag{8}$$

$$= r_{ij}\, \Phi_n(0)\, F^{test}(t)$$

The exciton energy $\mathcal{E}_n$ is counted from the sublevel distance $E_{ei} + E_{hj}$. Assuming stationarity, the time dependence is governed by the test frequency ω and multiples of the pump frequency ω_p. We expand into a Fourier series according to

$$P_{ijn}(t) = \sum_{s=-\infty}^{\infty} P_{ijn}^{s}\, e^{-i(\omega + s\omega_p)t} \tag{9}$$

For the test field the rotating wave approximation has been applied. The vector potential of the pump field is chosen as

$$\mathbf{A}(t) = -\frac{2\mathbf{F}}{\omega_p} \sin(\omega_p t) \tag{10}$$

Comparing coefficients results in a matrix equation with the multiple index ijns

$$\left[E_{ei} + E_{hj} + \mathcal{E}_n - \hbar\omega - i\Gamma_{ij} - s\hbar\omega_p + \frac{e^2 F_x^2}{\mu\omega_p^2} \right] P_{ijn}^{s} - \tag{11}$$

$$\frac{e^2 F_x^2}{2\mu\omega_p^2} \left(P_{ijn}^{s+2} + P_{ijn}^{s-2} \right) + \frac{\hbar e F_x}{\mu\omega_p} \sum_m (n|\partial_x|m) \left(P_{ijm}^{s+1} - P_{ijm}^{s-1} \right)$$

$$- eF_z \left[\sum_k (ei|z|ek) \left(P_{kjn}^{s+1} + P_{kjn}^{s-1} \right) - \sum_k (hj|z|hk) \left(P_{ikn}^{s+1} + P_{ikn}^{s-1} \right) \right]$$

$$= r_{ij}\, \Phi_n(0)\, \delta_{s,0}\, F^{test}$$

A phenomenological dephasing $i\Gamma_{ij}$ has been introduced here.

The quadratic field term $\Delta = e^2 F_x^2 / \mu\omega_p^2$ gives a constant blue shift and comes out to be the dominant Stark effect for x-polarization, being far away from any resonance. It can be traced back to the temporal average of the $A^2(t)$ term in the starting equation. In atomic physics, it is known as the "ponderomotive threshold shift" (Javainen et al. 1988).

For the experimental situation with the CO_2 laser frequency tuned close to the electron sublevel transition e1–e2, we pick up explicitly the coefficients $P^0_{e1,h1,n}$ and $P^1_{e2,h1,n}$ which have a resonant energy denominator if probed at $\hbar\omega \approx E_{e1} + E_{h1} + \mathcal{E}_n$. The 2x2-system is easily solved

$$P^0_{e1,h1,n} = \frac{r_{e1,h1}\,\Phi_n(0)\,F^{test}}{E_{e1} + E_{h1} + \mathcal{E}_n - \hbar\omega - i\Gamma_1 + \Delta - Z^{NR} - \dfrac{Z^2_{12}}{E_{e2} + E_{h1} + \mathcal{E}_n - \hbar\omega - \hbar\omega_p - i\Gamma_2 + \Delta}} \tag{12}$$

with
$$Z_{12} = eF_z\,(e1|z|e2) \tag{13}$$

An additional non-resonant shift in z-polarization has been included using second order perturbation theory (Zimmermann 1990)

$$\begin{aligned} Z^{NR} = {} & (eF_z)^2 * \\ & * \left[\sum_i (e1|z|ei)^2 \left(\frac{1-\delta_{i,2}}{E_{ei} - E_{e1} - \hbar\omega_p} + \frac{1}{E_{ei} - E_{e1} + \hbar\omega_p} \right) \right. \\ & \left. + \sum_i (h1|z|hi)^2 \left(\frac{1}{E_{hi} - E_{h1} - \hbar\omega_p} + \frac{1}{E_{hi} - E_{h1} + \hbar\omega_p} \right) \right] \end{aligned} \tag{14}$$

The resonant term is excluded. If summing the complete Z-expression for $\hbar\omega_p \to \infty$ a shift $e^2 F_z^2 / \mu\omega_p^2$ is recovered for this polarization, too. However, as already mentioned, the limit is not reached here.

The susceptibility as measured in the transmission of the test beam is related to the coefficient P^0 (no pump photons) via

$$\chi(\omega)\,F^{test} = \sum_{ijn} r_{ij}\,\Phi_n(0)\,P^0_{ijn} \tag{15}$$

P^0 has a two-pole structure which can be used to express the *nonlinear* susceptibility by a weighted superposition of the *linear* susceptibility

$$\chi_{lin}(z) = \sum_{ijn} \frac{r^2_{ij}\,\Phi^2_n(0)}{E_{ei} + E_{hi} + \mathcal{E}_n - \hbar z} \tag{16}$$

Defining the laser detuning as $\delta = \hbar\omega_p - E_{e2} + E_{e1}$ the Stark-shifted energies and weights are given as ($\tilde{\delta} = \delta + i(\Gamma_2 - \Gamma_1)$)

$$W_{1,2} = \frac{1}{2}\left(\tilde{\delta} \pm \sqrt{\tilde{\delta}^2 + 4Z^2_{12}} \right) \tag{17}$$

$$u_{1,2} = \frac{1}{2}\left(1 \mp \frac{\tilde{\delta}}{\sqrt{\tilde{\delta}^2 + 4Z^2_{12}}} \right)$$

and combine to the final expression for the nonlinear susceptibility

$$\chi(\omega) = \sum_{l=1,2} u_l\,\chi_{lin}\left(\hbar\omega + i\Gamma_1 - \Delta + Z^{NR} + W_l\right) \tag{18}$$

The linear susceptibility comprises bound exciton lines and an enhanced continuum, here characterized by an effective quantum defect q (Zimmermann 1990) which describes well the intermediate situation between 3D and 2D. The exciton energies are given by $\mathcal{E}_n = -R_y / (n-q)^2$. Additionally, the contributions from e1 - HH1 and e1 - LH1 are superimposed. Dipole matrix elements and sublevel energies are calculated for the given quantum well.

4. EXPERIMENTAL RESULTS AND DISCUSSION

In Fig. 3 we present experimental results for the x-Stark effect as a function of

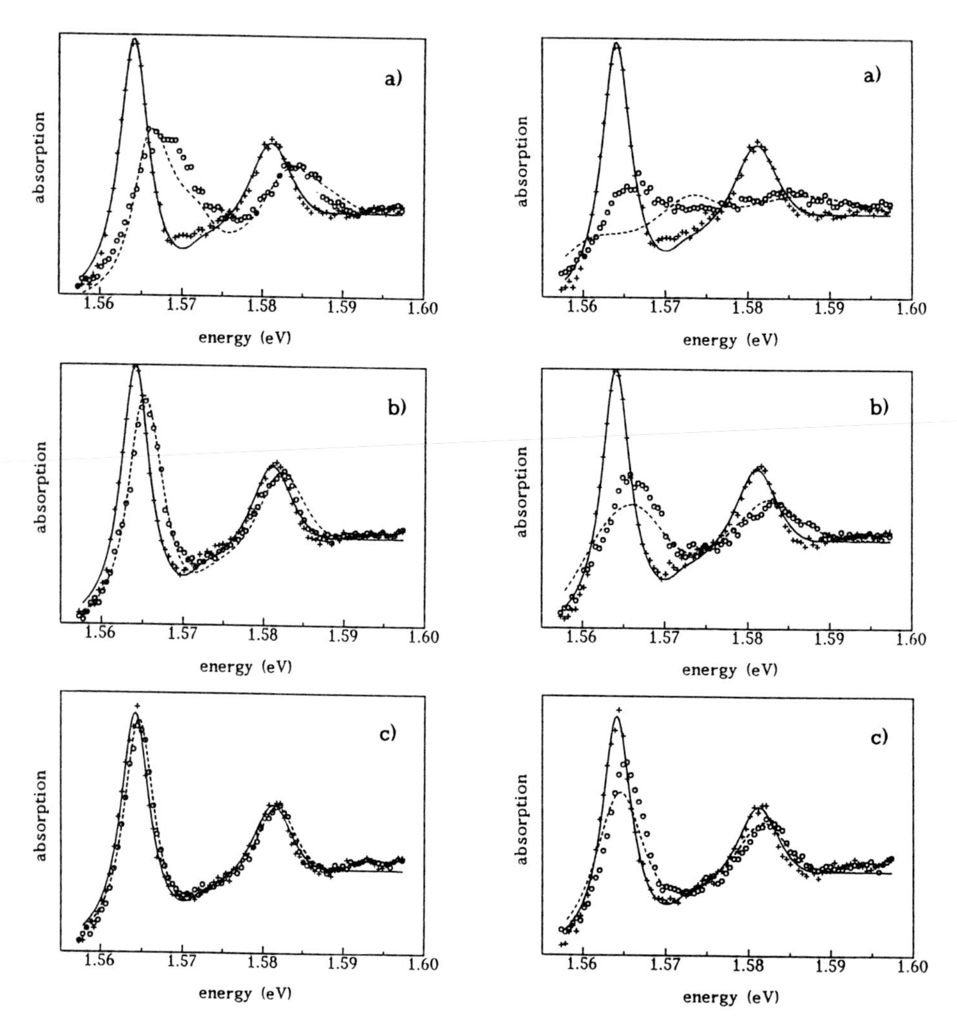

Fig.3. Intensity dependence of the x-Stark effect. Crosses, linear absorption; circles, x-spectrum; lines, theoretical fits, a) $I_0 = 144\ \mathrm{MW/cm^2}$ b) $I_0 = 51\ \mathrm{MW/cm^2}$ c) $I_0 = 22\ \mathrm{MW/cm^2}$.

Fig.4. Intensity dependence of the z/y Stark effect. Crosses, linear absorption; circles, x-spectrum; lines, theoretical fits; using the same intensities as in Fig. 3; detuning $\delta = 0.3\ \mathrm{meV}$; $\Gamma_2 = 4\,\Gamma_1$.

the CO_2 laser intensity together with a fit of the theory derived in chapter 3. The exciton parameters are chosen to describe the linear spectrum. We get $q = 0.35$ and $E_{e1} + E_{h1} = 1.574$ eV, $\Gamma_1 = 2.1$ meV for the heavy hole exciton and $E_{e1} + E_{h1} = 1.591$ eV, $\Gamma_1 = 3.5$ meV for the light hole exciton. The only remaining fit parameter for the x-spectrum is then the pump intensity which is proportional to F_0^2 outside the sample. The x-spectrum shows the expected blue shift and an additional broadening which we attribute to the inhomogeneous field distribution along the growth direction because of multiple reflections in the well. In the analysis we have taken Fabry-Perot modulations into account. We see for all intensities a good quantitative agreement of the theory with the experimental results. The maximum experimental intensity (Fig. 3a) was attenuated by a factor of 0.37 (b) and 0.14 (c) respectively which agrees well with the fitted values. The corresponding spectra for the z/y Stark effect are shown in Fig.4. For the theoretical fit of the z/y-spectrum there is no free parameter besides the damping Γ_2. The three spectra in Fig. 4 show intensity dependent measurements with a detuning $\delta = 0.3$ meV, whereas the spectrum shown in Fig. 5 is taken with a larger detuning of $\delta = 17.8$ meV. The experimental results show only a weak dependence on the detuning δ. The agreement with the theory is rather poor particularly at higher intensities.

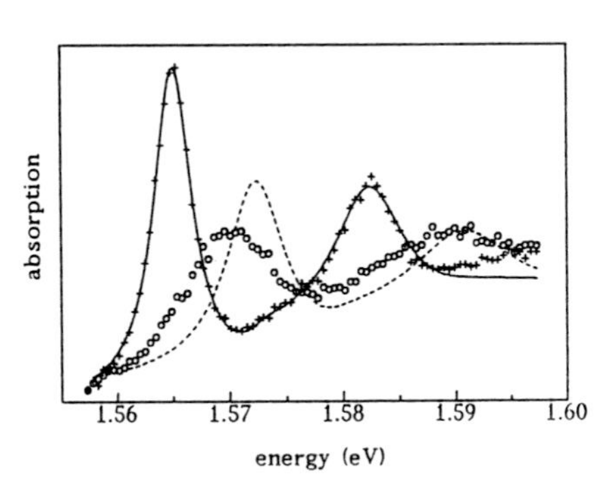

Fig. 5. Spectrum of the z/y-Stark effect. $I_0 = 228$ MW/cm^2, detuning $\delta = 17.8$ meV.

In conclusion, we have presented experimental data and a theoretical modelling of the intraband optical Stark effect in MQW. For the x-polarization of the pump the agreement was satisfying after inclusion of the inhomogeneous field distribution within the MQW. We stress that any nonlinear optical experiment is confronted with this problem since a good MQW sample exhibits nearly perfect parallel end faces. The field modulation is even stronger for a free standing sample compared to our sample which is sandwiched between ZnSe (n = 2.4). For the z-polarized Stark effect the agreement is relatively poor. In particular the expected resonance feature at zero detuning is not seen in the measured data. Further investigations are necessary to clarify this point.

5. ACKNOWLEDGEMENTS

The supply of samples by G. Khitrova (Optical Science Center, University of Arizona, Tuscon, USA) is gratefully appreciated. We acknowledge the financial support of this project by the Deutsche Forschungsgemeinschaft and the Volkswagen-Stiftung.

6. REFERENCES

Autler S.H. and Townes C.H. 1955 Phys.Rev. 100 703

Bassani F., Forney J.J. and Quattropani A. 1977 Phys. Rev. Lett. 39 1070

Bonch-Bruevich A. M. and Khodovoi V. A. 1968 Sov. Phys. Usp. 10 637

Catalano I.M., Cingolani A., Cingolani R., Lepore M. and Ploog K. 1989 Phys. Rev. B40 1312

Cingolani R. and Ploog K. 1991 Advances in Physics to be published

Fröhlich D., Nöthe A. and Reimann K. 1985 Phys. Rev. Lett. 55 1335

Fröhlich D., Wille R., Schlapp W. and Weimann G. 1987 Phys. Rev. Lett. 59 1748

Fröhlich D., Wille R., Schlapp W. and Weimann G. 1988 Phys. Rev. Lett. 61 1878

Fröhlich D., Neumann Ch., Uebbing B. and Wille R. 1990a phys. stat. sol. (b) 159 297

Fröhlich D., Neumann Ch., Uebbing B., Zimmermann R. and Khitrova G. 1990b Proceed. 20th International Conference of Physics of Semiconductors, ICPS 90, Thessaloniki 2 1214

Javainen J., Eberly J.H. and Su Q. 1988 Phys. Rev. B38 3430

Knox W.H., Chemla D.S., Müller D.A.B., Stark J.B. and Schmitt-Rink S. 1989 Phys. Rev. Lett. 62 1189

Mysyrowicz A., Hulin D., Antonetti A., Migus A., Masselink W.T. and Morkoc H. 1986 Phys. Rev. Lett. 56 2748

Nithisoontorn M., Unterrainer K., Michaelis S., Sawaki N., Gornik E. and Kano H. 1989 Phys. Rev. Lett. 62 3078

Pasquarello A. and Quattropani A. 1988 Phys. Rev. B38 6206

Pasquarello A. and Quattropani A. 1990a Phys. Rev. B41 12728 ; 1990b Phys. Rev. B42 9073 ; 1991 Superlattic. and Microstruct. 9 157

Schmitt-Rink S., Chemla D.S. and Miller D.A.B. 1989 Advances in Physics 38 89

Tai K., Mysyrowicz A., Fisher R.J., Slusher R.E. and Cho A.Y. 1989 Phys. Rev. Lett. 62 1784

Von Lehmen A., Chemla D.S., Zucker J.E. and Heritage J.P. 1986 Optics lett. 11 609

Zimmermann R. 1990 Adv. Solid State Physics 30 295

Inst. Phys. Conf. Ser. No 123
Paper presented at the International Meeting on Optics of Excitons in Confined Systems, Giardini Naxos, Italy, 1991

Electronic structure and optical properties of [111]-oriented strained-layer superlattices

Christian Mailhiot

Physics Department, Lawrence Livermore National Laboratory, Livermore, CA 94550, USA

ABSTRACT: We discuss a unique property of strained-layer superlattices not found in bulk semiconductors: large internal electric fields are generated by the piezoelectric effect and lead to novel electronic and optical properties. These internal strain-induced piezoelectric fields are largest for strained-layer superlattices grown along the [111] axis and typically reach a magnitude of 10^5 V/cm for a lattice mismatch of approximately 1 %. In [111]-oriented strained-layer superlattices, we show that the presence of internal piezoelectric fields leads to large nonlinear, electro- and piezo-optical effects.

1. INTRODUCTION

For specific growth orientations, strained-layer superlattices (SLS) grown from piezoelectric materials, such as III-V zincblende structure semiconductors, exhibit novel physical properties not present in neither bulk materials or lattice-matched superlattices. Lattice-mismatched-induced strain in SLS can generate internal electric polarization fields due to the piezoelectric effect (Smith 1986). These polarization fields can generate internal electric fields. The symmetry of the strain tensor determines whether or not polarization fields are generated by the internal strain. This symmetry is fixed by the growth axis of the superlattice. For superlattices grown from zincblende structure materials with a [001] growth axis, only diagonal strain components are non-zero. Since diagonal strain components do not induce an electric polarization vector in zincblende structure materials, piezoelectric effects do *not* occur in these widely studied systems. However, for SLS made from zincblende structure materials grown along any orientation other than [001], the off-diagonal strain components do not vanish and internal polarization fields are generated.

The orientation of the internal polarization fields depends on the growth axis of the superlattice. The growth of SLS along the [111] orientation represents a particularly interesting case because the polarization vectors are oriented along the [111] growth axis. The polarization charge has opposite sign on the two types of interfaces and, as a result, generates an internal electric field with opposite polarity in the two constituent materials. The magnitude of these internal electric fields can be very large, typically exceeding 10^5 V/cm for lattice mismatches of the order of 1 % (Smith 1986). These large internal strain-induced electric fields substantially modify the electronic structure (Mailhiot and Smith 1986a, Mailhiot and Smith 1987a) and optical properties (Smith and Mailhiot 1987) of the superlattice. By changing the superlattice subband energy levels and wavefunctions, internal strain-induced electric fields are responsible for modifying optical transition energies and oscillator strengths. The magnitude of these changes is second-order in the magnitude of the electric field (second-order Stark effect). The internal electric fields can be externally modulated by photoabsorption (as a result of the screening of photogenerated carriers), the application of an external electric field or the application of an external stress. As a result of these modulations, large nonlinear, electro- and piezo-optic effects occur.

2. INTERNAL PIEZOELECTRIC FIELDS

Zincblende-structure semiconductor are piezoelectric materials. Off-diagonal components of the strain induce a polarization field proportional to the off-diagonal strain component (Cady 1946) and given by

$$P_i^s = 2e_{14}\epsilon_{jk}, \quad (i,\ j,\ k \text{ cyclic}) \tag{1}$$

where $\mathbf{P}^s$ is the induced polarization, e_{14} is the piezoelectric constant and ϵ_{jk} is a symmetrized strain component. Diagonal strains (ϵ_{ii}) do not induce a polarization in these materials (i.e., $e_{11} = 0$), however. A SLS grown along the [001] orientation will only induce diagonal strains . However, off-diagonal strain components will be generated for any other growth orientation. Consequently, [001]-oriented SLS will not generate internal strain-induced polarization fields, but SLS grown along any other orientation will generate internal polarization fields.

The strain-induced polarization $\mathbf{P}^s$ is determined by Eq.(1) from a knowledge of the lattice-mismatch induced strain components ϵ_{ij} in the materials forming the superlattice. The polarization vector $\mathbf{P}^s$ is constant in each material and changes abruptly (in magnitude and direction) at the superlattice interfaces. Because one of the superlattice material is under biaxial tension and the other material is under biaxial compression, the polarization vector $\mathbf{P}^s$ changes sign at the interface provided that the sign of the piezoelectric constant e_{14} is the same in the two materials forming the superlattice. For a [111] growth axis, the polarization $\mathbf{P}^s$ is oriented along the growth axis; for a [110] growth axis, the polarization $\mathbf{P}^s$ lies in the plane of the superlattice interface; for a [001] growth axis, the polarization $\mathbf{P}^s$ vanishes. For a general growth axis, the polarization $\mathbf{P}^s$ has components both along and normal to the growth axis (Smith and Mailhiot 1988). The strain-induced polarization $\mathbf{P}^s$ generates electric fields $\mathbf{E}$ and/or displacement fields $\mathbf{D}$. General expressions for strain-induced electric and displacement fields result from the application of standard electrostatic relations (Smith and Mailhiot 1988).

As an illustrative example, we consider a GaAs / $Ga_{0.80}In_{0.20}As$ SLS in which the GaAs and GaInAs layers have the same thickness. This SLS exhibits a 1.4 % lattice mismatch. In Fig. 1, we indicate the strain-induced electric field parallel to the SLS growth axis ($\mathbf{E}_\perp$) in the $Ga_{0.80}In_{0.20}As$ layers as a function of the growth axis. The growth axis is defined by its polar coordinates (θ, φ) and we show results as a function of the polar angle θ for a fixed value of azimuthal angle, $\varphi = \pi/4$. Inspection of Fig. 1 reveals that $\mathbf{E}_\perp = 0$ for a [100] growth axis ($\theta = 0$), has a broad maximum centered around the [111] growth axis and vanishes for a [011] orientation. The strain-induced electric fields $\mathbf{E}_\perp$ shown in Fig. 1 modify the superlattice energy band diagram and, consequently, alter the superlattice electronic structure and its optical properties.

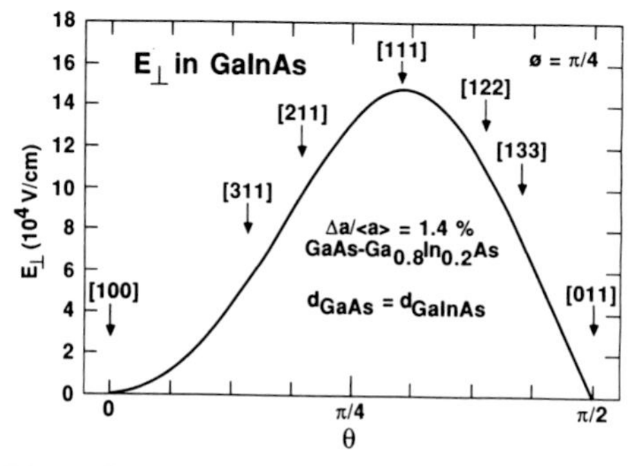

Fig. 1: Strain-induced electric field parallel to the superlattice growth axis ($\mathbf{E}_\perp$) in the $Ga_{0.80}In_{0.20}As$ layers of a GaAs / $Ga_{0.80}In_{0.20}As$ SLS as a function of the growth axis. Results are shown as a function of the polar angle θ for a fixed value of azimuthal angle, $\varphi = \pi/4$.

We emphasize that the strain-induced *electric fields* only have a component ($\mathbf{E}_\perp$) along the growth axis of the SLS. There is no component of the strain-induced electric field lying in the plane of the superlattice interfaces (i.e., $\mathbf{E}_{||} \equiv 0$) even if the growth axis is off the [111] orientation. Components of the polarization $\mathbf{P}^s$ lying in the plane of the superlattice interfaces ($\mathbf{P}^s_{||}$) do *not* generate electric fields. Conversely, strain-induced *displacement fields* only have a component ($\mathbf{D}_{||}$) lying in the plane of the superlattice interfaces. There is no component of the strain-induced displacement field along the growth axis of the SLS (i.e., $\mathbf{D}_\perp \equiv 0$).

3. EFFECTS OF STRAIN-INDUCED INTERNAL PIEZO-ELECTRIC FIELDS ON THE PROPERTIES OF SLS

In this section, we illustrate the effects of strain-generated internal electric fields on the electronic structure and optical properties of SLS. We consider a [111]-oriented $Ga_{0.47}In_{0.53}As$ /$Al_{0.70}In_{0.30}As$ SLS. This system is a type I SLS in which electrons and holes are confined within the $Ga_{0.47}In_{0.53}As$ quantum well layers. The confining $Ga_{0.47}In_{0.53}As$ layers are under biaxial compression with a relative lattice mismatch of 1.5 %. The theoretical method which serves as the basis for the calculation of the electronic structure and optical properties of the SLS has been presented elsewhere (Smith and Mailhiot 1986, Mailhiot and Smith 1986b).

3.1 Nonlinear Optical Properties

Superlattice electronic energy levels and wavefunctions are modified by the presence of internal strain-generated electric fields. The changes in the superlattice wavefunctions lead to changes of the optical matrix elements and to a screening of the strain-induced electric field by photogenerated electron-hole pairs. We now discuss the nonlinear optical response of a [111]-oriented SLS caused by the screening of the internal strain-induced electric fields by photogenerated carriers.

In Fig. 2(a), we show calculated energy levels of the conduction (c_1), heavy-hole (hh_1, hh_2, hh_3) and light-hole (lh_1) subbands at the center of the superlattice Brillouin zone as a function of superlattice layer thickness. The thickness (N_b) of the $Al_{0.70}In_{0.30}As$ layers is twice the thickness (M_a) of the $Ga_{0.47}In_{0.53}As$ confining layers. The strain-induced internal electric field has a magnitude of 1.41×10^5 V/cm in the $Ga_{0.47}In_{0.53}As$ layers and half this value in the $Al_{0.70}In_{0.30}As$ layers. Results of calculations both including and neglecting the internal strain-induced electric fields are shown in Fig. 2(a). In bulk $Ga_{0.47}In_{0.53}As$ under biaxial compression, the heavy-hole band is raised with respect to the light-hole band at the center of the bulk Brillouin zone.

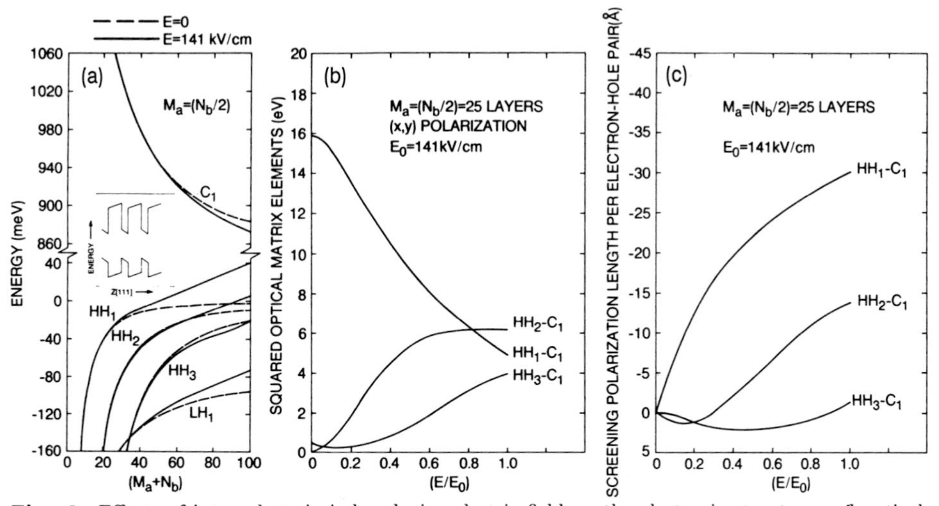

Fig. 2: Effects of internal strain-induced piezoelectric field on the electronic structure and optical properties of a [111]-oriented SLS. *(a)* Superlattice energy levels as a function of repeat distance calculated with (solid line) and without (dashed line) the internal strain-induced fields. *(b)* Squared optical matrix elements and *(c)* electron-hole dipole screening length as a function of the internal field magnitude for a [111]-oriented SLS consisting of $M_a = 25$ layers of $Ga_{0.47}In_{0.53}As$ alternating with $N_b = 50$ layers of $Al_{0.70}In_{0.30}As$.

Inspection of Fig. 2(a) reveals that Stark shifts are largest for heavy particles. The magnitude of the Stark shifts increases with layer thickness because the electrostatic potential drop across the superlattice layers increases linearly with layer thickness (for constant strain-induced electric field). Internal strain-induced electric fields shift the conduction band states to lower energy and the valence band states to higher energy, thus reducing the SLS band gap.

The internal strain-induced electric fields modify the superlattice wavefunctions and, therefore, alter the optical properties of [111]-oriented SLS. Calculated squared optical matrix elements for the first heavy-hole-to-conduction-band transitions (labelled $hh_1 - c_1$, $hh_2 - c_1$, $hh_3 - c_1$) are indicated in Fig. 2(b) as a function of the internal electric field in the $Ga_{0.47}In_{0.53}As$ quantum well layers. The SLS consists of $M_a = 25$ layers of $Ga_{0.47}In_{0.53}As$ alternating with $N_b = 50$ layers of $Al_{0.70}In_{0.30}As$. The polarization of the incoming light lies in the plane of the superlattice interfaces (x, y). The maximum field shown (E_0) is the unscreened value of the internal strain-induced electric field. The $hh_1 - c_1$ transition, which is strongly allowed at zero field, is suppressed by the internal field. The $hh_2 - c_1$ and $hh_3 - c_1$ transitions are very weak at zero field, but the internal field increases their strength so that the three transitions have comparable oscillator strengths at the unscreened value of the field $E = E_0$.

Electrons and holes are spatially displaced by the internal electric field. We now discuss the screening polarization fields originating from the field-induced spatial separation of free carriers along the [111] axis of the SLS. The screening polarization field $P_{sc}^{n,n'}(\rho, E)$ is a function of the free-carrier density ρ in the subbands n and n' and of the electric field E. Calculated screening polarization lengths of an electron-hole pair (for $hh_1 - c_1$, $hh_2 - c_1$, $hh_3 - c_1$) are indicated in Fig. 2(c) as a function of the internal field in the $Ga_{0.47}In_{0.53}As$ SLS layers. The negative sign in Fig. 2(c) indicates that the carrier-induced dipole moment opposes the strain-generated internal electric field. The value of the *screened* field E in the quantum wells is given by

$$E\left[1 - \frac{P_{sc}^{hh_1,c_1}(\rho, E)}{\kappa E}\right] = E_0, \qquad (2)$$

A density of $1.2 \times 10^{17} cm^{-3}$ electron-hole pairs (holes in the hh_1 state) is required to reduce the magnitude of the unscreened internal field by 10 %. This screening effect originates predominantly from the distortion of the heavy-hole wavefunction by the field.

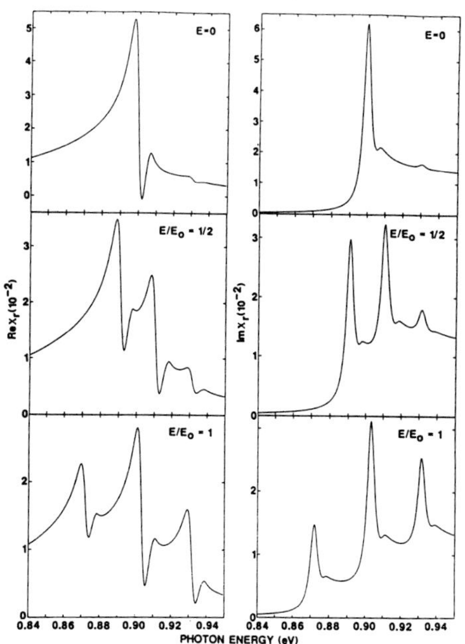

Fig. 3: Real (left) and imaginary (right) parts of the resonant susceptibility χ_r as a function of photon energy calculated at three values of the internal electric fields: the unscreened value ($E = E_0$) , half the unscreened value ($E = E_0/2$) and zero field ($E = 0$) for a [111]-oriented SLS consisting of $M_a = 25$ layers of $Ga_{0.47}In_{0.53}As$ alternating with $N_b = 50$ layers of $Al_{0.70}In_{0.30}As$.

Because the internal strain-induced electric fields change the superlattice energy levels and wavefunctions, they modify the optical properties of the superlattice. We now consider the ($M_a = 25$, $N_b = 50$) $Ga_{0.47}In_{0.53}As$ / $Al_{0.70}In_{0.30}As$ SLS discussed above and examine calculations of the resonant contribution to the susceptibility from near-band-edge transitions. Transitions from the highest three heavy-hole subbands (hh_1, hh_2, hh_3) and the lowest conduction subband (c_1) are included.

Figure 3 shows the calculated real and imaginary parts of the resonant susceptibility χ_r as a function of photon energy for three values of the internal electric fields: the unscreened field ($E = E_0$), half the unscreened field ($E = E_0/2$) and zero field ($E = 0$). At zero field, $hh_1 - c_1$ transitions dominate the spectrum and a weak $hh_3 - c_1$ exciton transition is superimposed on the $hh_1 - c_1$ continuum. As indicated in Fig. 3, the presence of internal strain-induced electric fields significantly changes the optical properties of [111]-oriented SLS. These internal electric fields can be externally modulated. An example of such modulation is the screening of the internal strain-induced fields due to photogenerated carriers produced by across band-gap optical absorption. The electron-hole pairs formed by this absorption process screen, and thus reduce, the magnitude of the internal fields. This screening effect is responsible for the nonlinear optical response of [111]-oriented SLS.

3.2 Electro-optical Properties

Materials whose optical properties can be modulated by the application of an external electric field are very valuable in the area of optoelectronics. Recently, there has been considerable interest in the electro-optic properties of lattice-matched superlattices because rather large second-order electro-optic coefficients occur in these materials (Miller et al 1984, 1985). Here we predict very large ***first-order*** electro-optic coefficients in [111]-oriented SLS (Mailhiot and Smith 1988a).

As discussed above, the internal strain-induced electric fields significantly change the electronic structure and optical properties of the superlattice. These changes are second-order in the magnitude of the electric field, as is usually the case with the Stark effect. We discuss below situations where the internal strain-induced electric fields are being modulated by the application of an external electric field. If the magnitude of the external field is small compared to that of the internal field, variations of the electronic and optical properties of [111]-oriented SLS due to the external field will be *linear* in the magnitude of the external field. Because the response of [111]-oriented SLS is linear in the magnitude of the external field, a blue shift (i.e. a shift towards higher energies) of the exciton transition energies will result if the direction of the external electric field is opposite to that of the internal strain-generated electric field in the quantum wells of the superlattice. Consequently, [111]-oriented SLS afford the possibility of exhibiting blue shifts as well as the normal red shifts (i.e. a shift towards lower energies) observed in quantum-confined Stark effect.

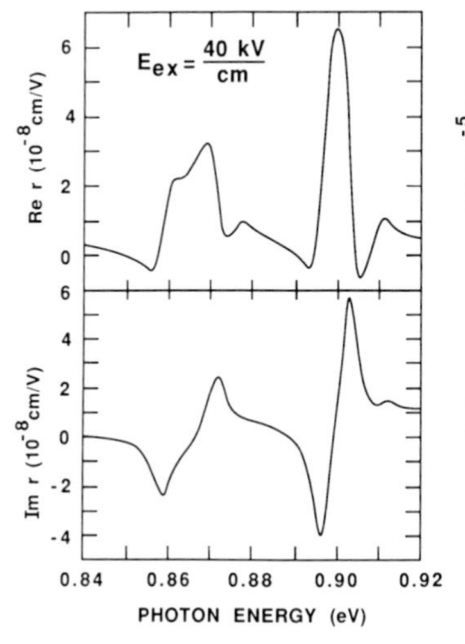

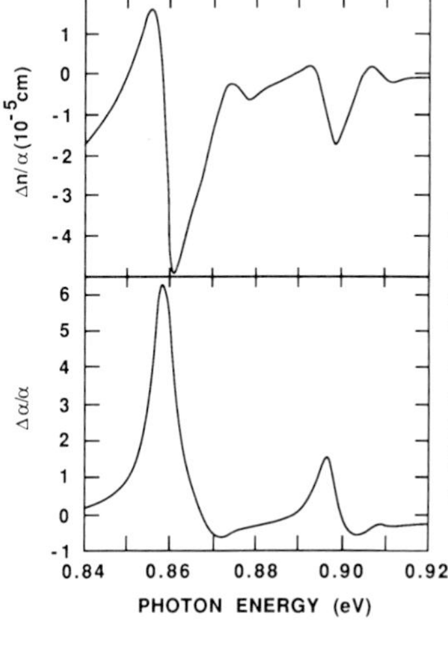

Fig. 4: Calculations of electro-optic coefficients as a function of photon energy for a [111]-oriented SLS consisting of $M_a = 25$ layers of $Ga_{0.47}In_{0.53}As$ alternating with $N_b = 50$ layers of $Al_{0.70}In_{0.30}As$.

Calculations of electro-optic coefficients for the $Ga_{0.47}In_{0.53}As$ / $Al_{0.70}In_{0.30}As$ SLS described above are indicated in Fig. 4. The coefficient r is defined as

$$n(E_{ex}) - n(E_{ex} = 0) \equiv -\left(\frac{1}{2}\right) |n(E_{ex} = 0)|^3 r E_{ex}, \tag{3}$$

where n is the index of refraction and E_{ex} is the externally applied electric field. For many electro-optic applications, the change in refractive index (Δn) or absorption coefficient ($\Delta\alpha$) divided by the zero-field absorption coefficient is an important figure of merit. Therefore, Fig. 4 also shows these quantities for an externally applied field of E_{ex} = 40 kV/cm as a function of photon energy. The electro-optic coefficients shown in Fig. 4 for [111]-oriented SLS can reach very large magnitude. As a comparison, let us mention that $r \approx 10^{-9}$ cm/V in potassium dihydrogen phosphate (KDP) (Kaminow et al 1971).

3.3 Piezo-optical Properties

The application of an external stress on a lattice-matched superlattice causes modifications of the electronic structure because of deformation potential effects (Mailhiot and Smith 1987b, Jagannath et al 1986). An additional effect arises in [111]-oriented SLS where the application of an external stress also causes a variation of the internal piezoelectric fields. As a result, the electronic structure of a [111]-oriented SLS is modified by the application of an external stress due to the combination of two effects: (1) deformation potential effects arise because of the variation of the internal strain and (2) Stark effects arise because the internal piezoelectric fields are modulated in accordance to the strain variations (Mailhiot and Smith 1988b).

Piezoelectric fields can be generated by an external strain (see Section 4) both for conventional [001]-oriented superlattices (lattice-matched or lattice-mismatched) and for lattice-matched superlattices where the piezoelectric fields vanish without the presence of external strain. These external-stress-induced electric fields produce modifications of the electronic structure of the superlattice which are quadratic in the magnitude of the field. For the magnitude of the strains typically considered, these modifications are rather small in the absence of built-in electric fields (i.e., with no internal strain). However, if large built-in electric fields are present (as in the case of [111]-oriented SLS), the modifications of the electronic structure caused by external-stress-induced electric fields are quite significant because they correspond to modulations of a quadratic effect about a large bias point. Theoretical analyses of this piezo-optical effect have been presented elsewhere (Mailhiot and Smith 1988b).

4. EXPERIMENTAL SITUATION

In this section, we briefly review recent measurements which have confirmed the presence of large internal strain-generated electric fields in [111]-oriented SLS. The first experimental observation of the presence of internal piezoelectric fields in [111]-oriented SLS was provided by a comparison of the optical absorption properties between GaAs / $Ga_{1-x}In_xAs$ SLS grown along the [001] and [111] orientation (Laurich et al 1989). Optical absorption and photoluminescence experiments on these SLS which differ only by their growth orientation revealed qualitative differences which were attributed, on the basis of theoretical analyses, to the presence of internal piezoelectric fields in the [111]-oriented SLS. More recently, the existence of internal strain-induced electric field in [111]-oriented SLS has been directly demonstrated by the observation of a blue shift in the photoconductivity spectrum of a GaAs / $Ga_{0.90}In_{0.10}As$ / GaAs $p-i-n$ diode under reverse bias operation (Caridi et al 1990, Goosen et al 1990). A schematic representation of the GaAs / $Ga_{0.90}In_{0.10}As$ / GaAs $p-i-n$ structure is shown in Fig. 5 along with measured photocurrent spectra. Under reverse bias operation, photoconductivity measurements reveal a *blue* shift of the $Ga_{0.90}In_{0.10}As$ quantum well electroabsorption peak. This observation is in contrast with the red shifts normally observed due to the quantum-confined Stark effect. Moreover, a theoretical analysis of the measured blue shifts indicate that the magnitude of the internal strain-induced piezoelectric field is given by Eq. (1).

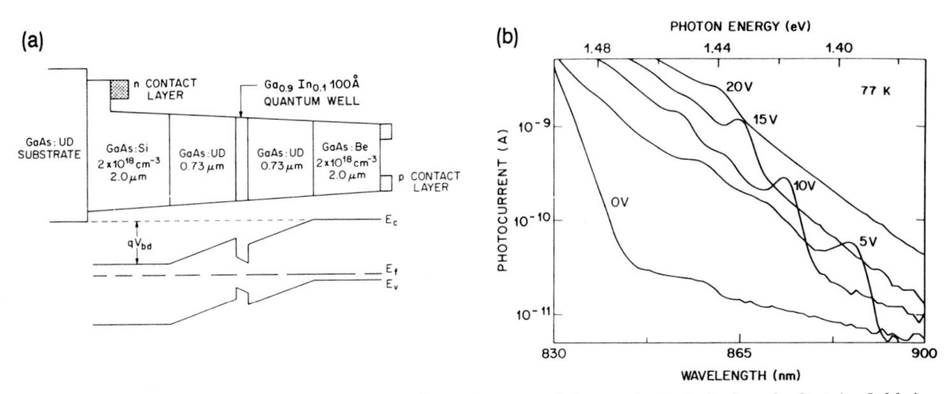

Fig. 5: Experimental demonstration of the existence of internal strain-induced electric field in [111]-oriented SLS by the observation of a blue shift in the photoconductivity spectrum of a GaAs / $Ga_{0.90}In_{0.10}As$ / GaAs $p-i-n$ diode under reverse bias operation. *(a)*: Schematic representation of the structure and energy band diagram showing an internal strain-induced electric field in the quantum well opposing the field in the diode. *(b)*: Measured photocurrent spectra under reverse bias operation for various reverse bias voltages. (Reproduced by permission from Caridi et al 1990).

The application of large compressive uniaxial stresses along the [100] and [110] direction have served as the basis for the determination of novel piezo-optical effects in [001]-oriented SLS (Qiang et al 1991). In these experiments, the application of large compressive uniaxial stresses along the [110] direction of a [001]-oriented GaAs / $Ga_{0.79}In_{0.21}As$ SLS produced a large red shift of photoreflectance peaks and a significant increase of several "symmetry-forbidden" heavy-hole-to-conduction optical transitions. These observations are absent when the compressive uniaxial stress is applied along the [100] axis. These results are interpreted in terms of the generation of internal piezoelectric fields along the [001] growth axis of the SLS when the compressive uniaxial stress is applied along the [110] direction where off-diagonal strains are induced. However, piezoelectric fields are absent when the compressive uniaxial stress is applied along the [100] axis.

5. CONCLUSION

Strained-layer superlattices grown from piezoelectric materials exhibit novel physical properties. The lattice-mismatch-induced internal strain present within the SLS layers generate large internal polarization fields by the piezoelectric effect. The orientation of these internal polarization fields depends on the components of the strain tensor and, consequently, on the growth orientation of the SLS. Piezoelectric polarization fields vanish for the usual case of SLS grown along the [001] axis. In contrast, the case of [111]-oriented SLS is particularly interesting because the internal piezoelectric polarization fields are parallel to the growth axis of the SLS and they generate internal electric fields whose magnitude typically exceed 10^5 V/cm for lattice mismatches of the order of 1 %. These large internal electric fields modify the electronic structure and optical properties of the SLS. Modulation of these internal strain-generated electric fields by photogenerated carriers, external electric field and external stress lead to large nonlinear optical, electro-optical and piezo-optical responses in [111]-oriented SLS. Experimental measurements have unambiguously verified the existence of internal piezoelectric fields in [111]-oriented SLS and demonstrated their modulation properties.

6. ACKNOWLEDGMENTS

This work was performed at Lawrence Livermore National Laboratory under the auspices of the U.S. Department of Energy under contract no. W-7405-ENG-48.

REFERENCES

Cady W F 1946 Piezoelectricity (New York : McGraw-Hill) p. 192
Caridi E A, Chang T Y, Goossen K W and Eastman L F 1990 Appl. Phys. Lett. 56 659
Goossen K W, Caridi E A, Chang T Y Stark J B, Miller D A B and Morgan R A 1990 Appl. Phys. Lett. 56 715
Jaganath C, Koteles E S, Lee J, Chen Y J, Elman B S and Chi J Y 1986 Phys. Rev. B 34, 7027
Kaminow I P and Turner E H 1971 Handbook of Lasers (Cleveland: Chemical Rubber) p. 452
Laurich B K, Elcess K, Fonstad C G, Beery J G, Mailhiot C and Smith D L 1989 Phys. Rev. Lett. 62 649
Mailhiot C and Smith D L 1986a J. Vac. Sci. Technol. B 4 996
Mailhiot C and Smith D L 1986b Phys. Rev. B 33 8360
Mailhiot C and Smith D L 1987a Phys. Rev. B 35 1242
Mailhiot C and Smith D L 1987b Phys. Rev. B 36 2942
Mailhiot C and Smith D L 1988a Phys. Rev. B 37 10 415
Mailhiot C and Smith D L 1988b Phys. Rev. B 38 5520
Miller D A B, Chemla D S, Damen T C, Gossard A C and Wiegmann W 1984 Phys. Rev. Lett. 53 2173
Miller D A B, Chemla D S, Damen T C, Gossard A C and Wiegmann W 1985 Phys. Rev. B 32 1043
Qiang H, Pollak F H, Mailhiot C, Pettit G D and Woodall J M, 1991 Phys. Rev. B xx, yyyy
Smith D L 1986 Solid State Commun. 57 919
Smith D L and Mailhiot C 1986 Phys. Rev. B 33 8345
Smith D L and Mailhiot C 1987 Phys. Rev. Lett. 58 1264
Smith D L and Mailhiot C 1988 J. Appl. Phys. 63 2717

Inst. Phys. Conf. Ser. No 123
Paper presented at the International Meeting on Optics of Excitons in Confined Systems, Giardini Naxos, Italy, 1991

Two-photon spectroscopy of CdS quantum dots

K. I. Kang, B. P. McGinnis, Sandalphon, Y. Z. Hu, S. W. Koch, N. Peyghambarian

Optical Sciences Center and Physics Department, University of Arizona, Tucson, AZ 85721

A. Mysyrowicz

Ecole polytechnique, ENSTA, Laboratoire d'Optique Appliquée, Palaiseau, France

L. C. Liu and S. H. Risbud

Mechanical, Aeronautical, and Material Engineering, University of California, Davis, CA 95616

ABSTRACT: Quantum-confined electron-hole transitions of CdS quantum dots are studied using one- and two-photon absorption spectroscopy. Nearly energetically degenerate one- and two-photon resonances are observed. The spectra are analyzed using Luttinger theory and numerical matrix diagonalization. The results indicate significant modifications of the quantum dot valence states as compared to the three-dimensional case.

1. INTRODUCTION

In this paper, we present experimental evidence for confinement-induced valence-band mixing by measuring and comparing one- and two-photon absorption spectra of CdS quantum dots in glass. The combination of one- and two-photon absorption spectra allow a direct comparison of electron-hole pair states with total angular momentum zero and one. We find that our experimental results can be explained by a model introduced by Xia (1989) and Vahala and Sercel (1990) that takes into account valence-band mixing caused by the spherical confining potential of small quantum dots.

2. EXPERIMENTAL RESULTS

We used samples containing CdS microcrystallites with a known size distribution (Liu and Risbud, 1990) in a glass matrix. The samples were mounted onto a cold finger inside a closed cycle helium cryostat that cooled the samples to approximately 10 K. The sample with the smallest microcrystallites had average radii of 0.9 ± 0.7 nm, while the sample with the largest microcrystallites had average radii of 8.0 ± 0.7 nm; confinement effects are considered important for microcrystallite radii comparable to or smaller than the bulk exciton Bohr radius, which is $\simeq 3$ nm for CdS.

Optical characterizations consisted of both one- and two-photon absorption measurements for each sample. One-photon spectra were obtained through direct transmission measurements or by excitation spectroscopy. Two-photon spectra were obtained through two-photon excitation spectroscopy where the photoluminescence (PL) intensity is detected as the excitation frequency is varied at constant input intensity excitation. Most of the PL from our samples was emitted in a broad band, peaking near 800 nm. These PL spectra were observed to be independent of the excitation photon energy, allowing us to monitor a constant band of PL while tuning the excitation source to obtain the two-photon absorption spectrum. The validity of this technique was checked by obtaining one-photon absorption spectra through PL excitation, comparing it to absorption spectra obtained from transmission measurements, and verifying that the two techniques give identical results.

Spectra were obtained with constant excitation pulse energies, discriminated to within 5% of the mean. The laser pulse temporal profiles were digitized for every tuning and stored for later analysis of the two-photon absorption spectra. We verified that the collected PL follows the expected quadratic dependence on pump energy for the two-photon spectra.

Examples of the one- and two-photon spectra for two samples with different average quantum dot radii between 1 and 2 nm are shown in Fig. 1(a) and (b). It is apparent that the one- and two-photon absorption peaks occur at the same energies. Similar spectra were obtained for other quantum dot samples with different average sizes.

3. DISCUSSION OF RESULTS

To analyze the data, we performed a theoretical analysis, which includes mixing of the heavy- and light-hole valence bands induced by the spherical confining potential. In the numerical calculation, we choose the eigenstates of the square of the total pair angular momentum and its z-component as the basis into which the Hamiltonian is expanded and diagonalized. The Coulomb energy is included, as discussed by Hu et al. (1990a and 1990b). The numerically computed one- and two-photon absorption spectra from this model using parameters appropriate for cubic CdS are shown in Fig. 1(c). The location of the vertical lines represents the energetic positions of the transitions and their height is their relative oscillator strengths. There are several predicted new resonances that satisfy the selection rules of total angular momentum, including quantum confinement and Coulomb interaction. The solid and dashed curves show the broadened one- and two-photon absorption spectra. The broadening originates from intrinsic effects as well as from the size distribution of the quantum dots in the sample. We see that the resonances merge and form an absorption curve that agrees well with the experimental data.

The near degeneracy of the one- and two-photon transition energies observed here results from the fact that the two lowest energy-hole states are roughly degenerate for the case of CdS quantum dots.

4. CONCLUSIONS

In conclusion, spectroscopic analysis of one- and two-photon absorption spectra of CdS quantum dots indicates that the quantized states of the holes develop from a mixture of the bulk light- and heavy-hole valence bands. Models that only describe uncoupled valence bands in the parabolic band approximation do not accurately describe our experimental spectra. A good fit to the experimental results is obtained when the mixing of the light-and heavy-hole valence bands due to the spherical confining potential is included in the Luttinger Hamiltonian calculations. Further improvements in the production of quantum dots should reveal the detailed spectrum of two-photon resonances predicted with this model.

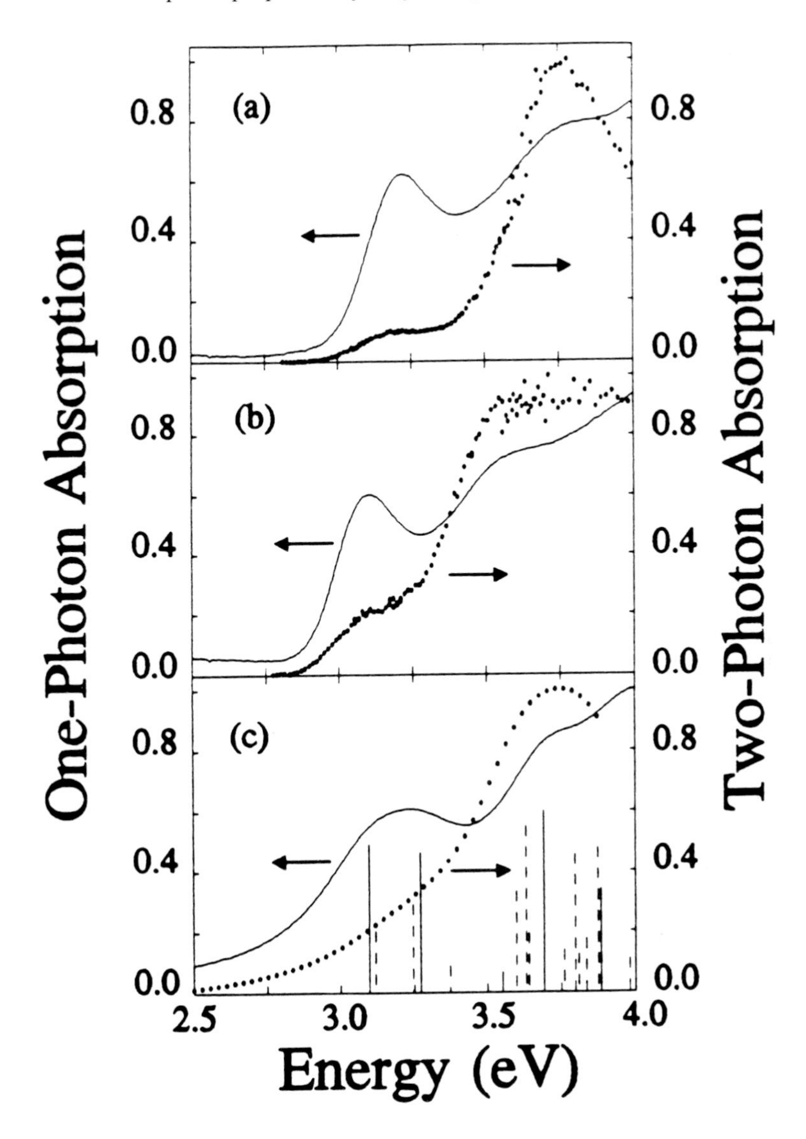

Fig. 1. (a) Experimental results of one-photon (solid line) and two-photon (dots) absorption spectra for a quantum dot sample heat treated at 640°C for 1 hour. No observable differences are seen in the transition energies between the one- and two-photon spectra. (b) *Similar* experimental results for a quantum dot sample heat treated at 640°C for 3 hours. (c) *Calculations* of one-photon (solid line) and two-photon (dots) absorption spectra based on the Luttinger Hamiltonian. The vertical lines are calculated with no broadening, while the dashed and solid curves represent spectra for a broadening of $\gamma = 8E_R$,where E_R is the bulk exciton binding energy of 27 meV. The magnitude of a vertical line represents the absorption strength. The other parameters used in the calculations are $R/a_B = 0.5$, with $\gamma_1 = 2.97$ and $\mu = 0.75$. The experimental data of Fig. 1(a) and (b) should be compared with Fig. 1(c).

5. ACKNOWLEDGEMENTS

The authors would like to acknowledge support from NSF, Joint Services Optical Program, AFOSR, ARO, SDI/ONR, and the Optical Circuitry Cooperative of the University of Arizona.

6. REFERENCES

Hu Y Z, Koch S W, Lindberg M, Peyghambarian N, Pollock E L, and Abraham F F 1990a, Phys. Rev. Lett. **64** 1805.
Hu Y Z, Lindberg M , and Koch S W 1990b, Phys. Rev. B **42** 1713.
Liu L C and Risbud S H 1990, J. Appl. Phys. **68** 28.
Vahala K J and Sercel P C 1990, Phys. Rev. Lett. **65** 239.
Xia J B 1989, Phys. Rev. B **40** 8500.

Inst. Phys. Conf. Ser. No 123
Paper presented at the International Meeting on Optics of Excitons in Confined Systems, Giardini Naxos, Italy, 1991

Electronic structure of ZnSe/ZnS superlattice studied by two-photon absorption spectroscopy

Fujio Minami, Kouji Yoshida, Jan Gregus and Kuon Inoue
Research Institute of Applied Electricity, Hokkaido University,
Sapporo 060, Japan

Hiroshi Fujiyasu
Department of Electronics, Faculty of Engineering, Shizuoka University,
Hamamatsu 432, Japan

ABSTRACT: We have studied the electronic structure of a strained-layer superlattice composed of 20 Å ZnSe/20 Å ZnS by using two-photon absorption spectroscopy. The exciton binding energy is found to be 41 meV, exhibiting a blue shift of 21 meV as compared to the bulk value due to the exciton confinement effect. Based on the observed subband energies, we derive a conduction-band offset of ~70 meV and a valence-band offset of ~785 meV.

1. Introduction

ZnSe-based structures are becoming more and more important as prospects for utilizing this compound in a variety of device applications. The room temperature direct bandgap of ZnSe at 2.7 eV makes this semiconductor especially attractive for optoelectronic device applications operating in the blue spectral region. In particular, ZnSe/ZnS superlattices are potentially useful for such devices as blue-emitting diodes and blue injection lasers. As compared to the case of GaAs/AlAs superlattices, however, less background knowledge is available for this material. In this paper, we apply two-photon absorption spectroscopy (TPA) to the study of electronic structure of a ZnSe/ZnS strained-layer superlattice (SLS). The TPA provides detailed information about the fundamental electronic structures of this superlattice, which cannot be obtained by means of linear spectroscopic techniques. We determine the energy of the $1S$-$2P$ splittings of heavy-hole (hh) and light-hole (lh) excitons, exciton binding energies, and conduction- and valence-band offsets in the ZnSe/ZnS superlattice.

2. Theoretical background

The TPA in bulk ZnSe has been shown to result predominantly from transition terms which are described in the framework of a two-band model (Mahr 1975). In this model, allowed two-photon transitions occur to final P-exciton states via discrete S-exciton states as intermediates. The lowest-energy final state is then the $2P$ exciton. In ZnSe/ZnS superlattices, one must distinguish between $\mathbf{E}\|\mathrm{z}$ and $\mathbf{E}\perp\mathrm{z}$ polarization configurations because of the different selection rules (Tai *et al* 1989). For $\mathbf{E}\perp\mathrm{z}$, the situation is not so different from the bulk case. The onset of TPA occurs at $2\hbar\omega$ equal to the $hh1$-$c1$ band edge. We use hereinafter the notation hhi-cj and lhi-cj for the optical transitions, where hh (lh) refers to the heavy (light) hole, and i (j) is the hole (electron) subband. Including the exciton effect, the lowest-energy state reached is the $2P$ state of the $hh1$-$c1$ excitons as in the bulk case. For $\mathbf{E}\|\mathbf{z}$, the TPA spectra differ greatly from those of the bulk crystal, reflecting subband quantization. The onset of two-photon transition occurs at the $lh2$-$c1$ or $lh1$-$c2$ transition, which is higher in energy than the $hh1$-$c1$ transition. (The $hh2$-$c1$ transition is forbidden because for $\mathbf{E}\|\mathbf{z}$ the interband transition is not allowed between heavy-hole and conduction bands.)

In ZnSe/ZnS SLS grown on a [001]-oriented substrate, the strain tensor ϵ_{ij}, corresponding to ZnSe or ZnS layers, are given by (Van de Walle 1989, Shahzad *et al* 1990) $\epsilon_{xx}=\epsilon_{yy}=(a_{\|}-a_0)/a_0$, $\epsilon_{zz}=-2\epsilon_{xx}C_{12}/C_{11}$, $\epsilon_{xy}=\epsilon_{yz}=\epsilon_{zx}=0$, where $a_{\|}$ is the in-plane lattice constant of the superlattice, a_0 is the lattice constant of bulk ZnSe or ZnS, and C_{ij} are the elastic constants. Due to the in-plane biaxial strain, the fourfold Γ_8 valence band splits into two doubly degenerate bands, i.e., hh ($J_z=\pm 3/2$) and lh ($J_z=\pm 1/2$) bands. As a function of the biaxial strain ϵ_{xx}, the energy separations between the conduction and valence bands are given by (Van de Walle 1989, Shahzad *etal* 1990)

$$\Delta E_{hh} = [2a(1-C_{12}/C_{11})-b(1+2C_{12}/C_{11})]\epsilon_{xx}, \qquad (1)$$

$$\Delta E_{lh} \simeq [2a(1-C_{12}/C_{11})+b(1+2C_{12}/C_{11})]\epsilon_{xx}, \qquad (2)$$

where a and b refer respectively to the hydrostatic and shear deformation potentials appropriate to tetragonal distortion. In addition to the strain-induced renormalization of the energy bands, electron and hole confinement also causes the transition energies to move higher energies. This confinement energies are calculated by using the Kronig-Penney model (Bastard 1981).

3. Experimental

The ZnSe/ZnS superlattice studied here was grown on a (001) GaAs substrate without buffer layers by hot wall epitaxy (HWE). The sample consisted of 1200 periods of 20 Å ZnSe and 20 Å ZnS layers. The incident laser beam was propagated normal to the epitaxial growth direction, $\mathbf{z}$, so that either the $\mathbf{E}\|\mathbf{z}$ or $\mathbf{E}\perp\mathbf{z}$ polarization configuration was possible (Minami *et al* 1991). The samples were directly immersed into liquid helium pumped to 2K. The excitation source was a home-made Ti:sapphire laser pumped by a cw Q-switched, intracavity frequency-doubled Nd:YAG laser operating at 6 kHz. The Ti:sapphire laser had a tuning range from 800 to 1000 nm, a pulse duration of 30 ns and an average output power of ~100 mW. The tuning accuracy of the Ti:sapphire laser was about 1 Å, corresponding to ~0.3 meV uncertainty in the energy determination of $2\hbar\omega$. The TPA signals were monitored via the luminescence from $1S$ free excitons (two-photon excitation spectroscopy). The luminescence was dispersed with a monochromator and detected by a gated optical-multichannel detector. The spectral resolution of the present measurement is mainly limited by the linewidth of the tunable Ti:sapphire laser.

4. Results and Discussions

The photoluminescence (PL) and one-photon PL excitation (PLE) spectra at 2K are shown in Fig. 1. Two peaks at 2.915 and 2.987 eV in the PLE spectrum correspond to $1S$ $hh1$-$c1$ and $lh1$-$c1$ free-exciton transitions. The free-exciton peaks exhibit a blue shift from the bulk value (~2.802 eV). The higher energy peak in PL, which is consistent in energy with the PLE onset, can be attributed to the $1S$ $hh1$-$c1$ free-exciton transition. The Stokes shift between the PL and PLE peaks may reflect inhomogeneities in layer thicknesses. The one-photon and two-photon PLE spectra were obtained by monitoring the PL intensity of this peak.

In Fig. 2, we show the TPA spectra obtained from our sample. For experimental points indicated by open circles, the polarization vector of the laser beam is chosen parallel to $\mathbf{z}$. The filled circles are obtained for $\mathbf{E}\perp\mathbf{z}$. In the $\mathbf{E}\perp\mathbf{z}$ configuration, the TPA process involves final

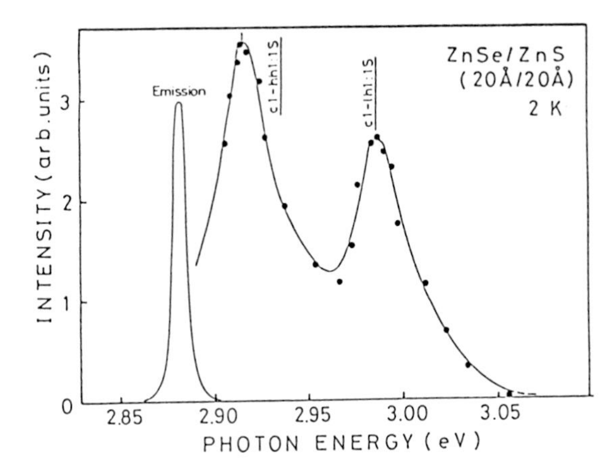

Fig. 1. Low-temperature photoluminescence (PL) and one-photon PL excitation spectra of a ZnSe/ZnS superlattice with 20 Å ZnSe and 20 Å ZnS layers.

$2P$ excited states for the $hh1$- and $lh1$-$c1$ excitonic transitions. It is noted that the one-photon-allowed $1S$ states of the $hh1$- and $lh1$-$c1$ excitons are absent in the two-photon spectrum. From the measured energy splitting between the $1S$ and $2P$ excitons, we can estimate the exciton binding energy. The binding energy is found to be 41 meV. There is a significant increase in the binding energy in our sample, as compared to that in bulk ZnSe (~20 meV). This increase will be caused by the exciton confinement effect. An appreciable difference in the binding-energy between the hh and lh excitons (Greene 1984, Matsuura and Shinozuka 1984) cannot be detected in the present sample.

For $\mathbf{E}\|\mathbf{z}$, the TPA enhancement is observed around $2\hbar\omega$=3.05 eV, which has a resemblance to the one-photon exciton absorption feature. In view of the polarization selection-rules and the small conduction-band offset between ZnSe and ZnS reported earlier (Shahzad *et al* 1990, Yamada *et al* 1991), this feature may be assign to the transition between the $lh1$ and the conduction-band continuum states. If we assume that the exciton binding energy of this complex is the same as that of the bulk ZnSe, the energy separation between the top of the well and the $c1$ level can be estimated as ~40 meV. Further, by using the square-well model in the

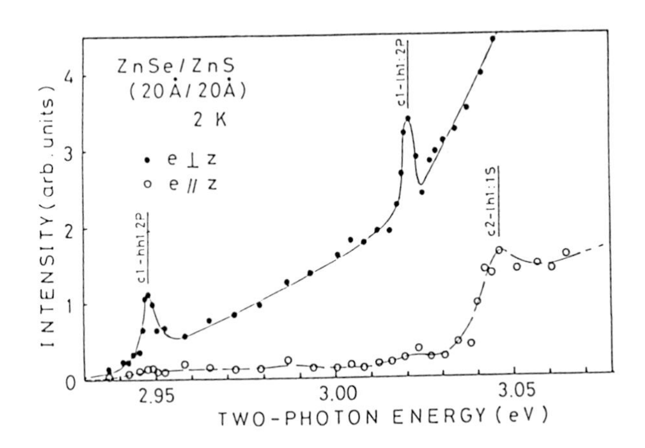

Fig. 2. Two-photon absorption spectra of the ZnSe/ZnS superlattice for the $\mathbf{E}\perp\mathbf{z}$ (filled circles) and $\mathbf{E}\|\mathbf{z}$ (open circles) configurations.

effective-mass approximation, we can calculate the conduction-band offset as ~70 meV. The TPA allowed *lh*2-*c*1 transition is not observed in the present energy region.

A careful comparison between the observed subband splittings and those from the Kronig-Penney calculations using $a_{||}$ as an adjusting parameter reveals the band lineups in the superlattice. The lineups we propose are illustrated in Fig. 3. The strain-induce splitting between the *hh* and *lh* becomes ~20 meV in ZnSe layers. It is thus found that the observed *hh*-*lh* splitting (~70 meV) is attributable mainly to quantum confinement effects. The discontinuities in the *hh* and *lh* bands at the interface are 910 and 660 meV, respectively. The discontinuity in the center of gravity of the *hh* and *lh* bands is 785 meV. The in-plane lattice constant $a_{||}$ is 5.62 Å and ϵ_{xx} is calculated to be 0.5%, which agrees approximately with the value of 1% estimated from Raman scattering. The lattice constant $a_{||}$ is larger than that (~5.51 Å) predicted in the case of a free standing superlattice (Van de Walle 1989, Shahzad *et al* 1990). This means that the $a_{||}$ is still affected by the substrate even in thick samples.

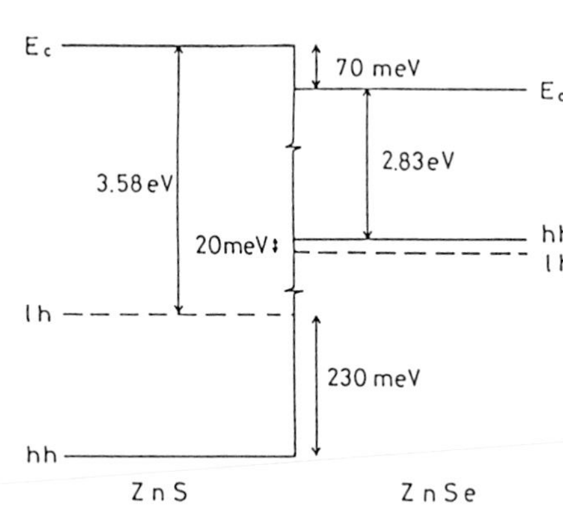

Fig. 3. Band lineups of the ZnSe/ZnS strained-layer superlattice. The heavy- and light-hole valence bands are represented by solid and dashed lines, respectively.

5. Concluding Remarks

We have shown that two-photon absorption spectroscopy can yield crucial information on electronic structures of ZnSe/ZnS strained-layer superlattices. This method has improved the accuracy in determination of exciton binding energy and band offsets in the strained-layer superlattices. It should be interesting to apply this method for other strained-layer superlattices, such as CdS/ZnS superlattices.

REFERENCES

Bastard G 1981 *Phys.Rev* **B24** 5693-5702
Greene R L 1984 *Phys.Rev.* **B29** 1807-1812
Mahr H 1975 *Quantum Electronics*, edited by H. Rabin and C.L. Tang, vol I (Academic Press, New York) p.285.
Matsuura M and Shinozuka Y 1984 *J.Phys.Soc.Japan* **53** 3138-3145
Minami F, Kato Y, Yoshida K, Inoue K and Era K 1991 *Appl.Phys.Lett.* **59** 712-714
Shahzad K, Olego D J, Van de Walle C G and Cammack D A 1990 *J.Luminescence* **46** 109-136
Tai K, Mysyrowicz A, Fischer R J, Slusher R E and Cho A Y 1989 *Phys.Rev.Lett.* **62** 1784-1787
Van de Walle C G 1989 *Phys.Rev.* **B39** 1871-1883
Yamada Y, Masumoto Y, Taguchi T and Takemura K 1991 *Phys.Rev.* **B44** 1801-1805

Inst. Phys. Conf. Ser. No 123
Paper presented at the International Meeting on Optics of Excitons in Confined Systems, Giardini Naxos, Italy, 1991

Two-photon-absorption spectroscopy of magnetically confined excitons in bulk GaAs

J S Michaelis [a], K Unterrainer [a], M Nithisoontorn [a], E Gornik [b], E Bauser [c]

[a] Institut für Experimentalphysik, Universität Innsbruck, A-6020 Innsbruck, Austria
[b] Walter-Schottky-Institut, TU München, D-8046 Garching, Germany
[c] Max-Plank-Institut für Festkörperforschung, D-7000 Stuttgart 80, Germany

ABSTRACT: We have studied the two-photon-absorption in bulk GaAs using a high sensitivity photoconductive technique. For the first time the 2P exciton could be observed without magnetic field. From the separation of the 1S and the 2P exciton we directly deduce the binding energy of the exciton. By applying a magnetic field the Zeeman splitting of the 2P-exciton could be observed even for low magnetic fields.

1. INTRODUCTION

Nonlinear absorption spectroscopy is a powerful tool for the investigation of optical and electronic properties of crystalline solids. A great number of two-photon absorption (TPA) experiments on bulk semiconductors like ZnSe /1/, ZnTe, CdTe /2/ and GaAs /2-4/ have shown that this method can provide additional information to one-photon absorption (OPA) measurements. Considering interband absorption mechanisms in zincblende semiconductors, the different selection rules due to the absorption of two photons result in optical transitions to P-exciton states which cannot be observed in OPA spectra, whereas transitions to S excitons are forbidden in TPA. Up to now the 2P exciton could not be resolved in the magnetic field-free case, because the binding energy of the 2P exciton in GaAs is very low (around 1 meV).

By applying a magnetic field, a strong enhancement of the P-exciton oscillator strength due to the compression of the exciton wave function in the presence of a magnetic field is observed /4/. Furthermore the 2P-exciton line splits due to the Zeeman effect. This splitting was first investigated by Neumann et al. /2,4/ in GaAs for intermediate magnetic fields B between 3 T and 6.5 T. For the low magnetic field range (B < 3 T), where $\gamma < 1$ ($\gamma = \hbar\omega_c / 2R_y$ with the cyclotron frequency $\hbar\omega_c$ and the effective Rydberg energy of the exciton R_y) there are no data available to describe the evolution of the exciton lines in a magnetic field.

Here we present experimental data of two-photon-magnetoabsorption (TPMA) in bulk GaAs with applied magnetic fields between 0 T and 7 T. The spectra show clear peaks representing transitions to both 1S and 2P excitons. The 2P magnetoexciton shows a Zeeman splitting into three lines which is observed down to 0.5 T.

2. EXPERIMENTAL RESULTS

The TPA experiments on bulk GaAs have been performed using the same setup as reported earlier by K. Unterrainer et al. /5/. The sample consisted of a 57 μm GaAs layer with a carrier mobility of 200.000 cm^2/Vs at 77 K (n =1.4x10^{13} cm^{-3}), grown by liquid phase epitaxy on a Si-doped GaAs substrate. Ohmic contacts were made by evaporating Au on the surface and alloying at 450°C for 4 min. The absorption processes at 2.2 K have been investigated by measuring the photoconductive response of the sample on intense near infrared light with a photon energy of $2\hbar\omega \approx E_g$, where E_g is the gap energy of GaAs (1.519 eV at 4.2 K). Magnetic fields between 0 T and 7 T in Faraday configuration have been applied.

In Fig. 1 the TPA spectra of the sample for different magnetic fields at T = 2.2 K are shown. The photoconductive signal is plotted versus the two-photon energy $2\hbar\omega$. At B = 0 T the spectrum exhibits a strong increase of absorption at 1.518 eV with a clear maximum at 1.5193 eV. Furthermore a strong excitonic contribution is observed at 1.5159 eV. To explain all the observed features it is necessary to consider the TPA parity selection rules in bulk GaAs. Taking into account a p-like Γ_8-valence band as the initial state, two-photon-allowed (Δl = 0, ±2) interband transitions lead to final p- or f-states. Therefore the lowest energy two-photon-allowed transition is that to the 2P exciton /6/. Following a hydrogenic exciton model with a reduced effective exciton mass $\mu = 0.058\ m_o$, where $\mu = m_e m_h/(m_e+m_h)$, m_e = 0.067 m_o, $m_h = 0.45\ m_o$, and a static dielectric constant ε = 12.4 for GaAs /7/, the 2P-exciton level is separated from the conduction band edge by only 1.2 meV. This small energy difference together with the rapidly decreasing oscillator strength of higher exciton states tends to smear out the 2P-exciton contribution and has prohibited the observation of the 2P exciton at zero magnetic field up to now. Therefore the peak at 1.5193 ± 0.0001 eV in the TPA spectrum in Fig.1 exhibits the first observation of the 2P exciton in bulk GaAs at zero magnetic field.

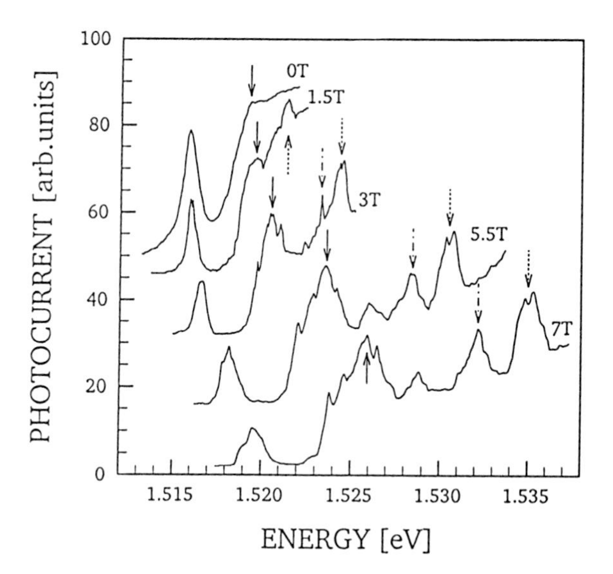

Fig. 1
Two-photon-magnetoabsorption spectra of the sample at magnetic fields between 0 T and 7 T in Faraday configuration. The different 2P-exciton transitions are indicated by arrows: solid for $2P^-$, dash-dotted for $2P^o$ and dotted for $2P^+$.

The very high quality of the sample ($\mu = 200.000\ cm^2/Vs$ at 77 K) combined with a high resolution photoconductivity technique allowed to resolve the excitonic features of the spectrum by using a data acquisition system and digitally averaging over a great number of TPA spectra. The identification of the 2P exciton is perfectly confirmed by our TPMA measurements, as reported later.

The additional peak at 1.5159 ± 0.0001 eV with $\Gamma_{FWHM} = 1.1$ meV is attributed to the ($\Delta l = 1$) 1S-exciton transition. With respect to the strong oscillator strength of this peak compared to the 2P exciton indicating a different absorption mechanism we explain the presence of transitions to S excitons in our TPA spectra by an additional nonlinear process due to the high intensity of the near infrared light source. Jha /8/ proposed that a second harmonic wave (SHW) can be generated inside a solid which is illuminated by an intense laser beam. For the case where the energy of the SHW is equal to or greater than the fundamental gap of the solid, it can be absorbed by the solid itself. Recently, the observation of 1S excitons in TPA spectra of $GaAs/Al_xGa_{1-x}As$ quantum wells was explained by this idea /9/. We adopt this idea for the case of bulk GaAs.

Extracting the separation energy between the 1S and 2P exciton from the data, one obtains $E(1S\text{-}2P) = 3.5 \pm 0.1$ meV. Calculating the binding energy E(1S) of the exciton ground state by $E(1S) = 4/3\ E(1S\text{-}2P)$, one obtains $E(1S) = 4.6 \pm 0.1$ meV. This is a relative high value compared to earlier reported data /2/, but is in good agreement with the theoretical value of 4.4 meV, calculated by Baldereschi and Lipari /10/.

The TPA measurements on bulk GaAs have been extended to study the influence of a magnetic field in Faraday configuration on the excitonic transitions. Following the structure of the 1S exciton from 0 T to 7 T, the width of the peak increases and at the same time the peak position shifts to higher energy. Next we want to discuss the evolution of the 2P exciton with increasing magnetic field. Taking into account a hydrogenic exciton model to describe the exciton in a magnetic field, one expects a splitting of the 2P exciton due to the linear Zeeman effect into three components, denoted by $2P^-$, $2P^o$ and $2P^+$, according to the different orientation of the angular momentum of the exciton with respect to the z direction, the direction of the magnetic field. As can be seen from Fig. 1, the zero field 2P exciton line at 1.5193 eV splits in fact into these three components. At the same time the oscillator strength of the P-exciton lines increases strongly with increasing field in comparison to the background contribution of the continuous absorption above the band gap. This effect is larger for the $2P^o$ exciton than for the $2P^{\pm}$ excitons. The splitting of the exciton line is resolved even for a magnetic field $B = 0.5$ T. At magnetic fields higher than 3 T an additional fine structure of the 2P-exciton lines is observable which will be discussed in detail elsewhere.

In Fig. 2 the observed transition energies of the 1S exciton and the 2P excitons have been plotted as a function of magnetic field. As can be seen from the plot the dependence of the transition energies on the magnetic field is strongly nonlinear. The experimental data points are fitted by second order polynomials. The difference between the experimental data points and the fitting curve is smaller than 2% which confirms the quadratic dependence of the transition energies. In addition the splitting is nonsymmetric with a shift of the $2P^o$ exciton towards the $2P^+$ exciton due to the envelope-hole coupling of the exciton which was also reported by Neumann et al. /2/.

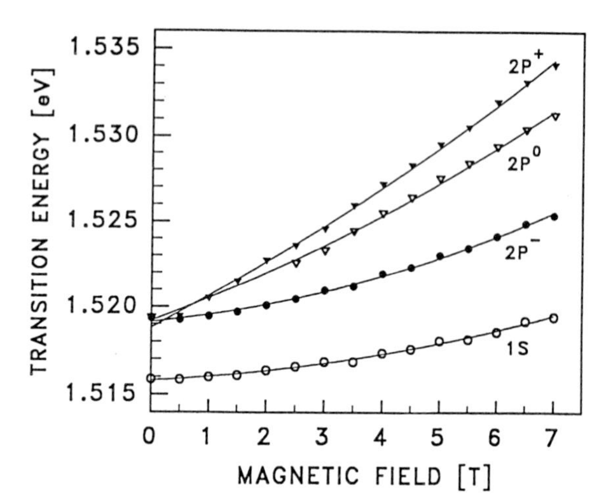

Fig. 2
1S- and 2P-exciton transition energies as a function of magnetic field.

3. CONCLUSION

In conclusion, we have presented experimental results of TPMA in bulk GaAs using the photoconductivity technique. It has been shown that both the forbidden 1S exciton and the allowed 2P exciton can be observed at zero magnetic field, thus allowing the direct determination of the exciton binding energy to be 4.6 meV. The data give an evidence for two different absorption mechanisms; one is the TPA and the other is the absorption of the SHW, which is generated inside the sample. We have shown that a magnetic field lifts the degeneracy of the 2P-exciton multiplet, giving raise to a Zeeman-splitting of the 2P-exciton line. The results show that two-photon-magnetoabsorption spectroscopy can yield crucial information on the exciton fine structure in bulk GaAs beyond the conventional experimental techniques.

This work was partially supported by the Bundesministerium für Wissenschaft und Forschung, Zl. 71.048/2-25/90.

REFERENCES

/1/ H.W. Hölscher, A. Nöthe, and Ch. Uihlein, Phys. Rev. B **31**, 2379 (1985)
/2/ Ch. Neumann, A. Nöthe, and N.O. Lipari, Phys. Rev B **37**, 922 (1988)
/3/ D.G. Seiler, C.L. Littler, and D. Heiman, J. Appl. Phys. **57**, 2191 (1985)
/4/ Ch. Neumann and A. Nöthe, Europhys. Lett. **4**, 351 (1987)
/5/ K. Unterrainer, M. Nithisoontorn, M. Helm, A. Köck, E. Gornik, and E. Bauser, Proc. "19th. Int. Conf. on the Physics of Semiconductors" , Ed. W. Zawadzki, Institute of Physics, Polish Academy of Science, p. 1331 (1989)
/6/ J.P. van der Ziel, Phys. Rev. B **15**, 2775 (1977)
/7/ J.S. Blakemore, J. Appl. Phys. **53**, R123 (1982)
/8/ S.S. Jha, Phys. Rev. **145**, 500 (1966)
/9/ M. Nithisoontorn, K. Unterrainer, S. Michaelis, N. Sawaki, E. Gornik, and H. Kano, Phys. Rev. Lett. **62**, 3078 (1989)
/10/ A. Baldereschi and N.O. Lipari, Phys. Rev. B **3**, 439 (1971)

Inst. Phys. Conf. Ser. No 123
Paper presented at the International Meeting on Optics of Excitons in Confined Systems, Giardini Naxos, Italy, 1991

Transient resonant Rayleigh scattering from excitons in semiconductor quantum well structures

H. Stolz, D. Schwarze, W. von der Osten and G. Weimann*

Fachbereich Physik, Universität–GH, D-4790 Paderborn, FRG
*Walter Schottky Institut, TU München, D-8046 Garching, FRG

ABSTRACT: By means of classical scattering theory we demonstrate that resonant Rayleigh scattering from excitons in GaAs/AlGaAs quantum wells after pulsed excitation decays with the coherence time. Using picosecond spectroscopy, thereby exploiting the selection rules, the coherence time and energy relaxation time across the inhomogeneously broadened ($n = 1$, $e - hh$) exciton band are determined.

1 INTRODUCTION

Electronic states in solids with spatial static disorder, quite analogous to a multi-atom two-level system (Loudon 1986), give rise to resonant Rayleigh scattering (RRS), i. e. an elastic contribution to the scattering spectrum which becomes resonantly enhanced for excitation in the electronic transition. As first demonstrated by Hegarty et al. (1982), this process occurs for excitons in quantum well (QW) structures that exhibit disorder due to compositional fluctuations at the interfaces and in the barriers. While these investigations were restricted to cw measurements, it was only recently that RRS could be shown to exhibit a finite decay time and to enable, as a linear optical method, the study of the exciton dynamical behaviour (Stolz et al 1991).

Applying scattering theory based on Maxwell's equations, we demonstrate that the temporal decay of RRS following short pulse excitation provides the coherence time of the exciton. Experimentally, by exploiting selection rules using polarised light (see Fig. 1), the coherent and incoherent (or hot luminescence like) contributions to the intensity can be discriminated. From the different time behaviour of these processes, all relaxation times of the exciton system can be determined and the portions of the exciton homogeneous linewidth due to inelastic and elastic scattering separated.

2 MECHANISM OF RESONANT RAYLEIGH SCATTERING

The elastically scattered intensity for short pulse excitation can be derived from a classical treatment of Rayleigh scattering. Using the previously discussed model to describe exciton motion in heterostructures (Stolz et al 1989), the QW is regarded to consist of small regions in which free motion of the two-dimensional exciton (Bohr radius a_B^{2d}) is possible. Their lateral extension, determined by the coherence length of the electron and hole Bloch states ($\xi_c > a_B^{2d}$), is finite because of disorder-induced scattering. Each region corresponds to an effective wellwidth and, hence, effective potential for the electronic states causing local variations in exciton resonance frequency $\omega_0(\vec{r})$. These result in the inhomogeneous broadening of the optical transition and, because of the strong dispersion near resonance, in spatial fluctuations in (frequency dependent) susceptibility $\chi_S(\vec{r},\omega)$ across the sample.

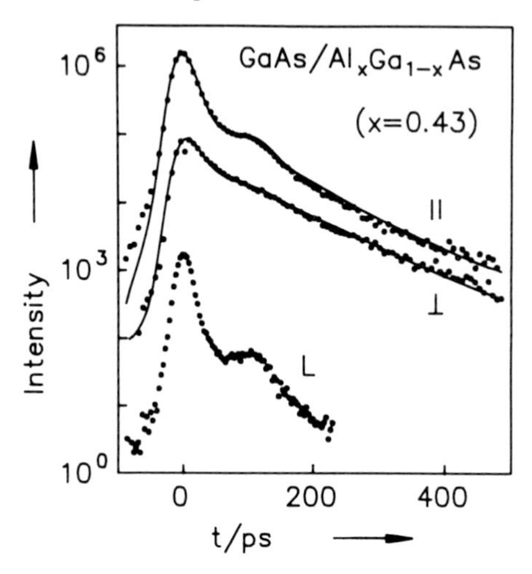

Fig. 1. Transient resonant Rayleigh scattering in a GaAs/Al$_x$Ga$_{1-x}$As ($x = 0.43$) quantum well at 2 K. The well and barrier widths are $L_z = 9$ nm and $L_b = 105$ nm, respectively. Excitation in the exciton absorption peak with light polarised along [110] ($I_{\|}$) and [$\bar{1}$10] ($I_{\perp}$). The full lines are fits to the experimental data (points); for a clearer representation the curves are shifted along the ordinate. The slow decay at longer times is due to thermal reactivation of localised into free exciton states. L: system response to the laser pulse far off resonance.

Being the source for the scattered field, the time dependent dielectric polarisation may be written as time integral

$$\vec{P}_S(\vec{r},t) \propto \int_{-\infty}^{\infty} dt'\, \chi_S(\vec{r},t')\, \vec{E}_0(\vec{r},t-t')\,. \tag{1}$$

Here $\chi_S(\vec{r},t)$ represents the Fourier–transform of $\chi_S(\vec{r},\omega)$ (background susceptibility neglected, $\xi_c < \lambda$) and $E_0(\vec{r},t)$ the (laser) light field incident at frequency ω_L. Considering a plane wave with amplitude $A_0(t)$, the scattered electric field can be calculated by Maxwell's equations. In far-field (i. e. scattering volume small compared to the distance of the observation point) it results in a wave at frequency $\omega_S = \omega_L$ with the scattering amplitude given by the component of the spatial Fourier transform of the susceptibility at the scattering wavevector as usual. Assuming $\chi_S(\vec{r},\omega)$ to have Lorentzian structure with a homogeneous linewidth $\delta E_{hom} = 2\hbar/\tau_{coh}$

$$\chi_S(\vec{r},\omega) = \frac{\omega_p}{\omega_0(\vec{r}) - \omega - i(1/\tau_{coh})} \tag{2}$$

(ω_p: strength of dielectric response, τ_{coh}: coherence time), then by renouncing the spectral information it is straightforward to derive the scattered intensity detected at distance R. For excitation within the inhomogeneously broadened exciton transition one obtains

$$I(\vec{R},t) \propto \int_0^{\infty} dt'\, |A_0(t-t'-\frac{R}{c})|^2\, e^{-2t'/\tau_{coh}} \tag{3}$$

which for δ-pulse excitation results in intensity decaying with time constant $2/\tau_{coh}$. In order to derive eq. (3) and to perform the necessary integration over the space coordinate, the susceptibility correlation function $\langle \chi_S^*(\vec{r},\omega)\, \chi_S(\vec{r},\omega)\rangle$ is taken of Gaussian form and unequal zero only within range ξ_c of the scattering regions. As in case of a multi-atom system where the light waves from the individual atoms exhibit a random distribution of relative phases, in the QW the total intensity is the incoherent sum of intensities scattered from the small regions described above.

3 EXPERIMENTAL RESULTS AND ANALYSIS

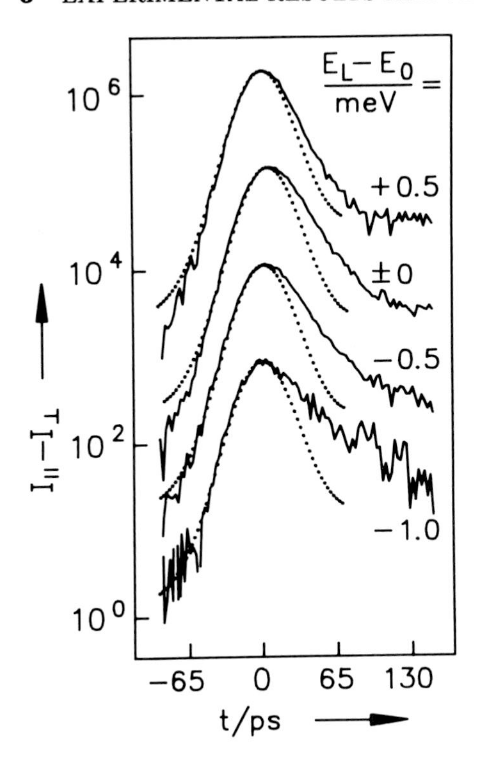

Fig. 2. Time decay of resonance Rayleigh scattering at 2K. The normalised intensity difference for the two polarisations is plotted vs. time. Excitation with energy E_L around the exciton resonance at E_0. Dotted curve: system response to the laser pulse off resonance.

Experiments with picosecond time resolution at the lowest ($n = 1, e - hh$) exciton in various $GaAs/Al_xGa_{1-x}As$ quantum wells (Stolz et al 1991) clearly reveal the finite delay of the elastic process (see Fig. 1 and Fig. 2 for an example). The criterion to distinguish between coherent Rayleigh scattering (RS) and incoherent hot luminescence (HL) is based on the degeneracy and polarisation behaviour of the exciton. In accordance with the D_{2d} symmetry of the QW, the exciton (symmetry Γ_5) is doubly degenerate having transition moments along [110] and [$\bar{1}$10]. Excitation with light polarised along one of these directions (for details of scattering geometry see Stolz et al (1989)) prepares the exciton in a state with well-defined polarisation and phase. By elastic scattering (characterised by rate $1/T_2'$) subsequently it may decay into the degenerate state having different polarisation so that the polarisation of the emitted light loses correlation with that of the incident light. This implies $I_\perp$ to be HL, while the intensity $I_\parallel - I_\perp$ remaining in parallel polarisation is due to coherent RS. Introducing the energy relaxation time T_1 to describe the inelastic processes, the analysis by means of rate equations for population (Schwarze 1991) (like a more elaborate quantum mechanical theory that takes into account the excitation and the detection process (Stolz 1991)) finally gives for the corresponding time dependent intensities

$$I^{RS} = I_\parallel - I_\perp \sim e^{-2t/\tau_{coh}} \qquad (4)$$

$$I^{HL} = I_\perp \sim e^{-t/T_1} - e^{-2t/\tau_{coh}}. \qquad (5)$$

As usual, the relaxation times are related to each other and to the homogeneous linewidth by

$$\delta E_{hom} = 2\hbar/\tau_{coh} = 2\hbar(1/2T_1 + 1/T_2'). \qquad (6)$$

Acc. to eq. (4), the decay of the intensity difference plotted in Fig. 2 directly provides τ_{coh}.

The intensity $I_\perp$ measured in perpendicular polarisation (see e. g. Fig. 1) then serves to separately obtain the energy relaxation time T_1 and, through eq. (6), the pure dephasing time T_2'. Calculated with a corresponding set of relaxation times determined in that way, the full lines in Fig. 1 represent fits to the experimental data (after convolution with the measured system response to the laser pulse) showing very good agreement.

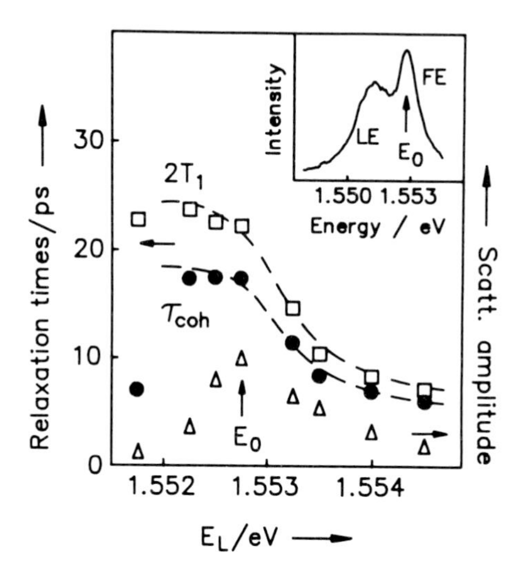

Fig. 3. Coherence time and energy relaxation time in the exciton region derived from the experimental data. The dashed lines are guidelines for the eye. Also shown is the scattering amplitude (Δ). Inset: free and localised exciton spectrum (FE, LE) at 2K.

As already obvious from Fig. 2, τ_{coh} depends on excitation photon energy. Fig. 3 represents data together with the corresponding energy-relaxation times extracted from a systematic study of polarised RRS across the inhomogeneously broadened exciton transition. Due to the increasing number of relaxation channels, τ_{coh} and T_1 decrease at higher energies. The values found for τ_{coh} (between 20 and 6 ps which is presently the detection limit of our set-up) are in good agreement with the results obtained by nonlinear techniques (for references see Stolz et al (1991)). In particular they are consistent with our assumption of non-coherence of the spatially separated exciton states (section 2), as the dephasing connected with the inhomogeneous broadening would result in much faster decay time.

Comparison of the data in Fig. 3 with eq. (6) suggests that the exciton coherence is predominantly destroyed by inelastic processes. The linear dependence of δE_{hom} on temperature that we find in our measurements suggests acoustic phonon scattering to be the dominant mechanism. Pure dephasing contributes with about 25% to the homogeneous linewidth nearly independent of exciton energy. On the low energy side of the free exciton (FE), localised exciton states (LE) lead to a more complex time behaviour not discussed here.

ACKNOWLEDGEMENT: The authors are grateful to the Deutsche Forschungsgemeinschaft for support of the project.

REFERENCES

Hegarty J, Sturge M D, Weisbuch C, Gossard A C, and Wiegmann W 1982 Phys. Rev. Lett. 49 930

Loudon R 1986 The Quantum Theory of Light, 2nd edition (Oxford: Clarendon Press)

Schwarze D. 1991 Thesis Universität Paderborn (to be published)

Stolz H, Schwarze D, von der Osten W and Weimann G 1989 Superlattices and Microstructures 6 271

Stolz H, Schwarze D, von der Osten W and Weimann G 1991 Superlattices and Microstructures 9 511

Stolz H 1991 Festkörperprobleme/Advances in Solid State Physics XXXI, ed. U Rößler (Braunschweig: Pergamon, Vieweg)

Inst. Phys. Conf. Ser. No 123
Paper presented at the International Meeting on Optics of Excitons in Confined Systems, Giardini Naxos, Italy, 1991

Many-particle effects on four-wave-mixing signals in quantum-well systems

F. Jahnke and W. Schäfer*

Institut für Physik, Hochschule Güstrow, O-2600 Gustrow, FRG
*Forschungszentrum Jülich, HLRZ, W-5170 Jülich, FRG

ABSTRACT: Recent experimental and theoretical investigations of Four-Wave-Mixing (FWM) signals in GaAs/AlGaAs MQW's have shown that earlier concepts based on isolated two-level systems failed to describe novel features, resulting from strong polarization interactions. Using the semiconductor density-matrix equations, which treat the Coulomb-interaction within the unrestricted Hartree-Fock approximation, we investigate time-resolved and time-integrated FWM-signals in MQW's. We show that a photon-echo, which is expected for inhomogeneously broadened two-level systems, can not be obtained as consequence of many-body effects.

1. INTRODUCTION

Transient nonlinear laser spectroscopy methods provide a powerful tool to study fundamental interaction processes in solids. Especially transient four-wave-mixing (FWM) spectroscopy has been used recently to investigate the loss of coherence of optically excited electron-hole-pairs in semiconductor MQW's (see e.g. Schultheis et al 1986). Usually FWM-signals were interpreted in terms of optical Bloch-equations for two-level systems which are treated perturbationally up to the third order in the fields (Yajima et al 1979). Then the *time-integrated* FWM-signal decays exponentially for increasing delay between the applied pulses and gives a direct measure for the polarization dephasing time (T_2). Only recently shortcomings of this usually adopted picture of FWM where demonstrated (Leo et al 1990 and Stafford et al 1990). Novel features result from the strong polarization interactions in semiconductors due to many-body Coulomb interaction. For low excitation-intensities the time-integrated FWM-signal was measured at time delays for which the two-level theory predicts no signal at all. At higher field intensities a measured nonexponential decay of the signal indicates the braekdown of perturbative treatment. It was shown by Wegener et al (1990) that this behaviour can be described by the semiconductor Bloch-equations, which treat the Coulomb interaction within the unrestricted Hartree-Fock approximation (see Binder et al 1991).

In addition to time-integrated FWM-signals *time-resolved* signals yield more detailed informations about the influence of many-body effects. The two-level theory predicts a free polarization decay for homogeneously broadened systems and a photon echo in the case of inhomogeneous broadening. Strong polarization interaction, however, causes a signal which is smeared out by more than one order of magnitude in comparison with the photon echo already at low excitation-intensities, for which a perturbative treatment (χ_3) is valid. At high intensities the FWM-signal contains two components, a prompt decay similar to the usual photon echo and a long wing. We show that the duration of the latter contains information about the strength of polarization interactions.

2. SEMICONDUCTOR BLOCH-EQUATIONS

The optical response of a laser puls that drives a coherent interband polarization $\Psi_{\mathbf{k}}$ and carrier occupation $f_{\mathbf{k}}$ is gouverned by the semiconductor Bloch-equations

$$i\hbar\frac{\partial}{\partial t}\Psi_{\mathbf{k}} = \left[\varepsilon_{\mathbf{k}} - 2\sum_{\mathbf{k'}} V_{\mathbf{k}-\mathbf{k'}}\, f_{\mathbf{k'}} - \frac{i\hbar}{T_2}\right]\Psi_{\mathbf{k}} - \left(1 - 2f_{\mathbf{k}}\right)\left[\mu E + \sum_{\mathbf{k'}} V_{\mathbf{k}-\mathbf{k'}}\,\Psi_{\mathbf{k'}}\right]$$

$$\frac{\partial}{\partial t} f_{\mathbf{k}} = -\frac{f_{\mathbf{k}} - f_{\mathbf{k}}^0}{T_1} + \frac{2}{\hbar}\,\mathrm{Im}\left\{\left[\mu E + \sum_{\mathbf{k'}} V_{\mathbf{k}-\mathbf{k'}}\,\Psi_{\mathbf{k'}}\right]\Psi_{\mathbf{k}}^{*}\right\}.$$

They extend the Bloch equations of independent ensembles of two-level systems by the socalled exchange terms which couples states with different wave vectors **k** by the Coulomb interaction $V_{\mathbf{k}-\mathbf{k'}}$. This leads to a renormalization of the one-particle energies $\varepsilon_{\mathbf{k}}$ and the Rabi-energy μE. The interaction processes are usually assumed as instantaneous, thus leading to a Lorentzian absorption lineshape and an exponential decay of the polarization for an independent two-level system. However, in semiconductors the driving terms are governed not only by the the external field E, but also by an internal one $\propto \sum V_{\mathbf{k}-\mathbf{k'}}\,\Psi_{\mathbf{k'}}$, which is determined by the interaction processes in the system. This internal field is responsible for an interaction induced signal which may be enforced also after the external pulse has passed. Thus the shape of the signal is strongly influenced by the interaction processes in the system which leads to a nonexponential behaviour, despite of a constant T_2 approach.

In the following we will consider the simplest case of two-pulse self-diffraction. The laser pulses propagating in the directions $\mathbf{k}_1$ and $\mathbf{k}_2$ interfere in the sample and generate a transient grating, as long as their time delay is of the order of the polarization dephasing time. The lowest order of a self-diffracted signal occures for the direction $2\mathbf{k}_2 - \mathbf{k}_1$. The evaluation of semiconductor Bloch-equations for this geometry was discussed recently by Binder et al (1991) and will be applied for the calculation of time-resolved FWM-signals in a MQW system of 100 Å well thickness.

3. RESULTS AND DISCUSSION

We will investigate the case of nonresonant excitation (2ryd below the excitonic resonance) with two 100fs pulses and a dephasing time of 500fs. On this time-scale scattering is particle number conserving. Further, for excitation energies below or in the vicinity of the excitonic resonance the generated one-particle distributions differ not much from quasiequilibrium distributions and will be redistributed only slightly. Hence, longitudinal relaxation processes can be neglected for the carrier distribution and for the population modulation resulting from the interference of both signals.

Time-resolved FWM-signals for low excitation intensity and different time delay between the two pulses are shown in Fig. 1a. In contrast to the two-level theory, which predicts for positive delay (probe precedes the pump puls) a photon echo at the twiced delay time decaying with the inverse inhomogeneous linewidth of the system, we obtain a strongly broadened signal. As the already discussed contribution of the internal field is limited by the dephasing time and not by the pulse duration, the total source contribution in the semiconductor Bloch-equations becomes broadened in time thus leading to the broadened diffracted signal.

At higher intensity (Fig. 1b), for which perturbation theory fails, the time-resolved signal shows for positive and zero delay two components. The first signal appears in the vicinity of $t = 0$ and decays nearly with the exciting pulse. The second delayed signal is much broader and decays within some ps. Note, that the position of this second signal depends only weakly on the delay between the two pulses. The magnitude of the delayed signal, however, depends definitely on the strength of the exciton-exciton interaction, which is given in the kinetic equation for the diffracted signal by:

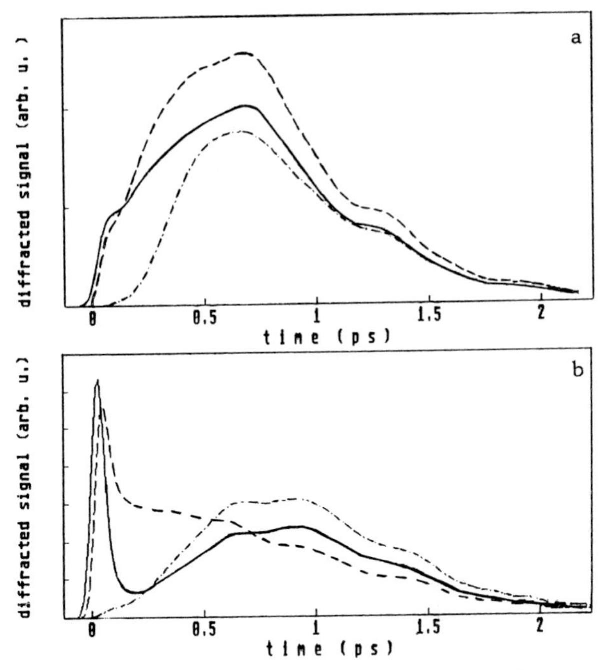

Fig. 1 a,b. Time resolved FWM-signal for different time-delay $\Delta = 0.1$ ps (full line), $\Delta = 0$ (dashed line), $\Delta = -0.1$ ps (dashed dotted line) and maximum puls energy $\Omega_R^0 = 0.3$ ryd (a), $\Omega_R^0 = 1.5$ ryd (b).

$$I_{ex-ex} = \delta n_{\mathbf{k}} \sum_{\mathbf{k'}} V_{\mathbf{k}-\mathbf{k'}} \Psi^*_{\mathbf{k'}} - \Psi^*_{\mathbf{k}} \sum_{\mathbf{k'}} V_{\mathbf{k}-\mathbf{k'}} \delta n_{\mathbf{k'}}$$

where $\delta n_{\mathbf{k}}$ is the population modulation. For comparison we have reduced in Fig. 2 the coupling strength of the Coulomb interaction in I_{ex-ex}. The remaining first signal corresponds to the usual photon-echo of inhomogeneously broadened two-level systems, which is however enhanced and smeared out due to the many-body Coulomb interaction. In the case of completely vanishing Coulomb interaction (Fig.3) the signal is more than five orders of magnitude smaller because the uncorrelated density of states is drastically reduced and the Coulomb enhancement of the Rabi-energy is absent. Furthermore, a symmetrical echo-signal will be obtained for a statistical ensemble average only. For nonresonant short pulse excitation, the average over the excited one

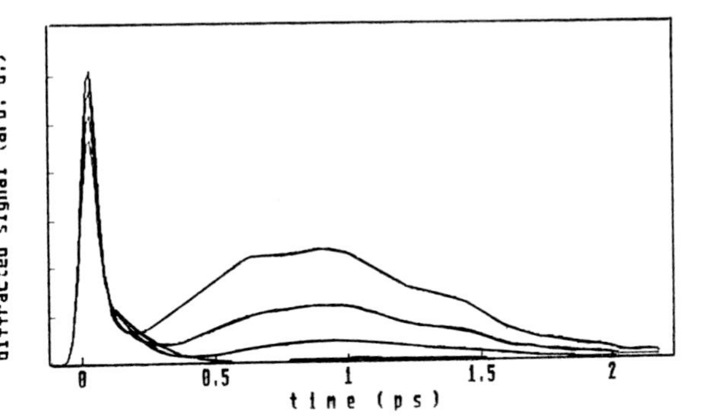

Fig. 2. Time-resolved FWM-signal for $\Omega_R^0 = 1.5$ ryd, delay $\Delta = 0.1$ ps and coupling strength of exciton-exciton interaction varying between 100%, 75%, 50% and 25% (top to bottom).

particle energies, however, does not correspond to a statistical average, due to missing localization of the weight function. Thus one obtains for increasing detuning below the resonances already without - Coulomb interaction a modified signal with a prompt decay near t=0. With Coulomb interaction this signal will be enhanced and broadened, and mixed with a second signal arising from exciton-exciton interaction. This result sheds new light on the interpretation of recent measurements of time-resolved stimulated FWM-signals in GaAs/AlGaAs MQW's by Webb et al (1991), where a prompt and a delayed signal were observed.

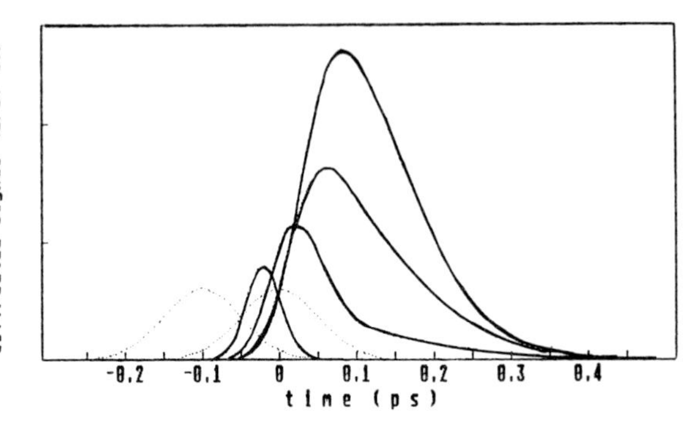

Fig. 3. Probe and pump pulses (dotted lines) and time-resolved FWM-signal for vanishing Coulomb-interaction with $\Omega_R^0=0.3$ ryd, $\Delta=0.1$ps and various detuning $\omega_o=0,-1.5,-3.5,-10$ ryd (top to bottom). The normalization constants in units of 10^5 are 1,2,10,50.

As a consequence of many-body effects in time-resolved FWM-signals, the time-integrated signal shows nonexponential decay at higher intensities (Fig. 4) and a temporal hole in the vicinity of zero delay, which was found earlier by Leo et al (1990). The latter results from the different saturation behaviour of the prompt and delayed signal.

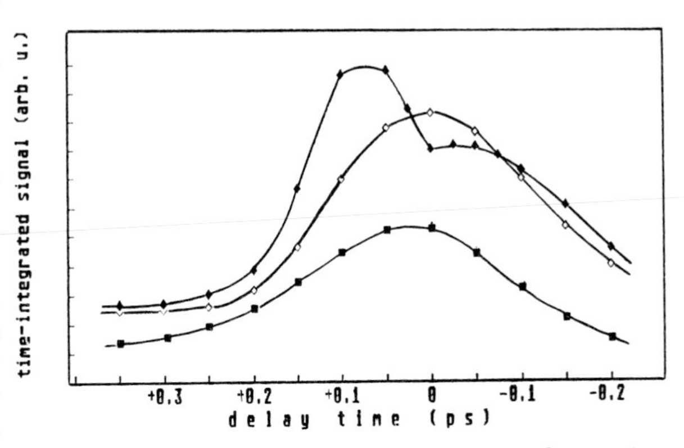

Fig. 4. Time-integrated FWM-signal for Rabi-energies $\Omega_R=1.5$, 0.9, 0.3 ryd (top to bottom).

Concluding, we have shown that in semiconductor MQW's FWM-signals will be considerably modified in comparison to inhomogeneously broadened two-level systems due to Coulomb-enhancement of the Rabi-energy and anharmonic exciton-exciton interaction. Especially at high excitation intensities these processes are responsible for different contributions to the time-resolved signal and nonmonotoneous behaviour of the time-integrated signal.

REFERENCES:

Binder R, Koch S W, Lindberg M, Schäfer W and Jahnke F 1991 Phys. Rev. B43 6520

Leo K, Wegener M, Shah J, Chemla D S, Göbel E O, Damen T C, Schmitt-Rink S, and Schäfer W 1990 Phys. Rev. Lett. 65 1340

Schultheis L, Honold A, Kuhl J, Köhler K, and Tu C W 1986 Phys Rev B34 9027

Stafford C, Schmitt-Rink S and Schäfer W 1990 Phys. Rev. B41 10000

Webb M D, Cundiff S T and Steel D G 1991 Phys. Rev. Lett. 66 934

Wegener M, Chemla D S, Schmitt-Rink S and Schäfer W 1991 Phys. Rev. A42 5675

Yajima T and Taira Y 1979 J. Phys. Soc. Japan. 47 1620

Inst. Phys. Conf. Ser. No 123
Paper presented at the International Meeting on Optics of Excitons in Confined Systems, Giardini Naxos, Italy, 1991

Time-resolved and time-integrated four-wave-mixing signals of two-dimensional magneto-excitons

M. Hartmann, W. Schäfer, S. Schmitt-Rink [×]

Forschungszentrum Jülich, HLRZ, D-5170 Jülich, FRG
[×]Fachbereich Physik, Philipps-Universität Marburg, D-3550 Marburg, FRG.

ABSTRACT: Two-dimensional excitons in a perpendicular magnetic field represent a quantum-confined system with properties varying from two-dimensional at low magnetic fields to quasi-zero-dimensional at high magnetic fields. Their nonlinear optical properties depend sensitively on the strength of the magnetic field. We present results of a systematic numerical study of time-resolved and time-integrated four-wave-mixing signals for various sets of the relevant parameters characterizing the system.

1. INTRODUCTION

It is well known that optical and transport properties are essentially determined by the dimension of the system. Linear and nonlinear optical properties in the vicinity of the absorption edge depend crucially on the degree of quantum confinement of e-h pairs that occur if relevant length scales of the system become of the same order of magnitude as the bulk Bohr radius. Especially quasi-two-dimensional quantum well structures and quasi-zero-dimensional quantum dots have attracted much attention in this context (Hanamura (1988), Banyai et al. (1988), Schmitt-Rink et al. (1989)).
Two-dimensional (2D) excitons in a perpendicular magnetic field represent a system that is well-suited to study the transition from two-dimensional to zero-dimensional behaviour in low and high magnetic fields, respectively. Extensive studies of nonlinear optical spectra of these systems were performed both, theoretically and experimentally (Stafford et al. (1990), Stark et al. (1990)). One essential result of these investigations is that - concerning nonlinear optical spectra - 1s excitons in strong magnetic fields behave like noninteracting particles. This unique behaviour results from the fact that only inter-Landau-level Coulomb interaction contributes to the exciton-exciton interaction. In this numerical study we investigate the role played by these interactions in the context of time-resolved and time-integrated four-wave-mixing (FWM) signals.

2. NONLINEAR OPTICAL RESPONSE

Optical properties of 2D Coulomb systems in perpendicular magnetic fields can be appropriately described within the unrestricted Hartree-Fock theory. This approximation becomes exact in the strong field limit and allows to treat the effects of external laser fields and the Coulomb interaction on equal footing. The corresponding density matrix equations have been derived earlier (Stafford et al. (1990)). They have the same structure as the well-known semiconductor Bloch equations (see e.g. Binder et. al. (1991)), however, k-vectors are replaced by Landau-level indices n. Here we will only briefly discuss the structure of the system of equations which is obtained for the simplest four-wave-mixing geometry: A strong pump field E(t) propagates in direction $\vec{k}_2$ and a weak probe field $\delta E(t)$ propagates in the direction $\vec{k}_1$. The exciting field gives rise to a renormalized 'ground state', characterized by one-particle distributions $f_n(t)$ and polarizations $\psi_n(t)$, while the linear response to the probe field $\delta\psi_n^+(t)$ contains the information about the renormalized 'excitation spectrum'. Starting in third order in the fields, one obtains a diffracted polarization $\delta\psi_n^-(t)$ resulting from the diffraction of the pump off the grating generated by the interference with the probe and propagating in the direction $2\vec{k}_2 - k_1$. From this diffracted field one obtains the time-dependent FWM-signal as $S(T, t) = |\Sigma\delta\psi_n^-(t)|^2$, where T denotes the time delay between pump and probe. The equation determining the diffracted signal reads

$$(-i\frac{\partial}{\partial t} - i\gamma - \lambda(2n-1) + 2\Sigma_n)\delta\psi_n^- + (1-2f_n)\delta\Omega_n^- = 2\delta f_n\Omega_n^\times - 2\psi_n^\times\delta\Sigma_n, \tag{1.1}$$

where γ is a transverse (phenomenological) damping constant. The inhomogeneity of (1.1), i.e. the source of the diffracted signal contains two contributions. The first term which depends on the renormalized Rabi-frequency $\Omega_n = \Sigma V_{nn'}\psi_{n'} + \mu E$ describes not only exciton-photon but also exciton-exciton interaction, as does the second one which depends on the probe-induced self-energy $\delta\Sigma_n = \Sigma V_{nn'}\delta f_{n'}$. These nonlinearities of (1.1) are due to the coherent population modulation δf_n while the incoherent population f_n takes into account the Pauli exclusion principle. The former dominates in the FWM experiment if the probe proceeds the pump (negative time delay) while the latter dominates at positive time delays. The complete system of equations determining the diffracted signal includes those for the quantities f_n, δf_n, ψ_n, and $\delta\psi_n^+$ (for details see Schäfer et al. (1990)).
In the limit of vanishing magnetic field $V_{nn'}$ reduces to the 2D Coulomb potential and the system of equations describes the nonlinear optics of 2D excitons. With increasing magnetic field excitons become more and more confined to a region which is characterized by the magnetic length $l = (c/eH)^{1/2}$. A convenient measure for the degree of confinement is the dimensionless parameter $\lambda = \omega_c/2E_o$, which is the ratio of magnetic and Coulombic zero-point energies, with the e-h pair cyclotron frequency $\omega_c = eH/mc$ and the 3D excitonic Rydberg E_0.

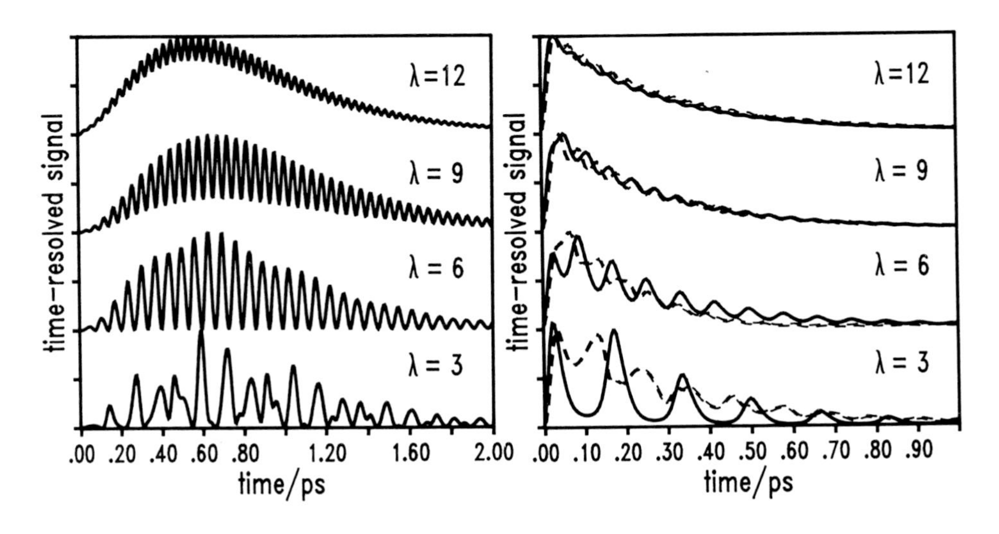

Fig.1. Time-resolved FWM signal for various magnetic field strengths and Gaussian pump and probe pulses of 25 fs duration centered in between the 1s and 2s exciton states.

Fig.2. Time-resolved FWM signal for the same parameters as in Fig.1.but without exciton-exciton interaction (dashed) and without any Coulomb interaction (full).

3. NUMERICAL RESULTS AND DISCUSSION

Impulsive excitation in ultra-short pump and probe experiments allows even at high magnetic fields to excite several magneto-exciton states simultaneously. For this reason one expects pronounced quantum beats in the time-integrated signal as was demonstrated earlier (Stafford et al. (1990)). In addition one expects an oscillating structure of the time-resolved signal which occurs due to the interference of polarizations with different frequencies. Figs.1-4 show the time-resolved signal for GaAs parameters, zero-time delay, a transverse relaxation time of 500fs and a Rabi-frequency of $0.01E_0$ corresponding to the low-intensity limit. The excitation with a Gaussian pump pulse of 25 fs duration is centered in between the 1s and 2s exciton states in order to excite at least two states simultaneously (but closer to the 1s-state). Fig.1 shows for medium magnetic fields a time-resolved signal which is characterized by rather irregular oscillations and a decay of the enveloppe on a time scale of about 2 ps. The irregular oscillations result from the interference of the polarizations of more than two magneto-exciton levels coupled by the excitation pulse. At higher magnetic field only 1s and 2s magneto-excitons contribute and the frequency of oscillations is equal to the level spacing. With further increasing field the amplitude of the oscillations decreases as consequence of the diminishing off-diagonal matrix elements of the Coulomb potential. The expectation that for very high magnetic fields ($\lambda = 12$) the system behaves like a two-level atom is obviously not fullfilled as consequence of the Coulomb interaction. If the latter is neglected completely one obtains the results shown by the full lines in Fig.2. Besides the already discussed oscillations due to the coupling of the higher Landau levels by the pulse, the envelope in this case shows the typical behaviour expected for two-level systems: A monotoneous decay proportional to $\exp(-2\gamma t)$. Nearly the same results are obtained (dashed lines) if exciton-exciton interaction is neglected and only the Coulomb interaction between e-h pairs is taken into account. Comparison with Fig.1 shows that the essential features of the signal are dominated by the exciton-exciton interaction which prohibits a simple interpretation in terms of two-level systems even at high magnetic fields. Details of the signal depend crucially on parameters characterizing the excitation process. Fig.3 shows the dependence of the time-resolved signal for $\lambda = 3$ on the duration of the excitation pulse. Decreasing spectral width with increasing pulse duration leads to decreasing occupation of higher states and finally only the lowest state is excited. As a consequence the oscillations vanish completely.

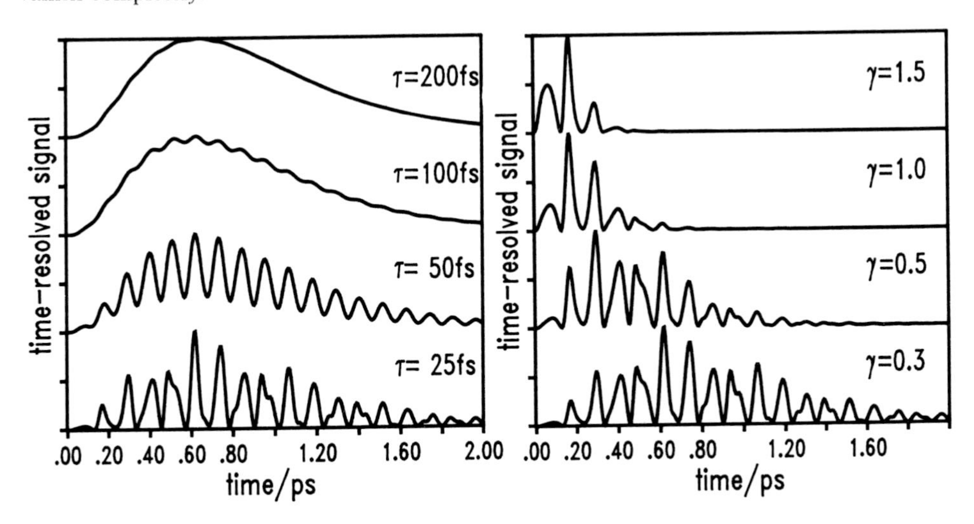

Fig.3. Time-resolved FWM signal for different pump pulse durations and $\lambda = 3$. Other parameters as in Fig.1.

Fig.4. Time-resolved FWM signal for various dephasing time and $\lambda = 3$. Other parameters as in Fig.1.

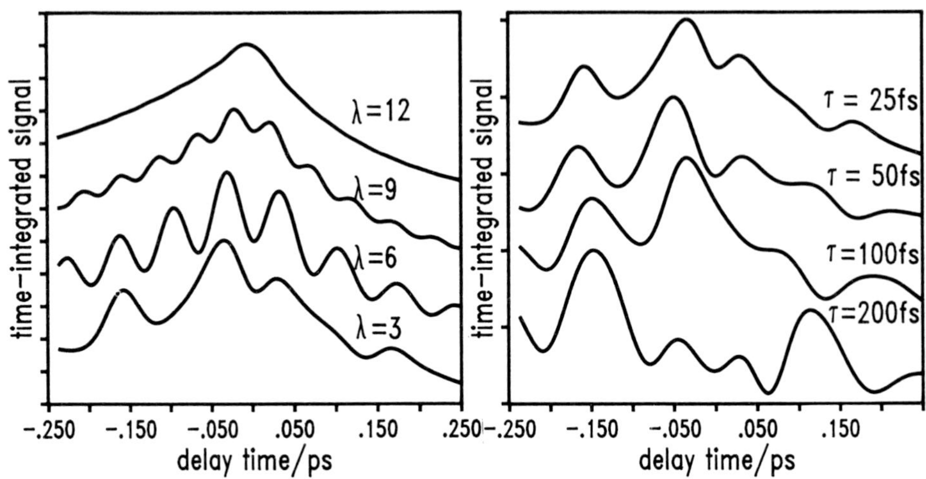

Fig.5. Time-integrated FWM signal for various magnetic field strength and high density excitation. Other parameters as in Fig.1.

Fig.6. Time-integrated FWM signal for different pump pulse duration, $\lambda = 3$ and high density excitation. Other parameters as in Fig.1.

Another essential parameter determining the lineshape of the signal is the transverse relaxation time. As illustrated in Fig.4 the decay of the envelope function is determined by this time, however, no simple exponential decay is obtained.
The time-integrated signals reflect again the novel features due to the Coulomb interaction, in comparison with the physics of two-level systems. Fig.5 shows the time-integrated signal calculated for a Rabi frequency of $1E_0$ and a pulse duration of 25 fs for different magnetic field strengths. For $\lambda = 12$ the signal is essentially the same as for bulk or MQW-systems: At negative time delay it behaves like $\exp(2\gamma t)$ while for positive time delay a decay proportional to $\exp(-4\gamma t)$ is obtained due to the Coulomb interaction (Stafford et al. (1990)). With decreasing magnetic field one obtains pronounced quantum beats with the frequency $E_{1s} - E_{2s}$. Varying the pulse duration and keeping the other parameters fixed the lineshape of the signal becomes more complex due to the contributions of higher order nonlinearities which come into play for larger pulse duration at a fixed Rabi frequency. For the longest pulse of 200 fs the signal shows a pronounced dip in the vicinity of zero delay. Similar behaviour was found in bulk and MQW-systems both, experimentally and theoretically and can be attributed to higher order nonlinearities as consequence of the exciton-exciton interaction (Wegener et al. (1989)).

ACKNOWLEDGMENTS

One of us (M. H.) would like to thank the Volkswagen Stiftung for financial support.

REFERENCES

Banyai L., Hu Y.Z., Lindberg M., Koch S. W. 1988 Phys. Rev. B 38 8142
Binder R., Koch S. W., Lindberg M., Schäfer W., Jahnke F. 1991 Phys. Rev. B 43 6520
Hanamura E. 1988 Phys. Rev. B 37 1273
Schäfer W., Schmitt-Rink S., Stafford C. (1990) SPIE vol. 12/80 High Speed Phenomena in Photonic Materials and Optical Bistability, ed. D. Jäger p. 24
Schmitt-Rink S., Chemla D. S., Miller D. A. B. 1989 Adv. Phys. 38 89
Stafford C., Schmitt-Rink S., Schäfer W. 1990 Phys. Rev. B 41 10000 and references therein
Stark J. B., Knox W. H., Chemla D. S., Schäfer W., Schmitt-Rink S., Stafford C. 1990 Phys. Rev. Lett. 65 3033
Wegener M., Chemla D. S., Schmitt-Rink S., Schäfer W. 1990 Phys. Rev. A 42 5675

Inst. Phys. Conf. Ser. No 123
Paper presented at the International Meeting on Optics of Excitons in Confined Systems, Giardini Naxos, Italy, 1991

Linear and nonlinear optical properties of $Al_xGa_{1-x}As/AlAs$ multiple-quantum-wells at room temperature

[a]M Dabbicco, [b]R Cingolani, [a]M Ferrara, [c]K Ploog and [c]A Fisher

[a] *Unita' GNEQP, Dipartimento di Fisica dell'Universita', I-70126 Bari, Italy*
[b] *Dipartimento di Scienze dei Materiali dell'Universita', I-73100 Lecce, Italy*
[c] *Max-Planck-Institut für Festkörperforschung, D-7000 Stuttgart, Germany*

ABSTRACT: Nonlinear optical behavior has been observed in ternary alloy $Al_{0.37}Ga_{0.63}As/AlAs$ multiple-quantum-well structures both in absorption and emission spectra. The bleaching of the n =1 and n = 2 heavy-hole exciton absorption has been observed by nanoseconds pump-and-probe spectroscopy. The relative variation of the absorption coefficient $((-\Delta\alpha)/\alpha_0)$ are about 15% and 8%, respectively. The value of the saturation density deduced from our spectra is in good agreement with the theoretical predictions. We have also observed stimulated emission by electron-hole plasma recombination under the same experimental conditions. The influence of the staggered band alignment on the observed processes is discussed.

1 INTRODUCTION

The investigation of optical nonlinearities is a powerful tool to determine the fundamental properties of semiconductors (Haug 1988). Same attention has recently been paid to the nonlinear behavior of excitons in ternary alloy multiple-quantum-well (MQW) structures because of their potential application in optoelectronic and photonic devices (Feldmann 1990, Sasaki 1990).
To our knowledge excitonic nonlinearities have not been studied under quasi-stationary conditions (nanosecond range) in this material system.

In this paper we report a comprehensive study of the linear and nonlinear absorption processes and of the electron-hole plasma (EHP) formation in type-II $Al_xGa_{1-x}As/AlAs$ ternary alloy MQW structures, investigated at room temperature by nanosecond high excitation pump-and-probe (P&P) and photoluminescence (PL) experiments, respectively. We focuse our attention on the results obtained in a 25 periods $Al_{0.37}Ga_{0.63}As/AlAs$ MQW structure grown by conventional solid source molecular beam epitaxy (MBE) onto a (100) GaAs substrate. Its structural parameters, assessed by means of double-crystal x-ray diffraction are $L_{Al_xGa_{1-x}As}$ = 193 Å, L_{AlAs} = 103 Å, x = 0.37. To perform transmission P&P experiments the substrate was removed by selective chemical etching with the sam-

ple face mounted on a sapphire plate. The broad fluorescence of a DCM dye was used as probe beam. Both the sample and the dye were pumped by the second harmonic of a Nd:YAG laser. The maximum pump power before focusing was about 800 μW per pulse with a 10 Hz repetition rate and a pulse width of 10 ns. The transmitted light was dispersed by a 0.6 m spectrometer and collected by a photomultiplier followed by a box-car integrator with computer acquisition. Great care has been taken in subtracting any luminescence contribution induced by the pump in the transmission spectra.
Using the same experimental apparatus we performed PL experiments in near back-scattering configuration. All measurements have been performed at room temperature.

2 ABSORPTION NONLINEARITIES

In Fig. 1 we show the differential absorption for two different pump intensities.

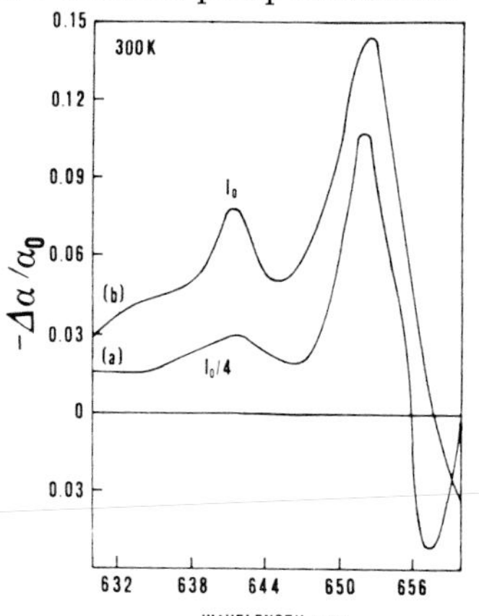

Fig. 1 Differential room temperature absorption spectra at two different pump-beam powers (I_0 = 400 kW/cm^2): $I_0/4$ (a) and I_0 (b). The two peaks are due to the excitonic absorption saturation.

At a pump power of 100 kW/cm^2 only the fundamental exciton absorption has been saturated. Rising the pump beam intensity we have saturated also the n = 2 exciton absorption. The relative variation of the absorption coefficient is $(-\Delta\alpha)/\alpha_0 \simeq 15\%$ and 8% for the n = 1 and n = 2 exciton, respectively.
From our spectra we can estimate the nonlinear absorption cross section σ_{eh} according to the relation

$$\alpha = \alpha_0 - \sigma_{eh} n \tag{1}$$

where n is the volume density of the photogenerated electron- hole plasma in non-resonant excitation condition. The value of n is given by the relation

$$n = \phi \alpha_0 \tau_{eff} \tag{2}$$

where ϕ is the photon flux, and τ_{eff} the carrier lifetime. As for the lifetime we have to take into account the staggered band alignment of our sample. Owing to the Al-content and the valence-band discontinuity, our sample exibits type-II band alignment. The excited carriers at the $Al_xGa_{1-x}As$ Γ-point thus relax down to the lowest lying energy levels at the AlAs X-point minimum. This additional "escape-channel" is much more effective in reducing the actually photogenerated carrier density given by eq. (2) than the pure radiative lifetime. To account for the different processes occurring in reducing the carrier density, we assume an effective lifetime of the Γ states, given by

$$\frac{1}{\tau_{eff}} = \frac{1}{\tau_{\Gamma-X}} + \frac{1}{\tau_{rad}} \tag{3}$$

where the $\tau_{\Gamma-X}$ is the electron intervalley transfer time from the $Al_xGa_{1-x}As$ Γ-point to the AlAs X-point minima, and τ_{rad} is the intrinsic radiative lifetime of the direct excitons in the ternary alloy. Unlike the case of type-I MQWs, here the most important contribution to the carrier loss is given by the Γ-X electron transfer. Feldmann (1990) have measured the value of $\tau_{\Gamma-X} \simeq 2$ ps at room temperature, about two orders of magnitude shorter than the purely radiative lifetime (Andreani 1991). Thus we can resonably assume that $\tau_{eff} \simeq \tau_{\Gamma-X}$. Using these values we obtain from eq. (2) $n \simeq 2.3\text{x}10^{16}$ cm^{-3} and a nonlinear absorption cross section of about 10^{-13} and $7\text{x}10^{-14} cm^2$ for the n = 1 and n = 2 excitons, respectively.

To compare our experimental findings with the theoretical prediction we have calculated the saturation density also according to the model of Schmitt-Rink (1985). In the limit of high temperatures, both considering the phase space filling and the exchange contribution, the saturation density is proportional to exciton binding energy (E_{1s}) and inversely proportional to the excition Bohr radius ($a_{2D}{}^2$). We determined $E_{1s} \simeq 15$ meV by fitting the linear absorption spectrum with the model proposed by Chemla (1984) including the n = 1 heavy- and light-hole excitons, the continuum of the n = 1 heavy-hole exciton and also the n = 2 heavy-hole exciton contributions. Once known E_{1s} we find $a_{2D} \simeq 40$ Å. Using these values we obtain a theoretical saturation density of $N_s \simeq 1.5\text{x}10^{10}$ cm^{-2} derived from the curve of Fig. 1 of ref. (Schmitt-Rink 1985), at a E_{1s}/kT value of 0.57. This is consistent with the experimentally determined value $n_s \simeq 6.6\text{x}10^{10}$ cm^{-2}, obtained by scaling the eq. (2) for L_{AlGaAs}, at which both the n = 1 and n = 2 excitons are bleached.

3 ELECTRON-HOLE PLASMA EMISSION

In Fig. 2 we report room temperature PL spectra obtained under high photogeneration rate.

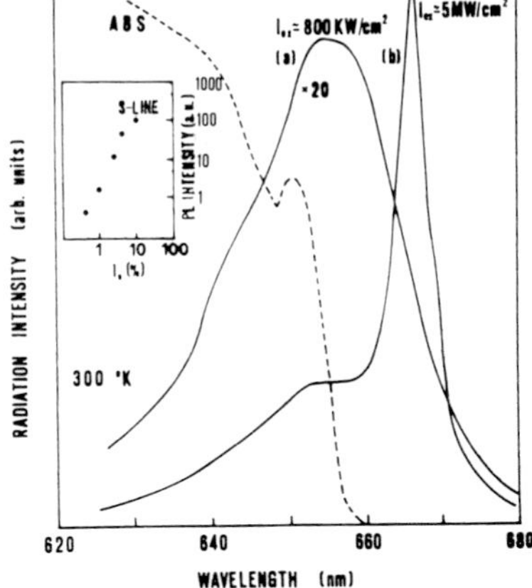

Fig. 2 Absorption (dashed line) and photoluminescence (solid lines) at two different excitation intensities: 800 kW/cm², below the stimulation threshold (a); 5 MW/cm², above the stimulation threshold (b). In the inset is reported the maximum peak value of the S line versus the relative excitation intensity (100% = 50 MW/cm²).

At the lowest excitation power used in this PL experiment, the spontaneous emission spectrum shows a broad band (FWHM $\simeq$ 60 meV) centered around 1.895 eV, red-shifted by about 10 meV with respect to the n = 1 ABS peak. By increasing the exciting power (I_{exc}) we first observe the saturation of the n = 1 subband related emission and the progressive filling of the higher energy states. At a threshold power about 1.5 MW/cm² a sharp stimulated emission (FWHM $\simeq$ 18 meV) sets in around 1.86 eV (S line). Further increase of the excitation intensity causes a small shift of the S line to lower energy and a superlinear growth of the emission intensity ($I_S \simeq I^2_{exc}$) as shown in the inset of the Fig. 3. Preliminary measurements at lower temperature show a similar behavior with two features: i) the threshold of the S line reduces by about one order of magnitude; ii) the energy splitting between the spontaneous and the stimulated emissions decreases monotonically down to 12 meV at 80 °K. Such spectral behavior is expected for the EHP stimulated recombination already studied in GaAs/$AlGa_xAs_{1-x}$ MQW (Cingolani 1990).

Owing to the Al-content and well width, our sample exibits a type-II band alignment and is also near the ternary-alloy Γ-X crossover. To properly describe the emission line-shapes when the stimulated emission occurs, time resolved experiments are necessary which are actually underway in this laboratory.

4 CONCLUSIONS

We have observed the excitonic absorption saturation in type-II $Al_xGa_{1-x}As$/AlAs ternary alloy MQW structures at room temperature under quasi-stationary non-equilibrium conditions. Pulsed nanosecond pump-and-probe experiments have revealed a negative differential absorption of about 15% and 8% for the n = 1 and n = 2 exciton, respectively, at a pump density of 400 kW/cm^2. The experimentally found saturation density agrees with the theoretical prediction only if the effective carrier loss rate in the type-II MQW structure is taken into account. This finding demonstrates the relevance of the staggered alignment in the photogeneration process and in determing the nonlinear properties of MQWs. At much higher excitation power (2 MW/cm^2) a stimulated emission occurs due to the electron-hole plasma recombination.

AKNOWLEDGMENTS

This work was partially supported by the Research National Council of Italy (under the special project "Tecnologie Elettroottiche") and by the Bundesministerium für Forschung und Technologie of the Federal Republic of Germany.

REFERENCES

Andreani L C 1991 Solid State Commun. **77**, 641
Cingolani R, Ploog K, Cingolani A, Moro C and Ferrara M 1990 Phys. Rev. B **42**, 2893
Chemla D S, Miller D A B, Smith P W, Gossard A C and Wiegmann W 1984 IEEE J. Quantum Electron. QE-**20**, 265
Feldmann J, Nunnenkamp J, Peter G, Göbel E, Kuhl J, Ploog K, Dawson P and Foxon C T 1990 Phys. Rev. B **42**, 5809
Haug H 1988 *Optical nonlinearities and instabilities in semiconductors* (San Diego: Academic Press)
Sasaki F, Mishina T and Masumoto Y 1990 Phys. Rev. B **42**, 11426
Schmitt-Rink S, Chemla D S and Miller D A B 1985 Phys. Rev. B **32**, 6601

Inst. Phys. Conf. Ser. No 123
Paper presented at the International Meeting on Optics of Excitons in Confined Systems, Giardini Naxos, Italy, 1991

High density excitation in quantum confined systems

D.S. Chemla
Physics Department, UC Berkeley and Material Sciences Division, Lawrence Berkeley Laboratory

ABSTRACT: We present: i) investigations of the effect of the exciton-exciton interaction on the line shape of time resolved four wave mixing resonant with quasi-2D excitons in quantum wells. ii) studies of the evolution of the nonlinear optical response of QW as the quasi-2D electronic states are further confined into quasi-0D by a strong magnetic field perpendicular to the quantum wells.

1) INTRODUCTION

Exciton-exciton interaction dominates the optical response of semiconductors at high excitation, i.e. when the average distance between photogenerated (real or virtual) excitons becomes of the order of the exciton Bohr Radius, a_o. Because excitons are composite particles made of charged Fermions their interaction is very sensitive to dimensionality through the Pauli exclusion and Coulomb interaction. In this lecture we present investigations of exciton-exciton interaction in low dimensionality semiconductor structures. In the first Section we discuss how exciton-exciton interaction strongly affects the line shape of time resolved four-wave mixing near quasi-two dimensional exciton resonances in semiconductor quantum wells (QW). In the second Section we present investigations of the evolution of the nonlinear optical response of QW as the quasi-two-dimensional electronic states are further confined into quasi-zero-dimensions by a strong magnetic field perpendicular to the plane of the QWs.

2) EFFECTS OF EXCITON-EXCITON INTERACTION ON TIME RESOVED FOUR WAVE MIXING.

In the density matrix description of quantum mechanical systems the effect of random perturbations and damping is usually accounted for by introducing phenomenological longitudinal and transverse relaxation terms, T_1 and T_2, in the equations of motion. These damping parameters are extremely difficult to calculate from first principles and therefore they are often deduced from experiments. A very powerful method to study the dephasing process in the optical response is time-resolved four-wave mixing (FWM). In this technique, two pulsed laser beams with wave-vectors k_1 and k_2 interfere in a sample to produce a diffracted beam in the direction $k_3 = 2k_2 - k_1$, as sketched in the inset of Fig. 1. The magnitude of the diffracted signal in the direction k_3 is then recorded as a function of the time-delay, $T = t_2 - t_1$, between a pulse of beam-2 and a pulse of beam-1. The results are usually interpreted in terms of the two level-model of Yajima and Taira [1] to determine the time, T_2, during which the system has not experienced incoherent scattering. In the context of semiconductor physics, important information on incoherent scattering of excitons (X) by acoustic phonons, impurities, other X and free carriers have been obtained by time-resolved

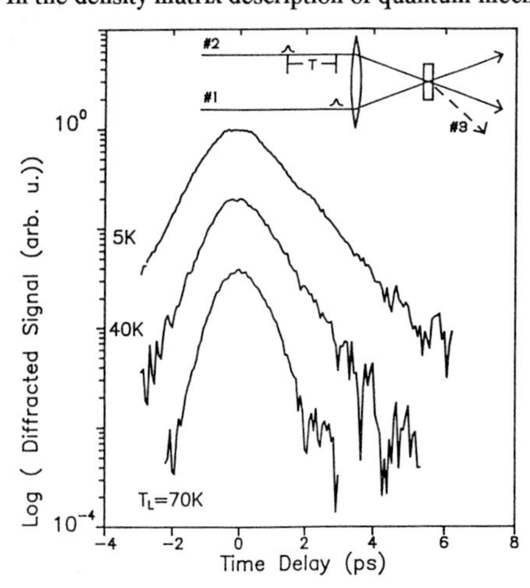

Figure 1 : DFWM signal in GaAs quantum well vs time delay for 3 lattice temperatures. Inset: Experimental configuration.

FWM experiments in GaAs quantum wells (QW) [2,3]. In all these cases, however, the experiments were interpreted in terms of the independent two-level model. This model predicts that the signal profile is identically zero at negative time delay, $T < 0$, and shows a step like onset, around $T = 0$, limited by the duration of the laser pulses followed by an [$\exp(-2T/T_2)$] exponential decay for positive time delay. In a set of experiments performed at low temperature on two types of QW structures, GaAs/GaAlAs-QWs (L_z = 170Å, $E_g \approx$ 1.5eV) and InGaAs/InAlAs-QWs (L_z = 200Å, $E_g \approx$ 0.8eV) [4,5] we have observed that the time delay dependence of the FWM signal is in qualitative disagreement with that predicted by the independent two-level model.

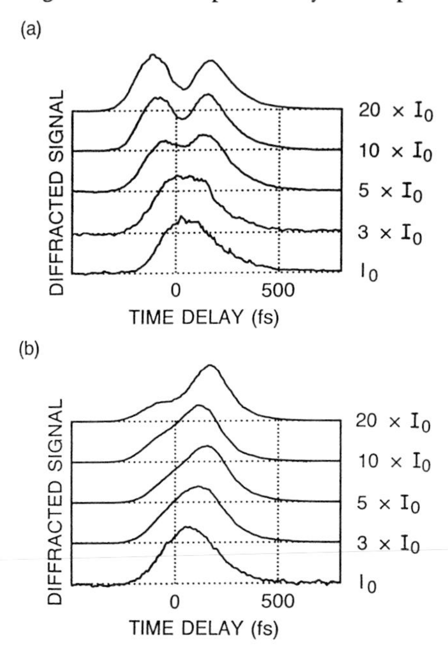

Figure 2 : DFWM signal in InGaAs quantum wells as a lattice temperature of 10K and for 5 excitation intensities ($I_o \approx$ 3MWcm^{-2}).
a) 10meV detuning and b) 6meV detuning.

Fig. 1, for example shows the FWM signal vs the time delay T obtained at very low excitation intensty ($\approx$ 160kWcm^{-2}) about 3meV below the heavy hole exciton resonance of the GaAs/GaAlAs-QW sample for three temperatures, 5K, 40K and 70K [4]. The signal is non-zero for negative time delays. In fact, the profile consists of exponantially rising and decaying wings. The time constant of the rising wing is exactly half of that of the decaying wing. As a function of temperature the two time constants get shorter but always in the same 1:2 ratio, as long as dephasing by phonons scattering has not reduced the coherence below the experimental resolution [4]. The low temperature value, T_2 = 1.15ps, which one would deduce from the $T > 0$ portion of the signal profile is in satisfactory agreement with earlier experiments on the same material system [2,3]. In Fig. 2(a) and 2(b), the FWM signals obtained on a InGaAs/InAlAs-QW sample for 10meV and 6meV detuning respectively are plotted as the intensity of both beams is scaled up simultaneously from I_o = 3MWcm^{-2} to $20 \times I_o$. As the incident intensity is increased we observe the evolution of two different features. First the strength of the signal for negative time delays increases. Then, the single maximum time-profile at low intensities gradually evolves into a lineshape that exhibits two distinct maxima in time for the 10meV detuning case, and a strong asymmetry for the 6meV detuning case.

For the 10meV detuning experiment, the highest intensity ($20 \times I_o \approx$ 60MWcm^{-2}) corresponds to a regime of significant saturation of the resonance, where the exciton density is approximately $N_X \approx 10^{11}$cm^{-2}, comparable to the saturation density of the Xs. It is remarkable that in this regime it is still possible to resolve a decay time, $\approx$ 80fs, for positive time delays. At this point the excitonic line has broadened due to the high exciton density. Assuming a homogeneous line yields: $T_2 \approx$ 160fs. In both cases, particular care has been taken to check that the $T < 0$ siganl is not an experimental artifact, for example by measuring the time reverse profile in the direction $k_4 = 2k_1 - k_2$ and by accurately determining the origin of time delay, $T = 0$. Furthermore, experiments performed on numerous samples including InGaAs/InP-QWs with different barrier materials have demonstrated that this behavior is general.

Optical nonlinearities in semiconductors originate from both exciton-exciton (X-X) interaction and anharmonicities in the exciton-photon (X-hν) interaction. Recently, following Schmitt-Rink et al. [6,7] theories based on the unrestricted Hartree-Fock approximation have been used successfully to describe nonlinear optical effects in the vicinity of the gap [8,12]. Conceptually this approach is a direct extension of Anderson's theory of collective excitations in superconductors [13]. We consider a two parabolic-band model, the wavevector, k, indexes the conduction and valence band levels. We neglect the photon momentum. The optically connected conduction and valence band levels form a set of two-level systems labelled by k. The density matrix of the semiconductor breaks into 2×2 blocks:

$$\hat{n}_k(t) = \begin{bmatrix} n_{ck}(t) & \psi_k(t) \\ \psi_k^*(t) & n_{vk}(t) \end{bmatrix} \tag{1}$$

$n_{c,vk}(t)$ are the populations in the conduction and the valence bands and $\psi_k(t)$ is the pair amplitude. The density matrix obeys the Liouville equation,

$$\frac{\partial}{\partial t} \hat{n}_k(t) = -i[\, \hat{\varepsilon}_k(t) \, , \, \hat{n}_k(t) \,] + \frac{\partial}{\partial t} \hat{n}_k(t) \,|_{\,relax} \tag{2}$$

where $\hat{\varepsilon}_k(t)$ is the energy matrix. The difference between this BCS-model of the semiconductor and a collection of independent two-level systems is that the Coulomb interaction, $V_{k,k'}$, couples the various levels. Hence in the presence of a strong laser field, E(t), (pump) propagating in the direction k_2, the energy matrix is:

$$\hat{\varepsilon}_k(t) = \begin{bmatrix} \varepsilon^o_{ck} & -\mu_k E(t) \\ -\mu_k^* E^*(t) & \varepsilon^o_{vk} \end{bmatrix} - \sum_{k'} V_{k,k'} \, \hat{n}_{k'} \tag{3}$$

In this equation $\mu_k E$ is the Rabi frequency which describes the coupling of the two-level system at k with the pump field. The conduction and valence band energies are: $\varepsilon^o_{ck} = E^o_g/2 + \hbar^2 k^2/2m_e$ and $\varepsilon^o_{vk} = -E^o_g/2 - \hbar^2 k^2/2m_h + \sum_{k'} V_{k,k'}$, where E^o_g is the bare band gap. As usual, the energy separation is: $\varepsilon^o_{ck}-\varepsilon^o_{vk} = E_g + \hbar^2 k^2/2m$ as a function of the dressed band gap, $E_g = E^o_g + \sum_{k'} V_{k,k'}$, and the reduced mass $m^{-1} = m_e^{-1} + m_h^{-1}$. As compared to the independent two-level systems, the physics is modified by the Coulomb force in two ways. First, the conduction and valence band energies are renormalized,

$$\varepsilon^o_{jk} \rightarrow \varepsilon_{jk} = \varepsilon^o_{jk} - \sum_{k'} V_{k,k'} \, n_{jk} \tag{4a}$$

with j = e or h. Second the coupling expressed by the Rabi frequency is modified according to:

$$\mu_k E \rightarrow \Delta_k = \mu_k E + \sum_{k'} V_{k,k'} \psi_{k'} \tag{4b}$$

This modification translates the fact that the optically connected e-h levels at k do not experience the applied field, $\mu_k E$. Rather, they see the self-consistent "local field", Δ_k, which is the sum of the applied field and the "molecular" field, due to all the other e-h levels at different wavevector k' [6,7]. The last term of Eq. (2) describes the coupling to the thermal reservoir and hence relaxation. In the usual manner it may be approximated by transverse and longitudinal relaxation rates, $\gamma_2 = T_2^{-1}$ and $\gamma_1 = T_1^{-1}$. In a FWM experiment a "weak" probe field $\delta E(t)$ propagating in the direction k_1 interferes with the pump field to generate a polarization which radiates in the direction $k_3 = 2k_2 - k_1$. The linear response to the probe beam is obtained by linearizing Eq. 2 with respect to $\delta E(t)$, with the precaution that the change in the matrix density, $\delta\hat{n}_k(t)$, induced by the probe field, is treated self-consistently. The total polarization induced by the probe is:

$$\delta P(t) = \sum_k \mu_k^* \delta\psi_k \tag{5}$$

and the FWM signal is given by the part of $\delta P(t)$ which propagates in the direction k_3. It is instructive to write explicitly the expressions of the probe induced changes in the pair amplitude in the electron-hole representation i.e. $n_{ck} \rightarrow n_k$ and $n_{vk} \rightarrow 1 - n_k$.

$$i\frac{\partial}{\partial t}\delta\psi_k + [i\gamma_2 - (E_g + \frac{\hbar^2 k^2}{2m})]\delta\psi_k + \sum_{k'} V_{k,k'}\delta\psi_{k'} =$$

$$-(1 - 2n_k)\mu_k\delta E + 2\delta n_k \mu_k E \tag{6}$$

$$+ \sum_{k'} 2V_{k,k'}\left[(\delta\psi_{k'} n_k - \delta\psi_k n_{k'}) + (\psi_{k'}\delta n_k - \psi_k \delta n_{k'})\right]$$

This is a driven time dependent Wannier Equation which deserves several comments. Let us note that in the first order, where $n_k = \delta n_k = 0$, the driving term becomes constant and one recovers the form of the Wannier equation often used when nonlinear effects are neglected.

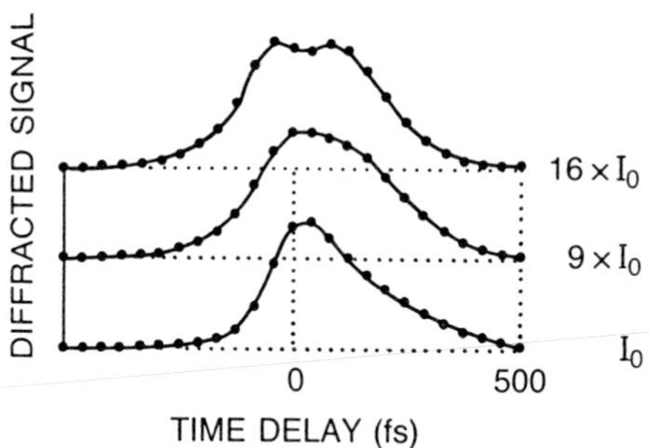

Figure 3 : Calculated DFWM signal. The points are results of numerical integration to all orders of the excitonic quantum kinetic equations.

The case of noninteracting e-h pairs is obtained by setting the Coulomb interaction, $V_{k,k'}$, equal to zero. Then Eq. (6) reduces to the usual Bloch equations. This establishes a direct connection between the BCS description of the band edge excitations, [6,12] the more conventional theories of excitons in semiconductors, [14] and two level-models [1]. We note also that in the further limiting cases Eq. (6), after some manipulations, takes the form of a Ginzburg-Landau equation. The driving terms are interesting in their own right. The first line expresses the coupling to the field corrected for Pauli exclusion, i.e., the X-hν nonlinearity. The terms of the second line are proportional to $V_{k,k'}$. They express X-X interaction and are thus absent for independent two-level systems. On the diagonal, $k = k'$, they vanish indentically.

The X-X interaction terms are responsible of specific features of excitonic effects in the nonlinear response. As compared to the X-hν terms which are proportional to the applied fields, they are proportional to the coherent and incoherent populations and pair amplitudes. Hence they have a very different temporal evolution which, as shown bellow, explains the negative time delay signal observed experimentally. Let us consider the case of excitating and probing with ultrashort pulses. In the independent two-level system description, the probe field sets up a first-order polarization which interferes with the pump field to create a second-order grating. Then the pump field is scattered from this grating in the direction k_3, thus producing the signal. Obviously if the pump pulses are over before the grating is set, ($T < 0$), there is no signal. For excitons, however, because of the interaction terms a grating can be formed by the probe-induced and pump-induced polarizations. The pump-induced polarization then can scatter off this grating in the k_3-direction. Therefore, a signal can be generated by this process for negative time delay, $T < 0$. Since polarizations are involved twice in this case, the time constant of the $T < 0$ signal is $T_2/4$ as compared to the $T_2/2$ time constant of the signal that originates from the scattering of the pump field. To verify this interpretation and treat the problem in its full generality we have solved numerically Eq. (6) with parameters corresponding to our experiments. In particular we have accounted for the finite pulse duration (110fs FWHM). The results for three different intensities are presented in Fig. 3 and compare very satisfactorily with experiment. At low intensity the time-integrated signal decays as expected, with a time constant $2/T_2$ for positive time-delay. For negative time-delay it exhibits a rising edge with a $4/T_2$ time constant. As the excitation increases, the positive time-delay signal becomes more pronounced and a time profile with two distinct maxima emerges due to saturation at the maximum overlap time, $T \approx 0$, as seen experimentally.

3) NONLINEAR OPTICAL RESPONSE OF QUASI-ZERO DIMENSIONAL MAGNETO-EXCITONS.

In QW, X are confined in one direction by the crystal potential. Attempts to make quasi-one-dimensional and quasi-zero-dimensional structures have been plagued by the difficulty of obtaining defect-free samples and of eliminating size fluctuations.

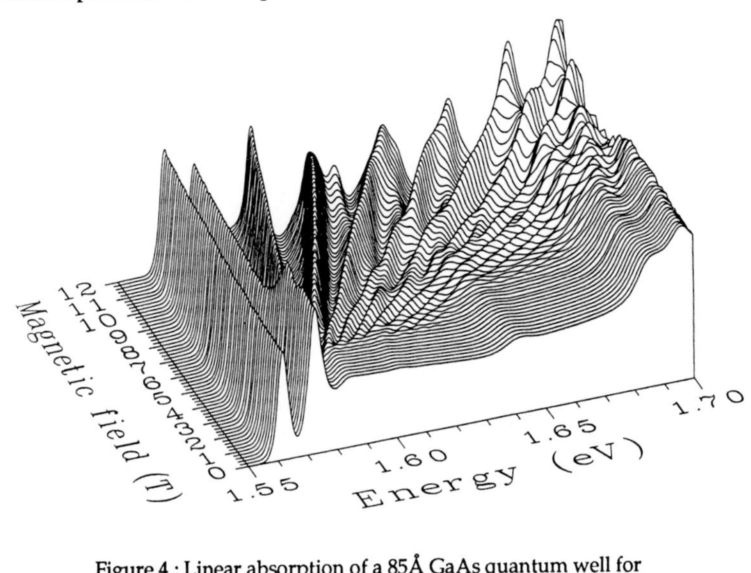

Figure 4 : Linear absorption of a 85Å GaAs quantum well for σ_- polarized light and B = 0 -> 12 Tesla.

Further confinement within the QW plane, however, can be obtained by applying a large perpendicular magnetic field, H. In addition to the Coulomb potential, the electron-hole (e-h) pairs then experience the quadratic potential imposed by the magnetic field. The e-h states in the presence of the two potentials are called magneto-excitons (MX). As shown below, the relative strength of Coulomb and magnetic energies can be measured by the ratio of the magnetic and Coulombic zero-point energies i.e. the dimensionless parameter $\lambda = (a_o/l_c)^2 = \omega_c/2R_y$, where $l_c = (c/eH)^{1/2}$ is the cyclotron radius, ω_c is the cyclotron frequency, and R_y is the material Rydberg ($\hbar = 1$). By varying the magnetic field the dimensionality of the MX states can thus be tuned continuously from quasi-two dimensions ($\lambda << 1$) to quasi-zero dimensions ($\lambda >> 1$) in materials with excellent quality and uniformity. Futhermore, using circular polarization the spin dependence of the MX-MX interaction can be explored as well. The theory of MX linear optical properties of has been discussed by numerous authors [15-19]. Within the effective mass approximation the Schrödinger equation for photoexcited 2D e-h pairs is:

$$[\frac{1}{2m_e}(p+\frac{e}{c}A)^2 + \frac{1}{2m_h}(p-\frac{e}{c}A)^2 - \frac{e^2}{\varepsilon_o r}]\,\psi_\alpha = E_\alpha\,\psi_\alpha \tag{7}$$

Here r is the e-h relative coordinate which, for optically active s-like MX ($l_z = 0$), is governed by the total potential $V(r) = [\,(\lambda r/2)^2 - 2/r]\,R_y$, which consists of the sum of the Coulomb potential and the quadratic potential imposed by the magnetic field. It is easy to see how the relative strength of two potentials is measured by λ. For $H = 0$, Eq. (7) yields the ususal 2D-X spectrum and, for noninteracting pairs, $\varepsilon_o^{-1} = 0$, it yields the Landau Levels (LL).

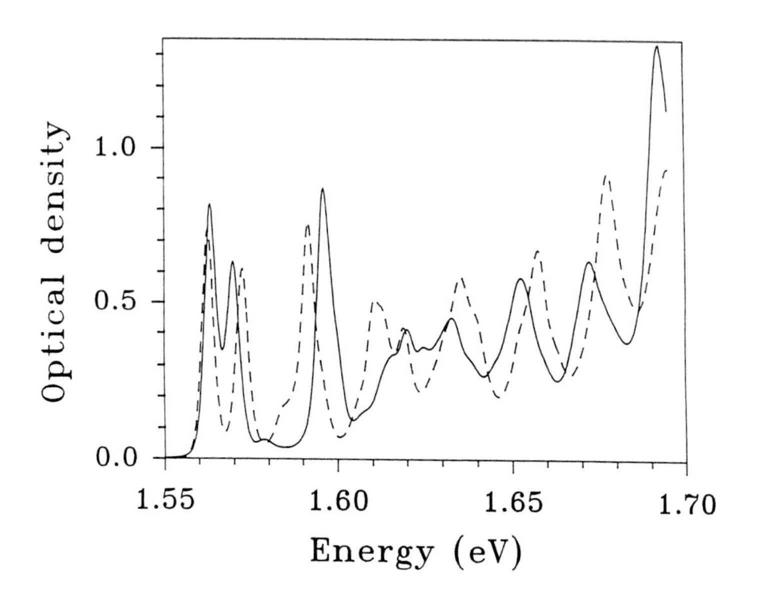

Figure 5 : Comparison of the B=12 Tesla spectra for σ_- (dashed) and σ_+ (solid) polarized light.

As H→0 all MX extrapolate to the bound 2D-X states. Fig. 4 shows the raw experimental linear absorption spectra of a high quality L_z = 85Å GaAs/AlGaAs-QW structure at 4K, taken with σ_- polarized light as the magnetic field is tuned from H = 0→12T (λ = 0→3). One sees very nicely the transition from the 2D behavior at low fields to 0D at high field. The way the LLs originate from the bound 2D-X states is also clearly displayed. Fig. 5 compares the 12T-linear absorption for σ_- (dashed line) and σ_+ (solid line). They are consistent with the QW selection rules at the Γ-point of the Brillouin zone. At such a high field the confinement is so strong that the absorption strength is almost zero between the 1s and 2s MX, indicating that the MX are almost diagonal in the LL basis. These spectra are in good agreement with the theoretical calculations of Ref. [19]. Experiments comparing the nonlinear optical responses at H = 0 and H = 12T were performed at 4K using the pump/probe technique with a time resolution of 100fs on L_z = 85Å and L_z = 70Å GaAs/AlGaAs-QW structures [20-22]. In these experiments a strong, narrow-band ultrashort pump pulse excites the sample and a weak, broad-band ultrashort probe pulse measures the transmisson of the sample at various time delays. The differential transmission spectra (DTS) give the normalized change of transmission when the pump is off and on, $\Delta T/T = (T_{on} - T_{off})/T_{off}$. In the small signal regime this is direcly proportional to the change in absorption, $1 >> \Delta T/T \approx -\Delta\alpha l$. The most straightforward way of interpreting such experiments is to consider that the pump creates (real or virtual) populations of MX (i.e.linear superposition of excited states), and the DTS measures the change of absorption due to the presence of these populations. The insets of Fig. (6a) and (6e) show the linear absorption spectra of the sample whith and without field, H = 12T. They also display the pump spectra for excitation resonant on the 1s-X and on the 1s-MX and 2s-MXs. The DTS obtained with a linear polarized pump resonant with the 1s-X at H = 0 is shown in Fig. (6b). It is remarkably dissimilar from the DTS obtained with pump resonant with the 1s-MX, Fig.(6c), or with 2s-MX, Fig. (6d) [20-22]. Creation of 1s-X causes both saturation and blue-shift of the 1s-X transition in agreement with theory [23] and as seen earlier [24]. This shift originates from X-X exchange interaction and can be interpreted as a hard core repulsion

[23]. On the contrary when 1s-MX are created at H = 12T, only the saturation of the 1s-Mx is seen, whereas the blue-shift is almost completely quenched.

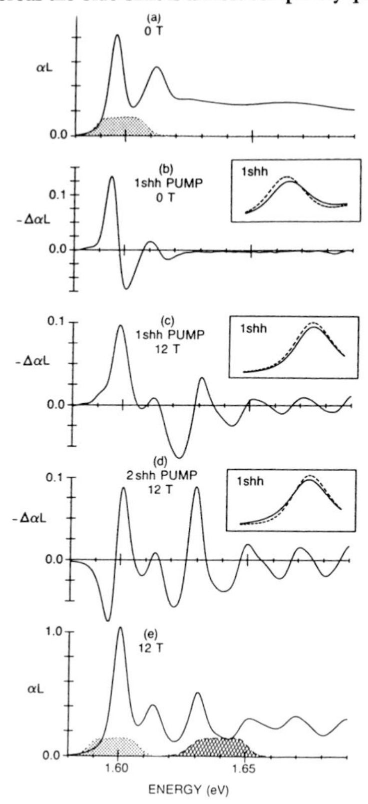

Figure 6 : Linear and differential absorption spectra of a 70Å GaAs quantum well for resonant excitation.

Simultaneously, the 2s-MX exhibit a red-shift and the higher lying MX show both red-shift and broadening. The later features are, of course, absent at H = 0. The creation of 2s-MXs causes a red-shift and saturation of the 1s-MX, whereas the signal at the 2s-MX is almost symmetric, indicating saturation and some broadenening but no shift of the resonance. The higher MXs react to a population of 2s-MXs essentially as they react to 1s-MXs. The theoretical discussion of the nonlinear response of MXs can be developed following the method presented in Section I [25]. Since at high magnetic field the MX look very much like the well known LL, one can look for solutions of Eq. (7) by expanding the MX wavefunction, $\psi_\alpha(r)$, on the LL orbitals, ϕ_{ν,l_z}. For the optically active s-like MX, (l_z = 0), the expansion coefficients, $\phi_\alpha(\nu)$, satisfy [25]:

$$\sum_{\nu'} [\, E_\nu \, \delta_{\nu,\nu'} - V_{\nu,\nu'} \,] \, \phi_\alpha(\nu') = E_\alpha \, \phi_\alpha(\nu) \qquad (8)$$

where $E_\nu = (2\nu\text{-}1)\lambda$, ($\nu = <\nu'$) and $V_{\nu,\nu'}$ are the LL-Coulomb matrix elements [25]. The linear susceptibility is:

$$\chi^{(1)}(\omega) = \frac{\lambda}{2\pi} \sum_\alpha \frac{|\, \mu \sum_\nu \phi_\alpha(\nu) \,|^2}{[E_\alpha - \omega - i\gamma_2]} \qquad (9)$$

Here and in the following, the units of energy and length are R_y and a_o.

In Eq. (9), μ is the interband matrix element. It is worth noting that $\lambda/2\pi$ is the LL degeneracy. Calculations of $\text{Im}\chi^{(1)}(\omega)$, in the pure 2D parabolic band case using a base of 1700 LL for $\lambda = 0.25 \rightarrow 8$ and $\gamma_2 = 0.7$ are in good agreement with the trends seen experimentally [25]; i.e. the 1s-MX exhibits a small diamagnetic shift while the higher MX disperse strongly. At each MX the confinement increases the oscillator strength as gaps appear between them, thus insuring the conservation of the total absorption strength.

To calculate the nonlinear optical response, approximations similar to those introduced in Section I must be applied. They are more appropriate in this case since they become exact in the extreme magnetic limit, $\lambda \rightarrow \infty$ [26-29]. In particular, at equilibrium they yield the exact and remarkable result that an ensemble of spin-polarized MX behave like a gas of noninteracting point-like Bosons [26,27]. In the LL-orbital basis the Liouville equation, Eq. (2), yields coupled equations for the LL populations, n_ν, and pair amplitude, ψ_ν that are analogous to those already encountered in Section I, such as Eq. (6). All the mathematics follows directly from the simple change of indices $k \rightarrow \nu$ [25]. The Coulomb interaction renormalizes the transition energies and the Rabi frequency again. For example, the expressions corresponding to Eq. (4a) and (4b) are:

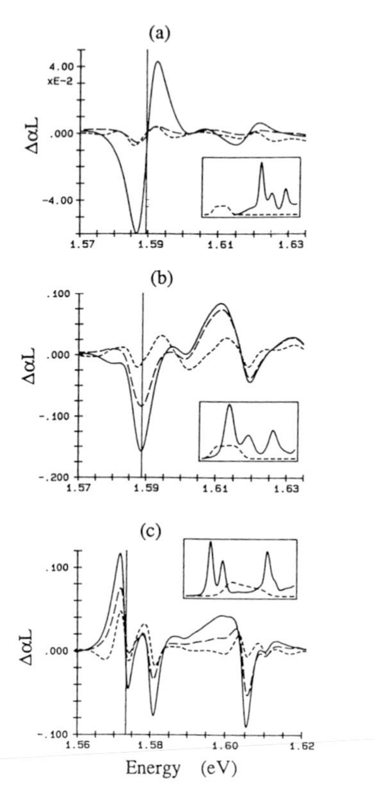

Figure 7 : Linear and differential absorption spectra for below, at and above resonance excitation.

$$E_\nu \rightarrow E_\nu - 2\sum_{\nu'} V_{\nu,\nu'}\, n_{\nu'} \qquad (10a)$$

and

$$\mu E \rightarrow \mu E + \sum_{\nu'} V_{\nu,\nu'}\, \psi_{\nu'} \qquad (10b)$$

As in Section I, it is found that the MX-MX interaction cancels for $\nu = \nu'$, giving a clue to the nonlinear optical response. For small field, MX are composed of many LL and therefore there is a strong interaction between all MX states whenever one MX-state is photoexcited. At high field the MXs become more and more diagonal in the LL basis, the interaction within a MX state vanishes but persist between different MX states. The residual interaction is mostly due to the self-energy corrections and is attractive. This explains the high field disappearance of the 1s-MX blue-shift found experimentally. The attractive inter-MX self-energy correction explains the experimental red shift of the 2s-MX induced by a population of photo excited 1s-MXs. Numerical solution of the Liouville equation on a basis of 1000-LL shows that the Coulomb correlations are indeed completely quenched in the extreme magnetic limit [25] in agreeent with the equilibrium exact result [26,27]. At high field MXs behave like two-level systems and their nonlinear optical response becomes dominated by Pauli exlusion i.e. MX-hν. We now discuss how the theory explains the nonlinear response of MX for the various excitation configurations.

The DTS obtained for below, at and above the 1s-MX resonance excitation are shown in Fig.(7). For comparison Fig. (8) displays the theoretical nonlinear absorption spectra, $\mathrm{Im}\chi^{(3)}(\omega)$, calculated using 500 LL. In the experimental spectra the solid lines correspond to the pump and probe arriving simultaneously on the sample, the short-dashed and long-dashed lines correspond to the pump arriving 600fs after and 600fs before the probe respectively. The insets show the position of the pump spectrum relative to that of the sample in each case. The theoretical spectra are caculated for 2D parabolic bands and therefore do not show the light-hole exciton series [25]. For below gap excitation, Fig. (7a), one sees a Stark shift to the blue of the 1s-MX and the 2s-MX that compares very nicely to the theoretical spectrum shown in Fig. (8a). For excitation at the 1s-MX, Fig. (7b), only bleaching of the 1s-MX is seen, without shift, whereas the 2s-MX exhibits a red shift as already discussed. This behavior is again in good agreement with the corresponding theoretical spectrum of Fig. (8b). The window of transparence seen in the σ_+absorption spectrum of the 85Å sample between the 1s-MX and the 2s-MX (see Fig. 5), gives a unique opportinity to investigate the response of a solid where the absorption spectrum resembles that of a molecules. The DTS obtained for excitation in this window is shown in Fig. (7c). The 1s-MX response is a red shift plus an induced saturation, in fact very similar to the response of the 2s-MX. The transient part of the 1s-MX response is essensitally a strong red Stark shift opposite in sign to that seen for below resonance excitation. This behavior is also seen in the theoretical spectrum of Fig. (8c). This overall comparison demonstrates that the essensitals of the physics are well captured by the BCS theory of MX [21,25].

4) CONCLUSION

We have shown that exciton-exciton interaction produces new nonlinear optical effects in time-resolved four wave mixing. At low densities it gives rise to a signal at negative time delay for which the time constant is two times smaller than that of the usual positive time delay signal. At high densities the lineshapes become complicated and can exhibit two maxima. We have also developed a manybody theory that incorporates on the same footing interactions with the photon-field and the Coulomb interaction. It accounts correctly for all experimental observations.

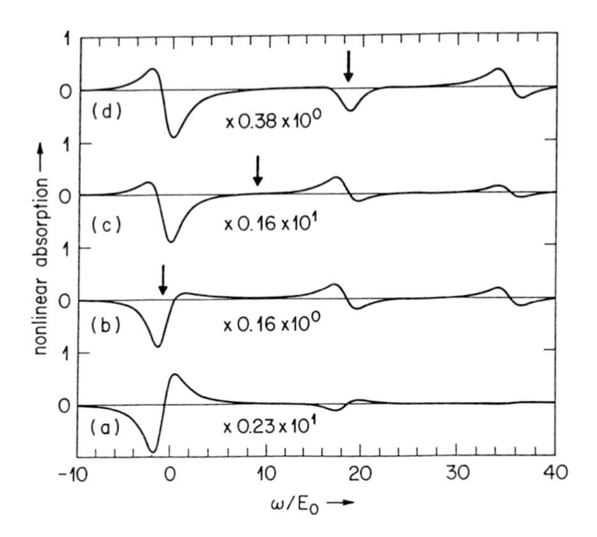

Figure 8 : Theoretical nonlinear absorption spectra for excitation indicated by the arrows.

We have also explored the femtosecond dynamics of the nonlinear optical response of magneto-excitons, as the quasi-2D quantum well electronic states are further confined in quasi-0D by a strong magnetic field. It was found, in agreement with theory, that at high field the Coulomb correlation between magneto-excitons in the same state is quenched. The correlation, however, persists between magneto-excitons in different states giving strong nonlinear responses. These results also show that a gas of magneto-excitons is a unique two-component manybody system that behaves very differently from the one-component systems such as electron gas in the fractional quantum Hall effect or Wigner crystal.

ACKNOWLEGEMENT: This work was performed in collaboration with J.B. Stark, S. Schmitt-Rink, W.H. Knox, and W. Schäfer.
The work of D.S. Chemla, was supported by the Director, Office of Energy Research, Office of Basic Energy Sciences, Division of Materials Sciences of the US Department of Energy under Contract No. DE-AC03-76SF00098.

REFERENCES

1. T. Yajima,and Yoichi Taira, J. Phys. Soc. Jap. 47, 1620 (1979)
2. L. Schultheis, J. Kuhl, A. Honold and T.W. Tu, Phys. Rev. Lett. 57, 1635, and 1797 (1986)
3. A. Honold, L. Schultheis, J. Kuhl and C. W. Tu, Appl. Phys. Lett 52, 2105 (1988)
4. K. Leo, M. Wegener. J. Shah, D.S. Chemla, E.O. Göbel, T.C. Damen, S. Schmitt-Rink and W. Schäfer, Phys. Rev. Lett. 65, 1340 (1990)
5. M. Wegener, D.S. Chemla, S. Schmitt-Rink and W. Schäfer, Phys. Rev. A42, 5675 (1990)
6. S. Schmitt-Rink, D.S. Chemla, Phys. Rev. Lett. 57, 2752 (1986).
7. S. Schmitt-Rink, D.S. Chemla, H. Haug, Phys. Rev. Lett. B37, 941 (1988).
8. W. Schäfer, Adv. Solid State Phys. 28, 63, (1988), W Schäfer, K.H. Schuldt, and R. Binder, Phys. Stat. Sol. B150, 407 (1988)
9. C. Ell, J.F. Muller, K. El Sayed, L. Banyai, and H. Haug, Phys. Stat. Sol. B150, 393 (1988) and Phys. Rev. Lett. 62, 304 (1989)
10. R. Zimmermann, Phys. Stat. Sol. B146, 545 (1988) and R. Zimmermann and M. Hartmann Phys. Stat. Sol. B150, 365 (1989)
11. I. Baslev, R. Zimmermann, and A. Stahl, Phys. Rev B40, 4095 (1989)
12. R. Binder, S.W. Koch, M. Lindberg, N. Peyghambarian, and, W. Schäfer, Phys. Rev. Lett. 65, 899 (1990)
13. P.W. Anderson, Phys. Rev. 112, 1900 (1958)
14. R. S. Knox, Solid State Physics Suplement-5, Academic Press, NY (1963)
15. O. Akimoto, and H. Hasegawa, J. Phys. Soc. Jnp., 22, 181 (1967)
16. M. Shinada, and K. Tanaka, J. Phys. Soc. Jnp., 29,1258 (1970)
17. S.R.E. Yang, and L.J. Sham, Phys. Rev. Lett. 58, 2598 (1987)
18. G.E.W. Bauer, and T. Ando, Phys. Rev. B38, 6015 (1988)
19. H. Chu, and Y.C. Chang, Phys. Rev. B40, 5497 (1989)
20. J.B. Stark, W.H. Knox, D.S. Chemla, W. Schäfer, S. Schmitt-Rink and C. Stafford, Phys. Rev. Lett. 65, 3033 (1990)
21. S. Schmitt-Rink, J.B. Stark, W.H. Knox, D.S. Chemla, W. Schäfer, Appl. Phys. B in press
22. J.B. Stark, W.H. Knox, D.S. Chemla, S. Schmitt-Rink, W. Schäfer, to be published
23. S. Schmitt-Rink, D.S. Chemla, and D.A.B. Miller, Phys. Rev. B32, 6601 (1985)
24. N. Peyghambariam, H.M. Gibbs, J.L. Jewell, A. Antonetti, A. Migus, D. Hulin, and A. Mysyrowicz, Phys. Rev. 53, 2433 (1984)
25. C. Stafford, S. Schmitt-Rink, W. Schäfer, Phys. Rev. B41, 10,000 (1990-I)
26. I.V. Lerner and Yu. E. Lozovik, Zh. Eksp. Teor. Fiz. 80, 1488 (1981) [Sov. Phys. JEPT 53, 763 (1981)]
27. D. Paquet, T.M. Rice, and K. Ueda, Phys. Rev. B32, 5208 (1985)
28. C. Kallin, and B.I. Halperin, Phys. rev. B30, 5655 (1984)
29. A.B. Dzyubenko, and, Yu. E. Lozovik, J. Phys. A24, 415 (1991)

Inst. Phys. Conf. Ser. No 123
Paper presented at the International Meeting on Optics of Excitons in Confined Systems, Giardini Naxos, Italy, 1991

Excitonic correlations in confined electron–hole magneto-plasmas

Gerrit E.W. Bauer

Philips Research Laboratories, 5600 JA Eindhoven, The Netherlands

ABSTRACT: The theory of the magneto-optical properties of metallic electron-hole plasmas in quantum wells is reviewed, emphasizing recent efforts to include screening and excitonic effects. The comparison of numerical calculations using different approximation schemes with magneto-luminescence experiments by Butov et al (1990,1991c) demonstrates the importance of excitons even at high plasma densities and provides evidence of dynamic screening effects.

1. INTRODUCTION

The attractive interaction between the carriers in electron-hole plasmas has been predicted to cause condensation of bound electron-hole pairs (excitons) into an *excitonic insulator* state (Keldysh and Koslov 1968, Halperin and Rice 1968, Comte and Nozieres 1982). Unfortunately, these theoretical expectations have never been experimentally verified. New theoretical efforts in this direction on semiconductor quantum wells (Bauer 1990a,1991a) were motivated by the increased excitonic interaction in quasi-two-dimensional systems. Excitonic effects have been observed experimentally in magneto-luminescence spectra of highly excited GaAs/AlGaAs (Potemski et al 1988, 1990, Maan 1990) and lattice-matched InGaAs/InP (Butov et al 1990) quantum wells. Coupled quantum wells with applied electric fields which spatially separate electrons and holes (Fukuzawa et al 1990) offer the prospect of lower plasma temperatures and, ultimately, superconductivity (Lozovik and Yudson 1976). Another important result for confined electron-hole plasmas is the proof that in strictly two dimensions and in the high-magnetic-field limit Hartree-Fock theory is exact at zero temperature (Lerner and Lozovik 1981, Paquet et al 1985).

The optical spectra calculated by Bauer (1990a) suggest that an experimental discovery of the condensed phase could be possible via standard spectroscopic methods. The calculations were carried out for quantum wells in the Hartree-Fock approximation and include the effects of external magnetic fields (Bauer 1990a,1991a). Experiments have not yet yielded evidence of plasma condensation, presumably because of too high plasma temperatures. However, the excitonic correlations which are responsible for the condensation also exist above the critical temperature, just like paramagnetic fluctuations persist above the Curie temperature of a ferromagnet. At zero magnetic fields these correlations should be detectable in the optical, luminescence and absorption/gain spectra in terms of an enhancement of the oscillator strengths close to the chemical potential (Schmitt-Rink et al 1986), but have not yet been observed experimentally. On the other hand, the excitonic precursors of the condensation as predicted for magneto-optical spectra (Bauer 1990a) have been recently identified by a collaboration led by the Chernogolovka group (Butov et al 1990, 1991b, 1991c). Bychkov and Rashba (1990) subsequently proposed a simplified Hartree-Fock theory which incorporated the most important physical effects. However, the good agreement with experiments (Butov et al 1991a) will be shown below to be fortuitous. The Hartree-Fock theory of Stafford et al (1991) (see also Stark et al 1990, Rappen et al 1991) takes the coupling between magneto-excitons into account, but is limited to zero temperature and strictly two-dimensional systems.

The Hartree-Fock theory of the present author (Bauer 1990a, 1991a) has been extended recently to include the effect of screening and applied to modulation-doped quantum wells (Bauer 1991b). In the present article I would like to discuss in some detail this extended theory for metallic electron-hole plasmas in magnetic fields, which has been successful in describing recent magneto-luminescence experiments (Butov et al 1991c). The expressions for the optical spectra of metallic plasmas in the Hartree-Fock approximation are already published (Bauer 1990b) and details of the numerical computations will be provided elsewhere (Bauer 1991c). As far as relevant for the rest of the paper the modifications which have been implemented to include screening effects are presented in Section 2. The high-magnetic-field limit is dealt with in Section 3. In Section 4 a detailed comparison is made between calculations based on different approximation schemes and between theory and experiments, followed by the conclusions in Section 5.

2. SCREENING AND SELF-ENERGIES

In the following, a model of parabolic bands for electrons and holes is adopted, spin-splittings are neglected, and quasi-equilibrium of the carriers at a temperature T and magnetic field H normal to the interfaces is assumed. The single-particle energies of particle type a (electron or hole) are labelled by an integer Landau-level index n:

$$E_a(n) = E_a^{\text{sub}} + \frac{eH\hbar}{cm_a}(n + \frac{1}{2}) + \Sigma_a(n; E_{a,0}(n)) = E_{a,0}(n) + \Sigma_a(n; E_{a,0}(n)) \tag{1}$$

The energy argument in the self-energy $\Sigma_a(n; E_{a,0}(n))$ is the non-interacting single-particle (Landau-level) energy (Ando et al 1982). The random-phase approximation is chosen here as a reference, since it seems to be more reliable than the Hubbard approximation, especially at lower densities (Ando et al 1982). In this approximation, the self-energies are conveniently separated into a "screened-exchange" Σ_a^{SX} and "Coulomb-hole" Σ_a^{CH} contribution (Haug and Schmitt-Rink 1984), which in a magnetic field become (Katayama and Ando 1989):

$$\Sigma_a^{SX}(n; E) = \sum_m \mathrm{P} \int_{-\infty}^{\infty} \frac{d\omega}{\pi} f(\omega - \mu_a)\, V_s(n,m; E - \omega)\, \mathrm{Im}\, G_a^0(m; \omega), \tag{2}$$

$$\Sigma_a^{CH}(n; E) = -\sum_m \mathrm{P} \int_{-\infty}^{\infty} \frac{d\omega}{\pi} g(\omega)\, G_a^0(m; E - \omega)\, \mathrm{Im}\, V_s(n,m; \omega). \tag{3}$$

where $G_a^0(m; \omega) = [\hbar\omega - E_{a,0}(m)]^{-1}$ is the non-interacting single-particle Green's function and f and g are Fermion and Boson occupation numbers:

$$f_a(\omega) = \frac{1}{e^{\beta\omega} + 1}; \quad g(\omega) = \frac{1}{e^{\beta\omega} - 1}; \quad \beta = \frac{1}{k_B T}. \tag{4}$$

The (quasi-) chemical potentials μ_a have to be determined self-consistently using Eq. (1). The matrix elements of the Coulomb interaction read:

$$V_S(n,m;\omega) = \int \frac{qdq}{2\pi}\, V_S(q;\omega)\,|\,J_{nm}(q)\,|^2 \tag{5}$$

where

$$J_{nm}(q) = \sqrt{\frac{N!}{M!}}\, e^{-\ell^2 q^2/4} (\ell q\,/\sqrt{2}\,)^{M-N}\, L_M^{(M-N)}(\ell^2 q^2/2). \tag{6}$$

$L_n^{(\alpha)}$ are associated Laguerre polynomials, $M = \max(m,n)$, $N = \min(m,n)$ and $\ell = ch/eH$ is the magnetic length. The symbol P indicates that at simple poles of the integrand the principal part should be evaluated. The Coulomb interaction is screened by the free-particle (RPA) dielectric function $\varepsilon(q,\omega)$:

$$V_S(q,\omega) = \frac{2\pi e^2}{\kappa q}\, F(q)\, \frac{1}{\varepsilon(q,\omega)} \tag{7}$$

where κ is the dielectric constant of the semiconductor and $F(q)$ is the subband form factor. The calculations are simplified by employing the plasmon-pole approximation (Haug and Schmitt-Rink 1984, Katayama and Ando 1989), where the energy-loss function $\mathrm{Im}\{1/\varepsilon(q,\omega)\}$ is approximated by a single effective plasmon mode. The plasmon dispersion is fixed by sum rules in terms of $\varepsilon(q,0)$, which is also needed to screen the excitonic electron-hole interaction. Following Ando and Uemura (1974), the singular screening in magnetic fields is dealt with by including the effect of impurity broadening of the Landau levels in the self-consistent Born approximation. The screening contributions of the electron and hole systems are assumed to be independent. This approximation should be valid above the transition temperature and/or at high magnetic fields where the Landau-level splitting is larger than the exciton binding energy.

Even in the plasmon-pole approximation the calculations are rather cumbersome, and can be simplified further by neglecting all energy dependence of the dielectric function, which is the so-called quasi-static approximation (Haug and Schmitt-Rink 1984, Uenoyama and Sham 1989). The expressions for the self-energy then become

$$\Sigma_a^{SX}(n) \simeq -\sum_m f(E_a(m) - \mu_a)\, V_s(n,m;0) \tag{8}$$

$$\Sigma_a^{CH} \simeq \frac{1}{2}\int \frac{qdq}{2\pi}\,[\,V_S(q;0) - V_0(q)] \tag{9}$$

where use has been made of the completeness relation $\Sigma_m J_{nm}(q)^2 = 1$ and $V_0(q)$ is the unscreened Coulomb interaction obtained by setting $\varepsilon(q,\omega) \equiv 1$ in Eq. (7). Note that the Coulomb-hole self-energy became identical for all Landau levels and independent of magnetic field. The Hartree-Fock approximation is recovered by setting all $V_S \rightarrow V_0$. The Coulomb-hole part vanishes and Eq. (8) reduces to the bare exchange integral.

3. HIGH-MAGNETIC-FIELD LIMIT

At very high magnetic fields the couplings to higher Landau levels vanish and at zero temperature and in the strictly two-dimensional limit ($F(q) \equiv 1$) Hartree-Fock theory is known to become exact (Lerner and Lozovik 1981, Paquet et al 1985). In this limit the energy per optically active (zero total momentum) electron-hole pair (exciton) becomes

$$E_X = E_0 - V_0 \tag{10}$$

where $E_0 = E_{e,0}(0) + E_{h,0}(0)$ is the sum of the non-interacting Landau-level energies and $V_{0/S} = V_{0/S}(0,0)$ are the bare/screened Coulomb interaction matrix elements for the lowest Landau level. An approximate theory should approach the Hartree-Fock result Eq. (10) at high magnetic fields. The exciton energy in the plasmon pole approximation for the self-energies (Katayama and Ando 1989) is seen to satisfy Eq. (10), since

$$E_X = E_0 - (\nu_e + \nu_h)V_S + (V_S - V_0) - (1 - \nu_e - \nu_h)V_S, \tag{11}$$

where the terms on the right-hand side are the sum of the screened-exchange self-energies, the Coulomb-hole self-energies and the exciton interaction, respectively, and $\nu_{e/h}$ are the Landau-level filling factors. The screening effect is indeed seen to vanish identically. On the other hand, the self-energies in the quasi-static approximation (Uenoyama and Sham 1989) fail this test. The reason for this is found in the Coulomb-hole self-energy. The neglect of the dynamic interaction allows for the virtual excitation into excited states without having to pay for the kinetic energies. All Landau levels therefore contribute equally to the screening at all magnetic fields, which causes a serious overestimation of the screening effect when the fields are high.

At finite temperatures, mean-field theory is no longer exact, but it is not expected to be far off reality (Lerner and Lozovik 1981). Assuming an uncondensed ground state, the electron-hole Green's function is easily obtained from the Bethe-Salpeter equation (Bauer 1991a):

$$G_{eh}(\omega) = \frac{1 - \nu_e - \nu_h}{\hbar\omega - E_X} \tag{12}$$

The pole of G_{eh} is located at the transition and exciton energy Eq. (10). We can see from Eq. (12) that no temperature effects are expected in the gain/absorption spectra. In contrast, the luminescence spectrum

$$I^{\text{lum}}(\omega) = \frac{1}{2\pi\ell^2} \frac{1 - \nu_e - \nu_h}{e^{\beta(E_X - \mu_e - \mu_h)} - 1} \delta(\hbar\omega - E_X) \quad \text{for} \quad T > T_c. \tag{13}$$

is seen to diverge when the sum of the chemical potentials approaches the transition energy $\mu_e + \mu_h = E_X$, which is nothing but the criterion for Bose condensation of excitons. The critical temperature is easily calculated to be

$$k_B T_c = \frac{(1 - \nu_e - \nu_h)V_0}{\ln\,(\nu_e^{-1} - 1) + \ln\,(\nu_h^{-1} - 1)}. \tag{14}$$

In Fig. 1 this result, which for neutral plasmas has been obtained before by Kuramoto and Horie (1978) is plotted as a function of electron and hole density. Below the transition the luminescence intensity is temperature-independent:

$$I^{\text{lum}}(\omega) = \frac{1}{2\pi\ell^2} \min(\nu_e, \nu_h)\delta(\hbar\omega - E_X) \quad \text{for} \quad T \le T_c. \tag{15}$$

At zero magnetic fields the Coulomb interaction is responsible for a correlated motion between electrons and holes and thus the formation of bound pairs which

Fig. 1: Condensation transition temperature (Eq.(14)) of an electron-hole plasma within the magnetic quantum limit as a function of the electron and hole filling factors. The 9 contours are at constant T_c from 0.1 til 0.9 times the maximum critical temperature $V_0/(4k_B)$ which is reached at $\nu_e = \nu_h = 1/2$.

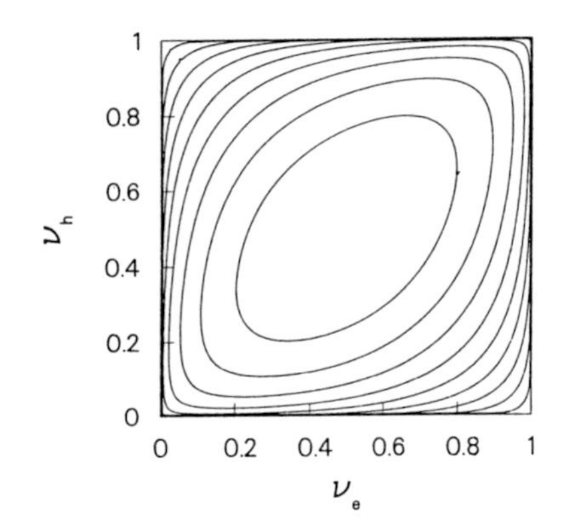

may condense and create a gap in the density of states. Optically active electron-hole pairs can be "bound" also by magnetic fields. At high magnetic fields these "excitons" are non-interacting and unmodified by the Coulomb interaction and temperature, at least in the Hartree-Fock approximation. In this limit only the entropy of the system is temperature dependent via the coherence of the phases of the individual excitons. When approaching T_c from above, the coherence area of the fluctuating dipoles increases in size and becomes macroscopic at the condensation temperature, which leads to the divergence in the luminescence intensity. In contrast, the condensed state decays only rather weakly because of strong reabsorption at the chemical potential.

Finally, it should be stressed that the above discussion of the plasma condensation is very incomplete, and it is by no means clear if the excitonic insulator can be created at all by high excitation. Already, the explanation of the low-temperature radiative lifetime of a single exciton requires the introduction of phonon scattering (Feldmann et al 1987) or polariton effects (Rappel et al 1988). A discussion of these complications is beyond the scope of the present paper.

4. COMPUTATIONAL RESULTS

The convergence of the expansion in Landau levels is uncritical at higher magnetic fields, and the results are fully converged with about 50 Landau levels in the basis at magnetic fields of 8.65 tesla, which is the maximum field at which Butov et al (1990) have carried out their experiments. The parameters $m_e = 0.05$, $m_h = 0.2$, $\kappa = 12.5$ for well and barrier material, and Gaussian subband envelope functions are adopted to take into account the effects of finite barrier heights. It is understood that non-parabolicity leads to an overestimation of the transition energies of higher Landau levels in the present parabolic band model.

Let us first look at the results of Hartree-Fock theory for different pair densities at intervals of $5{\cdot}10^{10}\mathrm{cm}^{-2}$ (Figs. 2-4). The computed luminescence at 10 K, which is below the (Hartree-Fock) phase-transition temperature for the discrete grid of densities considered here, is shown in Fig. 2. Very close to even occupation numbers the plasma is still normal: The strong reduction of the transition temperature for almost full or almost empty uppermost Landau levels can be understood from the discussion in Section 2. The lines with a strongly positive slope are a signature of condensation, but their oscillator strengths are very low at high magnetic fields. What is important

Fig. 2: Luminescence spectra of the electron-hole plasma in a 150 Å InGaAs/InP quantum well at 10K and a magnetic field of 8.65 T calculated in the Hartree-Fock approximation as a function of plasma density. The area of the dots is proportional to the luminescence intensities and the energy zero is the band gap of the bulk well material. The diamonds indicate the exciton energies at zero plasma density.

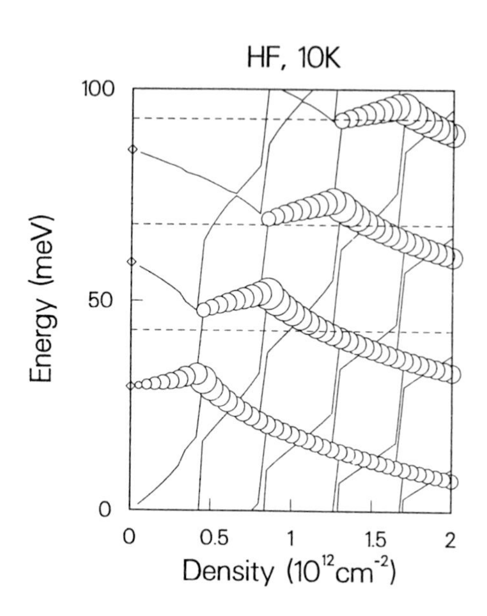

is the strong blue shift of the optical transitions between partially filled Landau levels. The exact cancellation of self-energy and vertex part, *i.e.*, constancy of E_X, is thus seen to be destroyed by the coupling to higher Landau levels (Lerner and Lozovik 1981, Paquet et al 1985). This is due to the modification of the wave functions by the Coulomb interaction, *i.e.* a genuine correlation effect. The coupling to higher Landau levels allows the carriers to come closer to each other than the simple Gaussian Landau-level wave functions would permit, which leads to an increased binding energy. On the other hand, this deformation does not have much effect on the exchange energy. The loss in the excitonic binding with increasing filling factor is then no longer fully compensated by the increasing exchange energy. Hence, a blue shift occurs which can alternatively be interpreted as a net repulsive interaction between Boson-like magneto-excitons (Butov et al 1991c). The results in the Hartree-Fock approximation for 50 and 100 K in Fig. 3 (top left) and 4 (left) emphasize the importance of temperature. At high temperatures the quenching of excitonic effects with density is reduced because the injected carriers are distributed over more Landau levels. The exchange energy is much less temperature-dependent, resulting in a small red shift with increasing density even at the Fermi level.

The theory of Bychkov and Rashba (1990) is recovered by setting all non-diagonal matrix elements to zero except those in the exchange energy, assuming zero temperature and taking the strictly two-dimensional quantum limit by eliminating the form factors. While indentifying correctly the underlying competition between excitonic and exchange effects, the quantitative agreement with the experiments (Butov 1991a) appears to be fortuitous due to the subtle cancellation of the effects of finite temperature, finite well width, and inter-Landau-level couplings at about 50 K.

Screening effects are included in the computational results presented in Figs. 3-4. In all cases considered here, the intra-Landau-level screening, which vanishes when $\Gamma \ll k_B T$, has been neglected. This approximation is justified because the effect of

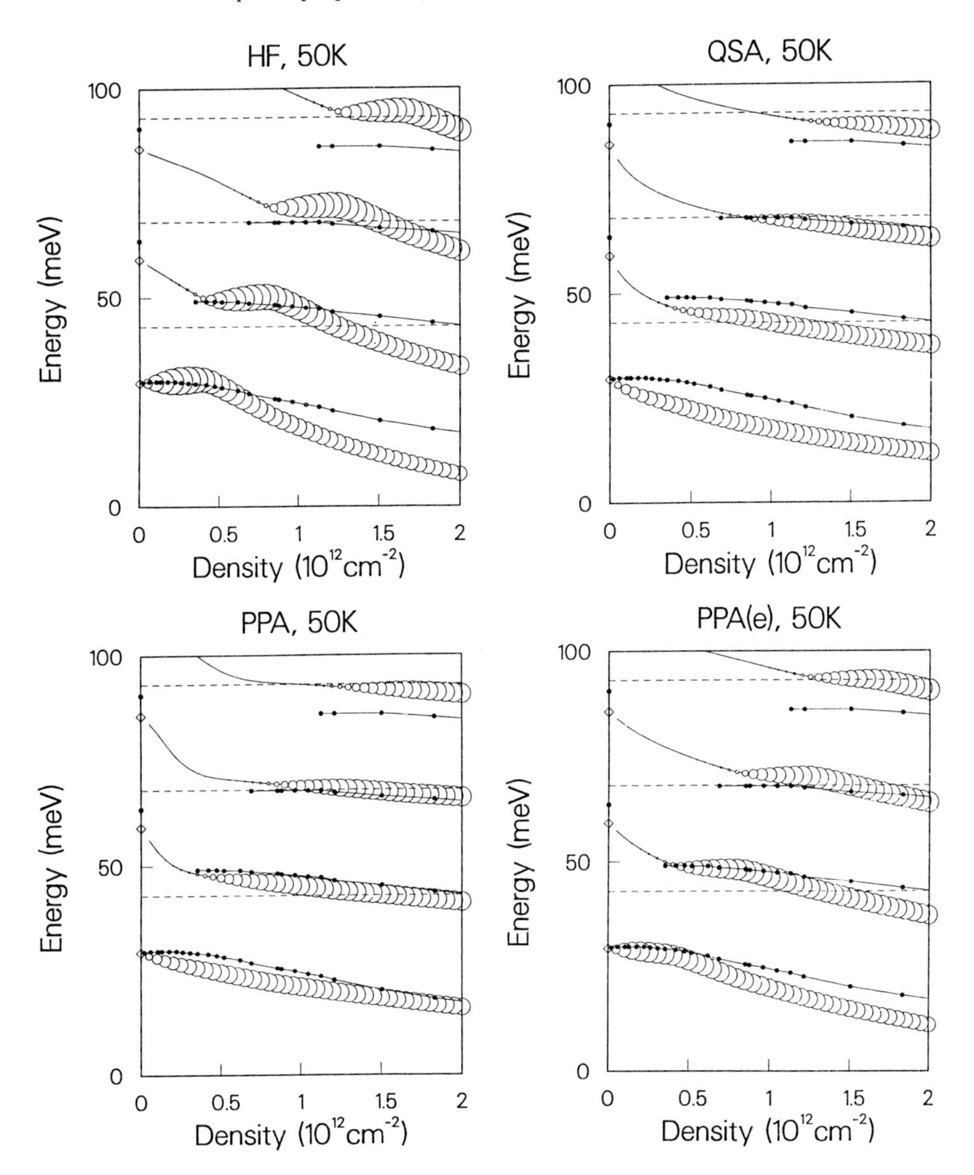

Fig. 3: Magnetoluminescence spectra of a 150 Å *InGaAs/InP quantum well as a function of electron-hole pair density at a plasma temperature of 50 K and magnetic field of 8.65 T, calculated in different approximations. The theoretical results (open dots) are compared with the data of Butov et al (1990) (filled dots) at about the same plasma temperature. A bulk gap of 796 meV has been substracted from the experiments. The area of the open dots is proportional to the calculated luminescence oscillator strengths. The diamonds indicate the exciton energies at zero plasma density. Results are given for the Hartree-Fock (top left), quasi-static (top right) and plasmon-pole (bottom) approximations. In the bottom right-hand figure the screening contribution of the hole system has been switched off (see text).*

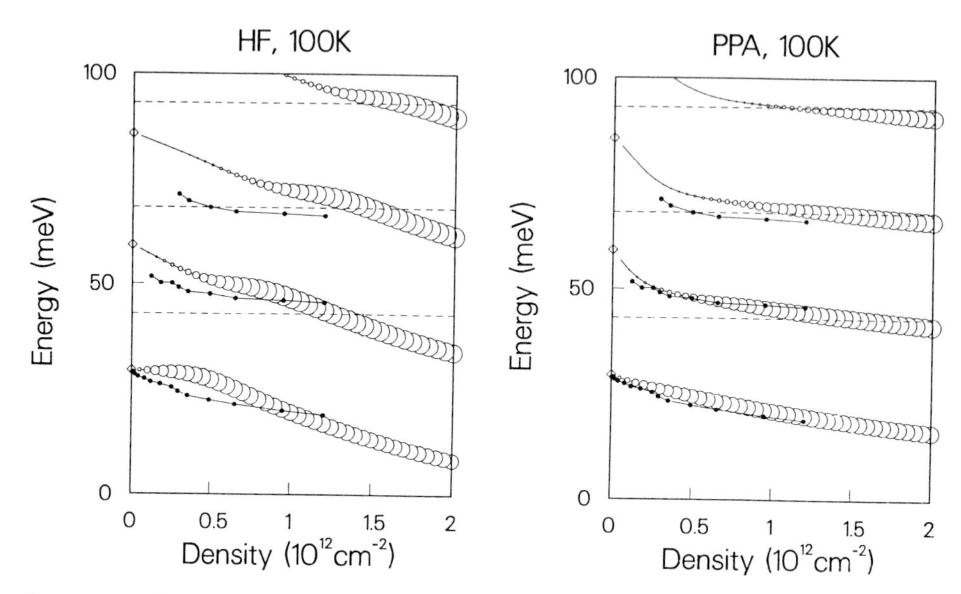

Fig. 4: As Fig. 3, but for a plasma temperature of 100K. The experiments have been carried out at a lattice temperature of 100K (Butov et al 1991c). Results are given for the Hartree-Fock (left) and plasmon-pole (right) approximations. A bulk gap of 788 meV has been substracted from the experiments, which reflects the thermal reduction of the bulk band gap.

intra-Landau-level screening is small even at low temperatures: The cancellation of Coulomb-hole and screened-exchange effects is even more complete for neutral plasmas than for modulation-doped quantum wells (Katayama and Ando 1989, Uenoyama and Sham 1989) and can be safely disregarded at 50 K.

The quasi-static approximation (Eqs. (8,9)) improves the agreement in comparison with experiments at high densities (Fig. 3, top right), but fails badly for low densities. This was to be expected from the discussion in Section 3. The results for the plasmon-pole approximation for 50 K and 100 K are given in Figs. 3 and 4. Especially in the high-density regime, the agreement is significantly improved compared to the Hartree-Fock approximation, emphasizing that the virtual inter-Landau-level polarizations cannot be neglected. On the other hand, the screening causes a red shift of the optical transitions at lower densities, thus disappointingly failing to reproduce the main experimental result. This overestimation of the van der Waals attraction between excitons (Nozieres and Comte 1982) can be explained as a dynamic screening effect as follows. The static screening approximation is equivalent to the assumption that the plasma responds infinitely swiftly to a density perturbation. In reality it will take a finité amount of time for the plasma charge to rush in and to screen the region of non-neutrality. These retardation effects are included approximately in the self-energies via the plasmon-pole approximation, but not in the excitonic electron-hole interaction. The introduction of dynamic screening effects into the Bethe-Salpeter equation is extremely difficult and is not attempted here, but the order of magnitude can be estimated easily. The retardation effects are largest for the hole gas: Due to the higher mass its response, as measured *e.g.* by the plasma frequency, is much slower than that of the electrons. A zero-order approximation to account for dynamic screening is introduced by retaining the static screening of the electrons (infinitely fast response), but assuming an infinitely slow response of the holes by setting the polarizability of the hole system equal to zero. In Fig. 3 (bottom right), we see that this approach gives very good results for partially occupied level energies while clearly improving on the Hartree-Fock results at high densities. For a plasma temperature of 100K the plasmon-pole approximation is clearly superior to the Hartree-Fock approximation. The screening contribution from the holes has become sufficiently small, though, and good agreement with experiment is obtained even without having to switch it off.

5. CONCLUSIONS

A detailed comparison between experiment and theory for the magneto-optical properties of high-density electron-hole plasmas has been carried out successfully and for the first time (Butov et al 1991c), identifying effects of excitons as well as dynamic screening. The agreement of experiment and theory is very satisfying and increases the confidence in the mean-field model (though the experiments by Potemski et al 1988,1990 disagree). The ultimate goal, *i.e.* the identification of the excitonic insulator, has not been achieved yet. In high magnetic fields the only effect of the condensation is the divergence of the luminescence at the transition temperature. More complicated phenomena are expected when the Landau-level splittings are smaller or of the order of the exciton binding. Since it seems doubtful whether the plasma temperature in simple quantum wells can be decreased below $\simeq 40K$ (Butov et al 1990), coupled quantum wells appear to be the most attractive candidates in the search for condensation phenomena.

I should like to acknowledge illuminating discussions with Leonid Butov on the two-dimensional electron-hole plasma in the high-magnetic-field limit.

REFERENCES

Ando T, Fowler A B and Stern F 1982 Rev. Mod. Phys. 54 437
Ando T and Uemura Y 1974 J. Phys. Soc. Japan 37 1044
Bauer G E W 1990a Surf. Science 229 374; Phys Rev Lett. 64 60
Bauer G E W 1990b Localization and Confinement of Electrons in Semiconductors, eds G. Bauer, F. Kuchar, and H. Heinrich, (Berlin: Springer)
Bauer G E W 1991a Condensed Systems of Low Dimensionality, ed L.J. Beeby, (New York: Plenum Press). Note the printing error in the denominators of Eqs. (18-19) where F(o) should not be squared. Th factor $1/\pi$ should be removed.
Bauer G E W 1991b Solid State Commun. 78 163; Surf. Science in press.
Bauer G E W 1991c Phys. Rev. B to be published
Butov L V, Kulakovskii V D, Forchel A and Grützmacher D 1990 Pis'ma Zh. Eksp. Teor. Fiz. 52 759; JETP Lett. 52 121
Butov L V, Kulakovskii V D and Rashba E I 1991a Pis'ma Zh. Eksp. Teor. Fiz. 53 104; JETP Lett. 53 108
Butov L V and Kulakovskii V D 1991b Pis'ma Zh. Eksp. Teor. Fiz. 53 444; JETP Lett. 53 466
Butov L V, Kulakovskii V D, Bauer G E W, Forchel A and Grützmacher D 1991c Phys. Rev. Lett. submitted.
Bychkov Yu A and Rashba E I 1990 Pis'ma Zh. Eksp. Teor. Fiz. 52 1209; JETP Lett. 52 624
Comte C and Nozieres P 1982 J. Phys. 43 1069

Fukuzawa T, Kano S S, Gustafson T K and Ogawa T 1990 Surf. Science 228 482
Halperin B I and Rice T M 1968 Solid State Physics 21 115
Haug H and Schmitt-Rink S 1984 Prog. Quantum Electron. 9 3
Feldmann J, Peter G, Göbel E O, Dawson P, Moore K, Foxon C T and Elliott R J 1987 Phys. Rev. Lett. 59 2337
Katayama S and Ando T 1989 Solid State Commun. 70 97
Keldysh L V and Koslov A N 1968 Zh. Eksp. Teor. Fiz. 54 978; Soviet Physics JETP 27 521
Kuramoto Y and Horie C 1978 Solid State Commun. 25 713
Lerner I V and Lozovik Yu E 1981 Zh. Eksp. Teor. Fiz. 80 1488; Sov. Phys. JETP 53 763
Lozovik Ye E and Yudson V I 1976 Zh. Eksp. Teor. Fiz. 71 738; Sov. Phys. JETP 44 389
Paquet D, Rice T M and Ueda K 1985 Phys Rev B 32 5208
Maan J C 1990 Localization and Confinement of Electrons in Semiconductors, eds Bauer G, Kuchar F and Heinrich H (Berlin: Springer)
Nozieres P and Comte C 1982 J. Phys. 43 1083
Potemski M, Maan J C, Ploog K and Weimann G 1988 Proc. 19th Int. Conf. on the Physics of Semiconductors, ed W. Zawadzki, (Warsaw: Polish Academy of Science)
Potemski M, Maan J C, Ploog K and Weimann G 1990 Solid State Commun. 75 185
Rappel W J, Feiner L F and Schuurmans M F H 1988 Phys. Rev. B 38 7874
Rappen T, Schröder J, Leiße J, Wegener M and Schäfer W 1991 to be published
Schmitt-Rink S, Ell C and Haug H 1986 Phys. Rev. B 33 1183
Stafford C, Schmitt-Rink S and Schäfer W 1990 Phys. Rev. B 41 10000
Stark J B, Knox W H, Chemla D S, Schäfer W, Schmitt-Rink S and Stafford C 1990 Phys. Rev. Lett. 65 3033
Uenoyama T and Sham L J 1989 Phys. Rev. B 39 11044

Paper presented at the International Meeting on Optics of Excitons in Confined Systems, Giardini Naxos, Italy, 1991

Optical excitation and Bose condensation of excitons in quantum confined systems

S Glutsch (a), F Bechstedt (a) and R Zimmermann (b)

(a) Friedrich–Schiller–Universität, Max–Wien–Platz 1, O-6900 Jena, Germany
(b) Zentralinstitut für Elektronenphysik, Hausvogteiplatz 5–7, O-1086 Berlin, Germany

ABSTRACT: The linear optical response of the excitons in low-dimensional semiconductor systems – quantum wells – pumped near the energy of the lowest exciton transition is studied under inclusion of many body effects. Generalized Bloch and gap equations are simultaneously solved. We focus our attention especially to three aspects: (i) time dependence, (ii) phase transition, and (iii) influence of dimensionality on the Coulomb interaction.

1. INTRODUCTION

New developments of subpicosecond spectroscopy have allowed to observe new coherent phenomena in semiconductor microstructures. One example is the discovery of the optical Stark effect in GaAs-GaAlAs quantum well (QW) structures. A first microscopic description of this phenomenon for bulk materials and QW structures explains the blueshift of the exciton lines for nonresonant pumping (Schmitt-Rink et al 1988). For pump frequencies near or above the exciton resonance the theory breaks down and drastic changes of the blueshift behaviour are predicted (Binder et al 1990, Bechstedt et al 1991). Even a Bose-like condensation of the virtually excited electron–hole pairs is predicted by Comte et al (1988) and Glutsch et al (1991).

In the present paper we study the quasi stationary and time dependent optical properties of a quasi-two-dimensional (quasi-$2D$) semiconductor structure pumped with laser pulses near the lowest exciton. Bloch and gap equation derived within the Hartree–Fock approximation and the collision-free limit are simultaneously solved. Results are compared with those for the $3D$ case (bulk) and $1D$ case (QW wire).

2. THEORETICAL BASIS

The Hamiltonian of interacting electrons (e) ($i = 1$) and holes (h) ($i = 2$) in the QW structure illuminated with a classical pump field $\vec{E}(t) = \vec{E}_p(t)\exp(-i\omega_p t) + \text{c.c.}$ takes the form

$$H = \sum_{i=1}^{2}\sum_{n=1}^{\infty}\sum_{\vec{k}_\perp} E_{in}(\vec{k}_\perp) c^\dagger_{in\vec{k}_\perp} c_{in\vec{k}_\perp} - \sum_{\substack{i,j=1\\ i\neq j}}^{2}\sum_{n=1}^{\infty}\sum_{\vec{k}_\perp} \vec{\mu}_{ij}\cdot\vec{E}(t) c^\dagger_{in\vec{k}_\perp} c_{jn\vec{k}_\perp}$$

$$+\frac{1}{2}\sum_{n_1n_2n_3n_4}\sum_{i,j=1}^{2}\sum_{\vec{k}_\perp\vec{k}'_\perp\vec{q}} V_{n_1n_2n_3n_4}(\vec{q}_\perp) c^\dagger_{in_1\vec{k}_\perp} c^\dagger_{jn_2\vec{k}'_\perp-\vec{q}_\perp} c_{jn_3\vec{k}'_\perp} c_{in_2\vec{k}_\perp-\vec{q}_\perp}, \quad (1)$$

$$V_{n_1,n_2,n_3,n_4}(\vec{q}_\perp) = \frac{2\pi e^2}{\varepsilon_0 q_\perp^2 A} \frac{4}{\pi^2} \int_0^\infty d\zeta \int_0^\infty d\zeta' e^{\frac{q_\perp d}{\pi}|\zeta-\zeta'|} \sin(n_1\zeta)\sin(n_2\zeta)\sin(n_3\zeta')\sin(n_4\zeta')$$

where the creation (annihilation) operator $c^\dagger_{in\vec{k}_\perp}$ ($c_{in\vec{k}_\perp}$) of an electron in an one-particle state $\left|in\vec{k}_\perp\right\rangle$ with the subband index n, a wavevector $\vec{k}_\perp$ perpendicular to the growth axis, and an energy $E_{in}(\vec{k}_\perp) = (-1)^{i+1}\left[E_g/2 + \varepsilon_{in} + \hbar^2 k_\perp^2/(2m_i)\right]$ (E_g - gap energy of semiconductor, ε_{in} - quantization energy, m_i - effective mass). $\vec{\mu}_{12}$ is the Bloch matrix element of the dipole allowed optical transition between valence and conduction states. The Coulomb potential $V(\vec{q}) = 4\pi e^2/(V\varepsilon_0 q^2)$ in the electron–electron interaction is screened by the background dielectric constant ε_0. d denotes the thickness of the well and A the normalization area.

The one-particle properties of the pumped system are studied within the density matrix formalism. Thereby, we assume the diagonality of the density matrix with respect to the subband indices. This approximation is valid in the limit of well thicknesses d smaller then the exciton radius. In Hartree–Fock approximation the resulting 2×2 electron–hole density matrix $n_{ij}(n,\vec{k}_\perp,t)$ $(i,j=1,2)$ satisfies the Heisenberg equation:

$$i\hbar\frac{\partial}{\partial t}\hat{n}(\vec{k}_\perp,t) = \left[\hat{E}(\vec{k}_\perp,t),\hat{n}(\vec{k}_\perp,t)\right]_- \quad . \tag{2}$$

These Bloch equation have to be solved self-consistently with gap equations for the energy matrix $E_{ij}(n,\vec{k}_\perp,t)$. With the particular form of the Hamiltonian (1) the gap equations take the form (Zimmermann 1990)

$$E_{in_1jn_2}(\vec{k}_\perp,t) = \delta_{n_1n_2}\begin{pmatrix} E_{1n_1}(\vec{k}_\perp) - \frac{1}{2}\hbar\omega_p & -\vec{\mu}_{12}\cdot\vec{E}_p(t) \\ -\vec{\mu}^*_{12}\cdot\vec{E}^*_p(t) & E_{2n_1}(\vec{k}_\perp) + \frac{1}{2}\hbar\omega_p \end{pmatrix} \tag{3}$$

$$-\sum_{n_3n_4}\sum_{\vec{q}_\perp} V_{n_1n_3n_2n_4}(\vec{k}_\perp - \vec{q}_\perp)\left[n_{in_3jn_4}(\vec{q}_\perp) - \delta_{ij}\delta_{i2}\delta_{n_3n_4}\right] .$$

The linear response of the pumped system to an optical probe beam can be evaluated from the Heisenberg equation (2) assuming a weak perturbation $\delta\vec{E}_p$ oscillating with frequency ω on the pump field $\vec{E}_p$ (Schmitt-Rink et al 1988). More direct, however, is the solution of the Bethe–Salpeter equation for the polarization function that is immediately related to the optical suszeptibility. Within the Hartree–Fock approximation and a twice application of the Bogoliubov transformation diagonalizing the matrix (3) the susceptibility can be calculated by inversion of an operator strongly related to the Hamiltonian (1) and the mentioned Bogoliubov transformation.

3. RESULTS

Results from the simultaneous solution of the equations of motion (2) and the gap equations (3) are plotted in Fig. 1. The time dependence of the amplitude of the exciting pump pulse during the switch on is assumed to be

$$\vec{E}_p(t) = \vec{E}_p\left\{\frac{1}{2} + \frac{1}{2}\mathrm{erf}\left[\left(\frac{t}{\tau} - 3\right)\right]\right\} \tag{4}$$

All quantities are expressed in terms of bulk excitonic units, the Rydberg energy R_{exc}, the Bohr radius a_{exc} and a characteristic time $\tau_{exc} = \hbar/R_{exc}$. In all explicit calculations we apply parameters of the heavy-hole exciton in GaAs. For numerical reasons only the lowest two subbands are considered. The steepness of the pulse is fixed at $\tau = 1.67\tau_{exc}$. In Fig. 1 we plot the total density of excited e–h pairs in a QW structure with thickness $d = 0.1a_{exc}$ versus time for fixed pump intensity $\vec{\mu}_{12}\cdot\vec{E}_p = 2\times10^{-2}R_{exc}$ and different pump detuning $\hbar\omega_p - E_g^*$ ($E_g^* = E_g+\varepsilon_{11}+\varepsilon_{21}$ - effective QW gap). The dashed line corresponds to off-resonant excitation whereas the solid

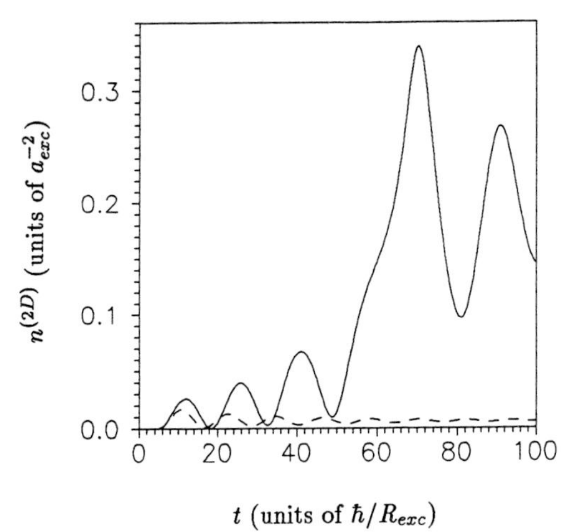

Fig. 1. Total density of excited e–h pairs.
Solid line - $\hbar\omega_p - E_g^* = -3R_{exc}$,
dashed line - $\hbar\omega_p - E_g^* = -4R_{exc}$

line describes pumping with frequencies above the exciton line. As in the $3D$ case (Glutsch et al 1991) the system exhibits Rabi-like oscillations after switching on the pump pulse. In the resonant case the system becomes unstable, i.e. the Rabi oscillations are strongly amplified by the Coulomb coupling of the e–h pairs. A transition into a new stable regime happens that is

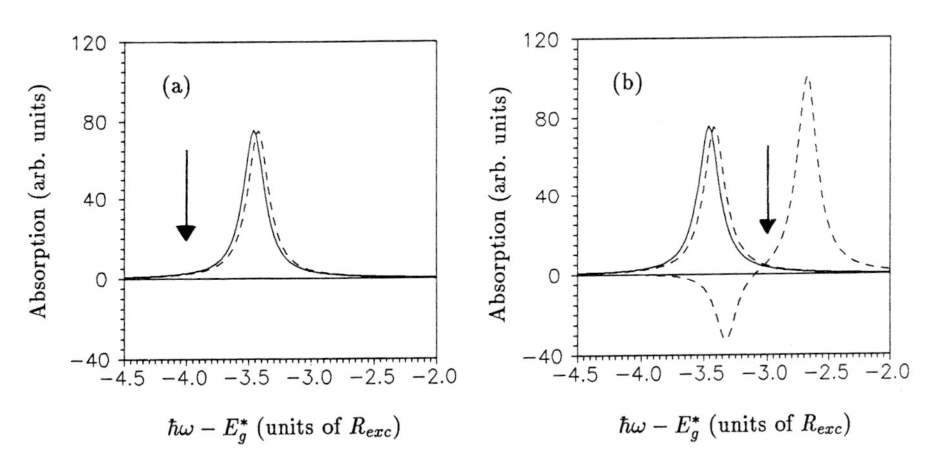

Fig. 2. Optical absorption versus probe detuning for $\hbar\omega_p - E_g^* = -4R_{exc}$ (a) and $\hbar\omega_p - E_g^* = -3R_{exc}$ (b). Solid line - $\vec{\mu}_{12}\cdot\vec{E}_p = 0$, dashed line - $\vec{\mu}_{12}\cdot\vec{E}_p = 2\times10^{-2}R_{exc}$.

characterized by a drastically Coulomb enhanced gap function $E_{12}(n, \vec{k}_\perp)$, i.e. effective Rabi frequency. Even if the pump field is switched off a finite pair density and polarization remain. This phenomenon is known as excitonic insulator (Comte et al 1988, Glutsch et al 1991).

The consequences of the behaviour described in Fig. 1 with respect to the optical properties of the pumped QW structure are shown in Fig. 2. The positions of the pump frequency are marked with arrows. For off-resonant pump excitation (a) the well-known results for the optical Stark effect (Schmitt-Rink et al 1988) are reproduced in the absorption spectra. The coherent excitation produces a blue shift. In the resonant case (b) we present two different absorption spectra with the stationary (higher pair density) and quasi-stationary (lower pair density) solutions (cf. Fig. 1). The metastable solution behaves like that of the off-resonant case. The stable solution related to the Bose-like condensate reveals increasing blue shift and oscillator strength. This excitonic insulator is characterized by a zero crossing just at the pump frequency ω_p with optical gain below.

4. SUMMARY

In the case of nonresonant pumping the direct solution of the Bloch equations for the density matrix confirms the dynamical Stark effect results known from the stationary treatment. After action of the pump pulse and some Rabi oscillations a finite virtual density of electron–hole pairs remains. It is responsible for the blueshift of the exciton absorption lines. On the other hand, in the resonant case drastic changes happen. After Rabi oscillations around a low-density value (which vanishes in the zero pump-intensity limit) the system jumps into a new stage with a pair density which does not vanish for zero intensity (driven Bose condensate). The accompanied absorption spectra are quite different. Zero absorption is observed at the pump frequency.

The particular behaviour described above is rather independent of the width of the microstructure. The same effects are observed in the bulk $3D$ case (Binder et al 1990, Glutsch et al 1991), the QW quasi-$2D$ case (this paper), as well as in the QW wire quasi-$1D$ case (Glutsch et al, to be published). As expected, the increasing binding energy and oscillator strength at reduced dimensionsinality lead to stronger nonlinearities.

5. REFERENCES

Schmitt-Rink S, Chemla D S and Haug H 1988 Phys. Rev. B37 941

Binder R, Koch S W, Lindberg M, Peyghambarian N and Schäfer W 1990 Phys. Rev. Lett. 65 899

Bechstedt F and Glutsch S 1991 Phys. Rev. B44 3638

Comte C and Mahler G 1988 Phys. Rev. B38 10517

Glutsch S, Bechstedt F and Zimmermann R (to be published)

Glutsch S and Zimmermann R 1991 Phys. Rev. B (accepted)

Zimmermann R 1990 Adv. in Solid State Phys. 30 295

Inst. Phys. Conf. Ser. No 123
Paper presented at the International Meeting on Optics of Excitons in Confined Systems, Giardini Naxos, Italy, 1991

Application of light-induced bandstructure modifications in heterostructures for light-activated switching

L M Weegels, E-J Vonken, J E M Haverkort, M R Leys and J H Wolter

University of Technology, Physics Department, P O Box 513, NL-5600 MB Eindhoven, The Netherlands

ABSTRACT: The experimentally observed blue shift of an excitonic recombination line originating from a 2-dimensional electron gas in a modulation doped heterostructure is explained by the mechanism of light-induced bandstructure modifications. This mechanism can be applied for light-activated switching in room temperature devices when a pseudomorphic heterostructure is specially designed and optimized.

1. INTRODUCTION

Recently, there has been an increasing interest in opto-electronic device applications. In particular, the Quantum Confined Stark Effect (QCSE) (Miller et al 1984) is employed in the self-electro optic effect device (SEED) (Miller et al 1988). The QCSE causes a red shift of the excitonic absorption resonances of multiple quantum well structures when an electric field perpendicular to the semiconductor layers is applied. In SEED devices utilizing Wannier-Stark localization (Mendez et al 1988, Bar-Joseph et al 1989) the absorption resonances are blue shifted. Here we propose a novel mechanism causing a blue shift based upon light-induced bandstructure modifications in heterostructures. It can be applied to an optical switch or an all-optical modulator. The mechanism is related to electrostatic effects caused by the build-up of photo-generated carriers. In a modulation doped heterostructure containing a 2-dimensional electron gas (2DEG) the bandstructure depends largely on the separation of the electrons in the 2DEG and their parent donors and on the formation of a depletion layer near the 2DEG. When the heterostructure is illuminated, the photo-generated electron-hole pairs are separated in the depletion field. The subsequent build-up of carriers modifies the bandstructure in such a way that the transition energies are blue-shifted. We have studied these light-induced bandstructure modifications in a modulation doped $GaAs/Al_xGa_{1-x}As$ heterostructure by means of low-temperature photo-luminescence (PL) spectroscopy (Weegels et al 1991a). We emphasize that this mechanism can be applied at room temperature when a pseudomorphic modulation doped $Al_xGa_{1-x}As/In_yGa_{1-y}As/GaAs$ heterostructure is specially designed and optimized. In this paper, we report an experimental study to elucidate the mechanism of light-induced bandstructure modifications. We describe a design of a pseudomorphic heterostructure which utilizes this mechanism at room temperature.

2. THE MECHANISM OF LIGHT-INDUCED BANDSTRUCTURE MODIFICATIONS

We have grown a specially designed modulation doped heterostructure containing a 2DEG in a 55 nm GaAs layer confined between a 25 period 5 nm GaAs/5 nm AlAs superlattice (SL) and a modulation doped $Al_{0.33}Ga_{0.67}As$ layer with a 10 nm spacer layer.The thin GaAs layer ensures a small overlap between electrons in the 2DEG and photo-generated holes near the SL barrier. This results in a finite probability of radiative recombination between the electrons in the 2DEG and the holes. In the low temperature PL spectrum (dashed line) of Fig. 1 we observe an intense excitonic recombination line in the spectral range between the bulk emission lines of (bound) excitons and electron to acceptor transitions. Previously, we have proven (Weegels et al 1991a) that this line originates from the 2DEG by changing the electron density by means of a bias voltage applied to a semi-transparent Schottky gate. The excitonic line is blue-shifted from 1.504 to 1.515 eV when the excitation intensity is increased by three orders of magnitude. In Fig. 1 the PL and PL excitation (PLE) spectra are shown at an intermediate excitation density, where the emission spectrum is dominated by the transition from the second subband of the 2DEG.

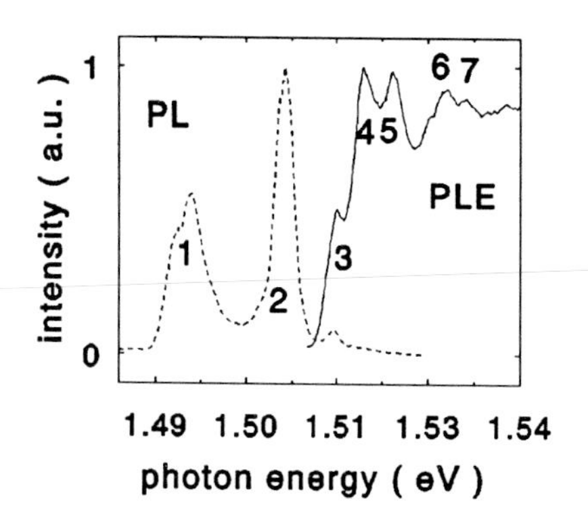

Fig. 1: Photoluminescence (dashed) and photoluminescence excitation (solid) spectra of the modulation doped heterostructure at 1.7 K. Labelling: 1. electron to acceptor transition, 2. recombination line due to the second subband of the 2DEG, 3. bulk excitons. For PLE the detection is set on the line labelled 2. The PLE spectrum shows transitions from the heavy and light hole subband to the third (resp. labelled 4 and 5) and fourth (resp. labelled 6 and 7) electron subband.

Further, in the PLE spectrum two doublets are well-resolved and arise from excitonic transitions from the heavy and light hole subbands to the third and fourth electron subbands. These excitonic transitions to the excited states of the 2DEG show a small blue shift of 1-2 meV with increasing excitation intensity. In Fig. 2 the experimentally observed transition energies in the PL and PLE spectra are shown versus the normalized excitation intensity.

The excitonic line in the PL spectrum is due to recombination of electrons in the 2DEG and photo-generated holes which remain near the 2DEG as a result of the electrostatic attraction due to the excess photo-generated electrons in the 2DEG. The blue shift of the transitions is related to electrostatic effects as a result of the build-up of photo-generated carriers tending to modify the bandstructure. In particular, two mechanisms can cause these bandstructure modifications:

(i) In the thin GaAs layer the photo-neutralization of acceptors gives rise to a flattening of the conduction and valence bands between the 2DEG and the SL. This process results in a potential difference equal to:

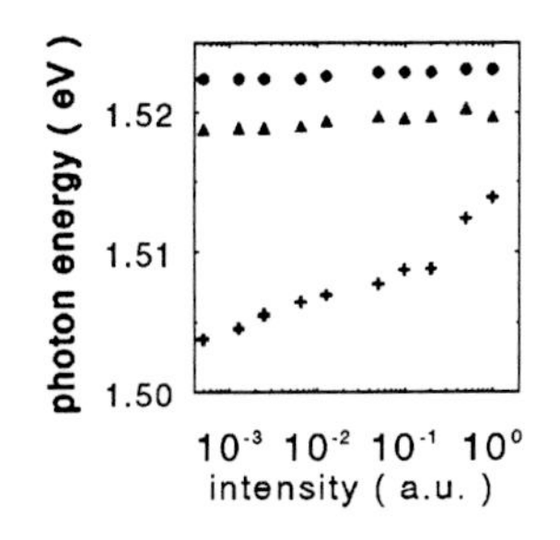

Fig. 2: Experimental observed transitions are shown versus the excitation intensity normalized to 90 W/cm^2. + second electron subband to heavy hole subband, ▲ first heavy hole subband to third electron subband and ● first light hole subband to third electron subband.

$$\Delta E = -\frac{e^2 N_A d_n^2}{2\epsilon_0 \epsilon_r} \tag{1}$$

where d_n is the width of the neutralized region, N_A the acceptor density, ϵ_0 the permittivity of free space, ϵ_r the relative dielectric permittivity, of GaAs and e the electron charge. In the case of unintentionally doped layers, this potential difference is too small to account for the observed blue shift.

(ii) A second mechanism is the separation of photo-generated electron-hole pairs in the depletion field in the thin GaAs layer. The build-up of excess electrons in the 2DEG and holes near the SL barrier is similar to charging a plate capacitor. In a first order approximation the shift of the bandminima near the SL with respect to the Fermi level of the 2DEG is given by the potential difference across the capacitor plates:

$$\Delta V = \frac{e N_{e-h} d}{\epsilon_0 \epsilon_r} \tag{2}$$

where d is the distance between the two carrier sheets and N_{e-h} the excitation density of electron-hole pairs. This potential difference is comparable to experimentally observed blue shift of 10 meV in the case of $d \approx 25$ nm and $N_{e-h} \approx 2\ 10^{10}$ cm^{-2}. Therefore, in our lightly p-doped samples the second mechanism is dominant.

3. APPLICATION OF LIGHT-INDUCED BANDSTRUCTURE MODIFICATIONS

At room temperature the mechanism of bandstructure modifications can be employed in all-optical modulators or optical switching devices when the absorption and blue shift of the excited states of the 2DEG are optimized by tailoring the design of the heterostructure. Proper operation implies that the absorption due to the lowest occupied electron subband is negligible compared to that of the unoccupied states of the 2DEG. Therefore the overlap between electrons in the 2DEG and the lowest hole subband should be as small as possible. Meanwhile the overlap between the excited states of the 2DEG and hole subbands should approach unity. In addition, the absorption edge of the unoccupied states should shift as a function of illumination.

This shift can be realized in a pseudomorphic n-$Al_xGa_{1-x}As/In_{0.13}Ga_{0.87}As$/GaAs modulation doped heterostructure. The excited states of the 2DEG are now confined in a GaAs layer of 40-50 nm and have a much better overlap with the hole wavefunction than the first electron subband which is confined in a 4 nm $In_{0.13}Ga_{0.87}As$ layer. The schematic band diagram of such a pseudomorphic heterostructure is shown in Fig. 3. The absorption edge of the dominant transition is blue shifted when the heterostructure is illuminated. By adding the intentional p-dopant in the GaAs layer the band bending in the GaAs has been substantially increased, leading to an increased blue shift of approximately 20 meV
As an example, we calculated the absorption spectra of such a pseudomorphic heterostructure in the dark and under illumination. For this purpose, we calculated the

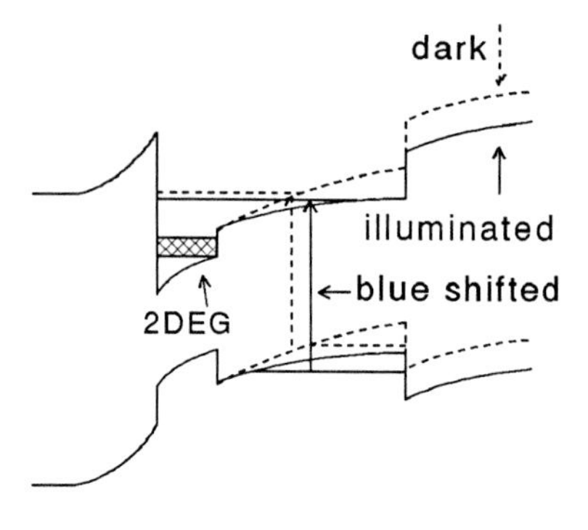

Fig. 3: Schematic band diagram of the pseudomorphic modulation doped heterostructure in the dark (dashed) and illuminated (solid). The 2DEG is defined by a thin $In_yGa_{1-y}As$ layer and the second electron subband is confined in the thin GaAs layer. When illuminated the transition energy of the second electron subband is blue shifted.

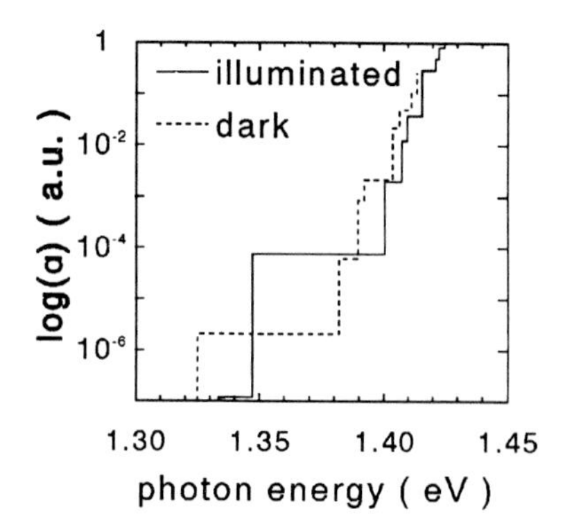

Fig. 4: Calculated absorption spectrum in the dark (dashed line) and illuminated (solid line) in the case of $In_{0.13}Ga_{0.87}As$ layer of 4 nm and a GaAs layer of 50 nm, which is p-doped at 10^{16} cm^{-3}.

bandstructure with a model which takes into account the modifications due to the photo-generated carriers (Weegels et al 1991b). In our model the subband structure for electrons and holes is calculated by solving the Poisson and Schrödinger equations self-consistently for room temperature. The GaAs layer is p-doped at 10^{16} cm^{-3}. The calculated absorption spectrum in the dark (solid line) and in the illuminated case (dashed line) are shown in Fig. 4. The calculations show that the transition energies to the second and third electron subbands shift 19 and 15 meV respectively. In addition, the absorption is enhanced in the illuminated case due to a larger overlap between the hole and electron states. The calculations predict an absorption contrast ratio of 27:1 around a photon energy of 1.395 eV. If this is experimentally realized, this blue shift mechanism provides attractive features for an all-optical modulator.

Bar-Joseph I, Goossen K W, Kuo J M, Kopf R F, Miller D A B and Chemla D S 1989 Appl. Phys. Lett. 55 340

Mendez E E, Agulló-Rueda F and Hong J M 1988 Phys. Rev. Lett. 60 2426

Miller D A B, Chemla D S, Damen T C, Gossard A C, Wiegmann W, Wood T H and Burrus C A 1984 Appl. Phys. Lett. 45 13

Miller D A B, Chemla D S and Smitt-Rink S 1988 Optical Nonlinearities and Instabilities in Semiconductors, ed Haug H (New York: Academic) pp 325-359

Weegels L M, Haverkort J E M, Leys M R and Wolter J H 1991a Superlattices and Microstructures 10 143

Weegels L M, Haverkort J E M, Leys M R and Wolter J H 1991b submitted to Phys. Rev. B

Inst. Phys. Conf. Ser. No 123
Paper presented at the International Meeting on Optics of Excitons in Confined Systems, Giardini Naxos, Italy, 1991

Low threshold current of multiple quantum well lasers by optimization of carrier capture efficiency

P W M Blom, J E M Haverkort and J H Wolter

Department of Physics, Eindhoven University of Technology,
P O Box 513, 5600 MB Eindhoven, The Netherlands

ABSTRACT: We have investigated the relevance of the carrier capture efficiency for the device characteristics of a multiple quantum well laser. It is demonstrated that the dependence of the threshold current on the structure parameters of the layers in the active region, which was reported in literature, is highly correlated to the electron capture efficiency. We explain this unexpected result by a carrier injection model in which a large carrier capture efficiency contributes to uniform pumping of the quantum wells and thus to a reduction of the threshold current.

1. INTRODUCTION

The performance of quantum well lasers is often suggested to be related to the injection efficiency of the carriers into the quantum well (Tsang 1981), but a detailed analysis is lacking. The first calculations of the carrier capture time were reported by Brum et al (1986). They predicted strong oscillations of the capture time as a function of well width from quantum mechanical calculations on a separate confinement heterostructure (SCH) single quantum well (SQW). In a recent study (Blom et al 1990) we expanded their model to SCH multiple quantum well structures (MQW) and we found the calculated capture time for the first time to be in agreement with experimental results obtained from subpicosecond time-resolved photoluminescence measurements. In addition, we demonstrated that the carrier capture efficiency in SCH- and graded-index SCH-MQW structures not only oscillates as a function of well width but in addition also oscillates as a function of barrier width between the wells (Blom et al 1991). We showed that the carrier capture efficiency can be improved by more than one order of magnitude by optimizing the dimensions and composition of the layers in the active region of SCH-MQW or GRINSCH-MQW structures.

2. COMPARISON OF THRESHOLD CURRENT DENSITY AND ELECTRON CAPTURE TIME

In this paper we discuss the relation between the carrier capture efficiency and MQW laser performance. We compare our calculated electron capture time with the experimental threshold currents reported by Tsang (1980,1981) for some SCH-MQW $GaAs/Al_xGa_{1-x}As$ lasers. Tsang systematically investigated the influence of the structure parameters such as well width (Tsang 1980), barrier width and composition of the barrier layers (Tsang 1981) on the threshold current density. In Fig. 1 the threshold current

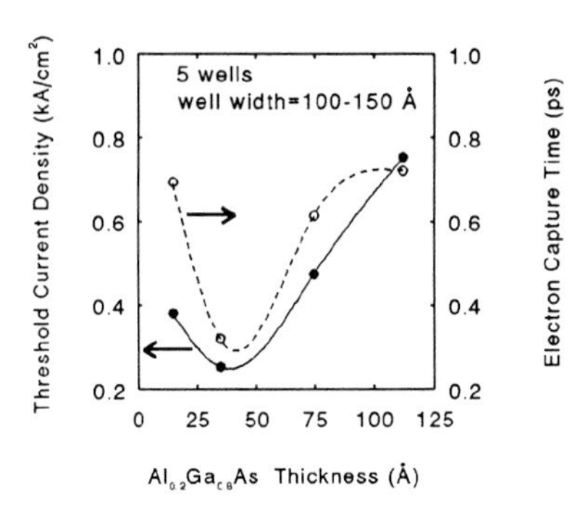

Fig. 1. Reported threshold current density (Tsang 1981, solid line) and electron capture time (dashed line) vs barrier width between the wells for a $GaAs/Al_xGa_{1-x}As$ SCH-MQW structure. The MQW structure consists of 5 wells, the well width ranges from 100 to 150 Å and the aluminum fraction x of the barrier layers is 0.2.

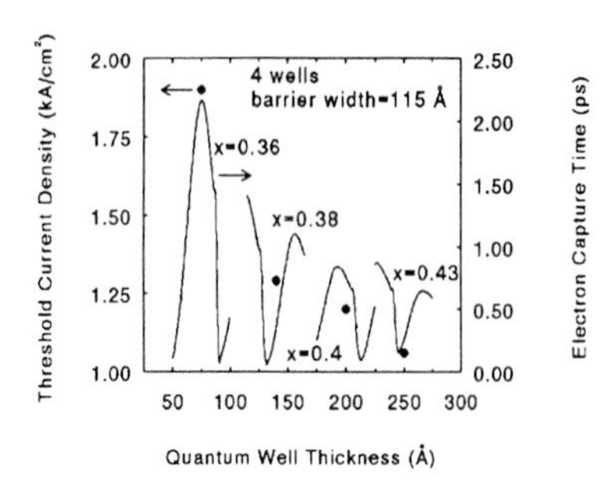

Fig. 2. Reported threshold current density (Tsang 1980, experimental points) and electron capture time (solid line) vs quantum well thickness for a $GaAs/Al_xGa_{1-x}As$ MQW structure with 8 wells. The thickness of the barrier layers is 115 Å and the composition ranges from x=0.36 to x=0.43.

density and the electron capture time are plotted in the same figure as functions of the barrier thickness between the wells for a $GaAs/Al_xGa_{1-x}As$ SCH-MQW structure with 5 wells. The well width ranges from 100 to 150 Å and the composition of the barrier layer is 0.2. It can be seen that the variations in the experimentally determined threshold current density and the calculated capture efficiency are in very good agreement. In particular, the minimum threshold current density matches with the maximum capture efficiency at a barrier width of 35 Å.

The dependence of the threshold current density and the electron capture time on the quantum well width is shown in Fig. 2 for a $GaAs/Al_xGa_{1-x}As$ MQW structure. The thickness of the barrier layers between the quantum wells in the MQW structure is 115 Å and the composition ranges from x=0.36 to x=0.43 for the different structures. We find that the threshold current density is well described by the predicted (Brum et al 1986, Blom et al 1990) oscillations of the carrier capture time as a function of well width. Of course, the scaling factor between the electron capture time and the threshold current is dependent on a number of structure parameters such as e.g. optical confinement. Therefore, the vertical scales in the figures are arbitrary. In Fig. 3 the threshold current density is compared with the electron capture time as a function of the composition of the $Al_xGa_{1-x}As$ barrier layers between the GaAs wells in a SCH-MQW structure with 8 wells. The barrier layer thickness ranges from 47 to 60 Å and the quantum well thickness from 142 to 167 Å. Again, it can be seen that the variations of the threshold current density and the electron capture time as a function of the composition are in very good agreement. The decrease of the capture time for aluminum fractions lower than 0.1 is caused by the fact that for these compositions the quantum wells are almost completely filled with carriers at an injection level of $1kA/cm^2$. This band filling reduces the number of available final states for a transition from an initial barrier state to a final bound state in the well, and thus increases the capture time. In the next session we discuss a model for explaining this correlation.

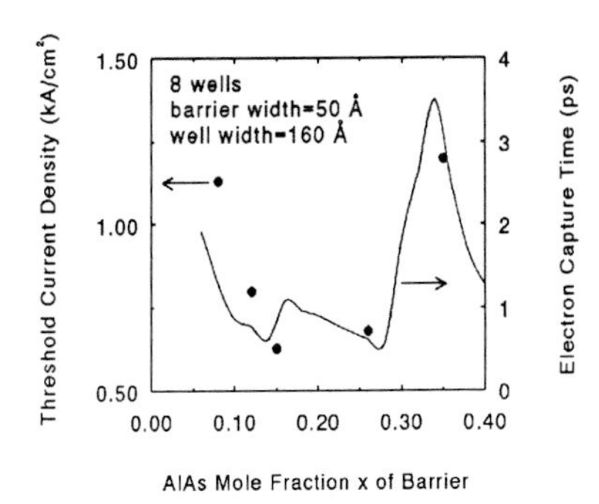

Fig. 3. Reported threshold current (Tsang 1981, experimental points) and electron capture time vs composition of the barrier layers x for a GaAs/$Al_xGa_{1-x}As$ SCH-MQW structure with 4 wells. The thickness of the barrier layers ranges from 47 to 60 Å and the quantum well thickness from 142 to 167 Å.

3. CARRIER DISTRIBUTION IN A MQW LASER

The loss rate of carriers in the barrier layers of a MQW structure is very small compared to the carrier capture rate, because the lifetime of the carriers in the barrier layers is of the order of nanoseconds while the capture time is in the picosecond regime. Consequently, this mechanism cannot explain the strong correlation between the threshold current density and the capture efficiency. An inefficient capture process only gives rise to significant additional losses in QW structures with very thick confinement layers of several thousands of ångströms, as was already pointed out by Nagle (1986). For these structures the capture time, which is proportional to the total structure width, is of the same order of magnitude as the lifetime of the carriers in the barrier layers.

In a laser structure, electrons are injected into the barrier layers from the n-cladding layer whereas holes are injected from the p-cladding layer, which gives rise to large drift fields and concentration gradients. A MQW laser structure with undoped active layers should be regarded as a p-i-n diode, as is shown schematically in Fig. 4. After injection, the electrons in the barrier states drift from the n-contact towards the p-contact ($x=0$) and the holes towards the n-contact ($x=L$). The number of electrons and holes which reach the p- respectively the n-contact is dependent on the electron and hole mobility and on the carrier capture efficiency at a certain bias current. So the capture time defines a mean free path length for carrier transport in the barrier states. Since the holes combine a large capture efficiency with a low mobility, resulting in a small mean free path, and the electrons have a high mobility combined with a low capture efficiency, resulting in a large mean free path, we expect that the accumulation of carriers at the p-contact is larger than at the n-contact. If, however, the electron capture efficiency is enlarged, more electrons will be captured directly after injection by the quantum wells located at the injecting n-contact. These electrons will also attract more holes to the n-contact as a result of the strong internal electric fields which tend to maintain charge neutrality in the structure. Therefore, an improved electron capture efficiency is expected to give rise to a more uniform distribution of the carriers in the active region of SCH-MQW laser structures and thus to a more uniform pumping of the quantum wells.

The carrier distribution can be calculated by using the Regional Approximation Method by Lampert and Schilling (1970). In this model the transport of carriers in the barrier layers is dominated by diffusion within one capture limited ambipolar free path length of the cladding layers and by drift in the barrier layers located in the middle of the structure.

In Fig. 4 the calculated carrier distribution in the barrier states of the active layers is shown for electron capture times of 0.1 and 1.0 ps respectively. As expected, an efficient carrier capture process improves the uniformity of the pumping of the quantum wells.

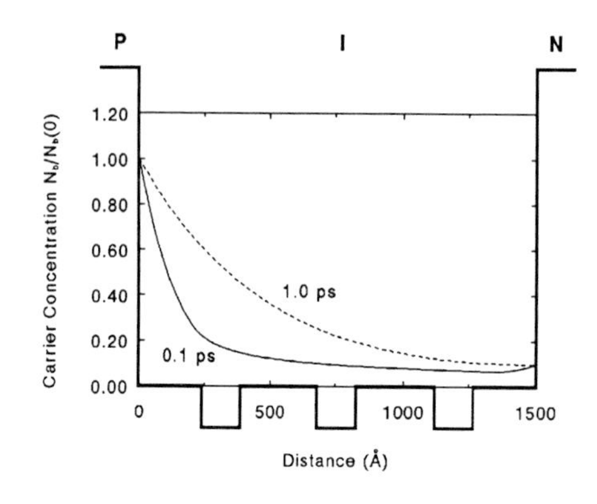

Fig. 4. Carrier distribution in a SCH-MQW laser at an injection current of 1 kA/cm^2 for capture times of 0.1 and 1.0 ps. The total width of the active layer is assumed to be 1500 Å, the net equilibrium density in the barrier layers is 3.10^{14} cm^{-3}.

For example, the carrier concentration reduces from x=250 Å to x=1250 Å by a factor 3 for a capture time of 0.1 ps and by a factor 5.5 for a capture time of 1.0 ps. In a MQW with nonuniform pumping, the spontaneous emission losses in the quantum wells which not yet have reached lasing threshold should be compensated for by the gain of the lasing quantum wells, which leads to a degradation of the threshold current. From a model of current injection in MQW lasers it was already proposed (Dutta 1983) that a uniform pumping of the quantum wells can be achieved by increasing the well-to-barrier ratio. This condition is met in our model, in which an increase of the well-to-barrier ratio gives rise to an increase of the carrier capture efficiency and thus to more uniform pumping conditions.

In conclusion, we demonstrate from a comparison of calculated electron capture efficiencies with reported threshold currents that the carrier capture efficiency is a very important parameter for the performance of a quantum well laser. As a possible mechanism for the strong correlation between electron capture efficiency and threshold current density we suggest that an efficient carrier capture process contributes to a more uniform pumping of the different quantum wells in a SCH-MQW laser.

Blom P W M, Mols R F, Haverkort J E M, Leys M R and Wolter J H 1990 Superlatt. and Microstruct. 7 319
Blom P W M, Haverkort J E M and Wolter J H 1991 Appl. Phys. Lett. 58 2767
Brum J A and Bastard G 1986 Phys. Rev. B 33 1420
Dutta N K 1983 IEEE J. Quantum Electron. QE-19 794
Lampert M A and Schilling R B 1970 Semiconductors and Semimetals 6, eds R K Willardson and A C Beer (New York: Academic Press) 1
Nagle J, Hersee S, Krakowski M, Weil T and Weisbuch C 1986 Appl. Phys. Lett. 49 1325
Tsang W T 1980 Appl. Phys. Lett. 38 204
Tsang W T 1981 Appl. Phys. Lett. 39 786

Inst. Phys. Conf. Ser. No 123
Paper presented at the International Meeting on Optics of Excitons in Confined Systems, Giardini Naxos, Italy, 1991

Experimental studies of the relaxation of magnetic moment in GaAs/GaAlAs quantum wells

V.A.Chitta+, E.C.F.da Silva+, D.Toet+, M.Potemski++, J.C.Maan+ and K.Ploog+++

+Max Planck Institut für Festkörperforschung,
Hochfeld Magnetlabor, BP 166 , F38042 Grenoble Cedex 9, France
++Service National des Champes Intenses,CNRS, BP 166, F38042 Grenoble Cedex 9, France
+++Max Planck Institut für Festkörperforschung,Postfach 800 665, 7000 Stuttgart 80, FRG

ABSTRACT: Measurements of the degree of polarization of the luminescence as a function of energy give information about the relaxation of the magnetic moments of the photo-excited carriers. A study of this effect in undoped quantum wells (QW) and modulation doped quantum wells (MDQW) as well as n-type MDQW with additional acceptors in magnetic fields both perpendicular and parallel to the layers, allows to analyze the different contributions of hole and electron relaxation. The experiments confirm the recent theoretical claim that the polarization of the luminescence in two dimensional systems is governed by the conservation of the hole parity rather than by the conservation of electron spin upon relaxation.

1. INTRODUCTION

In an optical pumping experiment carriers are optically excited with polarized light, thus allowing for a selective orientation of their magnetic moments, and the polarization of the recombining luminescence is measured. This technique has originally been applied to atomic spectroscopy [1] but later also to bulk semiconductors. A wide variety of phenomena such as spin relaxation, optically detected nuclear and electron spin resonance, exciton magnetic moment relaxation etc. for many different semiconductors have been studied. [2]. The application of these experimental techniques to QW (thin layers of semiconductors) is rather limited. Weisbuch et al. [3] used the technique to assign heavy and light hole transitions, shake-up of the Fermi sea was studied in doped samples of a special geometry [4], optically detected resonance was observed by several authors [5,6] and several polarization studies were reported [7,8]. Recent time resolved experiments at zero field [9] and steady state polarization measurements at high fields [10,11] together with a new theoretical analysis of older data [12,13] and studies of hole magnetic moment relaxation [13,14] in QW have produced results which have shown that the situation in these systems may be markedly different from what was assumed until recently. Here we want to present recent experimental results on different type of samples studied in the absence of magnetic fields, and in fields with different orientations with respect to the plane of the QW layer and the propagation direction of the light. In the next section we will briefly resume some generalities about optical pumping. Experimental details and results will be presented in sections 3,4 and 5 and we close with a discussion and analysis of those data.

2. POLARIZATION AND OPTICAL PUMPING

In fig. 1 we show the schematic in-plane dispersion relation of electrons and holes for a typical GaAs-GaAlAs QW. At zero magnetic field for the in-plane wavevector $k_{\parallel}=0$ the conduction band states are characterized by $|\pm1/2\rangle_e$ (spin-up and spin-down), the heavy hole band by $|\pm3/2\rangle_{hh}$, and the light hole band by $|\pm1/2\rangle_{lh}$ basis wavefunctions. Due to band mixing at finite wavevectors the holes have a mixed lh-hh character and their wavefunctions are a linear combination of the basis wavefunctions, but for simplicity we will ignore this complication at present.

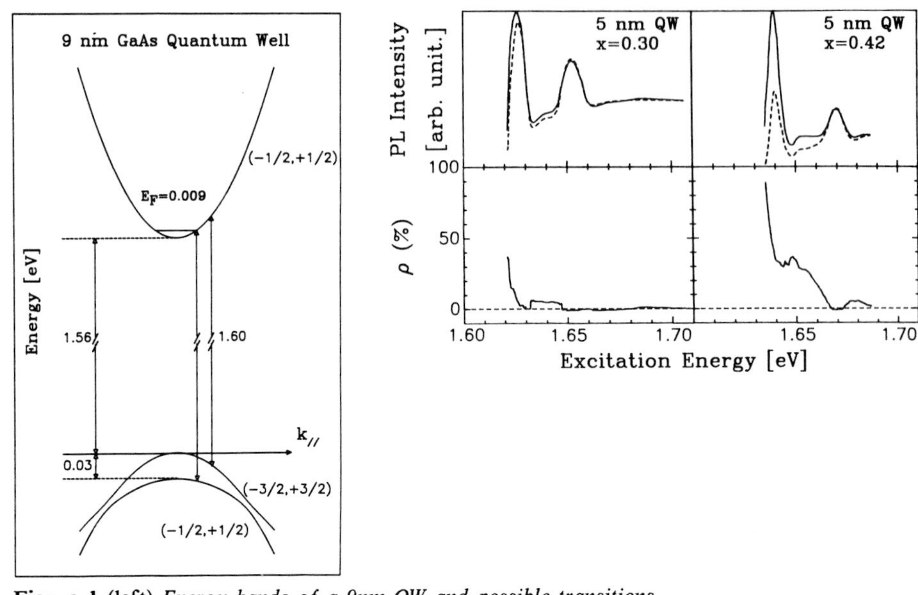

Figure 1 (left) *Energy bands of a 9nm QW and possible transitions.*
Figure 2 (right) *Photoluminescence excitation spectra of two different 5 nm undoped QW, exciting with σ^+ polarized light and detecting σ^+ polarized luminescence (solid lines) or detecting σ^- polarized luminescence (dashed lines). The degree of polarization ρ is shown in the lower part of the figure.*

Due to selection rules light with right circular polarization, σ^+, can induce the following interband transitions: $|-3/2\rangle_{hh} \rightarrow |-1/2\rangle_e$ and $|-1/2\rangle_{lh} \rightarrow |+1/2\rangle_e$, and similarly σ^- light can induce $|+3/2\rangle_{hh} \rightarrow |+1/2\rangle_e$ and $|+1/2\rangle_{lh} \rightarrow |-1/2\rangle_e$ transitions; the intensity of the transitions involving the hh states being three times larger than those with the lh. Since the momentum relaxation time is generally much shorter than the radiative recombination lifetime, the excited carriers will relax to the band edge before recombining under emission of a photon. The selection rules for recombination are the same as for excitation, therefore the intensity of the recombining light depends exclusively on the populations of electrons and holes in the respective magnetic moment states before recombination, which depend of the degree of relaxation of the magnetic moments of both electrons and holes.

Experimentally, the intensities I^+ and I^- of resp. the σ^+ and the σ^- polarized luminescence, exciting with either σ^+ or σ^- light, is measured. It is therefore customary to define the degree of polarization $\rho^\pm$ for $\sigma^\pm$ excitation respectively as:

$$\rho^\pm = \pm \frac{I^+ - I^-}{I^+ + I^-} \quad (+ \text{ for } \sigma^+ \text{ excitation, - for } \sigma^- \text{ excitation}) \tag{1}$$

In the case of QW the luminescence involves only the heavy hole $|+3/2\rangle$ and $|-3/2\rangle$ states which have the same matrix elements for recombination with the corresponding electrons. Therefore I^+ is determined by the number of holes (h^+) and electrons (n^+) in the states at the band edges between which recombination through the σ^+ channel is allowed (resp. the $|-3/2\rangle_{hh}$ and the $|-1/2\rangle_e$ states) and similarly for I^-. If n^+ and h^+ are equal the intensity can directly be evaluated. If they are different the intensity in the σ^+ channel depends on the smallest of n^+ and h^+ and similarly for n^- and h^- in the σ^- channel. The implicit assumption here is that the remaining carriers which do not find a partner of the right type will disappear through non-radiative processes. Under this conditions $\rho^\pm$ is given by [13]:

$$\rho^{\pm} = \pm \frac{\min(n^+,h^+)-\min(n^-,h^-)}{\min(n^+,h^+)+\min(n^-,h^-)} \tag{2}$$

where min(n,p) means the lowest of the two numbers. An implication of this formula is that the total luminescence intensity will be maximal when $n^+=h^+$ and $n^-=h^-$ and decreases when these numbers are very different. In this simple scheme, neglecting band mixing, (all states are pure) it can easily be seen that for energies lower than the light hole transitions, with one polarization, only n^+ and h^+ carriers are created. If these relax without loss of magnetic moment, the polarization is 100%, if both relax completely the polarization is zero; if one of the two relaxes completely and the other not, ρ=100% is found once again but the total luminescence intensity is halved with respect to the previous case. Furthermore for excitation above the lh-e transition with one polarization 75% spin-up and 25% spin down electrons are created and with full hole relaxation the degree of polarization should drop to 33% at this energies.

Band mixing is important for the hole states which at finite in plane wavevector are made of a linear combination of $|\pm 3/2\rangle_{hh}$ and $|\pm 1/2\rangle_{lh}$ states, with coefficients depending on energy and k-vector. The electrons remain almost pure $|\pm 1/2\rangle_e$ spin states. Therefore for excitation with σ^+ light at a finite wavevector a number of holes with a mixed character are created and at the same time a certain number of electrons both in spin-up and spin down; the ratio of these latter two densities depending on the degree of mixing of the hole states. The band mixing changes with wavevector, therefore a hole relaxing within its band changes its character, and the densities h^+ and h^- at the band edge depend on the mixing in the initial state and the relaxation. A full analysis of this problem [13] has revealed that the parity of the wave functions is the quantity which tends to be conserved during relaxation. This viewpoint has important consequences for the interpretation of polarization studies. Previous work [8] has almost invariably assumed that due to the mixed character of the hole states a full relaxation occurred, implying that the both $|\pm 3/2\rangle_{hh}$ states at the band edge were equally populated. The electrons are assumed to conserve at least partially their magnetic moment and therefore densities of the spin-up and spin-down populations at the band edge were thought to reflect the densities at which they were created. These densities in turn depend on the hole state from which they were excited. Therefore almost all polarization experiments were interpreted exclusively in terms of electron spin populations assuming full hole depolarization. Since this assumption has recently been put into question we report here experiments which are meant to distinguish between the concepts of relaxation of magnetic moment of electron or holes.

3. EXPERIMENTS AT ZERO MAGNETIC FIELD

3.1 Undoped samples

In this experiment, the intensity of the luminescence is measured as a function of the exciting laser energy (luminescence excitation spectroscopy). We adopt the following notation for the configurations: a (p1,p2) spectrum is a measurement of the luminescence intensity in p1 (σ^+ or σ^-) polarization for p2 (σ^+ or σ^-) polarization of the exciting light. In fig.2 we show (-,-) and a (+,-) spectra of two different GaAs QWs (20 5nm GaAs wells separated by 15 nm $Ga_{1-x}Al_xAs$ barriers (x=0.3) for sample 1 and three 5nm GaAs wells separated by 100nm $Ga_{1-x}Al_xAs$ barriers (x=0.42) for sample 2). The energy dependence of ρ is also shown. It can be seen that the polarization is maximal at the heavy hole energy, decreases with increasing energy, becomes zero at the light hole exciton energy, rises slightly at even higher energies and disappears at very high energies. The two samples have a globally similar behaviour but the amount of polarization is different. Similar behaviour has been reported also by other authors [8]. Originally [8] these results were explained in terms of electron relaxation. Upon excitation near the band edge only one electron spin polarization is created, which remains partially conserved during relaxation, thus giving rise to a polarized luminescence. Exciting above the heavy hole exciton 75% spin-up and 25% spin down electrons are created, which recombine with, presumably depolarized, holes explaining the reduction in polarization. However Uenoyama and

Sham [12,13] could account for exactly the same polarization behaviour in terms of conservation of parity of the holes. Therefore these results do not have a unique interpretation. It is important to note that ρ is strongly sample dependent. In fact in many samples which are apparently similar (similar linewidth, luminescence intensity and thickness) the amount of polarization is completely different, being absent in some cases. However, when polarization was measurable its energy dependence was always similar to the one shown in fig.2.

3.2 Modulation doped samples

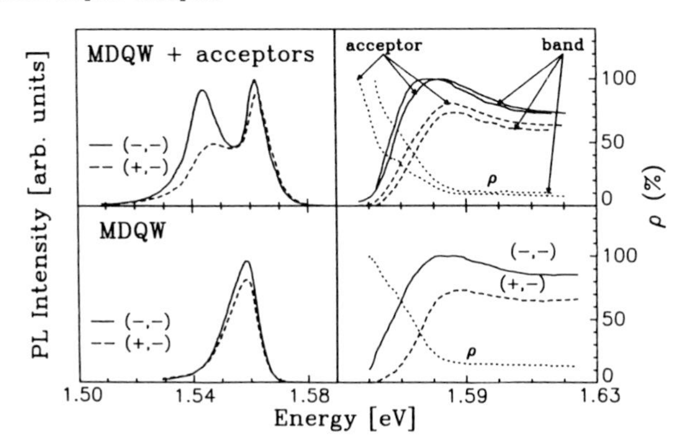

Figure 3 Left panel: *polarized luminescence spectra of a 9 nm MDQW (bottom) and a 9 nm MDQW with additional acceptors in the well (top). The excitation energy is 1.66 eV.* Right panel: *excitation spectra with the same polarization as the detected luminescence (dashed lines) and the opposite polarization (full lines) on the same samples. The degree of polarization ρ is given by the dotted line.*

In fig. 3 we show (-,-) and (+,-) excitation spectra together with the (-,-) and the (+,-) luminescence spectra when exciting at very high energies, for two different MDQW samples. These samples were grown under the same conditions as the undoped sample 1 in the previous section, with the same barriers. n-type MDQW and n-type MDQW with acceptors in the central part of the well, with well widths of 5 nm and 9 nm, were studied. Fig.3 shows the spectra of the 9 nm samples (sheet density n_S =2.8 10^{11} cm^{-2} for both samples). The low energy peak in the luminescence spectra on the higher part of the figure is due to the recombination of electrons with the acceptors. The excitation spectra on the free electron-hole transition in the (-,-) configuration show an onset in the luminescence when the exciting energy is above the Fermi energy, the excitation spectra in the (+,-) configuration show that the onset of the (-,+) luminescence occurs at much higher energies for (-,-) excitation. Reversing all + and - in the experiments gives the same results. Identical experiments on the free particle luminescence of the MD doped-acceptor sample give exactly the same outcome. These results cannot be explained in terms of electron relaxation, since the electron population is mainly extrinsic and thus there is a large amount of both spin-up and spin-down electrons available for recombination with the photocreated holes. On the other hand they are easily understood in terms of conservation of the hole parity. In the (-,-) experiments at low energies (-) heavy holes are created which relax to the band edge retaining their parity and then recombine with one of the available electrons under emission of a σ^- photon. In the (+,-) also (-) heavy holes are created but when they relax to the band edge retaining their parity they cannot recombine under emission of a σ^+ photon. Only if holes much deeper in the band are created where they are of a more mixed character they can recombine with one of the electrons with a σ^+ photon as is indeed observed. From the polarization curves is can be seen that the degree of ρ decreases steadily with increasing energy up to some energy after which it is constant. This energy corresponds to the the onset of absorption related to the light hole band, which agrees with theoretical predictions.[12,13]. The (-,-) and the (+,-) spectra taken on the acceptor related luminescence show the same overall features, however the degree of polarization is less (see fig. 3). Since the polarization is not zero it must be concluded that the acceptor states are polarized in the same manner as the photocreated hole which is captured by the acceptor. However the acceptor related luminescence has a much lower polarization implying that part of the hole magnetic moment is not conserved upon capture by acceptors.

4. EXPERIMENTS IN A PERPENDICULAR MAGNETIC FIELD

4.1 Undoped Quantum Wells

In a perpendicular magnetic field both electron and hole continuous dispersion curves split in discrete Landau levels. For the electrons these are given by $(N+1/2)\hbar\omega_c + g^*\mu B$ ($\omega_c = eB/m^*$ the cyclotron frequency in a field B for a particle with effective mass m^*, g^* the g-factor of GaAs and μ the Bohr magneton). Due to the complex valence band, the hole Landau levels are much more complicated. They have been the subject of extensive theoretical [15,16] and experimental studies [17] and although the energy spectra of interband transitions in a magnetic field in QW are very complicated, they are quantitatively well understood. Due to selection rules different sets of transitions can be optically excited in σ^+ and in σ^- polarization, which is a consequence of the breaking of time reversal symmetry when a field is applied. Therefore it is only meaningful to compare excitation spectra with the same exciting polarization (in which the same transitions are excited) but detected in different polarizations of the luminescence. In fig. 4 we show the excitation spectra of a 9 nm QW (sample 1) at 10 T, in both polarization of the exciting light. The σ^+ (resp.σ^-) excitation spectra is obtained by adding the (-,+) and the (+,+) spectra (resp. the (-,-) and the (+,-) spectra). In a magnetic field the luminescence at the band edge splits in two peaks, each peak with a different circular polarization, due to the Zeeman splitting of the electron and hole ground states. The intensity of the higher energy of this two peaks decreases with increasing field as a consequence of thermalization which tends to occupy only the lowest of the two. This leads to a natural difference of intensity of the (+,+) with respect to the (-,+) which has no relation to relaxation. To eliminate this trivial effect we normalized the two spectra to the same average intensity over the entire energy range. We then plotted the differences between the two spectra, which after this normalization are exclusively caused by specific relaxation effects. These differences are shown also in fig.4 and it can be seen that some excited states relax preferentially to the + ground state luminescence (positive polarization) and some others preferentially to the - luminescence (negative polarization). In the spectra shown here at 10 T, this trend becomes visible but it has been shown previously by Potemski et al.[10] that at higher fields this effect becomes very strong. By comparison with theory it was shown that the states which are seen to be strong in the σ^+ luminescence (excited either in - or + polarization) are the ones which involve the electron spin orientation allowed for the σ^+ luminescence and vice versa for the states seen preferentially in the σ^- luminescence. The phenomenon was interpreted in terms of the inhibition of electron spin relaxation due to the discrete nature of the spin states at high fields.

Uenoyama and Sham [18] demonstrated an equally convincing agreement on the basis of parity conservation of hole states without invoking conservation of magnetic moment for the electron spin. It seems therefore that two quite different models explain the data equally well. The reason that these apparently contradictory assumptions explain the data is the folllowing. Some hole states have the correct parity to excite an electron state with a well defined spin in a given polarization; when the hole relaxes to the band edge conserving its parity it can only recombine with an electron which has the proper parity, i.e. an electron that has the same magnetic moment orientation as the one which was originally excited.

4.2 Doped Quantum Wells

A possibility to distinguish the two reasonings mentioned in the previous paragraph is to do the same experiment in doped QWs. Due to the large density of electrons present in both spin polarizations, the spin polarization of the photoexcited electron is irrelevant for the polarization of

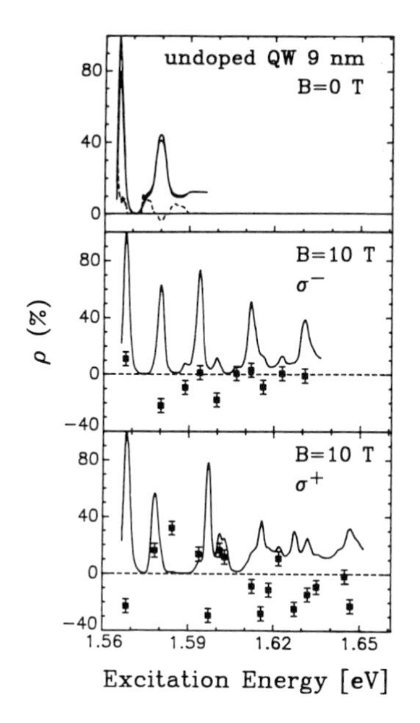

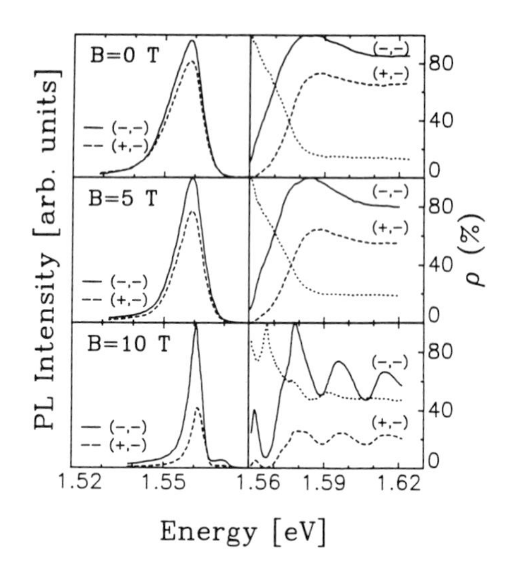

Figure 4 (left) Excitation spectra in σ^+ and σ^- excitation at 0 T and 10 T perpendicular magnetic fields of an undoped 9 nm QW. The data points show the degree of polarization of the different peaks after normalization of the average intensity of the σ^+ and σ^- luminescence intensity.

Figure 5 (right) Luminescence and excitation spectra with polarizations as indicated in the figures of a 9 nm MDQW at different values of the perpendicular magnetic field.

the recombining radiation. The experimental results on the same samples as described before are shown in fig.5 for (-,-) and (+,-) spectra, together with the polarization and the corresponding (+) and (-) spectra taken for excitation energies far from the band edge. A clear Landau level structure can be seen at high fields with some peaks apparently more pronounced in the (-,-) spectra than in the (+,-). However when ρ is calculated from the spectra no pronounced structure is seen, implying that we do not see any selective relaxation in these samples in a field, contrary to the undoped ones. However, these samples have much broader Landau level peaks than the undoped samples. Therefore the electron spin-splitting is not fully resolved and this could also explain the absence of selective relaxation, so that unfortunately no definite conclusion can be drawn from these data.

5. PARALLEL MAGNETIC FIELD

The polarization of the luminescence created by σ^+ or σ^- excitation at zero field is defined with respect to the propagation direction of the light. Therefore the electron and hole magnetic moments are aligned parallel or antiparallel to the direction of the light propagation. If a magnetic field is applied in the Faraday configuration all relevant directions i.e. field axis, k-vector of the light and growth axis of the QW are aligned. In a magnetic field parallel to the layer plane and perpendicular to the light propagation direction, the magnetic moment of electrons and holes will tend to be aligned with the field, i.e perpendicular to the sample growth axis and the magnetic field. The electrons, which posses pure spin states which are decoupled from the size quantization, are easily aligned along the field axis and therefore their polarization in the direction of the light propagation which is perpendicular to the field, will be lost. This effect is called the Hanle effect [2] and is often used to distinguish the spin relaxation time from the electron lifetime. The magnetic field needed to depolarize the electrons is determined by $\omega_S \tau_S$ where ω_S is the spin precession frequency and τ_S the spin relaxation time. For GaAs, and practically all bulk semiconductors, it is found experimentally that a magnetic field of 0.01 to 0.1 T perpendicular to the light propagation completely depolarizes the electron. In a QW, the holes are quantized along the component J_z of the orbital momentum parallel to the growth direction. A field parallel to the layer can therefore depolarize the hole population only when it succeeds to rotate the quantization direction. In order to accomplish this, the cyclotron energy for the holes must be comparable to the size quantization energy for which extremely high fields (>100 T) are needed.

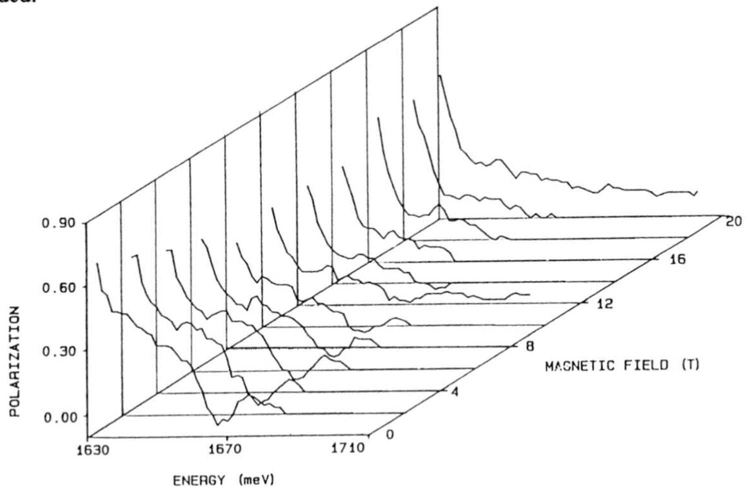

Figure 6 Degree of polarization as a function of energy of a 5 nm QW as a function of the magnetic field parallel to the layers and perpendicular to the direction of the light propagation.

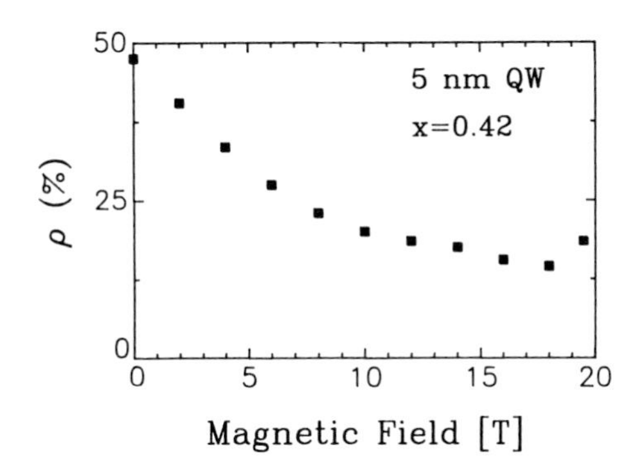

Figure 7 Degree of polarization of the heavy hole exciton as a function of the parallel magnetic field.

In fig. 6 we show the polarization of a the 5 nm sample as a function of the magnetic field. The polarization at the position of the heavy hole exciton is shown in fig. 7. It can be seen that the polarization diminishes from 50% to about 25% between 0 and 10T, but is never fully destroyed. This result clearly shows that the electron spin polarization cannot be responsible for the amount of polarization at zero field. If this were the case the luminescence would be completely depolarized in a field of 0.1 T as is experimentally observed for bulk GaAs. For reasons explained before the magnetic field cannot rotate the hole quantization axis of the orbital moment and therefore the hole population between the $|+3/2>_{hh}$ and $|-3/2>_{hh}$ states at the band edge remains conserved. The slight decrease of polarization with field can probably be explained by a slight admixture between the heavy hole and the light hole states induced by the field. Theoretical calculations [19] of the hole dispersion relation in a parallel magnetic field have indeed shown that this can be possible.

Conclusions

We have presented experimental results on different QW samples in different configurations in order to study relaxation phenomena of electrons and holes in QWs. All our results are consistent with the recent theoretical claim that the hole parity remains conserved upon relaxation. This new interpretation enabled the explanation of older results which were previously convincingly interpreted in terms of conservation of the magnetic moment of electrons. However, several of the new results presented here are not entirely consistent with the latter hypothesis, but are quite logically interpreted by conservation of the parity of the holes.

References

1. Proc. "Symposium Alfred Kastler", Paris 1975, [Ann. Phys. Paris] **10**, (1985)
2. "Optical Orientation", Modern problems in Condensed Matter Sciences, edited by F.Meier and B.P.Zakharchenya, North Holland, Amsterdam, (1984), Vol. **8**
3. C.Weisbuch, R.Dingle, A.C.Gossard and W.Wiegmann, Solid State Commun. **38**,709, (1981)
4. R.Sooryakumar, D.S.Chemla, A.Pinczuk, A.C.Gossard, W.Wiegmann, L.J.Sham, Phys. Rev. Lett. **58**, 1150, (1987)
5. G.P.Flinn, R.T.Harley, M.J.Snelling, A.C.Tropper and T.M.Kerr, J. of Luminescence **45**, 218, (1990)
6. G.R.Johnson, A. Kana-ah, B.C.Cavenett, M.S.Skolnick and S.J.Bass, Semicondn Sci. Technol. **2**, 182, (1987)
7. M.J.Snelling, A.S.Plaut, G.P.Flinn, A.C.Tropper, R.T.Harley and T.M.Kerr, J. of Luminescence, **45**, 208, (1990)
8. R.C.Miller and D.A.Kleinman, J. of Luminescence, **30**, 520, (1985)
9. M.R.Freeman, D.D.Awschalom, J.M.Hong and L.L.Chang, Phys. Rev. Lett. **64**, 2430, (1990)
10. M.Potemski, J.C.Maan, A.Fasolino, K.Ploog and G.Weimann, Phys. Rev. Lett. **63**, 2409, (1989)
11. V.A.Chitta, M.Potemski, J.C.Maan, A.Fasolino, K.Ploog and G.Weimann, J. Superlattices and Microstr. **9**, 303, (1991)
12. T.Uenoyama and L.J.Sham, Phys. Rev. Lett. **64**, 3070, (1990)
13. T.Uenoyama and L.J.Sham, Phys. Rev. **B42**, 7114, (1990)
14. R.Ferreira and G.Bastard, Phys. Rev. **B43**, 9687, (1991)
15. F.Ancilotto, A.Fasolino and J.C.Maan, Phys. Rev. **B88**, 1788, (1988)
16. G.E.W.Bauer and T.Ando, Phys. Rev. **B38**, 6017, (1988)
17. L.Vina, G.E.W.Bauer, M.Potemski, J.C.Maan, E.E.Mendez and W.I.Wang, Phys. Rev. **B41**, 10767, (1990)
18. T.Uenoyama and L.J.Sham, Proc. ICPS 20, ed. E.M.Anastassakis and J.D.Joannopoulos, p. 1286, World Scientific, Singapore, (1991)
19. A.Fasolino, G.Platero, M.Potemski, J.C.Maan, K.Ploog and G.Weimann, Surface Sci. to be published.

Inst. Phys. Conf. Ser. No 123
Paper presented at the International Meeting on Optics of Excitons in Confined Systems, Giardini Naxos, Italy, 1991

An exciton in a parabolic quantum dot and in a magnetic field

V Halonen

Department of Theoretical Physics, University of Oulu, SF-90570 Oulu 57, Finland

ABSTRACT: We have studied the properties of a two-dimensional hydrogenic exciton in a parabolic quantum dot and in an external magnetic field. The Hamiltonian of the system is written in terms of the center-of-mass and relative coordinates, and the effects of the cross term between center-of-mass and relative motions are investigated. The ground state and the lowest excited states of the system are calculated using the method of diagonalizing the Hamiltonian. The effects of magnetic field and parabolic confinement potential to the ground state and also to the excited states are studied.

1. INTRODUCTION

There has been a considerable interest on the magneto-optical properties of quasi-zero-dimensional electron systems, *i.e.* quantum dots (Sikorski and Merkt 1989, Maksym and Chakraborty 1990). The interesting role of electron correlations in these quantum confined systems have been demonstrated by the recent experimental (Demel *et al* 1990) and theoretical (Chakraborty *et al* 1991) studies. Theoretical works for quantum confined excitons (Bryant 1988) in quantum boxes (in the absence of a magnetic field) have shown the competing effects of quantum confinement and interparticle correlations. On the other hand, the magneto-optical properties of free excitons have been investigated theoretically some years ago (Shinada and Tanaka 1970, Paquet *et al* 1985). In this paper we present briefly some of our results for a two-dimensional exciton in a parabolically confined quantum dot subjected to a static external magnetic field.

2. THEORY

We consider a two-dimensional system of two oppositely charged particles confined by a parabolic potential and subjected to a static external magnetic field. The effective-mass Hamiltonian of this system is

$$\begin{aligned} H = &\frac{1}{2m_e}\left(-i\hbar\nabla_e - \frac{e}{c}\mathbf{A}_e\right)^2 + \frac{1}{2}m_e\omega_e^2 r_e^2 \\ &+ \frac{1}{2m_h}\left(-i\hbar\nabla_h + \frac{e}{c}\mathbf{A}_h\right)^2 + \frac{1}{2}m_h\omega_h^2 r_h^2 - \frac{e^2}{\varepsilon}\frac{1}{|\mathbf{r}_e - \mathbf{r}_h|}, \end{aligned} \tag{1}$$

where ε is the dielectric constant of the medium and m_e (m_h), $\mathbf{A}_e$ ($\mathbf{A}_h$) and $\hbar\omega_e$ ($\hbar\omega_h$) are the effective mass, the vector potential and the confinement potential energy of the electron (hole), respectively. We make a transformation from single particle coordinates to the center-of-mass (c.m.) coordinate $\mathbf{R} = (m_e\mathbf{r}_e + m_h\mathbf{r}_h)/(m_e + m_h)$ and to the relative coordinate $\mathbf{r} = \mathbf{r}_e - \mathbf{r}_h$. We also choose the symmetric gauge vector potentials as $\mathbf{A}_e = \frac{1}{2}\mathbf{B} \times (\mathbf{r}_e - \mathbf{r}_h)$ and $\mathbf{A}_h = -\frac{1}{2}\mathbf{B} \times (\mathbf{r}_e - \mathbf{r}_h)$. The Hamiltonian (1) can now be rewritten in the form

$$H = H_{cm} + H_{rel} + H_x, \tag{2}$$

where

$$\begin{aligned} H_{cm} &= -\frac{\hbar^2}{2M}\nabla^2_{cm} + \frac{1}{2}M\left[\frac{1}{M}\left(m_e\omega_e^2 + m_h\omega_h^2\right)\right]R^2, \\ H_{rel} &= -\frac{\hbar^2}{2\mu}\nabla^2_{rel} + \frac{i\hbar e}{2\mu c}\gamma\mathbf{B}\cdot\mathbf{r}\times\nabla_{rel} \\ &\quad + \frac{1}{2}\mu\left[\frac{e^2B^2}{4\mu^2c^2} + \frac{1}{M}\left(m_h\omega_e^2 + m_e\omega_h^2\right)\right]r^2 - \frac{e^2}{\varepsilon}\frac{1}{r}, \\ H_x &= \frac{i\hbar e}{Mc}\mathbf{B}\cdot\mathbf{r}\times\nabla_{cm} + \mu\left(\omega_e^2 - \omega_h^2\right)\mathbf{R}\cdot\mathbf{r}. \end{aligned} \tag{3}$$

Here we have adopted the usual notations: $M = m_e + m_h$, $\mu = m_em_h/M$, and $\gamma = (m_h - m_e)/M$. The terms of the Hamiltonian (2) are grouped so that H_{cm} depends only on the c.m. coordinate and H_{rel} depends only on the relative coordinate. There is also a cross term H_x which vanishes when $\mathbf{B} = 0$ and $\omega_e = \omega_h$.

The Hamiltonian H_{rel} can be separated into radial and angular parts. The Schrödinger equation for the radial part is

$$R'' + \frac{1}{r}R' + \left\{\frac{2\mu}{\hbar^2}\left(E + \frac{e^2}{\varepsilon}\frac{1}{r}\right) - \frac{\ell_{rel}^2}{r^2} - \left[\frac{e^2B^2}{4\hbar^2c^2} + \frac{\mu^2}{M\hbar^2}\left(m_h\omega_e^2 + m_e\omega_h^2\right)\right]r^2\right\}R = 0. \tag{4}$$

In this equation both the square of magnetic field and the square of confinement potential energy appear in the coefficient of the harmonic term. Therefore the energy eigenvalue E depends on the magnetic field and confinement potential energy exactly in the same way. The difference is that the external magnetic field couples to the relative angular momentum of the exciton ℓ_{rel}. This gives a contribution to the energy of the relative motion Hamiltonian H_{rel}

$$E_{rel} = E - \gamma\frac{eB}{2\mu c}\ell_{rel}. \tag{5}$$

We calculate the eigenfunctions and eigenvalues of the system using the method of numerical diagonalization of the Hamiltonian. We consider two possibilities. One is to expand the wave functions of the system in terms of the eigenfunctions of the noninteracting electron-hole pair, and the other is to use the eigenfunctions of $H_{cm} + H_{rel}$ as a basis functions. In general the latter approach gives a better convergence of the eigenvalues as a function of the number of basis states.

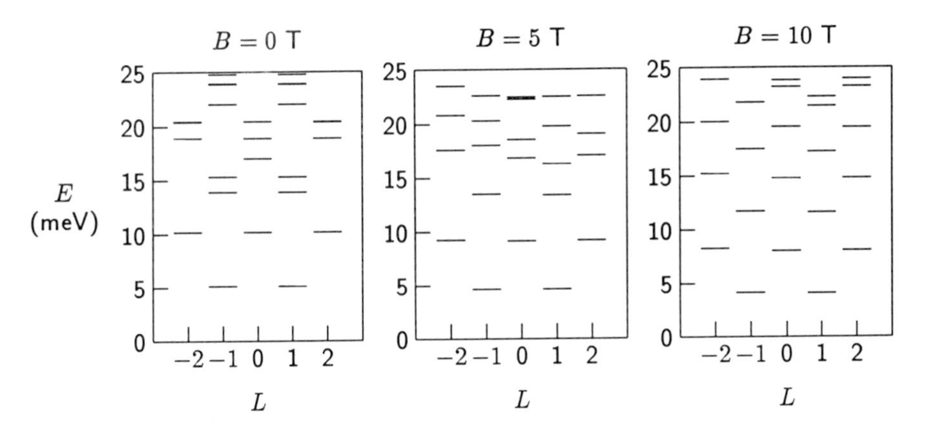

Fig. 1. Low-lying excitation energies of the light-hole ($m_h = 0.09m$) exciton as a function of the angular momentum L. The confinement potential energy is $\hbar\omega_e = \hbar\omega_h = 5$ meV. The basis consists 1000 noninteracting electron-hole pair states.

3. RESULTS AND DISCUSSION

We present here results which were calculated using GaAs material parameters, *i.e.*, dielectric constant $\varepsilon = 13.1$, electron effective mass $m_e = 0.067m$ and hole effective mass $m_h = 0.09m$ for light hole and $m_h = 0.377m$ for heavy holes. In Fig. 1 the lowest excitation energies of a light-hole exciton at various magnetic fields are shown. These results are obtained using the noninteracting electron-hole pair states as a basis functions. The excitation energies divide into two sets. One set behaves like c.m. excitations showing essentially no sign of magnetic field dependence while the other set can be interpreted to be almost pure excitations of the relative coordinate. This means that although the Hamiltonian doesn't separate into c.m. and relative parts the interaction between them must be, at least in this case, very small. In Fig. 2 we present the ground-state energy and also some of the lowest excitation energies of a light-hole exciton as a function of magnetic field. To obtain these results we have used the c.m. and relative motion separation approach. The effect of the cross term H_x is also illustrated, and as can be seen it is very small.

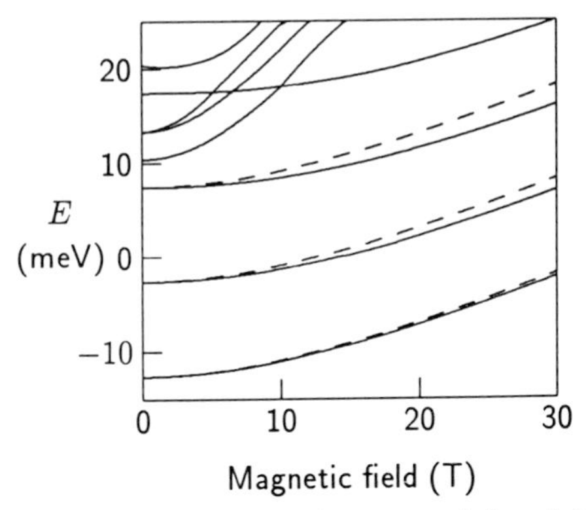

Fig. 2. The ground-state and low-lying excitation ($L = 0$) energies of a light-hole exciton (solid curves). Confinement potential energy is $\hbar\omega_e$, $\hbar\omega_h = 5$ meV. The dashed curves are calculated without the cross term H_x.

One of our aims is to investigate the electron-hole correlations as a function of the confinement potential energy and also as a function of the magnetic field. We compare the results obtained using both noninteracting electron-hole pair state basis and the c.m. and relative motion separation. The ground-state energy of a heavy-hole exciton

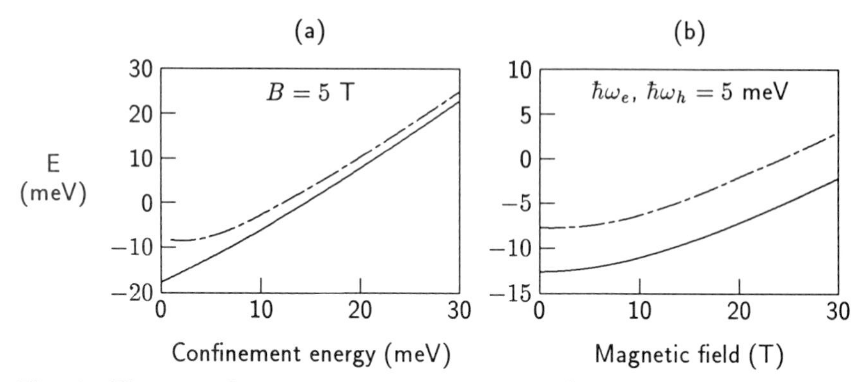

Fig. 3. The ground-state energy of a heavy-hole ($m_h = 0.377m$) exciton. The solid curves are calculated using c.m. and relative motion separation and the dot-dashed curves using noninteracting electron-hole pair state basis (100 basis states).

as a function of confinement potential energy is shown in Fig. 3(a). At low confinement potential energies the noninteracting electron-hole pair state basis approach needs a very large number of basis states to converge. The reason for this is that the electron-hole pair is strongly correlated due to the mutual Coulomb force. When the confinement potential energy is increased the noninteracting electron-hole pair states give a better discription of the system indicating the uncorrelated nature of the particles. In Fig. 3(b) the ground-state energy of a heavy-hole exciton as a function of magnetic field is shown. The two bases give approximately the same magnetic field dependence. This indicates that there is no change in correlations as a function of the magnetic field (at least up to 30 tesla).

In this paper we have present results for the ground-state and low-lying excitation energies of such a two-dimensional electron-hole system where both particles are confined by a parabolic potential. But it is also possible to leave the hole unconfined. It turns out, however, that there is no essential difference between these two choices. We have also studied other ground-state properties (electron-hole separation, optical absorption) of a parabolically confined exciton subjected to an external magnetic field. The results of these calculations will be published elsewhere (Halonen *et al* 1991).

ACKNOWLEDGMENTS

The work reported in this paper was partly done with Tapash Chakraborty, National Research Council, Canada. The author would also like to thank Pekka Pietiläinen for helpful discussions and encouragement.

REFERENCES

Bryant G W 1988 *Phys. Rev.* B**37**, 8763
Chakraborty Tapash, Halonen V and Pietiläinen P 1991 *Phys. Rev.* B**43**, 14289
Demel T, Heitmann D, Grambow P and Ploog K 1990 *Phys. Rev. Lett.* **64**, 788
Halonen V, Chakraborty Tapash and Pietiläinen P 1991 unpublished
Maksym P A and Chakraborty Tapash 1990 *Phys. Rev. Lett.* **65**, 108
Paquet D, Rice T M and Ueda K 1985 *Phys. Rev.* B**32**, 5208
Shinada M and Tanaka K 1970 *J. Phys. Soc. Jpn.* **21**, 1936
Sikorski Ch and Merkt U 1989 *Phys. Rev. Lett.* **62**, 2164

Inst. Phys. Conf. Ser. No 123
Paper presented at the International Meeting on Optics of Excitons in Confined Systems, Giardini Naxos, Italy, 1991

Two-dimensional exciton magnetic polaron in CdTe/(Cd,Mn)Te quantum wells

D R Yakovlev*, I N Uraltsev[1], W Ossau, G Landwehr, R N Bicknell-Tassius, A Waag, and S Schmeusser

Physikalisches Institut der Universität Würzburg, D-8700 Würzburg, Germany

[1] A.F.Ioffe Physico-Technical Institute, Academy of Sciences of the USSR, 194021 Leningrad, USSR

ABSTRACT: We report the investigation of two-dimensional magnetic polarons formed from excitons localized in very thin nonmagnetic CdTe quantum wells confined by $Cd_xMn_{1-x}Te$ semimagnetic barriers. Analysis of the Stokes shift of the exciton luminescence peak under resonance excitation, which is a dramatically decreasing function of temperature and magnetic field, allows us to determine the magnetic polaron energy. The polaron energy measured as a function of quantum well width is found to be 25 meV in 6 Å thick quantum wells and to vanish in wells thicker than 30 Å.

1. INTRODUCTION

The recent technological progress in molecular-beam epitaxy has led to the successful growth of high quality CdTe quantum wells (QW) confined by semimagnetic semiconductor barriers like $Cd_xMn_{1-x}Te$ (Ossau et al 1990) and made it possible to examine spin exchange interaction of two-dimensional carriers with the localized magnetic moments of the Mn ions in the barrier layers. As a consequence of the strong exchange interaction, a localized carrier or an exciton tends to organize ferromagnetically the spins of magnetic ions inside the barriers to form a magnetic polaron (MP) (Wolf 1988). Two types of MPs are expected in QWs resulting from different primary mechanisms of exciton localization in the well plane: i) MPs localized on a QW interface originating from excitons trapped by local potentials like that of a point defect or impurity; ii) two-dimensional exciton magnetic polarons (2D EMP) formed from excitons localized on extended defects like fluctuations of the QW width or of the Mn^{2+} spins. The former has been studied in thick $CdTe/Cd_xMn_{1-x}Te$ QWs both experimentally and theoretically (Nurmikko et al 1986, Gonsalves da Silva 1986, Wu et al 1986). The formation of the latter MP is expected to occur in thin QWs with a substantial extension of the exciton wavefunction into the barriers and to enhance the localization near both interfaces (Mauger et al 1988). A strong dependence of the MP energy on the QW width is expected in this case. 2D EMP formation has been recently observed in a 30 Å thick QW by resonance exciton luminescence spectroscopy (Yakovlev et al 1990).

In this paper we present the investigation of the 2D EMP problem in $CdTe/Cd_xMn_{1-x}Te$ quantum well structures. The exciton magnetic polarons formed from resonantly excited excitons are studied by low-temperature photoluminescence. We show that the 2D EMP energy is a dramatically decreasing function of the well width. It is comparable with the quasi-2D-exciton binding energy in very thin QWs and vanishes in wells thicker than 30 Å.

*) Alexander von Humboldt Fellow. Permanent address: A.F.Ioffe Physico-Technical Institute, Academy of Sciences of the USSR, 194021 Leningrad, USSR

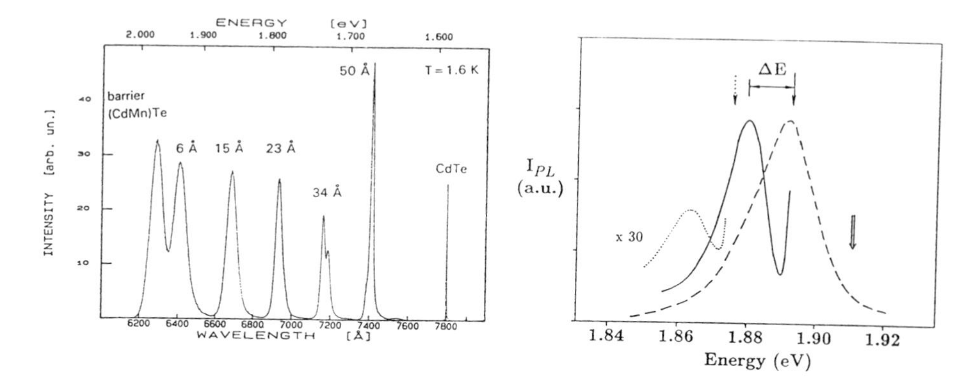

Fig.1. Photoluminescence spectrum of a $CdTe/Cd_{0.75}Mn_{0.25}Te$ structure with a set of SQWs with $L_z=6\div50$ Å separated by 1000 Å thick barriers. Excitation energy $\hbar\omega_{exc}=2.41$ eV; T= 1.6 K.

Fig.2. Photoluminescence spectra of a 12 Å thick SQW taken under resonance excitation in the localized exciton states with $\hbar\omega_{exc}=$ 1.892 eV (solid line) and 1.875 eV (dotted line). The localized exciton PL band, which was excited above the n=1 heavy-hole exciton energy (open arrow), is displayed by dashed line for comparison. T=1.6 K.

2. EXPERIMENT

$CdTe/Cd_{0.75}Mn_{0.25}Te$ single quantum well (SQW) structures were prepared by molecular-beam epitaxy on (100)-oriented CdTe substrates. The QW structures were grown on a 0.6 μm thick buffer layer of $Cd_{0.75}Mn_{0.25}Te$ and contain a set of SQWs which are separated by 1000 Å thick $Cd_{0.75}Mn_{0.25}Te$ barrier layers. Quantum wells with CdTe layer thickness L_z=6,12,15,23,28,34,40,50,60,100 and 300 Å were studied. A dye laser pumped by a cw argon-ion laser was used as an excitation source for photoluminescence (PL) and excitation PL spectroscopy. The samples were immersed in liquid or gaseous helium allowing to vary the sample temperature between 1.6 and 300 K. Magnetic fields up to 7.5 Tesla were applied perpendicular to the QW layers.

The luminescence spectrum of SQWs with $L_z=6\div50$ Å photoexcited above the band gap of the $Cd_{0.75}Mn_{0.25}Te$ barriers is shown in Fig.1. It consists of well-resolved peaks of the heavy-hole excitons in QWs and those of bound excitons in $Cd_xMn_{1-x}Te$ barriers and CdTe buffer layers. Both the red shift of the exciton peaks from the resonance energies of free heavy-hole excitons, measured from excitation PL and reflection spectra, and their linewidths are strongly decreasing functions of L_z. They are mainly caused by the exciton localization on monolayer fluctuations of the QW width (Yakovlev et al 1990) and to some extent by the magnetic polaron effect.

3. RESULTS AND DISCUSSION

Fig.2 displays PL spectra taken at T= 1.6 K under various excitation energies from a 12 Å thick QW. A wide, featureless PL band (dashed line) is observed under excitation above the n=1 heavy-hole exciton energy shown by an open arrow. For resonance excitation of localized excitons below the free exciton energy the PL spectrum reveals a narrower line which is well-separated from the excitation energy $\hbar\omega_{exc}$ (solid arrow), as the solid curve in Fig.2 shows. The lineshape of the PL and the Stokes shift from $\hbar\omega_{exc}$, $\Delta E=$ 13 meV, appear to be nearly independent of the energy of the selective excitation within the band of localized states. This fact is illustrated by a dotted-line

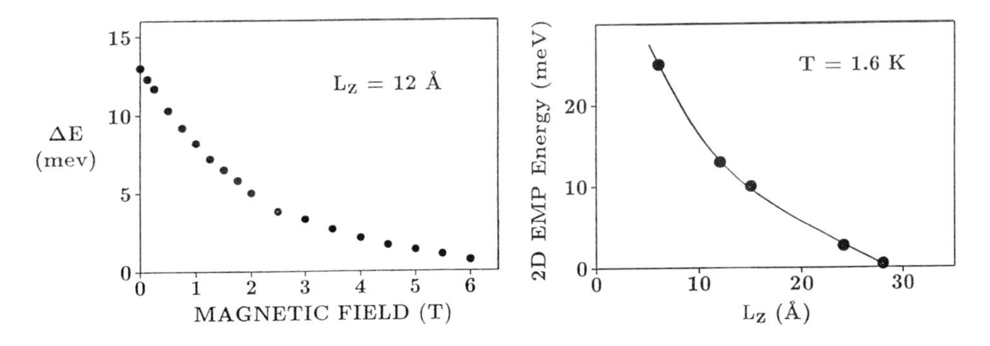

Fig.3. The Stokes shift ΔE of the PL peak from the resonance excitation as a function of magnetic field, T=1.6 K.

Fig.4. Energy of two-dimensional exciton magnetic polaron as a function of QW width in the $CdTe/Cd_{0.75}Mn_{0.25}Te$ SQWs, T=1.6 K. The solid line is given as a guide to the eye.

spectrum in Fig.2 which was excited at a lower energy (dotted arrow) corresponding to a tail of the localized exciton PL taken under nonresonant excitation. We interpret the observed Stokes shift of the PL peak from the laser excitation energy as resulting from magnetic polaron formation. Being resonantly excited in the localized states an exciton induces sizable magnetization within that part of its wavefunction which penetrates into the Mn-rich barrier regions, and decreases its energy due to an additional gain in the exciton-ion exchange energy. Suppression of the Stokes shift in the presence of magnetic fields shown for 12 Å thick QW in Fig.3 demonstrates the magnetic polaron character of this shift. The external magnetic field align the Mn^{2+} spins and the orientation of these spins by the exciton-ion exchange becomes insignificant.

One could suggest two possibilities to explain the Stokes shift ΔE: i) the MPs are formed from excitons and the whole ΔE should be attributed to the energy of MP formation; ii) excitons bound to impurities (donors or acceptors) form bound MPs. In the latter case ΔE is the sum of the exciton binding energy to the impurity and of the MP energy. The observed complete suppression of the Stokes shift at high magnetic fields argues against assignment of the PL peak to the bound MP recombination, as the exciton binding energy to an impurity could not be suppressed by a magnetic field. Thus, the whole value of the Stokes shift should be attributed to the MP energy.

The clear evidence for the two-dimensional character of MPs comes from the data of Fig.4. Here the energy of 2D EMP at 1.6 K is shown as a function of the quantum well width. The effective MP formation is found for QWs thinner than 30 Å. The PL peak for the 28 Å thick QW appears to be so close to the excitation energy that it hardly be resolved from the backscattered light. The MP energy is 25 mev in 6 Å thick QW and monotonically decreases for thicker QWs. The dramatic decrease of the 2D EMP energy with increasing L_z could be due to two effects: i) a decrease in the equilibrium MP energy and ii) an increase of the MP formation time. By use of picosecond time-resolved luminescence we have examined the dynamics of MP formation and conclude that the first reason is the principal one (Yakovlev et al 1992).

2D EMPs are formed mainly due to the exchange of hole spins with the spins of the Mn^{2+} ions, because the exchange integral for holes is 4 times larger than that for electrons (Gaj et al 1979). Our finding of the polaron shift in QWs thinner than twice the Bohr radius of heavy hole (~30 Å), in which the extension of the hole wave function into the semimagnetic barriers becomes significant (Gonsalves da Silva 1986), is an unequivocal experimental evidence for this suggestion. In a QW the hole is subjected to a net potential well consisting of two contributions of comparable values: the electron Coulomb potential and the valence band offset. The Coulomb potential enhances the hole localization in the center of the QW. As a result, the MP effect is quenched in thick QWs with a weak penetration of the hole wave function into the barrier layers. An enhancement of the MP effect is found in QWs of thicknesses comparable with the hole Bohr radius where an appreciable hole exchange with the Mn^{2+} spins is expected.

Goncalves da Silva (1986) has calculated the MP energy as a function of the position of the center of the hole wavefunction in the $CdTe/Cd_{0.75}Mn_{0.25}Te$ QW with $L_z=50$ Å. The polaron is expected to become stable if the hole is located closer than 15 Å from the heterointerface. Our finding of the 2D EMP formation in QWs with $L_z<30$ Å is not inconsistent with this prediction. It is noteworthy that the negligible polaron energy in QWs thicker than 30 Å proves the absence of interface localization of holes in our structures. It is not the case for the (111)-oriented $CdTe/Cd_{1-x}Mn_xTe$ QWs where interface localization of excitons provides significant overlap of its wavefunction with one of the barriers. This effect has been shown to result in the interface MP formation in thick QWs with $L_z>50$ Å (Nurmikko 1986).

In conclusion we have observed the two-dimensional magnetic polaron in thin nonmagnetic quantum wells confined by semimagnetic barriers. The polarons consist of excitons trapped, via exchange interaction, in magnetic potential wells created by locally aligning of the spins of the magnetic ions in the barrier layers near QW interfaces. Spin alignment of the magnetic ions is favored by the exchange interaction of that part of the exciton wavefunction, which penetrates into the Mn-rich barriers, but this effect is not the primary mechanism of localization. To observe the 2D EMP formation it is necessary to create excitons resonantly in localized states where their migration along the QW interfaces is restricted.

4. REFERENCES

Gaj J A, Planel R and Fishman G 1979 Solid State Commun.29 435
Goncalves da Silva C E T 1986 Phys.Rev.B 33 2923
Mauger A, Almeida N S and Mills D L 1988 Phys.Rev.B 38 1296
Nurmikko A V, Zhang X C, Chang S K, Kolodziejski L A, Gunshor R L and Datta S 1986 Surface Sci.170 665
Ossau W, Fischer S and Bicknell-Tassius R N 1990 J.Crystal Growth 101 905
Wolff P A 1988 Semiconductors and Semimetals, vol 25, eds J K Furdyna and J Kossut (London: Academic Press) ch 10,p 413
Wu J W, Nurmikko A V and Quinn J J 1986 Phys.Rev.B 34 1080
Yakovlev D R, Ossau W, Landwehr G, Bicknell-Tassius R N, Waag A, Uraltsev I N 1990 Solid State Commun.76 325
Yakovlev D R, Ossau W, Landwehr G, Bicknell-Tassius R N, Waag A, Schmeusser S, Uraltsev I N, Pohlmann A and Göbel E O 1992 J.Crystal Growth to be published

Inst. Phys. Conf. Ser. No 123
Paper presented at the International Meeting on Optics of Excitons in Confined Systems, Giardini Naxos, Italy, 1991

Asymmetry in the oscillator strengths of the optical transitions in a semiconductor superlattice under an electric field

P. Tronc

Laboratoire d'Optique. Ecole Supérieure de Physique et Chimie Industrielles
10 Rue Vauquelin 75231 Paris Cedex 05 - FRANCE

ABSTRACT : The oscillator strengths of the optical transitions in a semiconductor superlattice under an electric field parallel to the growth axis can be calculated using a perturbative model with Bloch envelope functions. The applied electric field and the indirect excitons both induce strength asymmetry between the oblique +p and -p transitions of the Wannier-Stark ladder. Features of the photocurrent spectra recorded at low temperature (for example the ranks of the transitions with the strongest aymmetry) can be accounted for by the present model in a very simple manner.

1. INTRODUCTION

Due to the lack of resonant coupling between adjacent quantum wells when an electric field is applied along the growth axis of a superlattice (SL) the miniband spectrum converges towards the evenly spaced "Stark ladder" (Bleuse et al. 1988). In a type I SL, the peak corresponding to the transition between a hole and an electron with a wavefunction whose maximum is p periods away from the hole state is labelled +p or -p (Fig. 1), the sign indicating that its energy is respectively greater or smaller than the energy of the peak corresponding to an electron and a hole localized in the same well. Photoluminescence (Mendez et al. 1988, Tronc et al. 1990) and photoconduction (Mendez et al. 1988, Agullo-Rueda et al. 1988) experiments show that the +p transitions are generally weaker than their -p counterparts. The ratio of the +p to -p transition intensities depends on the value of p and on the value of the applied electric field F. The photocurrent is proportional to the optical absorption in the structure while the photoluminescence spectra are strongly dependent on the radiative and non-radiative lifetimes of the carriers. It is widely known (Bastard 1988a) that the matrix element for an optical transition is proportional to the overlap of the electron and hole envelope functions.

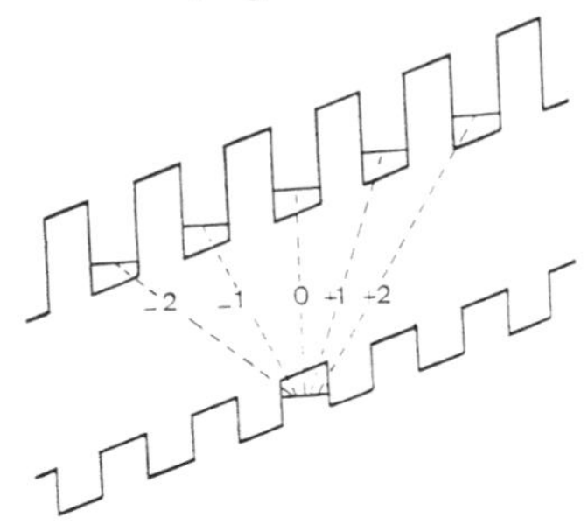

Fig. 1. Sketch of the conduction - and valence - band potential profiles in a type I SL under an electric field.

2. THE MODEL

We use a perturbative model operating with the superlattice envelope functions of the carriers introduced by Bastard (1988 b). The unperturbed superlattice envelope functions are Bloch functions $\psi_k(z)$, the allowed value of k being determined by use of cyclic boundary conditions. Let us consider a SL with an added perturbative potential Q(z). Moreover, if one considers the excitons, one notices that the system built from an electron centered on the 0th well and

a hole centered on the pth well is the mirror image of the system built from the same electron and a hole centered on the $(-p)^{th}$ well. The potential induced by the electron (hole) on the hole (electron) $f^{ph(e)}$ is changed into $f^{-ph(e)}$ when one goes from the first system to the second.
An important step is now to prove the validity of using a perturbative model to take into account the applied electric potential Q(z) = eFz and the excitons formation. We have seen that the applied electric field F localizes the carriers in the SL. In III-V SLs the heavy hole is strongly localized due to its large effective mass. If N is the number of wells over which the electron wavefunction is spread, N is limited because the potential drop over the whole spread of the electron wavefunction cannot exceed the electron first minibandwidth Δ_e (if not the probability for an electron to tunnel over N wells would be zero) :

$$NeFd \simeq \Delta_e \qquad (1)$$

N is odd from symmetry arguments. This value of N can be checked on the Bleuse's functions which provide a good approximation to the spread of the electron (Tronc et al. 1990). The voltage drop along N wells, being of the order of Δ_e, justifies the use of a perturbative model since Δ_e is generally widely smaller than the heights of the SL barriers for the electrons and the holes. Moreover the Rydbergs of the indirect excitons are small when compared to Δ_e (Blum et al. 1990) showing that the perturbation arising from the electron-hole interaction is widely weaker than the perturbation induced by the electric field. The spread of the electron wavefunction and hence of the electron-hole system being limited to N wells we have to use, to calculate the perturbed eigenenergies and the expansions of the perturbed eigenfunctions, the corresponding integration interval over z (of extension Nd), the allowed values of k being those of a periodic array of N wells. Indeed, after the electric field has been switched on, the electron wavefunction is the same whether the SL has only N wells or more. This drastically reduces the number of allowed values of k to N (including the O value) and drastically increases the energy difference between the states corresponding to two consecutive allowed values of k. The electric field being constant, the envelope function of a carrier centered on the nth well is deduced from the envelope function of the same carrier when centered on the $(n-p)^{th}$ by the pd translation. It is also clear that the $f^{ph(e)}$ potential is invariant under a md translation of the exciton along the z axis, m being an integer. The voltage drop along N wells being of the order of the first electron miniband Δ_e, makes it necessary, in a perturbative model, to diagonalize the electric field perturbation over, at least, the subspace corresponding to the first miniband both for the electrons and the holes. Higher minibands will not be considered because their difference in energy is large when compared to Δ_e, even for the holes. Moreover the contributions to the overlap of the electron and hole envelope functions is zero for the envelope functions from minibands with indexes of opposite parities and very weak for indexes of same parities but different from 1 (Bastard 1988a).

2.1. EFFECT OF THE ELECTRIC FIELD

To begin with, we first consider a SL with an applied electric field but without taking into account the electron-hole interaction. The bottom of the first electron miniband and the top of the heavy hole first miniband correspond to a k wavevector equal to zero ; both eigenenergies are not degenerate. The allowed value of k being very few and the energy difference between unperturbed eigenstates being large due to the localization (see above) it is assumed that the lowest perturbed eigenenergy corresponds to the perturbed state originating from the unperturbed one with k = 0. More precisely, it is assumed that the envelope functions of the electron and the hole which radiatively recombine are the Bloch envelope functions $|\chi_0^e(z)\rangle = |\psi_{k=0}^e(z)\rangle$ and $|\chi_0^h(z)\rangle = |\psi_{k=0}^h(z)\rangle$ perturbed by the electric potential Q(z). These perturbed functions are respectively written as $\lambda^{-1}|\chi_0^e + \chi_1^e\rangle$ and $\mu^{-1}|\chi_0^h + \chi_1^h\rangle$. $\chi_1^{e(h)}$ is expanded over the Bloch envelope function $\psi_{k\neq 0}^{e(h)}(z)$ and λ and μ are normalization coefficients.

The matrix element of the electric potential is :

$$q^{e(h)}_{k',k''} = <\psi^{e(h)}_{k'}(z) \mid Q^{e(h)}(z) \mid \psi^{e(h)}_{k''}(z)> \quad \text{with} \quad \psi^{e(h)}_{k}(z) = \exp(ikz)U^{e(h)}_{k}(z) \tag{2}$$

where $U^{e(h)}_{k}(z)$ is the periodic part of the electron (hole) Bloch envelope function. The potential of the barriers being even with respect to z, one gets :

$$U^{e(h)}_{-k}(z) = U^{e(h)}_{k}(-z) \tag{3}$$

and it can easily be shown that :

$$q^{e(h)}_{-k,0} = - q^{e(h)}_{k,0} \quad \text{therefore} \quad q^{e(h)}_{00} = 0 \tag{4}$$

The coefficient c_k of the expansion of $\chi_1(z)$ over $\psi_{k\neq 0}(z)$ is a function of the $q_{k,k}$ matrix elements. At the first order, c_k is $[E(0) - E(k)]^{-1}\, q_{0,k}$, and c_{-k} is equal to $-c_k$. But at the higher orders c_k has no defined parity when k is changed into -k. c_k can therefore be written as :

$$c_k = a_k + b_k \quad \text{with} \quad a_{-k} = -a_k \quad \text{and} \quad b_k = b_{-k} \tag{5}$$

The overlap for the $\pm p$ transition is calculated by operating the $\pm pd$ translation along the growth direction on the $\psi^{h}_{k=0}(z)$ function perturbed by the Q(z) potential.

The numerator of the overlap is then :

$$\begin{aligned} N_{\pm p} = {} & 1 + 2 \sum_{k>0} \cos kpd\left[(a^e_k)^* a^h_k + (b^e_k)^* b^h_k\right]\sigma_k \\ & \pm 2 \sum_{k>0} i\sin kpd\left[(b^e_k)^* a^h_k + (a^e_k)^* b^h_k\right]\sigma_k \end{aligned} \tag{6}$$

$$\text{with :} \quad |\lambda_{\pm p}|^2 = 1 + \sum_{k\neq 0} |a^e_k + b^e_k|^2 = |\lambda_o|^2, \; |\mu_{\pm p}|^2 = 1 + \sum_{k\neq 0} |a^h_k + b^h_k|^2 = |\mu_0|^2$$

$$<\psi^e_k(z) \mid \psi^h_{k'}(z)> = \sigma_k \delta_{kk'} \qquad \sigma_{-k} = \sigma_k \tag{7}$$

One goes from $<\chi^e \mid \chi^h>_p$ to $<\chi^e \mid \chi^h>_{-p}$ by changing a_k into $-a_k$. The main contribution comes from the smallest allowed values of $|k|$. It can be predicted that the rank of the transitions with the strongest asymmetry between the oscillator strengths of the +p and -p transitions are those which are such that $|\sin kpd| \approx 1$ and therefore $|\cos kpd| << 1$ where k has its lowest finite allowed value which is equal to $2\pi/Nd$. This leads to the criterion :

$$p \approx N/4 \tag{8}$$

which will be called "asymptotic criterion" for reasons appearing in 2.2. This relation provides an integer value for p on each side of N/4 (N is odd).

2.2. TAKING INTO ACCOUNT THE ELECTRON-HOLE INTERACTION

If the formation of excitons is now taken into account, the coefficient of the expansion of $\chi_1(z)$ over $\psi_{k\neq 0}(z)$ becomes d_k :

$$d_k = c_k + \varepsilon_k$$

ε_k is a function of the matrix elements of the electron hole interaction :

$$\varepsilon^{pe(h)}_{k',k''} = <\psi^{e(h)}_{k'}(z) \mid f^{pe(h)}(z) \mid \psi^{e(h)}_{k''}(z)> \tag{9}$$

The symmetry of the SL provides : $\varepsilon^{-pe(h)}_{-k',-k''} = \varepsilon^{pe(h)}_{k',k''}$ (10)

We have seen that the electron-hole interaction perturbation is weak when compared to the electric field one. It is the reason why we keep only the first order terms : $\varepsilon^{pe(h)}_{k} = [E(0) - E(k)]^{-1} \varepsilon^{pe(h)}_{k,0}$. The formulae from 2.1 are changed into :

$$
\begin{aligned}
N_{\pm p} = \; & 1 + \sum_{k\neq 0}\left[(\varepsilon_k^{pe})^*\varepsilon_k^{ph} + (b_k^e)^*\varepsilon_k^{ph} \pm (a_k^e)^*\varepsilon_k^{ph}\right]\sigma_k \\
& + \sum_{k>0} b_k^h\left[\cos kpd(\varepsilon_k^{pe} + \varepsilon_{-k}^{pe})^* + i\sin kpd(\varepsilon_k^{pe} - \varepsilon_{-k}^{pe})^*\right]\sigma_k \\
& \pm \sum_{k>0} a_k^h\left[\cos kpd(\varepsilon_k^{pe} - \varepsilon_{-k}^{pe})^* + i\sin kpd(\varepsilon_k^{pe} + \varepsilon_{-k}^{pe})^*\right]\sigma_k \\
& + 2\sum_{k>0}\left\{\cos kpd\left[(a_k^e)^* a_k^h + (b_k^e)^* b_k^h\right] \pm i\sin kpd\left[(b_k^e)^* a_k^h + (a_k^e)^* b_k^h\right]\right\}\sigma_k
\end{aligned}
\tag{11}
$$

and :

$$
\begin{aligned}
|\lambda_{\pm p}|^2 &= 1 + \sum_{k>0}\{|\varepsilon_{-k}^{pe} \mp a_k^e + b_k^e|^2 + |\varepsilon_k^{pe} \pm a_k^e + b_k^e|^2\} \\
|\mu_{\pm p}|^2 &= 1 + \sum_{k>0}\{|\varepsilon_{-k}^{ph} + \exp(-ikpd)(\mp a_k^h + b_k^h)|^2 \\
& + |\varepsilon_k^{ph} + \exp(ikpd)(\pm a_k^h + b_k^h)|^2\}
\end{aligned}
\tag{12}
$$

The criterion for the maximal asymmetry is more complicated because of the a_k^h coskpd term and because λ and μ are different for the +p and -p transitions. Nevertheless it can be seen that ε_0^p is real and that ε_k^p is a continuous function of k, making $(\varepsilon_k^{pe})^* - (\varepsilon_{-k}^{pe})^*$ to be zero at k = 0. For the smallest allowed values of |k|, $(\varepsilon_k^{pe})^* - (\varepsilon_{-k}^{pe})^*$ is therefore small when compared to $(\varepsilon_k^{pe})^* + (\varepsilon_{-k}^{pe})^*$. Moreover $f^p(z)$, which is small when compared to Q(z) even at p = 1, decreases with increasing values of p. The aymptotic criterion remains therefore still valid with perhaps some discrepancies for the first values of p.

3. DISCUSSION

If one considers the values of sinkpd and coskpd versus N for the smallest allowed finite values of |k|, in a type I SL and the features of the photocurrent spectra recorded at 5°K in the experiments by Agullo-Rueda et al. (1989) (the electron minibandwidth being assumed to be equal to 65 meV) one notices that the fit between the experimental results and the asymptotic criterion is good, except for the strong measured asymmetry between the +2 and -2 transitions when N=5, which does not correspond to a value of |sin 2kd| close to 1. It may be imagined (see 2.2) that the asymptotic criterion does not work well at N=5 because the difference between the two allowed values of k with the smallest finite modulus (i.e. k = 2π/5d and k = -2π/5d) is maximal at N=5 and equal to 4π/5d. The deviation from the asymptotic criterion can therefore be probably understood as the signature of indirect excitons. On the other hand one would expect that the asymmetry decreases when the Rydberg of the exciton decreases (i.e. for increasing values of N), which is not true from the measured photocurrent spectra (Agullo-Rueda et al. 1989). The strong asymmetry which does exist at large values of N can therefore be considered as the signature of the electric potential applied to the SL.

REFERENCES

Agullo-Rueda F, Mendez EE and Hong JM 1989 Phys. Rev. B40 1357

Bastard G 1988a, Wave Mechanics Applied to Semiconductor Heterostructures (Les Editions de Physique, les Ulis) pp. 246-250

Bastard G 1988b in ref. 1988a pp. 63-117

Bleuse J, Bastard G and Voisin P 1988 Phys. Rev. Lett. 60 220

Blum JA and Agullo-Rueda F 1990 Surf. Sci. 229 472

Mendez EE, Agullo-Rueda F and Hong JM 1988 Phys. Rev. Lett. 60 2426

Tronc P, Cabanel C, Palmier JF and Etienne B 1990 Solid State Commun. 75 825

Inst. Phys. Conf. Ser. No 123
Paper presented at the International Meeting on Optics of Excitons in Confined Systems, Giardini Naxos, Italy, 1991

Electric field effects on the electronic structure and radiative recombination of two-dimensional carriers in modulation-doped GaAs/AlGaAs heterostructures

Q X Zhao[1,2], P O Holtz[2], J P Bergman[2], B Monemar[2], T Lundstrom[2], M Sundaram[3], J L Merz[3] and A C Gossard[3]

1) Department of Physics, University of Trondheim - NTH, N-7034 Trondheim, Norway,
2) Department of Physics and Measurement Technology, Linköping University, S-581 83 Linköping, Sweden,
3) Center for Quantized Electronic Structures (QUEST), University of California, Santa Barbara, CA 93106, USA.

ABSTRACT: The radiative recombination of two-dimensional (2D) carriers at the interfaces of GaAs/AlGaAs one-side modulation doped double heterostructures has been investigated in specially designed structures grown by molecular beam epitaxy, with and without an external electric field applied perpendicular to the layers via a gate electrode. Several radiative transitions involving confined electron and holes states, such as band-to-band transitions, excitons and Fermi edge singularities, are observed. By applying an electric field across the structure, the position of the Fermi level can be deliberately altered within certain limits, which in turn provides a way to verify the interpretation of the observed transitions.

1. INTRODUCTION

Heterostructure samples of high quality can nowadays be fabricated by molecular beam epitaxy (MBE) and metal organic chemical vapor deposition (MOCVD) techniques. This fact has made it possible to study many fundamental properties related to two-dimensional (2D) carriers. It has been demonstrated in a number of recent investigations (Yuan et al 1985, Kukushkin et al 1988, Zhao et al 1989, Bergman et al 1991, Zhao et al 1990), that a specific photoluminescence (PL) band, the so-called H-band occurs close to the bandgap energy of GaAs in structures with a GaAs/AlGaAs heterointerface. The origin of this emission has been shown to be band-to-band transitions between confined electrons at the interface and holes in the valence band. We also report on the observation of a novel band close to the high energy limit of the H-band recombination interpreted as the Fermi-edge singularity (FES), which is enhanced in the presence of an applied electric field. The observation of a strong enhancement of the luminescence intensity for holes recombining with electrons close to the Fermi level in the 2D potential is the first manifestation of the FES in a heterostructure PL spectrum. The origin of the FES:s was first discussed by Mahan (1967) and have often been refereed to as Mahan excitons. The observation of the FES in optical experiments has earlier been reported for modulation doped quantum wells (QW) (Skolnick et al 1987, Chen et al 1990, Chen et al 1991,Lee et al 1987), where the FES:s appear due to the correlation between the sea of electrons and a single hole, which results in an enhancement of the oscillator strength for electrons close to the Fermi energy.

In this study, we present an optical study of modulation doped GaAs/AlGaAs heterostructures. Several radiative transitions involving confined electron and holes states, such as band-to-band transitions, excitons and Fermi edge singularities, are observed with and without an electric field as perturbation.

2. SAMPLES AND EXPERIMENTAL SETUP

The samples used in this study are grown by Molecular Beam Epitaxy (MBE) on either n^+ conducting or semiinsulating GaAs substrates, followed by a 10-period AlAs/GaAs superlattice (SL), an undoped GaAs layer with thickness d, a 20 nm undoped $Al_{0.35}Ga_{0.65}As$ spacer layer, an 80 nm Si doped (10^{18} cm^{-3}) $Al_{0.35}Ga_{0.65}As$ layer and finally a 5 nm GaAs cap layer. 2D electrons are confined in the notch potential at the interface between the undoped GaAs and the $Al_{0.35}Ga_{0.65}As$ spacer layer. We have studied samples, in which the thickness of the active GaAs layer (d) has been varied in the range from 400 to 1500 Å, but all spectra shown in this report originate from a heterostructure with a 500 Å wide GaAs layer. A semi-transparent metal gate consisting of 1nm Cr and 5nm Au was evaporated on top of the GaAs cap layer. To apply the electric field perpendicular to the hetero-interface between the GaAs and the $Al_{0.35}Ga_{0.65}As$ layer, thin electrical wires were contacted on both the gate metal and the back side of the GaAs substrate. In order to get a good bonding contact, 40 nm Au was used at the contact point on top of the semitransparent metal gate. The sample temperature could be continuously regulated down to 2.0 K. To eliminate heating effects due to the electric current, the electric field was pulsed, typically at 100Hz with a 10 percent duty cycle. The laser beam was also pulsed synchronously with the electrical bias pulse, by using an acousto-optical modulator. In this way, the current heating effects were minimized by choosing a proper length of the pulses. For the photoluminescence excitation (PLE) measurements a tunable Sapphire: Ti solid state laser, which covers the wavelength range from 700 to 1000 nm, was used as excitation source. A double-grating monochromator and a GaAs photomultiplier were used to disperse and detect the PL signals.

3. RESULTS AND DISCUSSION

The schematic band structures of the sample used for this study is shown in Fig.1. The electrons are confined in the notch potential formed at the interface between the active GaAs layer and the AlGaAs spacer layer. Due to the band bending across the active GaAs layer, the holes in the valence band are also quantized (Zhao et al 1991). From PL and PLE measurements it is verified that only the first electron subband is occupied at zero field. When a negative gate voltage is applied, the band bending across the active GaAs layer is reduced. The potential flattens out and the energy separation between the first and second electron subband decreases.

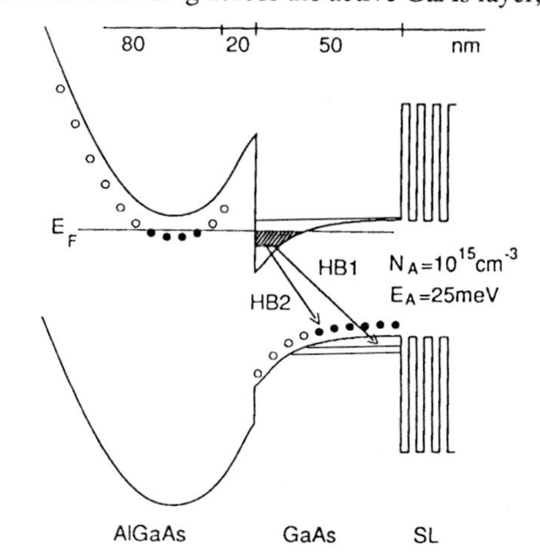

Fig.1 A schematic drawing of the potential of a typical hetero-interface. The two recombinations, HB1 and HB2, are indicated. Note that the energies indicated in this illustration are not on scale.

In Fig.2 a) a PL spectrum at 2.0 K with zero gate voltage is shown. To start at the lowest energy: The broad band at ~ 8350 Å is due to the recombination between electrons from the 2D gas and acceptor holes in the active layer, the so called H-band 2 (HB2 in Fig.2 a) (Zhao et al 1990). The emissions at ~ 8300 Å originate from GaAs bulk recombination: The free-to-bound (FB) and the donor acceptor (DA) pair transition. The band denoted HB1 in Fig.2 (H-band 1) is due to the transition between 2D electrons and holes from the valence band in the active layer. In Fig.2 b) the corresponding PLE spectrum of H-band 1 is shown. The two sharp peaks in the PLE spectrum are related to excited states of H-band, which have been assigned as excitons, related to CB2-HH1 and CB2-LH1, respectively (Zhao et al 1991).

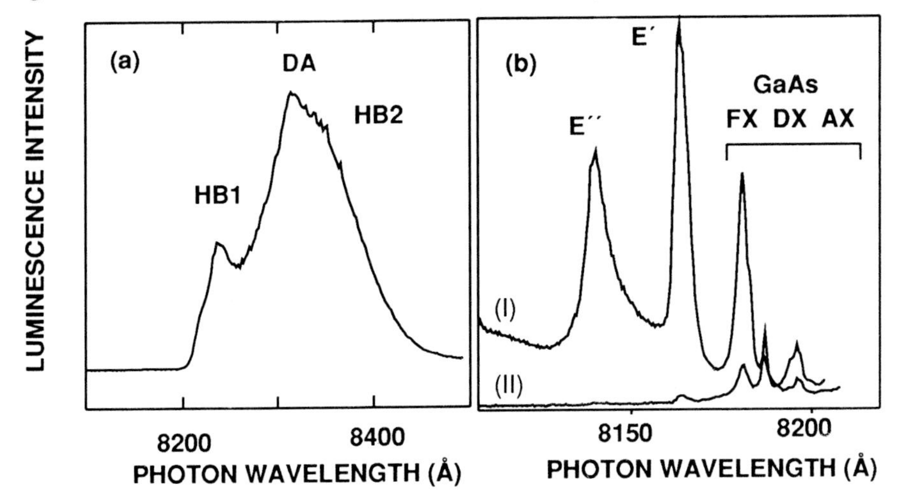

Fig.2 a) PL spectrum for a modulation doped GaAs/AlGaAs heterostructure measured at 2 K with excitation at 5145 Å. b) PLE spectra for the same sample as used for the PL spectrum measured at 2 K. The detection is (I) resonant with and (II) 3.7 meV above the so called HB1. E' and E" are exciton peaks related to CB2-HH1 and CB2-LH1, respectively.

To study the electric field effects on the radiative recombination, an electric field has been applied perpendicular to the layer. By varying the gate voltage, the energy separation between the Fermi edge and the first unoccupied electron subband can be changed in a controlled way, which in turn gives rise to striking many body effects close to the Fermi edge. Fig.3 shows PL spectra at different negative gate voltages. In these spectra, we see a new emission appearing at a position around 8180 Å, which becomes stronger with increasing electric field. It should also be noted that a peak at 8285 Å appears in the spectrum at the same time.

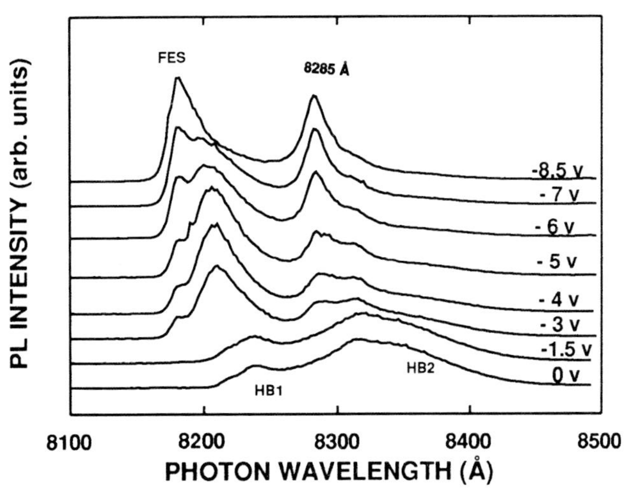

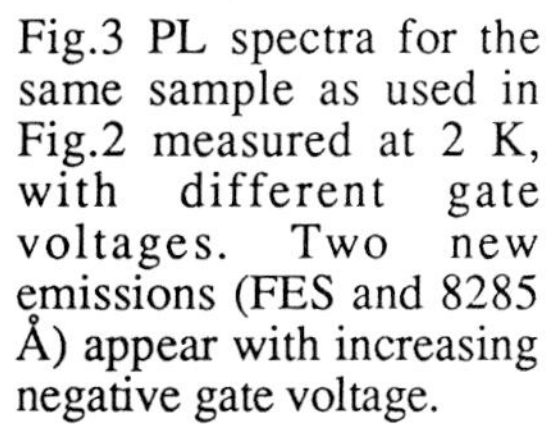

Fig.3 PL spectra for the same sample as used in Fig.2 measured at 2 K, with different gate voltages. Two new emissions (FES and 8285 Å) appear with increasing negative gate voltage.

For the novel transition at 8180 Å, tentatively interpreted as the FES, we first have to rule out alternative explanations : (1) the possibility of the free exciton in bulk GaAs can be excluded, since the emission disappears, when the top AlGaAs layer is etched away. (2) The possibility of transitions related to different excited subbands can be ruled out from pure energy considerations. The observed energy separation between the 8180 Å and E' (8164 Å) peaks is significantly less than expected for e.g. the n=2 and n=3 states. So we are left with the FES alternative. In previous studies (Chen et al 1990, Chen et al 1991), it is concluded that when the separation between the Fermi energy and the unoccupied electron subband is less than 5 meV, the Fermi-edge enhancements occur because the nearly resonant adjacent subband allows an efficient scattering path near k=0 for electrons at the Fermi energy in the context of the electron-hole interaction, thereby significantly increasing the optical oscillator strength. According to the theoretical calculation by Mueller (1990), the Fermi-edge enhancements depend on: (1) The energy separation between the Fermi energy and adjacent unoccupied electron subband; (2) The optical transition oscillator strength from electron subbands to holes; (3) Coulomb matrix element for electron-hole scattering between occupied and adjacent unoccupied electron subbands. In our measurements, the separation between first occupied and the second unoccupied electron subband can be reduced by the applied negative electric field, which gives rise to an increased electron-hole scattering, and as reduced band bending, which in turn results in an increasing optical transition oscillator strength between electrons and holes. These effects make k≠0 transition more favorable close to the Fermi-edge.

The interpretation of the transition at 8285 Å is at present uncertain. One possible candidate would be a correspondence to the FES at 8180 Å, but the hole would in this case originate from an acceptor in the active GaAs layer. Its nature is then similar to HB2 (Zhao et al 1990), which is due to the transition between a hole bound at acceptors and an electron at the bottom of the first electron subband.The separation between the FES and the 8285 Å band is about 19 meV, i.e. similar to the separation between HB1 and HB2. A tentative conclusion is therefore that the peak at 8285 Å is due to electrons at the Fermi level recombining with acceptors in the GaAs layer.

4. CONCLUSIONS

In summary we have presented optical studies on modulation doped GaAs/AlGaAs heterostructures. The transitions between a 2D electron and a hole in the valence band and a hole bound to an acceptor, respectively, are observed. We also report on the Fermi-edge singularities, which appear strongly when an electric field is applied across the structure.

5. REFERENCES

Bergman J P, Zhao Q X, Holtz P O, Monemar B, Sundaram M, Merz J L and Gossard A C 1991 Phys. Rev. B 43 4771

Chen W, Fritze M, Nurmikko A V, Achley D, Colvard C and Lee H 1990 Phys. Rev Lett. 64 2434

Chen W, Fritze M, Nurmikko, Hong M and Chang L L 1991 Phys. Rev. B 43 15738

Kukushkin I V, Klitzing K V and Ploog K 1988 Phys. Rev. B 37 8509

Lee J S, Iwasa Y and Miura N 1987 Semicond. Sci. Technol. 2 675

Mahan G D 1967 Phys. Rev. 153 882, 1967 Phys. Rev. Lett. 18 448

Mueller J F 1990 Phys. Rev. B 42 11189

Skolnick M S, Rorison J M, Nash K J, Mowbray D J, Tapster P R, Bass S J and Pitt A D 1987 Phys. Rev. Lett. 58 2130

Yuan Y R, Pudensi M A A, Vawter C A and Merz J L 1985 J. Appl. Phys. 58 397

Zhao Q X, Bergman P, Holtz P O, Monemar B, Hallin C, Sundaram M, Merz J L and Gossard A C 1989 Proceding of MRS fall Meeting, Boston, V136 337

Zhao Q X, Bergman J P, Holtz P O, Monemar B, Hallin C, Sundaram M, Merz J L and Gossard A C 1990 Semicond. Sci. Technol. 5 884

Zhao Q X, Fu Y, Holtz P O, Monemar B, Bergman J P, Chao K A, Sundaram M, Merz J L and Gossard A C 1991 Phys. Rev. B 43 5035

Inst. Phys. Conf. Ser. No 123
Paper presented at the International Meeting on Optics of Excitons in Confined Systems, Giardini Naxos, Italy, 1991

Theoretical and experimental studies on the binding energy of the shallow Zn acceptor in $Ga_xIn_{1-x}As/InP$ quantum wells

Al L Efros, A Kux, C Wetzel, B K Meyer, D Grützmacher[+] and A Kohl[+]

Physik Department E 16, TU Munich, James Franck Str., 8046 Garching, F.R.G.
[+]Institute of Semiconductors Electronic, RWTH Aachen, Sommerfeldstr., 5100 Aachen, F.R.G.

ABSTRACT In this work we study the behaviour of the shallow Zn acceptor in $Ga_xIn_{1-x}As$ Quantum Wells (1.5 nm<L_z<5 nm) lattice matched to InP. The experimental results as obtained by photoluminescence and optically detected impact ionisation to determine the binding energy of Zn are compared with a theoretical calculation. The binding energy increases from the 3D bulk value for smaller L_z, passes through a maximum for the 3 nm well width, but then has an unexpected minimum at L_z=1.5 nm.

1. INTRODUCTION

Photoluminescence (PL) is a very common technique to characterize semiconductor heterostuctures and Quantum Wells. It allows to separate between different radiative contributions, e.g. exciton or bound exciton decay or donor-acceptor in comparison with the free to bound transition. This is not always a simple task in GaInAs, due to the small binding energy of the defects. It requires e.g. temperature dependent studies. Also, the behaviour of the PL intensity as a function of the excitation light power can be useful to distinguish between the various decay processes (Götz et al 1983). Perturbation spectroscopy such as applying an external electric field or magnetic field, or thermally modulated PL (Gal et al 1991) can be used to gain more insight and to enhance the resolution. We have used a similar technique and studied PL under the influence of microwave irradiation (Meyer at al 1991). With this technique we determine the binding energy of the shallow Zn impurity in $Ga_xIn_{1-x}As$ lattice matched to InP as a function of quantum well thickness. We compare the experimental results with a theoretical calculation done in the framework of the effective mass theory.

2. EXPERIMENTAL

The samples were grown by LP-MOVPE. They were undoped, but contained Zn as a residual impurity (in the same reactor Zn doped laser active structures are grown and a pollution of QW structures grown in a subsequent run can hardly be avoided). The photoluminescence and microwave experiments were performed in a 12 GHz magnetic resonance spectrometer allowing for optical detection of impact ionisation (ODII). This technique is discribed by Wang et al 1989. The basic set-up can be found in Meyer at al 1991.

3. EXPERIMENTAL RESULT

In the microwave frequency range of 12 GHz the influence of microwave irradiation on the PL intensity gives rise to broad unresolved cyclotron resonance lines centered close to zero magnetic field (Booth et al 1985 , Pakulis et al 1987). The cyclotron resonance induced changes in the PL intensity are assigned to ODII. Free carriers, which are accelerated by the resonance gain energy and impact on bound excitons or neutral donors. Excitons and free electrons are released, therefore donor-bound exciton and donor-acceptor transitions show a decrease in PL intensity, whereas the free exciton and the free to bound acceptor transitions increase in PL intensity. This is now a well established mechanism in bulk and epitaxial samples. In narrow quantum wells (L_z<5nm) usually only electron to hole recombination within the confined subband states is observed. In this case ODII resolves the excitonic transitions due to monolayer fluctuations, which then have the same sign (see Figs.1 and 2) in the ODII spectrum (Meyer at al 1991) and correspond to a decrease in PL intensity. For impurity related transitions, such as free to bound transitons involving a shallow acceptor, a different sign is found i.e. that enhancement results from the impact ionisation. It is this selective mechanism we use to determine the binding energy of the Zn acceptor in narrow Quantum Wells.

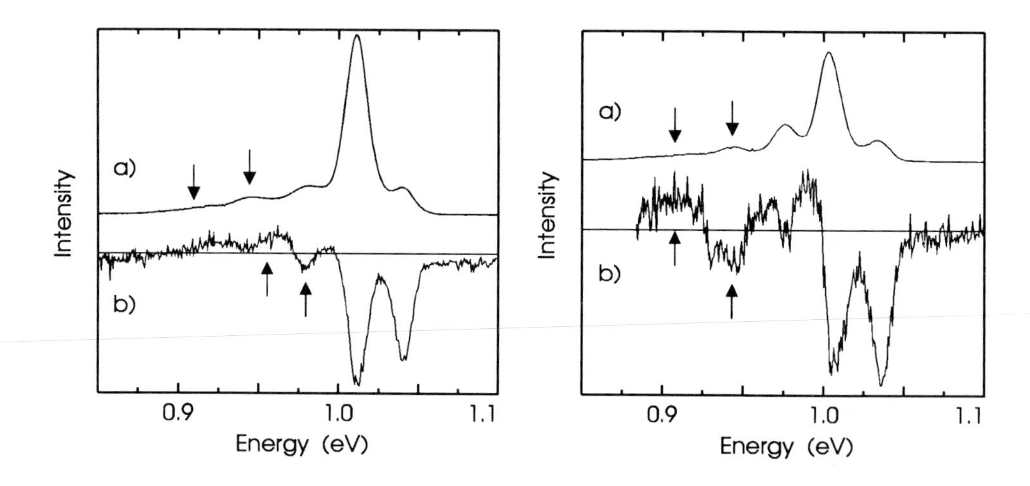

Fig.1 : PL (a) and ODII spectrum (b) for a 1.5 nm thick $Ga_{0.48}In_{0.52}As/InP$ QW

Fig.2 : PL (a) and ODII spectrum (b) for a 1.5 nm thick $Ga_{0.41}In_{0.59}As$ QW

The binding energy of the Zn acceptor in thick QWs (100 nm) is 22 meV, which is equal to the bulk value (3D) of $Ga_xIn_{1-x}As$. This is practically also the case for the 10 nm well (see Fig.3, full circle). The experimental value of the binding energy increases from the 3D bulk value for smaller L_z, passes through a maximum for L_z=3 nm, and then decreases for L_z=1.5 nm (see crosses in Fig. 3). Especially for the 1.5 nm wells PL studies alone were not sufficient. This is demonstrated in Figs.1 and 2, where from comparison of PL and ODII spectra an approximate value for the binding energy can be found (see arrows in Fig.1 and 2). Two compositions, x=0.48 and x=0.41, have been investigated. For both compositions we can expect that the binding energy is not greatly altered. The negative signals are caused by two excitonic transitions due to monolayer fluctuations of the QWs with 6 and 7 monolayers. Taking as a reference the respective lowest excitonic line and averaging over many samples, and taking into account the PL line positions and line shapes, we deduce as a upper and lower bound for the binding energy the value 24 ± 10 meV in the 1.5 nm well (crosses in Fig.3). The value for L_z=0, which corresponds to the acceptor in the barrier, is just the binding energy of Zn in InP, 43 meV (Fig.3, full circle).

4. THEORETICAL CALCULATIONS

The experimental results are rather unexpected, if compared to the dependence of the hydrogen-like donor binding energy in the QW of thickness L_z obtained by Bastard et al 1986. This dependence was calculated under the assumption that the electron has an effective mass equal to the heavy hole m_{hh}. This behaviour is shown in Fig. 3 by the dotted line. One can devide it into three regions. If L_z is larger than the Bohr radius of the impurity, a_B, the shift of the impurity levels from their bulk values E_B (Bohr energy) is exponentially small. The increase of the binding energy is determined by the displacment from the band edge due to the quantum size (QS) effect. When L_z is smaller than a_B, the impurity becomes 2D like. The impurity level follows the lowest QS subband and the binding energy approaches the 2D like hydrogen center, i.e. 4 E_B. In very narrow QWs, where the QS energy becomes comparable to the depth of the QW, a penetration of the wavefunction into the barrier restores the 3D behaviour again. As a result the binding energy decreases and approaches the bulk value in the barrier material (InP). It shows no additional minimum.

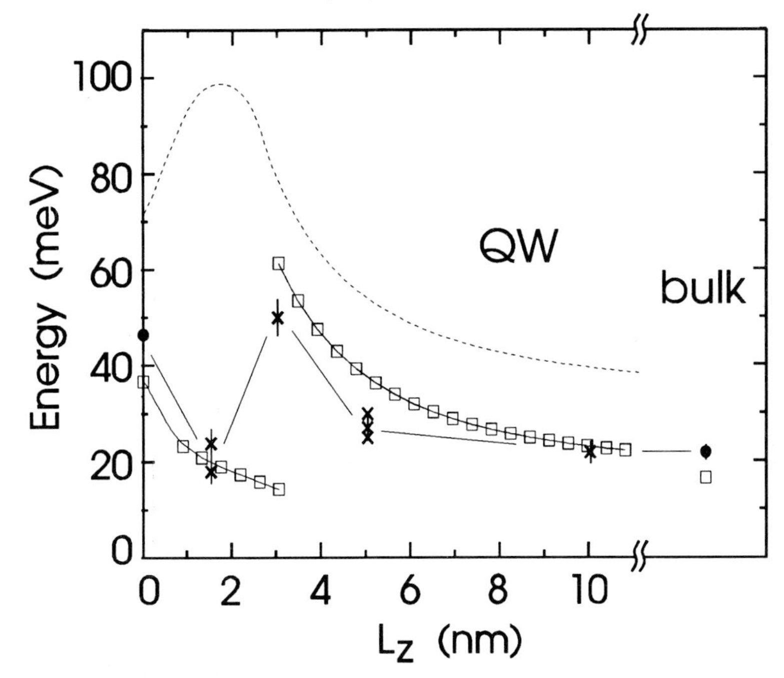

Fig. 3 : Experimental (crosses: this work, full circles: bulk values from literature) and theoretical values (open squares) for the Zn acceptor binding energy as a function of Quantum Well thickness

We connect the observed minimum of the acceptor binding energy for $L_z = 1.5$ nm to the small value of the longitudinal effective mass $m_\parallel$ of the lowest QS hole level. It is just of the order of the light hole effective mass m_{lh}. When m_{hh} is much larger than m_{lh}, the binding energy of a 3D acceptor is given by the following expression (Lipari et al 1970, Gel'mont et al 1972) :

$$E_{3D} = (4/9)\, m_{hh}\, e^2 / (2\, h^2 \epsilon^2) \qquad \text{(equ.1)}$$

where ϵ is the dielectric constant. For a pure 2D acceptor we find the expression

$$E_{2D} = 4\ m_{\parallel}\ e^2 / \ (2 h^2 \epsilon^2) \qquad \text{(equ.2)}$$

A comparison of equ.1 and equ.2 shows that E_{2D} becomes smaller as compared to the E_{3D} case, when $m_{\parallel} < m_{hh}/9$.

Our calculation of $m_{\parallel}$ taking into account the penetration of the wavefunction into the barrier gives $m_{\parallel} = 0.087$ for $L_z = 1.5$ nm.
The heavy hole masses in InP and $Ga_xIn_{1-x}As$ lattice matched to InP were taken to be $m_{hh} = 0.85$ and $m_{hh} = 0.465$, respectively. This means that the binding energy of the 2D acceptor in the QW should be smaller than the binding energy of the 3D acceptor in the barrier and hence larger than in the QW. The results of the corresponding calculations for the hydrogen-like acceptor binding energy are shown in Fig.3 (open squares). The theoretical calculation takes into account the perturbation of the binding energy for a 2D acceptor, due to the finite value of the QW thickness.
The position of the acceptor level changes exponentially weakly in the QWs, when the thickness L_z is larger than the light hole Bohr radius a_{lh} (Ivanov et al 1988), where

$$a_{lh} = h / \sqrt{2 m_{lh} E_{3D}}$$

In this range the increase of acceptor binding energy is described by the shift of the lowest QS level of holes.
We are aware of the limitations of our analytical calculation, nevertheless a comparison with experiment not only shows qualitative, but also quantitative agreement. We note that even with the relatively large error bar on the value for the binding energy of Zn in the QW with $L_z = 1.5$ nm, experiment and theory share a common trend and the minimum in the binding energy brings evidence for the existence of a pure 2D acceptor in our QW.

5. CONCLUSIONS

The binding energy of the shallow Zn acceptor in narrow $Ga_xIn_{1-x}As$/InP QWs has been determined by PL and ODII experiments. The unexpected result that the binding energy first goes through a minimum for 1.5 nm thick QWs, which is lower than the value found for the pure InP limit, has been supported by theoretical calculations.

References

Bastard G and Brum J A 1986 IEEE J. Quant Electr. **QE22** 1625
Booth I J and Schwerdtfeger 1985 Solid State Comm. **55** 817
Gal M, Xu Z Y, Green F and Usher B F 1991 Phys. Rev. **B43** 1546
Gel'mont B L and D'yakonov M I 1972 Sov. Phys. Semicond. **5** 1905
Götz K H, Bimberg D, Jürgensen H, Selders J, Solomonov A V, Glinskii G F and Razeghi 1983 J. Appl. Phys. **54** 4543
Ivanov M G, Merculov I A and Efros Al L 1988 Sov. Phys. Semicond. **22** 392
Lipari N O and Baldereschi 1970 Phys. Rev. Lett. **42** 1660
Meyer R, Hardtdegen H, Carius R, Grützmacher D, Stollenwerk M, Balk P, Kux A and Meyer B K 1991a J. Electr. Mat. to be publ.
Meyer B K, Wetzel C, Grützmacher D and Ömling P 1991b Mater. Sci. Engin. **B9** 293
Pakulis E J and Northrop G A 1987 Appl. Phys. Lett **50** 1672
Wang F P, Monemar B and Ahlström M 1989 Phys. Rev. **B39** 11195

Inst. Phys. Conf. Ser. No 123
Paper presented at the International Meeting on Optics of Excitons in Confined Systems, Giardini Naxos, Italy, 1991

Donor-bound excitons in CdTe/$Cd_{1-x}Zn_xTe$ quantum wells

K. Kheng, R. T. Cox, S. Tatarenko, F. Bassani, K. Saminadayar, N. Magnea

SPMM/Laboratoire de Physique des Semiconducteurs, CEN-Grenoble, B.P. 85X, 38041-Grenoble-Cedex, France
Laboratoire de Spectrométrie Physique, CNRS-Université J. Fourier, B.P. 87X, 38042-Saint Martin d'Hères-Cedex, France

ABSTRACT. Optical spectra for D^oX (excitons bound by neutral donors) are identified in emission and absorption spectra of indium doped CdTe/$Cd_{1-x}Zn_xTe$ quantum wells. The exciton localisation energy increases from the 3D value of 2.6meV to 5.1mev at the well-centres for well-width L_w= 95 Å. For one sample (L_w= 140 Å), two-electron satellites of D^oX were seen, yielding a donor-ionisation energy of 18.1 meV.

For bulk direct-gap semiconductors where free-exciton related luminescence is inhibited by polariton effects, the emission spectra are dominated by recombination transitions of D^oX and A^oX (excitons bound to neutral donors and neutral acceptors respectively). By contrast, free-exciton lines predominate in the spectra of semiconductor quantum wells. As a consequence, there is not a lot of information available about the effects of confinement on *bound* excitons, particularly for the very lightly bound species D^oX.

Our group has been developing a strained layer, II-VI semiconductor quantum well system consisting of CdTe wells with $Cd_{1-x}Zn_xTe$ ("CZT") alloy barriers (typical properties: E_g(CdTe)= 1.606eV; ΔE_g= 45 meV for x=0.08 with $\Delta Ec \approx 0.8 \Delta Eg$, depending on strain). We grow the structures by Molecular Beam Epitaxy (MBE) on $Cd_{1-x}Zn_xTe(x \approx 4\%)$ alloy substrates. This system appears interesting for studies of D^oX because the exciton localisation energy E_{loc} (difference between the energies of X and D^oX) is relatively large in CdTe: $E_{loc} \approx$ 3 meV in the bulk. We have started spectroscopy of D^oX in conjunction with a materials programme on Indium donor doping of CdTe and CZT during MBE growth.

We began by re-examining the luminescence of bound excitons in nominally undoped CdTe/CZT quantum wells and superlattices (the free exciton spectra have been reviewed at this conference, [Tuffigo 1992]). One sees nearly always an $\approx$1 meV wide line, called Y by [Mariette et al 1988] (also called B by [Tuffigo et al 1991] for superlattice samples), that lies some meV below the heavy-hole exciton line X (light-hole exciton states are split off by strain) . Because line Y is very weak compared to X in transmission and reflectivity spectra, it has been attributed to (heavy-hole) excitons trapped at "impurities or interface defects". One aim of the present work was to establish whether line Y corresponds to D^oX. In fact, closer study of line Y, particularly at low excitation power, showed that it has a complex shape whose appearance varies with well width, barrier-height and sample strain. The doping studies described below show that the principal component of Y, lying closest to X (we label this Y_1) is *not* D^oX. This major component remains unidentified. A component sometimes visible at lower energy, which we label Y_2, does increase on donor-doping, as now discussed.

We first doped *single* CdTe wells, introducing indium donors, using the technique described by [Bassani et al 1991a]. A "Y_2-type" line appears but, for single wells, it proved very difficult to adjust the doping to the relatively low donor concentrations ($\approx 10^{16}$ cm^{-3}) required to avoid broadening of the line [Bassani et al 1991b]. At low doping level, the donors in the well tend to be compensated by the large numbers of background acceptor impurities (Cu, Li, etc, in the 10^{15} cm^{-3} range) available in a depletion length extending far out into the CZT barriers. These difficulties led us to study doping of ***multiple*** quantum well (MQW) or superlattice samples of thickness one micron or greater and well-width $L_w \approx$ barrier-width L_b. Now the volume of doped material (the wells) and undoped material (the barriers) is about equal. The samples were nominally planar doped ("delta-doped") during a growth-interruption at each well-centre. SIMS (Secondary Ion Mass Spectrometry) measurements have shown that doping profiles sharp to within <50Å can be achieved in this way[Bassani et al 1991b].

Fig. 1(a) shows the luminescence of a doped MQW (L_w=L_b= 135 Å). There is now a very strong line in the "Y" spectrum region, completely dominating the free exciton line X. Capacitance-voltage (C-V) measurements are possible for a MQW structure and proved that

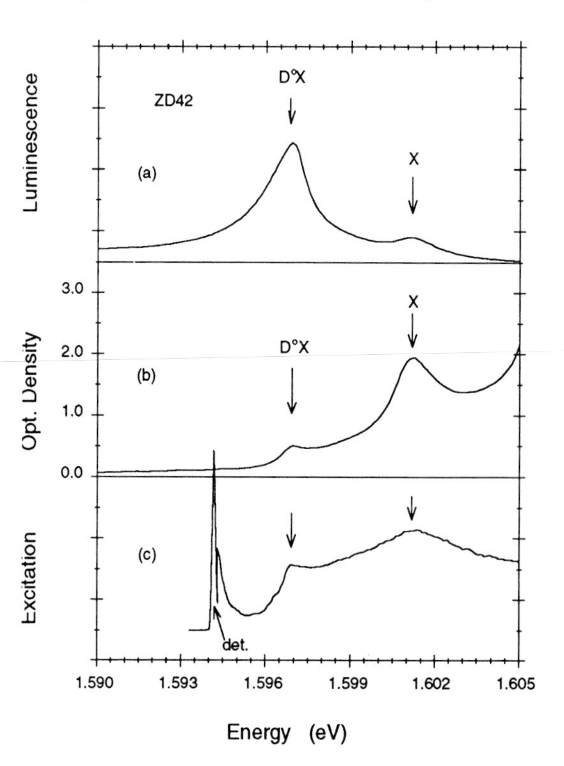

Figure.1:Spectra in the exciton region for MQW sample ZD42 (8000Å of 135ÅCdTe/135Å$Cd_{1-x}Zn_xTe$, x=8% on a CdTe/$Cd_{1-x}Zn_xTe$, x=3% substrate). The MQW is doped with indium at the well centres. X and D^oX label (heavy-hole) free and donor-bound excitons. (a) emission spectrum for argon laser excitation (b) optical density defined by $\log_{10}(I_{in}/I_{out})$; the exciton lines lie on the sharply rising substrate absorption.(c) luminescence-excitation spectrum detected at wavelength marked "det."in low energy wing of D^oX.

this MQW sample had been successfully doped n-type, with a volume-averaged neutral donor concentration of $\approx 4.\ 10^{16} cm^{-3}$. On this basis, we attribute the strong line in Fig. 1(a) to D^oX in the quantum wells.

The $Cd_{1-x}Zn_xTe$ substrate material with x≈3% is transparent to ≈1.605 eV and for L_w=135Å the confined excitons in the CdTe wells can be studied in transmission. Fig. 1(b) shows the absorption spectrum of the above MQW sample. The X and D^oX absorptions stand out clearly on the rising background of the (x=3%) substrate's absorption edge. Note in Fig. 1 that there is no obvious "Stokes shift" between absorption and luminescence for the transition peaks, an indication that the heterostructures are of high quality. Note also that, although the ratio of indium atoms to cadmium atoms in the wells is only 5. 10^{-6}, the D^oX absorption is within about an order of magnitude of the X absorption. This is the so-called "giant oscillator-strength" effect for D^oX, related to the large volume occupied by the ≈50Å radius donor wavefunctions. Fig. 1(c) also shows a photoluminescence excitation (PLE) spectrum monitored in the low energy wing of the D^oX emission. The PLE resembles the absorption spectrum, except that X is relatively weaker.

Donors D^o in a quantum well have a distribution of values of ionisation energy E_D, with a maximum value at the well-centre [Bastard 1980]. As concerns the excitonic species D^oX, little is known about the effect of confinement on the localisation energy E_{loc}, but we expect E_{loc} to increase with increasing E_D. Thus, we expect well-centre donors to have the highest localisation energy and so we attribute the sharp low-energy edge of the D^oX absorption and PLE, giving a marked peak at 1.597 eV (see Fig. 1), to donors at the well centres. The X-D^oX difference in the absorption spectrum then gives a very accurate value of E_{loc} at well-centre, namely 4.2 meV.

There is intensity above the main D^oX peak in Fig. 1, extending up to the free-exciton peak X. This may correspond to excitons bound at donors lying off the well-centres, that is it could be an indicator of impurity segregation on a scale unresolvable by SIMS. Or (more hopefully for the materials aspects), it may correspond to excited states of D^oX: for bulk CdTe crystals giving very narrow optical lines, four excited "rotator" states of D^oX can be distinguished between D^oX and X.

Another doped MQW (sample ZD51)grown with narrower wells, L_w= L_b=95Å, gave spectra similar to those of Fig. 1, with D^oX shifted ≈7 meV to higher energy. The well-centre value of E_{loc} (from the X-D^oX difference in PLE) increases to 5.1 meV. Fig. 2 plots the limited amount of data available to date for E_{loc} and compares with the 3D (indium) value. We have redefined the 3D value of [Francou et al 1990], taking E_{loc}(3D) to be E_t-E(D^oX)= 2.7 meV for indium. Here E_t is the transverse exciton energy = 1.5957 eV [Merle et al 1984], which is the limit of E(X) as $L_w \to \infty$ (in the centre-of-mass quantization description, see [Tuffigo 1992]).

Much information about energy levels of donors in bulk semiconductors has come from studies of "two-electron" recombination transitions, where the donor electron is promoted to an ns (or np) excited state during the recombination. For example, the energy difference between the two-electron transition $D^o{}_{1s}X \to D^o{}_{2s}$ and the normal ("one-electron") transition $D^o{}_{1s}X \to D^o{}_{1s}$ gives the 1s-2s energy difference. The two-electron transitions are typically 10^{-2} times weaker than the one-electron transition and for CdTe have been reported only for pure or relatively lightly doped crystals giving narrow luminescence lines [Dalbo et al 1989], [Francou et al 1990].

After an extensive search, we believe we have found such lines for one CdTe/CZT sample: an undoped MQW with L_w=L_b=140Å. This sample gave a strong, narrow "Y_2" type emission at the right position to be D^oX, 3.9 meV below X, see Fig. 3. We discuss the satellite lines labelled S_1 and S_2 , at 13.7 and 16.0 meV below Y_2. We propose that S_1 and S_2 correspond to two-electron transitions with the donor promoted to the 2s and 3s states respectively. We attribute the asymmetric lineshapes of S_1 and S_2 to the distribution of donor positions in the well, with the sharp low energy edges corresponding to the well centres.

In a well of width $L_w >> a_B$≈50Å, the 3D Rydberg formula: $E(n) = E_D/n^2$ appears more accurate than a 2D formula. Then, taking E_D= 4/3 x 13.7 meV from line S_1 or 9/8 x 16.0 meV from S_2 gives two, concordant estimates, 18.2 and 18.0 meV respectively, for E_D at the well centres. In bulk CdTe, the donor-ionisation energies range from 13.7 to 14.8 meV[Francou et al 1990]. For Ga and In (the most probable donors in our undoped MBE material) the 3D values are 13.9 and 14.1 meV respectively. The value 18.1 meV represents an enhancement of 29% for E_D at L_w= 140 Å. The D^oX localisation energy has increased from 2.6 to 3.9 mev at this well-thickness (Fig. 2), an ≈50% increase. Thus, E_{loc} is increasing faster than E_D, that is Haynes's rule for donor-bound excitons (proportionality between E_D and E_{loc}) does not apply in these quantum wells for the L_w dependence.

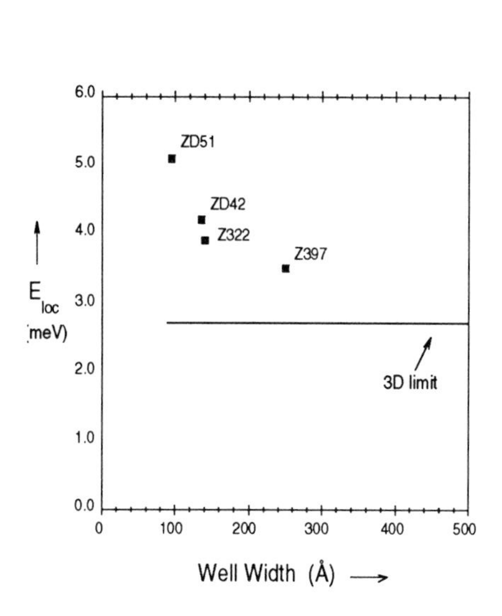

Figure 2: Available data for energy of localisation of an exciton on a donor in CdTe/ CZT(7-8%Zn) MQW samples. ZD51 and ZD42 are indium-doped; Z322 and Z397 are undoped MQWs giving a strong D^oX type line.The horizontal line at 2.7 meV is the energy of localisation of an exciton on an indium donor in bulk CdTe (referenced to E_t).

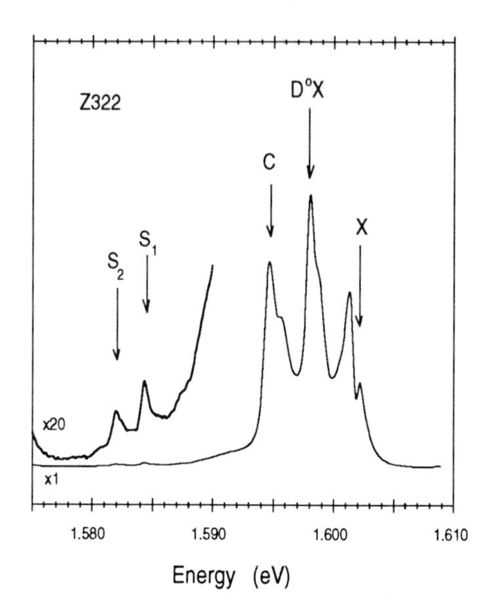

Figure 3: Luminescence spectrum of an undoped, 140Å CdTe /140Å CZT (8% Zn) MQW (sample Z322) giving a strong D^oX line. (The free exciton emission X shows a "reabsorption" dip; line C, not discussed here, is probably A^oX). Lines S_1 and S_2, shown at 20x increased gain, are assigned to two-electron recombinations that leave D^o in the 2s and 3s states respectively.

REFERENCES

Bassani F, Tatarenko S, Saminadayar K, Bleuse J, Magnea N and Pautrat J L 1991a Appl. Phys. Lett. 58 2651

Bassani F et al 1991b, J. Crystal Growth, in press (Proc.II-VI Conference, Tamano 1991)

Bastard G 1980, Phys. Rev B 24 4714

Dalbo F, Lentz G, Magnea N, Mariette H, Le Si Dang and Pautrat J L 1989, J. Appl. Phys. 66 1338

Francou J, Saminadayar K, and Pautrat J L 1990 Phys. Rev. B 41, 12035

Mariette H, Dalbo F, Magnea N, Lentz G and Tuffigo H 1988 Phys. Rev. B 38 12443

Merle J C, Sooryakuma R and Cardona 1984 Phys. Rev. B 30 3261

Tuffigo H 1992, this conference

Tuffigo H, Magnea N, Mariette H, Wasiela A and Merle d'Aubigné Y 1991 Phys. Rev. B 43 14629

Inst. Phys. Conf. Ser. No 123
Paper presented at the International Meeting on Optics of Excitons in Confined Systems, Giardini Naxos, Italy, 1991

Optical spectroscopy of bound excitons in AlGaAs/GaAs quantum wells

B Monemar, P O Holtz and J P Bergman

Department of Physics and Measurement Technology, Linköping University, S-581 83 Linköping, Sweden

C Harris and H Kalt

Max Planck Institut für Festkörperforschung, D-7000 Stuttgart 80, Germany,

M Sundaram, J L Merz and A C Gossard

Center for Quantized Electronic Structures (QUEST), University of California, Santa Barbara, CA 93016, US

K Köhler and T Schweizer

Fraunhofer Institut für Angewandte Festkörperphysik, D-7800 Freiburg, Germany

ABSTRACT: Bound excitons in AlGaAs/GaAs quantum wells, doped with donors or acceptors in the well, are discussed in relation to recent experimental data from optical spectroscopy. Results from two-particle spectroscopy, the influence of high doping levels, the interplay with exciton localization, as well as picosecond transient data are covered.

1. INTRODUCTION

Bound excitons (BE's) in quantum wells (QW's) were discovered about 10 years ago (Miller et al 1982), but detailed investigations comparable to those in bulk materials have only emerged during the last few years. Systematic studies of intentially doped QW's are needed for this purpose. In addition the impurities need to be introduced in a narrow region of the QW in order to get sharp line BE spectra, since the binding energies of impurities and BE's vary dramatically with the impurity position in the QW (Rune et al, 1991). In this paper we briefly summarize some recent experimental results for BE's in AlGaAs/GaAs QW's. Emphasis will be given to several aspects, such as two-hole transitions (THT's) for acceptors and two-electron transitions (TET's) for donors, spectroscopic evidence for localization of excitons in interface potentials interfering with BE spectra, and dynamic aspects of BE recombination.

2. SAMPLES AND EXPERIMENTAL

The samples used in this study were grown by MBE at 680 °C, without interruptions at the QW interfaces, and in most cases consist of multiple QW structures with 50 periods of alternating layers of GaAs QW's and 150 Å wide $Al_{0.3}Ga_{0.7}As$ barriers. The QW thicknesses and doping densities have been varied over a wide range. As excitation source in the cw photoluminescence (PL) measurements an Ar^+ ion laser was used, combined with a Titanium doped Sapphire solid-state laser for the selective PL (SPL) and PL excitation (PLE) measurements. Transient data were obtained either with a photon counting system or a streak camera, using tunable picosecond (ps) pulsed excitation.

3. SATELLITE SPECTRA AND BINDING ENERGIES

3.1. Donor-doped QW's

The samples used in this part of the study were selectively doped with Si in the central 20% of the QW with a concentration varied from 1×10^{16} to 3×10^{17} cm^{-3}. For moderate doping (low 10^{16} cm^{-3} range), only the FE and the exciton bound at the Si-donor are observed in the PL spectrum with above bandgap excitation. The BE binding energy for a 100 Å wide QW is 1.6 meV, in agreement with values reported by Liu et al (1988). When the excitation photon energy approaches the BE energy, a novel feature appears in the selective PL (SPL) spectrum, 10.6 meV below the BE (Holtz et al 1991a). This satellite is interpreted as the two electron transition (TET) of the BE, in which the Si donor is left in its excited 2s state after the recombination. The interpretation of the TET satellite is further supported by PLE measurements. The BE peak is strikingly enhanced in the PLE spectrum, when the TET peak is resonantly detected. Calculations on excited donor states have recently been reported, where different values for effective masses and dielectric constant in the well and the barrier were used (Fraizzioli et al 1990). A predicted value on the 1s - 2s energy separation of 10.6 meV was derived, i.e. in excellent agreement with our experimental results.

3.2. Acceptor-doped QW's

For the case of moderately doped QW's, more in depth spectroscopic investigations of the acceptor electronic structure have to date been made than for the donors. Acceptor transitions from the 1s ground state to both the excited 2s and 3s states have been observed in both SPL and Resonant Raman Scattering (RRS) experiments (Holtz et al 1989a,b). The dependence of these transition energies on the QW width and the acceptor position within the QW has been studied in a systematic way. In Fig 1 is shown an example of a THT spectrum observed with excitation resonant with the BE. Besides the strong 2s replica and the weaker 3s peak, 2p replica is also seen, allowing a determination of the 2s-2p splitting.

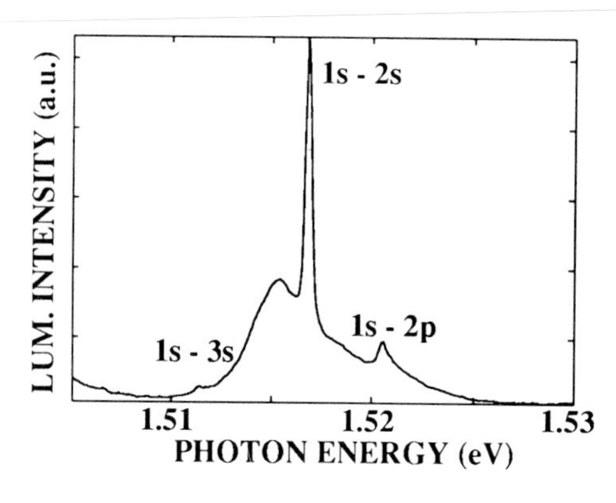

Fig. 1. Two hole spectrum at 2 K of a Be doped 150 Å MQW, exciting in the BE.

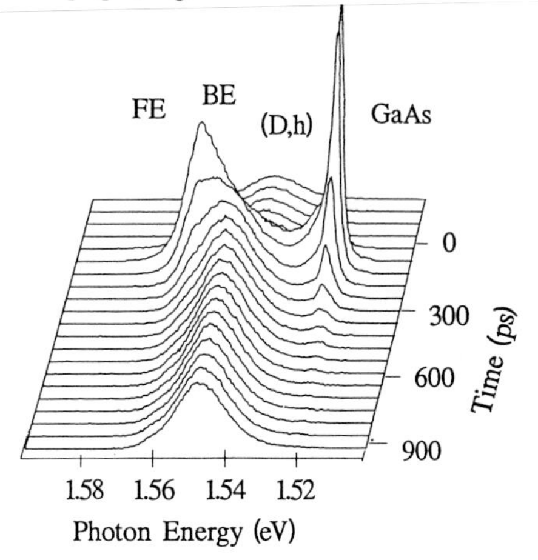

Fig. 2. Time resolved PL spectra at 5 K of a highly n-doped 100 Å QW.

4. HIGH DOPING LEVELS

High doping has also been studied for Si-doped single QW samples, up to doping levels above the degenerate limit. Above 5×10^{17} cm^{-3} the excitonic spectra are difficult to resolve in steady state PL measurements, due to very strong spectral transfer effects, as can be directly demonstrated in ps transient spectra (see Fig. 2). The FE is captured very rapidly into BE states (within tens of ps), and BE's in turn undergo further spectral diffusion down to more longlived lower energy spectra (dominating the time-integrated spectra), presumably due to the separate localization of electrons and holes in potential fluctuations due to the doping inhomogenieties present in the partially compensated samples. Analogous effects of potential fluctuations are observed in Si-doped bulk GaAs (Redfield et al 1970).

The PL spectra for acceptor doped QW's are also significantly altered by increasing doping levels. For a sample with a doping level above $3\text{x}10^{17}$ cm^{-3}, the intensity of the FE and, in particular, the BE with a binding energy of 4.2 meV is reduced, and instead a new and fairly broad (FWHM $\approx$ 3 meV) band downshifted about 7 meV in energy from the FE is gaining intensity (Fig. 3). Also the PLE spectrum is different from the corresponding spectrum of a moderately doped QW, in which the BE can be monitored only under certain circumstances (Holtz et al 1989b). In PLE a strong new band appears 5.0 meV below the FE_{hh} peak, thus the novel band is downshifted significantly more in PL. A plausible origin of this novel band is an exciton bound at two or more interacting acceptors. The PLE spectrum is expected to reflect the density of BE states, while the excitons could transfer to the locally lowest potential before the recombination observed in PL. Similar conversions from excitons bound at a single acceptor (BE) to excitons bound at interacting acceptors in more heavily doped material have earlier been reported only for the bulk case, e.g. InP and ZnTe (Molva and Magnea 1980). The potential binding the exciton increases due to pairing, as calculated as a function of acceptor concentration for bulk material (Molva and Magnea 1980).

5. INFLUENCE OF LOCALIZATION EFFECTS

In the case of non-interrupted growth (as exclusively studied in this work) it is well known that interface roughness gives rise to localization effects for the excitons. The excitons will experience stronger localization in the narrower QW's, so for QW widths of 50 Å the localization effects of FE's into localized potentials may easily dominate and obscure the capture process to impurities creating the BE's. This has a dramatic effect on the BE spectra in QW's. One observation is that the FE peak usually dominates the BE peak in time-integrated spectra even up to rather high doping levels, at least for thicknesses below 100 Å. This

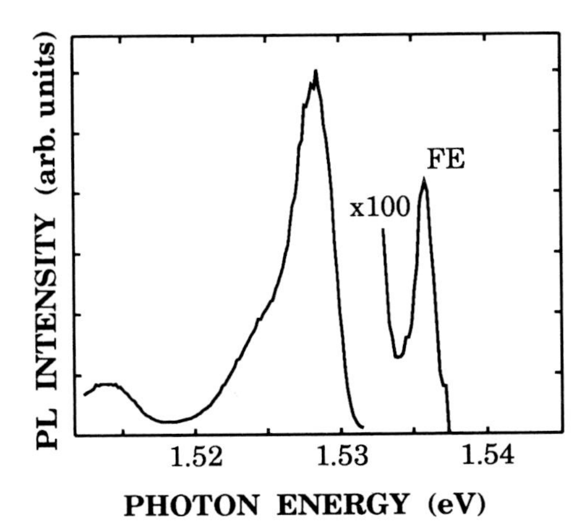

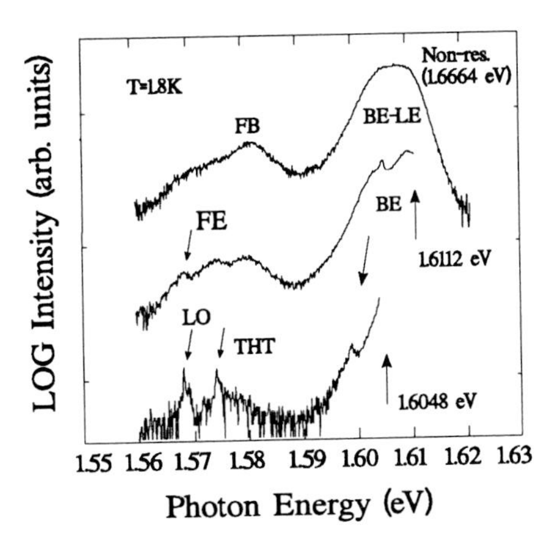

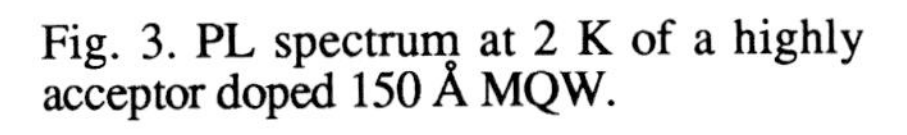

Fig. 3. PL spectrum at 2 K of a highly acceptor doped 150 Å MQW.

Fig. 4. PL spectra at 2 K with tunable cw excitation of a Be doped 50 Å MQW.

behaviour is in sharp contrast to bulk GaAs, where the BE peaks usually completely dominate the PL spectra at doping levels above say 10^{14} cm^{-3}. One reason for this difference could be that FE's in QW's are captured in the interface potentials and then do not form BE's. The FE spectra observed in PL at 2 K are then more naturally labelled as localized exciton (LE) peaks.

For narrow wells the interaction between localization potentials and the impurity potentials becomes strong, and the LE and BE spectra are strongly mixed and broadened in time integrated spectra, unless selective excitation is used. In Fig. 4 we show such spectra for a 50 Å MQW acceptor-doped sample, for different laser photon energies. For excitation above the FE energy, the LE and BE peaks are merged together. At lower excitation energy some LE's are resonantly excited, in which case capture into acceptors in the same localization potential gives rise to a sharp BE PL peak, allowing accurate determination of the BE binding energy (Monemar et al 1991a). This process is most apparent for excitation in the low energy part of the LE spectrum, corresponding to large area localization potentials. THT Raman replica as well as LO replica induced by localization are also observed. At higher photon energy these sharp replica are supplemented by strong corresponding PL processes occurring after hopping relaxation of the initially excited LE. Also, intrinsic THT replica related to recombination at LE's with excitation of localized holes occur (Holtz et al 1991b,Fig. 4).

6. KINETICS OF BE CAPTURE AND RECOMBINATION

These properties were studied with a streak camera, employing tunable ps pulsed excitation resonant with either the FE or the BE. Localization and also capture to the impurities can then be directly observed in the temporal development of the PL emission (Harris et al 1991). Thermalization occurs for the donor BE's in relation to the FE's already at 5 K, while the corresponding thermalization of the acceptor BE's to the FE (LE) do not occur until at 15-20 K, depending on well width. It has been concluded that the radiative recombination time for the BE decreases monotonically from a well width L_Z of 150 Å to L_Z = 50 Å, indicating that the BE takes advantage of the confinement to increase its oscillator strength in the same way as the FE (Monemar et al 1991b, Bergman et al 1991). The biexciton is present in most spectra obtained with pulsed excitation, at not too high doping levels. At high excitation densities the BE recombination kinetics are influenced by FE scattering processes (Harris et al 1991).

7. REFERENCES

J P Bergman, P O Holtz, B Monemar, M Sundaram, J L Merz and A C Gossard 1991 Phys Rev B 43 4795

S Fraizzioli, F Bassani and R Buczko 1990 Phys Rev B 41 5096

C I Harris, H Kalt, B Monemar, P O Holtz, J P Bergman, M Sundaram, J L Merz and A C Gossard 1991 Proc 16th Int Conf on Defects in Semiconductors, Bethlehem, USA, Materials Science Forum in press

P O Holtz, M Sundaram, R Simes, J L Merz, A C Gossard and J H English 1989a Phys Rev B 39 13293

P O Holtz, M Sundaram, K Doughty, J L Merz and A C Gossard 1989b Phys Rev B 40 12338

P O Holtz, B Monemar, M Sundaram, J L Merz and A C Gossard 1991a Phys Rev B in press

P O Holtz, H P Hjalmarson, M Sundaram, J L Merz and A C Gossard 1991b Superlattices and Microstructures 9 407

R C Miller, A C Gossard, W T Tsang and O Munteanu 1982 Phys Rev B 25 3871

B Monemar, P O Holtz, J P Bergman, C I Harris, H Kalt, M Sundaram, J L Merz and A C Gossard 1991a Surface Science in press

B Monemar, H Kalt, C I Harris, J P Bergman, P O Holtz, M Sundaram, J L Merz, A C Gossard, K Köhler and T Schweizer 1991b Superlattices and Microstructures 9 281

E Molva and N Magnea 1980 Phys Stat Sol (b) 102 475

D Redfield, J P Wittke and J I Pankove 1970 Phys Rev B 2 1830

X Liu, A Petrou, B D McCombe, J Ralston and G Wicks 1988 Phys Rev B 38 8522

G C Rune, P O Holtz, B Monemar, M Sundaram, J L Merz and A C Gossard 1991 Phys Rev B in press

Inst. Phys. Conf. Ser. No 123
Paper presented at the International Meeting on Optics of Excitons in Confined Systems, Giardini Naxos, Italy, 1991

Theory of two-dimensional carriers bound to ultrathin isoelectronic impurity intralayers

Kurt A. Mäder[a,b] and Alfonso Baldereschi[b,c]
[a]Laboratorium für Festkörperphysik, ETH Zürich, 8093 Zürich, [b]Institut de Physique Appliquée, EPF Lausanne, 1015 Lausanne, [c]IRRMA, PH Ecublens, 1015 Lausanne, and University of Trieste, Italy

A model to study the electronic properties of isoelectronic impurity intralayers in zincblende semiconductors is presented. It is based on an empirical tight-binding scheme and on a Koster–Slater approach to the short range perturbation. Results for Ga (001) intralayers in AlAs are given and compared to effective mass calculations.

1 Introduction

Isolated isoelectronic impurities in a semiconductor bind a hole (electron) only if their short-range potential is stronger than a certain threshold which depends on the valence (conduction) band dispersion of the host (Baldereschi 1973, Baldereschi and Hopfield 1972, Faulkner 1968). Due to their localized nature, such traps act as efficient recombination centres even in indirect gap materials, such as, *e. g.*, GaP:Bi (GaP:N). Using simple quantum-mechanical considerations we show that an atomic monolayer of isoelectronic impurities will *always* bind a hole (electron), regardless of the strength of the single-site short-range perturbation. Isoelectronic monolayers introduce in general two-dimensional (2D) bands in the energy gap of the host-material. Carriers in such 2D isoelectronic intralayer states may subsequently attract carriers of opposite charge and thus form 2D intralayer excitons. Recent experiments (Cingolani *et al.* 1990, Sato and Horikoshi 1990, Taira *et al.* 1988) indeed prove this to be the case for GaAs:$In_{Ga}(001)_1$, (where the indices indicate the crystallographic orientation of the isoelectronic intralayer and the subscript denotes the number of consecutive impurity planes), as well as for InP:$As_P(001)_n$ (Carlin *et al.* 1991). Prior interpretations by these authors of the observed photoluminescence spectra were made within an envelope-function scheme. Although the transition energies can be fitted surprisingly well by this method, it is not justified in the ultrathin quantum well limit, and atomistic models have to be used. For isoelectronic electron traps this has been done by Hjalmarson (1982) in the case of GaAs:$N_{As}(001)_1$, and Wilke and Hennig (1991) in the case of GaAs:$In_{Ga}(001)_n$. Taking the same empirical tight-binding (ETB) hamiltonian (Vogl *et al.* 1983) for the host as Hjalmarson (1982), we go beyond the on-site approximation, and include perturbation of nearest-neighbour interactions and treat both holes and electrons on the same footing. Furthermore we include spin-orbit interaction and strain effects in lattice-mismatched systems. This particular ETB model, also known as sp^3s^* model, has in the past been applied successfully to isolated deep traps (Hjalmarson *et al.* 1980). Once the host hamiltonian and the perturbation are specified, we construct the Green's function in the whole 2D Brillouin zone. We then determine hole (electron) binding energies, 2D dispersion relations, and wave function parameters as a function of n. The model is detailed in section 2, results for the AlAs:$Ga_{Al}(001)_n$ system are presented in section 3.

2 Description of the Model

The ETB hamiltonian (Vogl *et al.* 1983) is given in a basis of anion and cation s, p_x, p_y, p_z, s^* orbitals. The latter are excited states of the free atoms, and serve to mimic additional conduction bands. A nearest neighbour TB model is then sufficient to reproduce the bandstructures even of indirect energy-gap materials in good agreement with ***ab-initio*** calculations (Vogl *et al.* 1983). Inclusion of spin-orbit interaction is done as described by Chadi (1977), and the effects of biaxial strain on the TB parameters can be built in by empirical rules (Harrison 1981). The simplest approach to treat a perturbation of extremely short range is the one-site one-band Koster–Slater model (Koster and Slater 1954). The use of a TB crystal hamiltonian allows us to be more general, *i. e.*, to include many bands and to go beyond on-site perturbation. Without loss of generality we will outline our model for AlAs:$Ga_{Al}(001)_n$. If *all* ($n \to \infty$) cations Al in AlAs are replaced by Ga, the matrix elements of H are transformed into those of bulk GaAs, and the electronic spectrum is shifted with respect to AlAs by the valence band offset ΔE_v. Thus, in the limit $n \to \infty$, the perturbation U is simply given by $H(\mathrm{GaAs}) - H(\mathrm{AlAs}) + \Delta \mathrm{E_v}$. For finite n we proceed as follows: inside the GaAs intralayer bulklike GaAs behavior is assumed as in the case of infinite n discussed above, outside of the intralayer the perturbation is zero, and at the common-anion boundary layer we adopt an intermediate procedure by adding only a fraction λ of ΔE_v to the (diagonal) on-site matrix elements. λ is a free parameter of our model which can be used to adjust the results to experimental or theoretical data. Standard Green's function techniques (see , *e. g.*, Callaway 1974) are then used to solve the Schrödinger equation $(H+U)\Psi = E\Psi$ and to obtain binding energies and wave functions of bound states in the forbidden energy gap of the host, or resonant states within its bands. To exploit translational symmetry in the (001) plane, the perturbation U and the resolvent operator $R(E, \mathrm{k}) = [E - H(\mathrm{k})]^{-1}$ are both transformed to a "layer orbital" basis $|\mathbf{Q}lj\rangle$, where $\mathbf{Q} = (q_x, q_y)$ is a 2D wave vector, $l = s, x, y, z, s^*$ denotes the orbital character, and $j = 0, \pm1, \pm2, \pm3, \ldots$ labels the layers. The resolvent $R(E, \mathbf{k})$ expressed in this new basis defines the Green's function matrix $\mathcal{G}(E, \mathbf{Q})$ of a now *quasi* linear chain model with perturbing "coupling constants" $U_{jj',ll'}(\mathbf{Q})$. For $n = 1$ this can be visualized schematically in the following way:

$\ldots$	=	As	=	Al	=	As	=	Ga	=	As	=	Al	=	As	=	$\ldots$
$j =$		-3		-2		-1		0		1		2		3		
$U_{jj'}$						$U_{-1,-1}$	$U_{-1,0}$	$U_{0,0}$	$U_{0,1}$	$U_{1,1}$						

Bound states above the valence and below the conduction bands exist, if the secular equation

$$D(E, \mathbf{Q}) \equiv \det[\mathbf{1} + \mathcal{G}(E, \mathbf{Q})U(\mathbf{Q})] = 0 \tag{1}$$

has real roots $E_b(\mathbf{Q})$ in the (001) projected energy gap of the host, and thus the 2D band structure is completely determined by eq. (1). Resonances inside the projected bands occur if $\lim_{\eta\to 0} \mathrm{Re}[D(E_r + i\eta, \mathbf{Q})] = 0$. Since $U(\mathbf{Q})$ is non-zero only for the impurity layers the rank of the secular equation (1) is rather small for moderate n. In the one-band one-site approximation $\mathcal{G}(E, \mathbf{Q})$ is the Hilbert transform of the projected one-dimensional density of states at $\mathbf{Q}$, and therefore has a $\frac{1}{\sqrt{E}}$ singularity at the band edge of a TB s- or p-band. In this case eq. (1) reads $1 + \mathcal{G}(E, \mathbf{Q})J = 0$, where J is the on-site impurity matrix element, and has a solution for arbitrarily small J, with $E_b \propto J^2$. An impurity *layer* thus always has a bound state in this simplified case.

3 Results for AlAs:$Ga_{Al}(001)_n$

In this section we apply the above described model to the case of n layers of Ga substituting n Al (001) layers. It is chosen as a test case because (i) it is lattice-matched, (ii) due the indirect

nature of the AlAs band structure a rigorous treatment of the bound electrons is difficult by simpler schemes, such as the effective mass approximation (EMA). First, we need a reasonable value of the "boundary layer parameter" λ. To get an estimation of its magnitude we studied the variation of the Γ_1 conduction band energy of an $(AlAs)_2(GaAs)_2$ superlattice with λ, calculated within the same TB scheme as described in section 2, and compared with *ab-initio* results (Posternak *et al.* 1988). We conclude that λ is rather small ($\lambda \leq \frac{1}{4}$), although an exact quantitative determination is beyond the possibilities of this empirical model. For the present calculations we used $\lambda = \frac{1}{4}$, but one should keep in mind that this is probably an upper bound of the parameter.
For the bound holes we include the spin-orbit interaction in the ETB host hamiltonian and in the perturbation. The binding energies are then found by solving eq. (1) numerically. The results for $n = 1, \ldots, 7$ are presented in figure 1. As expected, there are always at least two

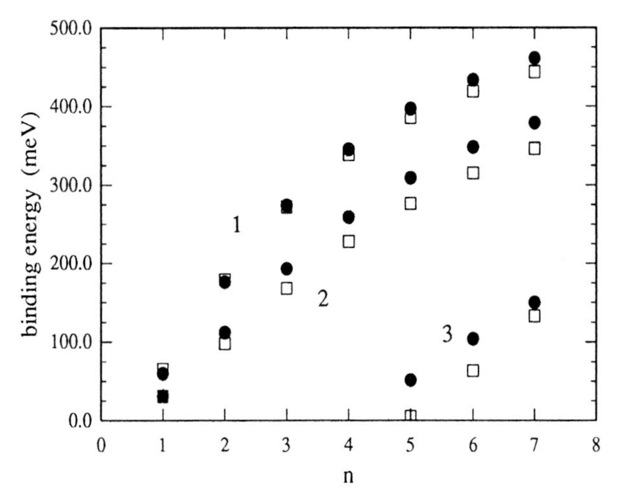

Figure 1: Binding energies of the confined holes in AlAs: $Ga_{Al}(001)_n$ as a function of n. The zero of energy is at the valence band maximum of AlAs. Full dots: present model, squares: EMA (Andreani 1991). States (1) are $\bar{\Gamma}_7$ (from heavy hole band), (2) are $\bar{\Gamma}_6$ (from light hole band), and (3) again $\bar{\Gamma}_7$.

bound states (each doubly degenerate), the higher stemming from the heavy hole band along Δ, the lower from the light hole band. The split-off band (lower by $\sim$ 275 meV at the Γ-point in AlAs) does not introduce a bound state for small n. Since the point group of the perturbed system is D_{2d}, the little group at $\bar{\Gamma}$ has two irreducible representations, labelled $\bar{\Gamma}_6$ and $\bar{\Gamma}_7$, respectively. In the figure we also give the results of a calculation in the effective mass approximation (Andreani 1991). Note the over-all agreement between the two models for the highest state, which should be best for thicker wells, so $\lambda = \frac{1}{4}$ might still be too great. If λ would be adjusted to reproduce the EMA values at $n \approx 7$, say, the present model would yield smaller binding energies in the ultra-thin limit, where the EMA is by construction not justified. The energy separation to the lower bound states predicted by our model is systematically smaller then in the EMA. Inclusion of non-parabolicity in the EMA calculation does not remove this difference, on the contrary, the light hole states tend to be even less bound (Andreani 1991). In our model all bands, along the whole length of the Δ-line in the 1st Brillouin zone, contribute to the Green's function at $\bar{\Gamma}$, unless prohibited by symmetry. This effect will have drastic consequences near the conduction band. AlAs has its conduction band minimum (CBM) at the X-point, whereas GaAs is a direct-gap material, and its lowest state at X lies higher in energy than the one of AlAs, assuming the conduction band offset to be 65% of the direct band gap difference. In the quantum well picture GaAs thus acts as a well at Γ and as a barrier at X. Since in our model full mixing along the Δ-line is warranted the additional states at $\bar{\Gamma}$ induced by the impurity layer have important contributions from Bloch states with $k_z \neq 0$. Our calculation (for the bound electron we neglect spin-orbit interaction) leads to the following conclusions: for small n there is a resonant state close to the projected CBM at Γ, and for

$n \geq n_c \approx 10$ there is a bound state below the projected CBM (which is 2.223 eV in our TB fit, and originates from a Δ-valley $\sim$ 15% off the X-point). The nature of these resonant and bound states, respectively, will be the object of a future study. Let us finally turn to the bound

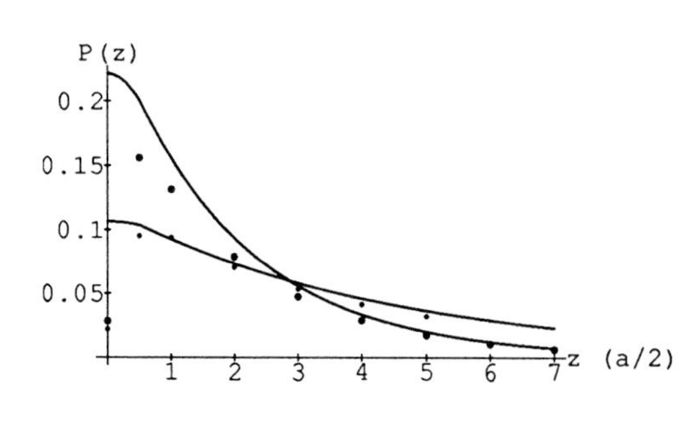

Figure 2: Probability per fcc monolayer as a function of z (in units of the impurity layer width $L = a/2$) of the bound holes at $\bar{\Gamma}$ for AlAs: $Ga_{Al}(001)_1$. The dots denote the weights of the layer orbitals as obtained by the present model. The curves give $|F(z)|^2$ of an EMA calculation (Andreani 1991). Large dots (upper curve at $z = 0$) heavy hole bound state, small dots (lower curve at $z = 0$) light hole bound state.

hole wave functions at $\bar{\Gamma}$. Since we work in a layer orbital basis the probabilities per layer (fig. 2) and decay lengths are easily obtained. For $n = 1$ we get for the heavy (light) hole state a decay length of 4.0 (7.4) fcc monolayers (ML), and a confinement of 18.4% (11.7%) (defined as the total probability inside the impurity layer). The EMA predicts 3.9 (8.6) ML, and 21.5% (10.5%), respectively.
To conclude we have presented a model for ultra-thin "quantum wells", or more precisely for isoelectronic impurity layers, which is valid even in the monolayer regime, and includes full mixing with all bands along the direction in the Brillouin zone defined by the impurity layer orientation.

References

Andreani L C 1991 private communication
Baldereschi A 1973 J. Luminescence **7** 79
Baldereschi A and Hopfield J J 1972 Phys. Rev. Lett. **28** 171
Callaway J 1974,1991 Quantum Theory of the Solid State (Boston: Academic Press) chapter 5
Carlin J F, Houdré R, Rudra A, and Ilegems M 1991 Proc. Int. Conference on Chemical Beam Epitaxy, Oxford, to be published in J. Cryst. Growth
Chadi D J, 1977 Phys. Rev. B **16** 790
Cingolani R, Brandt O, Tapfer L, Scamarcio G, La Rocca G C, and Ploog K 1990 Phys. Rev. B **42** 3209
Faulkner R A 1968 Phys. Rev. **175** 991
Harrison W A 1980 Electronic Structure and the Properties of Solids (San Francsisco: Freeman)
Hjalmarson H P 1982 J. Vac. Sci. Technol. **21** 524
Hjalmarson H P, Vogl P, Wolford D J, and Dow J D 1980 Phys. Rev. Lett. **44** 810
Koster G F and Slater J C 1954 Phys. Rev. **96** 1208
Posternak M, Baldereschi A, Massidda S, and Freeman A J 1988 Inst. Phys. Conf. Ser. **91** 537 and private communication
Sato M and Horikoshi Y 1990 Surface Science **228** 192
Taira K, Kawai H, Hase I, Kaneko K, and Watanabe N 1988 Appl. Phys. Lett. **53** 495
Vogl P, Hjalmarson H P and Dow J D 1983 Phys. Chem. Solids **44** 365
Wilke S and Hennig D 1991 Phys. Rev. B **43** 12470

Inst. Phys. Conf. Ser. No 123
Paper presented at the International Meeting on Optics of Excitons in Confined Systems, Giardini Naxos, Italy, 1991

Photoluminescence of a delta doping related exciton in GaAs:Si

J.C.M. Henning[*][+], Y.A.R.R. Kessener[*], P.M. Koenraad[+], M.R. Leys[+], W. van der Vleuten[+], J.H. Wolter[+] and A.M. Frens[**]

[*]Philips Research Laboratories, P.O. Box 80.000, 5600 JA Eindhoven, The Netherlands;
[+]Physics Department, University of Technology, P.O. Box 513, 5600 MB Eindhoven, The Netherlands
[**]Huygens Laboratory, University of Leiden, P.O. Box 9504, 2300 RA Leiden, The Netherlands

ABSTRACT: Photoluminescence (PL) spectra of Si delta–doped GaAs show a novel sharp feature "δ" at 1.4977 eV that is induced by the delta-doping procedure, but cannot be attributed to transitions involving conduction electron subbands. Its spectroscopic properties unambiguously prove the exciton character of "δ". A simple model, compatible with the experimental data is an exciton bound to a Si – As – Si complex, with the Si–Si pair axis aligned along $[1\bar{1}0]$. The (light) hole is strongly localised at the As site and has a magnetic axis along [001]. The electron behaves as a loosely bound effective mass particle.

1. INTRODUCTION

During the last decade interest has arisen in planar (delta) doping techniques which enable the incorporation of high concentrations of electrically active dopants in semiconductors in sheets of at most a few atomic layers in thickness (Schubert 1990). In order to avoid smearing out of the delta layer by diffusion or segregation the growth temperature has to be kept low. Since low substrate temperatures are known to lead to imperfect epitaxial crystal growth a detailed PL study of the defects is not out of order. To this end samples containing a single delta layer are ideally suited since radiative recombinations of electrons in conduction subbands (in n–doped material) with excitation induced holes cannot be expected: the holes tend to move away from the doped plane towards the flat-band part of the sample, thus reducing the transition probability. In this paper PL results are presented for MBE grown p–GaAs containing a single delta layer of Si dopants. An extensive account of this work is in the press (Henning et al. 1991).

2. EXPERIMENT

The samples were grown by MBE on semi-insulating (001) oriented GaAs substrates. First a layer of 2.5 μm thick undoped GaAs was grown at a rate of 1 μm/hr. Then the Ga shutter was closed and the surface was stabilised under As flux for 10 seconds. After this a Si flux of 3×10^{11} at/cm^3 was applied for a fixed amount of time. Finally a 1 μm GaAs top layer was grown, again at a rate of 1 μm/hr. The growth temperatures were chosen as 480, 530 or 620 ^{0}C. The Si concentrations used were either 2×10^{12} or 8×10^{12} cm^{-2}.

The low–T PL spectra show, in addition to the normal GaAs background spectrum, a number of sharp features with intensities that depend strongly on Si dope concentration and growth temperature. We restrict our discussion to one particular feature, called "δ", at $h\nu_f = 1.4977$ eV, that dominates at the lowest growth temperatures. We summarise the spectroscopic properties of "δ":

1. Its intensity $I(\delta)$ goes superlinearly with doping concentration N_d.
2. $I(\delta)$ decreases with increasing growth temperature t_s.
3. Its peak position $h\nu_f$ is independent of dope concentration N_d, temperature T and excitation density φ.
4. $I(\delta)\sim\varphi^{1.5}$ for low φ. Above $\varphi \simeq 300$ Wcm^{-2} saturation sets in.
5. The photoluminescence excitation spectrum, with the detector set at "δ" shows resonant cooling of hot electrons by LO phonon emission.
6. The PL linewidth is 0.5 meV, similar to that of a bulk bound exciton.
7. The peak position relative to the band edge is $E_g - h\nu_f = 21.3 \pm 0.5$ meV.
8. Thermal quenching of "δ" leads to an activation energy $E^* = 18 \pm 4$ meV.
9. The PL line "δ" is linearly polarised along $[1\bar{1}0]$ with $\pi = 0.5 \pm 0.1$; the $[1\bar{1}0]$ axis coincides with the long axis of the "oval defects".
10. Strain modulated photoluminescence (SMPL) leads to an isotropic deformation potential for "δ": $D_d + C_1 \simeq 10$ eV, which is practically identical with that of the bulk exciton. The improved resolution of the SMPL techique enables the observation of two local mode replicas of "δ", with $h\nu_{LM} =$ 5.2 meV and a Huang–Rhys factor $S = 0.25$ (fig. 1). The (stronger) first replica shows the same stress and polarisation behaviour as "δ".

11. Zeeman spectroscopy at low magnetic fields (B) gives an isotropic diamagnetic shift $\Delta h\nu_f = 0.20\ B^2$ (meV/T^2) identical with that of a bulk free exciton (Dingle 1973). At high B a spin splitting is observed, described by an axially symmetric effective g′ tensor: $g'[001] = 0 \pm 0.5$, $g'[110] = g'[1\bar{1}0] = \pm 1.4 \pm 0.2$.

3. DISCUSSION

Properties (3)–(6) provide circumstantial evidence for an exciton character of "δ". A definite proof is given by the diamagnetic shift (prop. 11), which shows that the electron-hole binding energy equals $E_b = 4$ meV. Combining this data with prop.(7) we obtain for the localisation energy $E_{loc} = 17$ meV, compatible with (8). The slopes of the $\Delta h\nu_f$ versus B curves at high B resemble those of the first Landau level. They are approximately isotropic, showing that the electron moves in 3 dimensions with an effective mass equal to the bulk value: $m^* = 0.067\ m_0$ and a concomitant Bohr radius of 14 nm. The hole, on the other hand, is strongly localised by a crystal field with [001] axial symmetry. Analysis of the effective g′ tensor (prop. 11) reveals that heavy and light holes are well separated by this crystal field and that the "δ" exciton belongs to the category (e,LH). Assuming $g_e = -0.44$ (Skolnick et al. 1976) our experimental data lead to $g_{lh} = -0.5 \pm 0.1$, in agreement with calculations by Kirkman et al. (1978) for light holes.

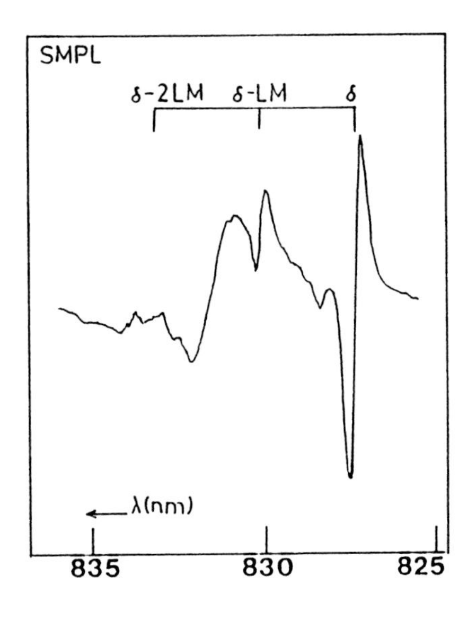

Fig. 1. SMPL spectrum of delta–doped GaAs:Si, $N_d = 8 \times 10^{12} cm^{-2}$; $t_s = 480\ ^0C$; T=4K; stress//[110]. The donor–acceptor line at 831 nm is insensitive to stress; this enables the emergence of the local mode sidebands of "δ".

The present data is insufficient to deduce an atomic model of the centre to which the "δ" exciton is bound. Magnetic resonance measurements are indispensible in this respect. Nevertheless, it is tempting to make some speculations. Property (1) suggests that more than one Si atom is involved. An attractive, simple model is an exciton bound to a Si–As–Si complex, with the Si–Si axis along the $[1\bar{1}0]$ direction. The hole is thought to be localised near the As site. The C_{2v} symmetry of this complex then naturally explains both the (quasi) axial [001] symmetry of the spin splitting, which is mainly determined by the hole wave–function, and the linear polarisation along the Si–Si bond axis $[1\bar{1}0]$, which is related to the antibonding and bonding character of the initial and final state wavefunctions (Henning et al. 1991). Adopting Chadi's (1987) model for the (2×4) reconstructed As surface, on which delta doping starts, the Si–Si pairs may be thought to occupy the "missing As–As dimer" sites, oriented along $[1\bar{1}0]$. The As is at a nearest–neighbour antisite (Ga) position in the next (001) layer. The complex as a whole is isoelectronic with respect to the lattice. This may explain both the large value of the localisation energy (Monemar 1988) and the self–compensation observed at high N_d and low t_s (Koenraad et al. 1990). The local vibration associated with the defect may then be identified with a symmetric bending mode. Its low frequency is due to the fact that the Si atoms are only weakly bound to the underlying layer. Bending mode energies of the order of 5 to 10 meV are indeed not uncommon for centres involving interstitial atoms (Bosomworth et al. 1970, Gislason et al. 1982).

4. REFERENCES

Bosomworth D R, Hayes W, Spray A R L and Watkins G D 1970 Proceed.Roy.Soc. London A317 133
Chadi D J 1987 J.Vac.Sci.Technol. A5 834
Dingle R 1973 Phys.Rev. B8 4627
Gislason H P, Monemar B, Holtz P O, Dean P J and Herbert D C 1982 J.Phys. C15 5467
Henning J C M, Kessener Y A R R, Koenraad P M, van der Vleuten W C, Wolter J H and Frens A M Semicond.Sci.Technol., in the press.
Kirkman R F, Stradling R A and Lin–Chung P J 1978 J.Phys. C11 419
Koenraad P M, Blom F A P, Langerak C J C M, Leys M R, Perenboom J A A J, Singleton J, Spermon S J R M, van der Vleuten W C, Voncken A P J and Wolter J H 1990 Semicond.Sci.Technol. 5 861
Monemar B 1988 CRC Critical Reviews in Solid State and Materials Sciences 15–2 111
Schubert E F 1990 J.Vac.Sci.Technol. A8 2960
Skolnick M S, Jain A K, Stradling R A, Leotin L, Ousset J C and Ashkenazy S J 1976 J.Phys. C9 2809

Subject Index

Author Index